6th Southwest Process Technology Conference 2014

Galveston, Texas, USA
9-10 October 2014

ISBN: 978-1-5108-1771-5

Printed from e-media with permission by:

Curran Associates, Inc.
57 Morehouse Lane
Red Hook, NY 12571

Printed by Curran Associates, Inc. (2016)

For permission requests, please contact AIChE
at the address below.

AIChE
120 Wall Street, FL 23
New York, NY 10005-4020

Phone: (800) 242-4363
Fax: (203) 775-5177

www.aiche.org

Additional copies of this publication are available from:

Curran Associates, Inc.
57 Morehouse Lane
Red Hook, NY 12571 USA
Phone: 845-758-0400
Fax: 845-758-2634
Email: curran@proceedings.com
Web: www.proceedings.com

TABLE OF CONTENTS

CONFERENCE KEYNOTE

CATALYSIS & REACTION ENGINEERING

ENERGY EFFICIENCY & ASSET OPTIMIZATION

THERMODYNAMICS & PROCESS SIMULATION

UPSTREAM OIL & GAS

DISTILLATION & SEPARATION

PROCESS SAFETY

PROFESSIONAL ETHICS

STUDENT PROGRAM

CHEMICAL PROCESS TECHNOLOGY

PROCESS CONTROL & OPTIMIZATION

PROCESS DEVELOPMENT & CHEMICAL MANUFACTURING

PETROLEUM REFINING TECHNOLOGY

TECHNICAL SESSIONS FOR YOUNG PROFESSIONALS

Wonders and Challenges of Process Research in the Chemical Industry

Rui Vogt Alves da Cruz
R&D Director
The Dow Chemical Company

October 9th 2014

Dow.com

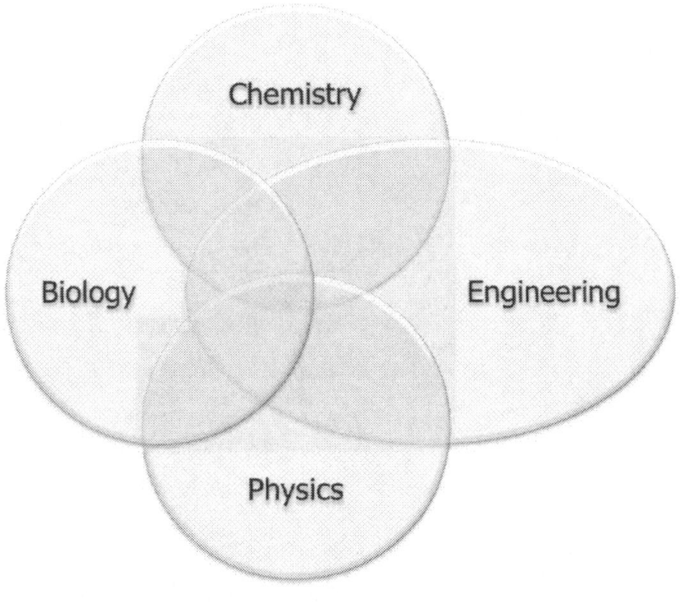

Challenging and Multidisciplinary

2

Large Scale Impact

Safer, more sustainable

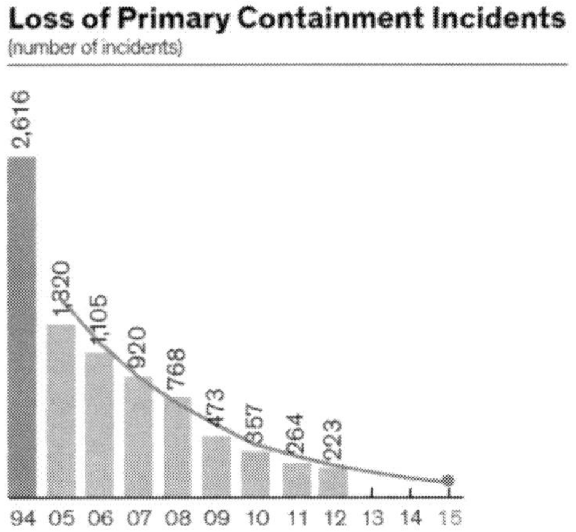

Equipment Design
Reaction Engineering
Yield Improvement
Energy Integration

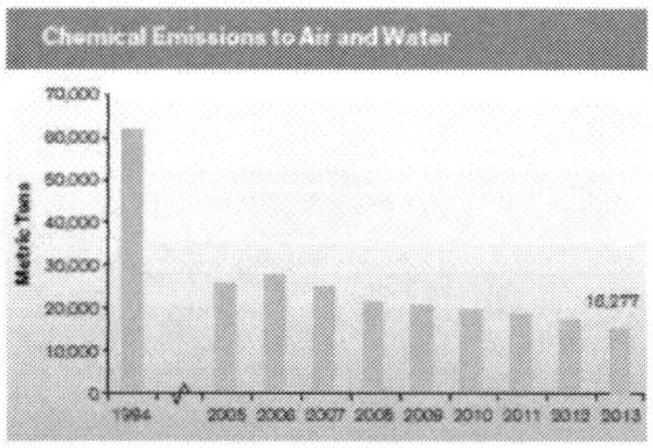

Dow Chemical 2013 Annual Sustainability Report

More Products, Cost Effective

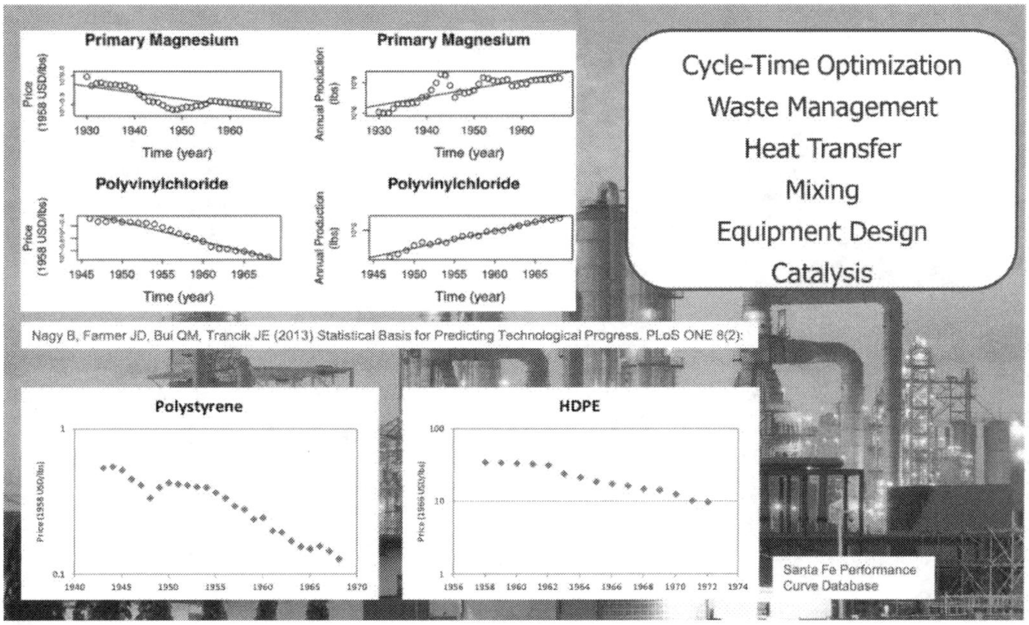

5

Enabling New Products

Scale-up
Route optimization
Mixing
Separations

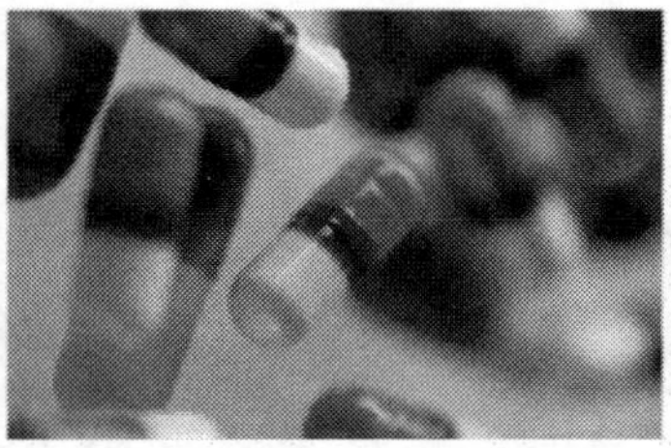

6

Synthesis of Ammonia

Abb. 4 Haber-Le Rossignol-Apparatur zur Ammoniaksynthese

Die Ammoniakfabrik der BASF in Oppau. Gemälde von Otto Bollhagen, 1914. Ludwigshafen, Unternehmensarchiv BASF

7

Polystyrene

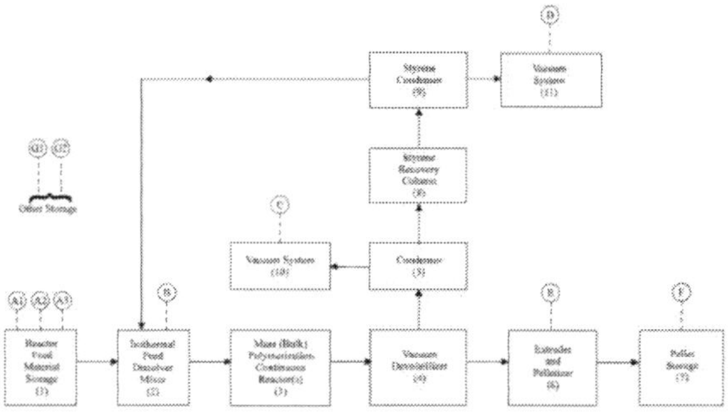

EPA AP 42, Fifth Edition, 1995

8

Defining Process Research

Breakthrough Process Research

Process Improvements and Plant Support

Process Development for New Products

9

Fundamental Challenges of Process Research

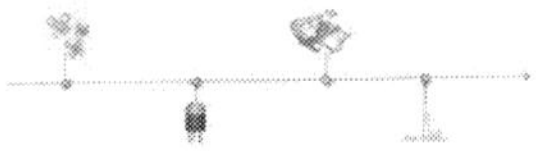

- Long development timelines

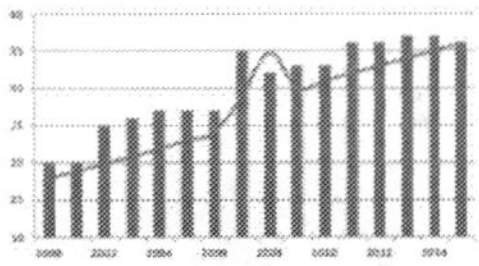

- Dependence on supply demand balances

- Investment levels

10

Getting Around – Best Practices

- Close engagement of business leadership and marketing
- Risk analysis and abatement planning
- Tracking of regulatory trends

11

Getting Around – Best Practices

- Prioritization of lab, modeling and piloting work
- Multipurpose piloting or piloting in the plant
- Multifunctional, independent evaluation of technology
- Right team

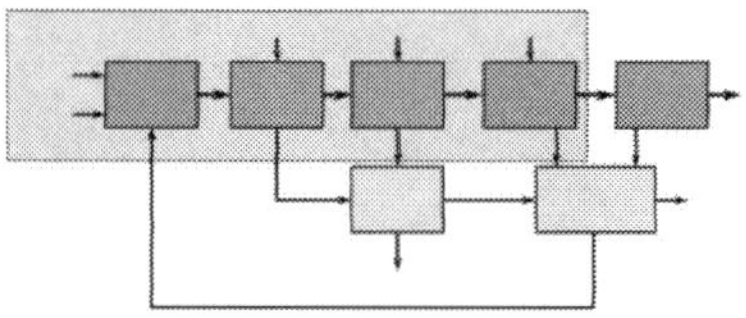

Zacharias, L., Painter, R. 8th Global Congress on Process Safety, 2012

12

Developing a Winning Process Research Team

Skills

Integration

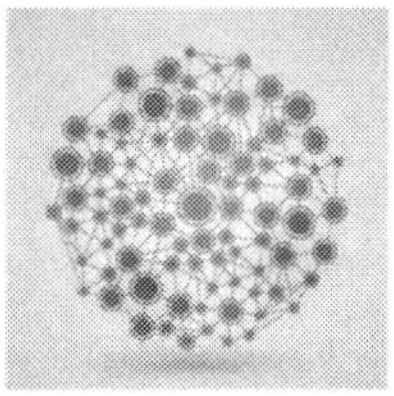

Talent Pipeline

Preparing the Next Generation

Skill	Average Relative Importance 1-5
UO: unit operations, transport phenomena, thermodynamics, separation processes *	4.6
RE: reaction engineering, catalysis, kinetics.	4.0
AM: analysis, modeling, simulation, process control *	4.0
MAT: materials, surface science, polymers *	3.2
BIO: biotechnology, medical and life sciences	2.1
NANO: nanotechnology and its applications	1.8

DOW

13

Future of Process Research

- Much higher sustainability standards
- Process intensification
- Soft sensors
- Biotechnology
- Faster development times
- Leveraging process research to application development

- **Always critical for competitive advantage**

DOW

14

Thank you

rvcruz@dow.com

15

Dewaxing Challenging Paraffinic Feeds in North America

Renata Szynkarczuk
Senior Technical Specialist
Criterion Catalysts & Technologies
Edmonton, Canada

Shell Global Solutions

CRITERION CATALYSTS & TECHNOLOGIES

AIChE Southwest Process Technology Conference

Dewaxing Challenging Paraffinic Feeds in North America
Economic/Technical Drivers of Catalytic Dewaxing

- Fuel Specifications
  - Cloud Point (CP)
  - Pour Point (PP)
  - Cold Filter Plugging Point (CFPP)
- Reduce Kerosene Blending
- Reduce Additive Usage
- Process "Waxy" (Paraffinic) Crudes
- Increase Feed Endpoint

AIChE Southwest Process Technology Conference

Dewaxing Challenging Paraffinic Feeds in North America
Cold Flow Specifications (1)

Cloud Point (CP) ASTM D-2500:
- CP: The temperature at which the first crystals of wax (n-paraffins, haze) appear when a sample is cooled down under standardized conditions
- CP: Intrinsically related to hydrocarbon composition
- CP: *Cannot* be significantly changed by kerosene blending/additives

Pour Point (PP) ASTM D-97:
- PP: The temperature at which a sample can just be poured (pumped) when cooled down under standardized conditions. This point is 3°C above the temperature at which sufficient wax has crystallized to prevent normal flow of the sample.
- PP: Gel-Like Structure Formation
- PP: *Can* be changed by kerosene blending/additives

2014 AIChE Conference Page 3

AIChE
Southwest Process Technology Conference

Dewaxing Challenging Paraffinic Feeds in North America
Cold Flow Specifications (2)

Cold Filter Plugging Point (CFPP) IP 309:
- CFPP: The temperature where a fixed volume of diesel blocks flow through a 45 μm metal gauze filter upon cooling
- CFPP: *Can* be modified by additives

Wax Formation Causes:
- N-Paraffin (Linear Paraffin) Crystallization
- Fuel N-Paraffin Concentrations
- Crude Source & Process Conditions

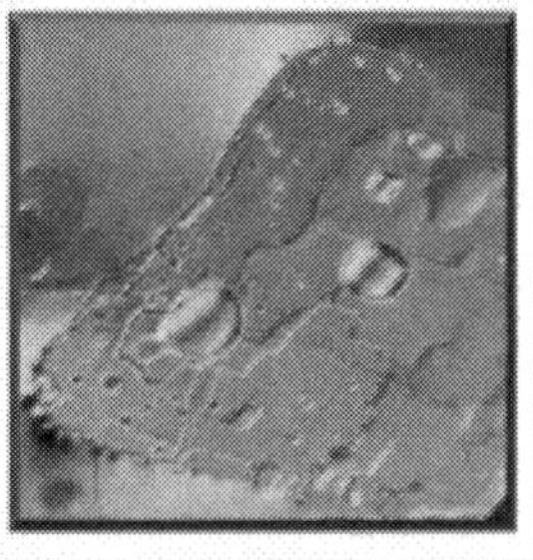

2014 AIChE Conference Page 4

AIChE
Southwest Process Technology Conference

Dewaxing Challenging Paraffinic Feeds in North America

Plugged Fuel Filter – Room Temperature

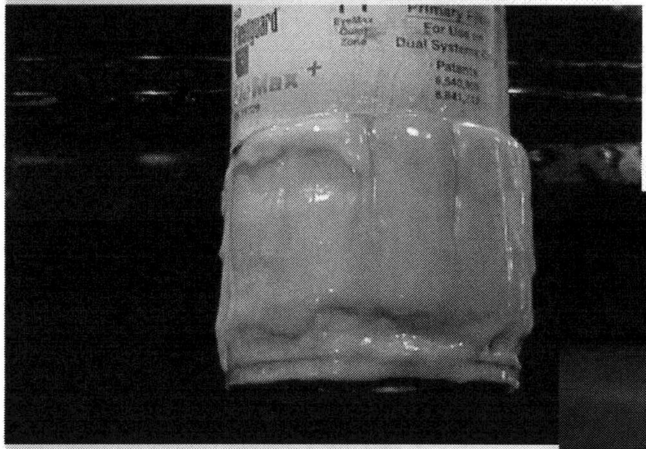

Courtesy ASTM

Dewaxing Challenging Paraffinic Feeds in North America

<u>N-Paraffins</u>: Melting Point vs. Carbon Number and Boiling Point Relationship

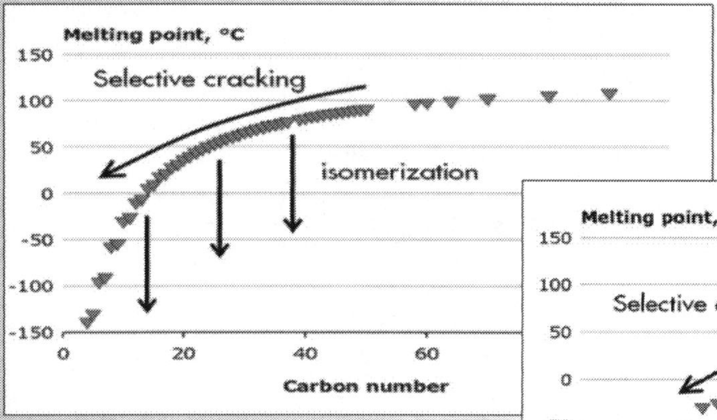

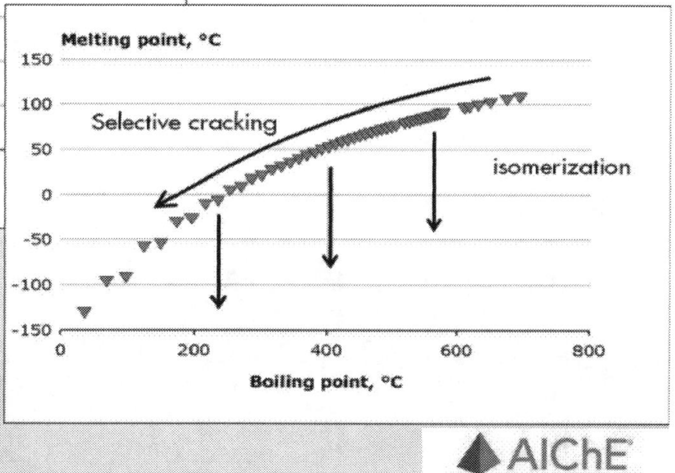

Dewaxing Challenging Paraffinic Feeds in North America
Influence of Molecular Structure on Properties (1)

Methyl Position ($C_{19}H_{40}$)	Melting Point (°C)
2-methyl-octadecane	+13.0
3-methyl-octadecane	+0.5
4-methyl-octadecane	-1.0
5-methyl-octadecane	-13.5
6-methyl-octadecane	-4.0
7-methyl-octadecane	-16.0
8-methyl-octadecane	-10.0
9-methyl-octadecane	-16.5

Formula ($C_{30}H_{62}$)	Melting Point (°C)
n-triacontane (n-$C_{30}H_{62}$)	+66
2,6,10,15,19,23-hexamethyltetracosane	-38

Molecular structure and degree of branching have significant impacts on melting point and cold flow properties.

Dewaxing Challenging Paraffinic Feeds in North America
Influence of Molecular Structure on Properties (2)

$C_{26}H_{54}$ Paraffin Isomer Pour Points

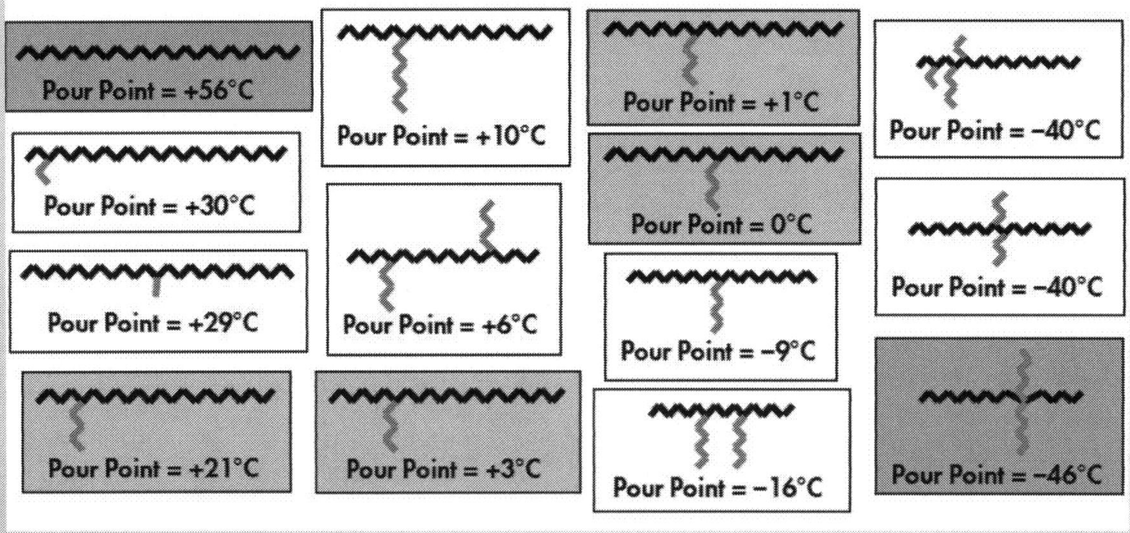

Dewaxing Challenging Paraffinic Feeds in North America

2xGC carbon - n-alkanes & isomerised alkanes in diesel boiling range (and extract)

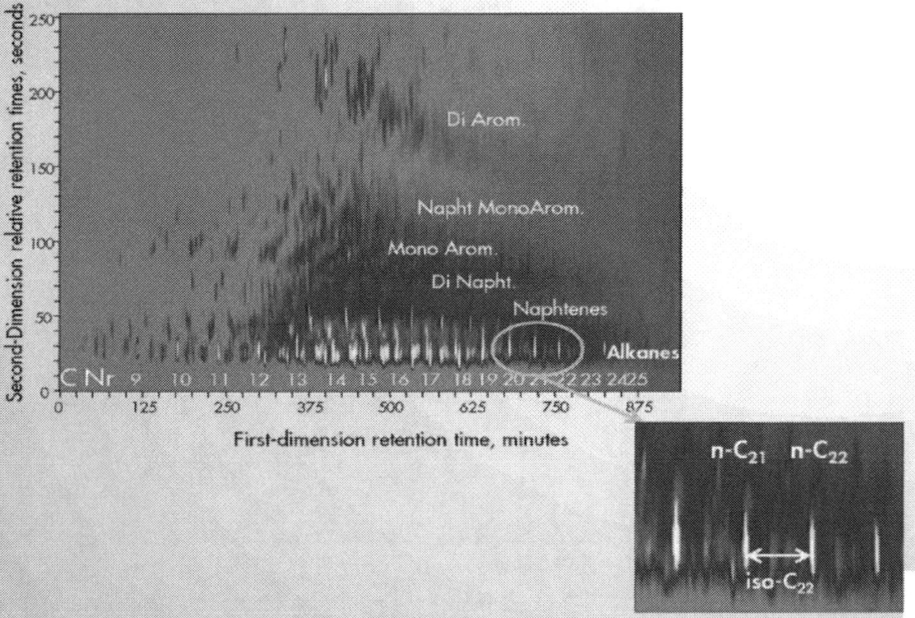

Dewaxing Challenging Paraffinic Feeds in North America

Influence of Boiling Range Distribution of n-Alkanes on Cold Flow Properties & Dewaxing (1).

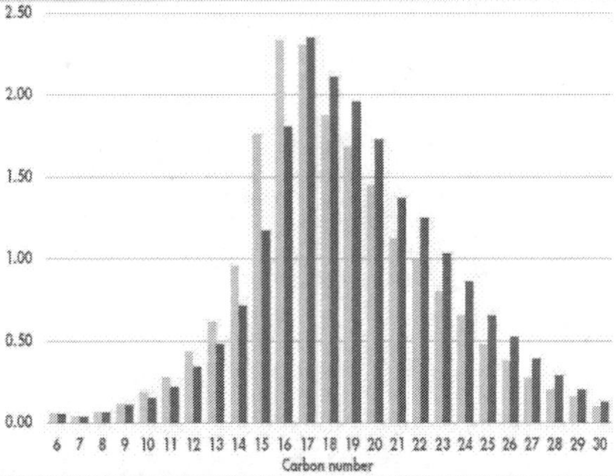

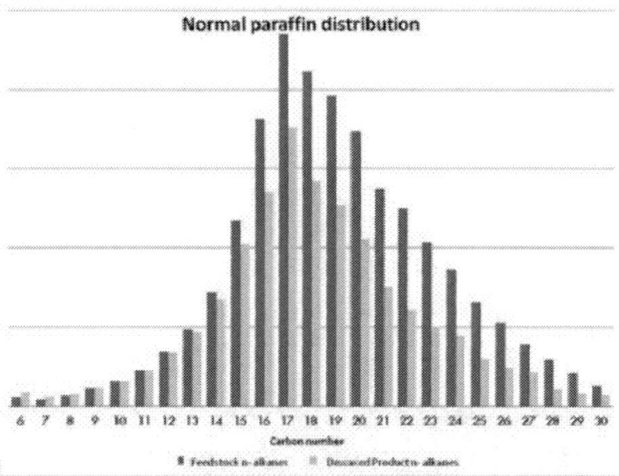

Dewaxing Challenging Paraffinic Feeds in North America

Influence of Boiling Range Distribution of n-Alkanes on Cold Flow Properties & Dewaxing (2).

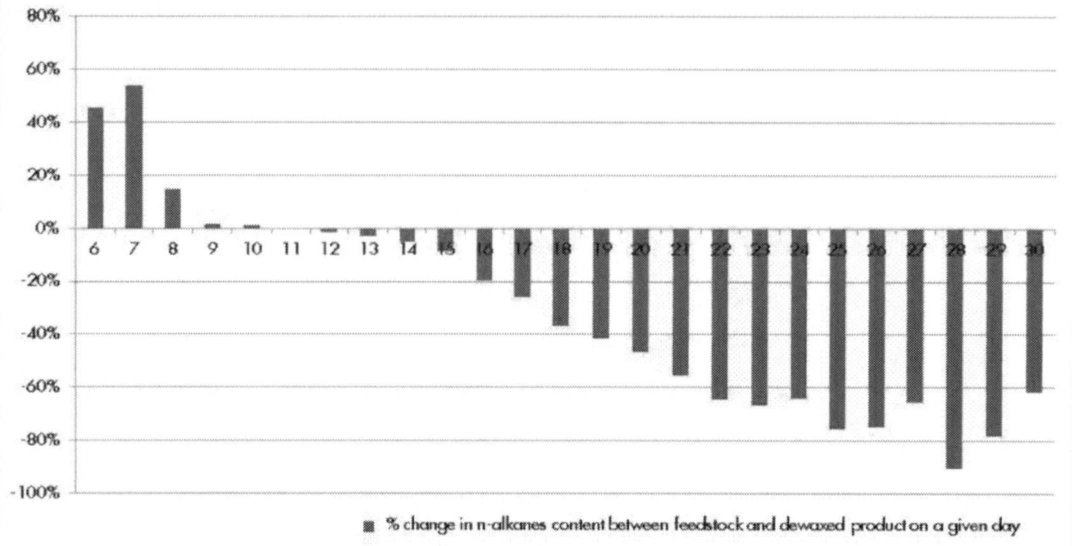

Dewaxing Challenging Paraffinic Feeds in North America

Shell GtL Base Oils Quality & main characteristics

Shell GtL Base Oils	4 cSt	5 cSt	8 cSt
Vk100 °C (cSt)	3.8-4.2	4.8-5.4	7.5-8.5
Vk40 °C (cSt)	-	-	-
Viscosity Index	135	145	150
VdCCS, -30 °C	1000	1860	5300
Pour Point, °C	< -30	< -24	< -15
Noack Volatility, %m	12	9	2
Flash Point, °C (D-93)	215	232	240

Dewaxing Challenging Paraffinic Feeds in North America

Heavy paraffins' mixture dewaxing for producing high-quality base oil for Shell lubricants

AIChE Southwest Process Technology Conference

2014 AIChE Conference Page 13

Dewaxing Challenging Paraffinic Feeds in North America

Reactant Shape Selectivity (1)

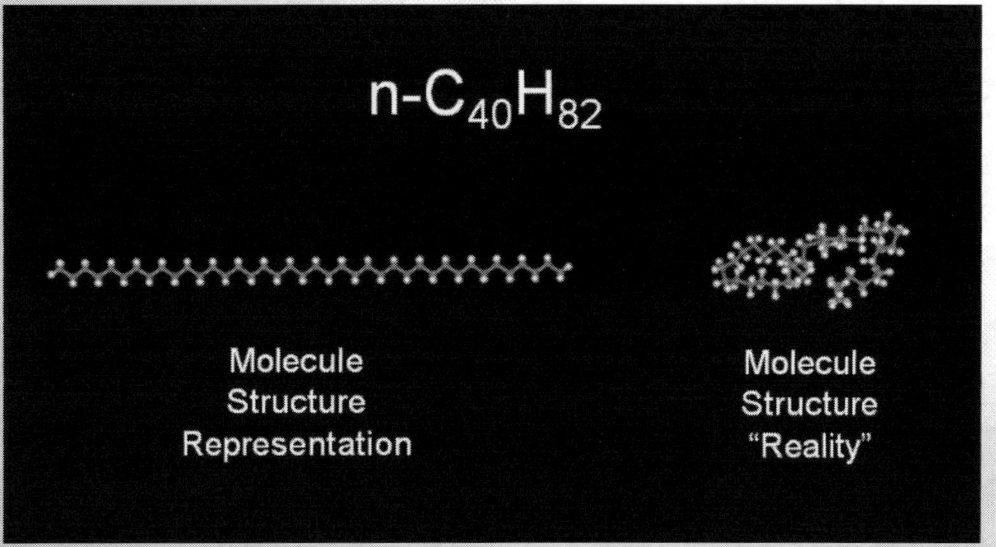

AIChE Southwest Process Technology Conference

2014 AIChE Conference Page 14

Dewaxing Challenging Paraffinic Feeds in North America

Reactant Shape Selectivity (2)

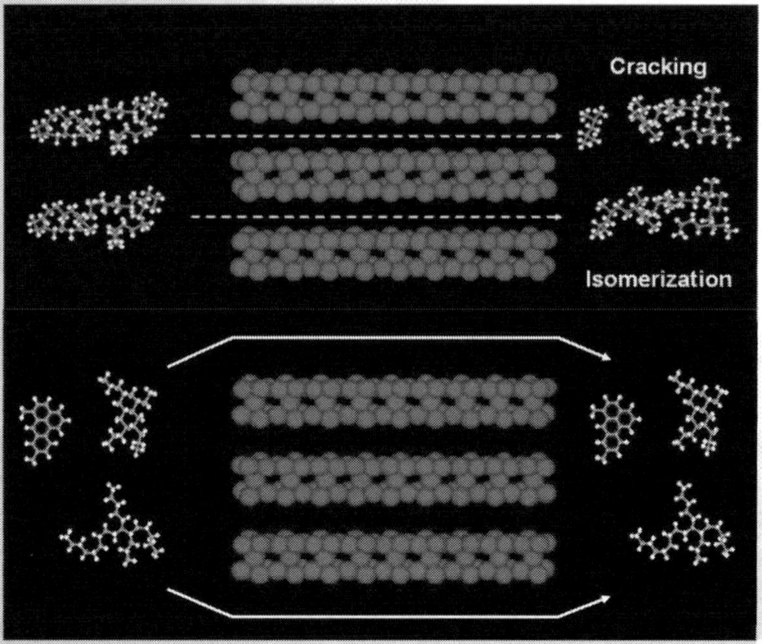

AIChE
Southwest Process Technology Conference

Dewaxing Challenging Paraffinic Feeds in North America

Reactant Shape Selectivity (3)

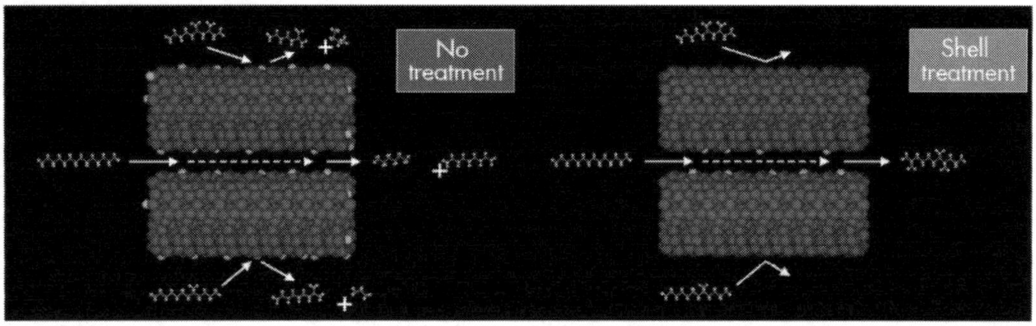

- Unique balance between isomerisation & selective cracking
- Highest selectivity in dewaxed products
- Longest cycle length
- Higher operational flexibility
- Lower dewaxing catalyst volume needed (synergy with HDS catalysts, nitrogen management)

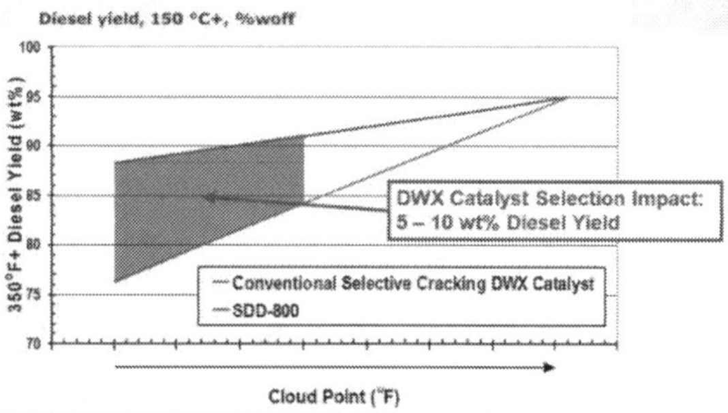

AIChE
Southwest Process Technology Conference

Dewaxing Challenging Paraffinic Feeds in North America

Summer-ULSD vs. Winter-Dewaxing Mode

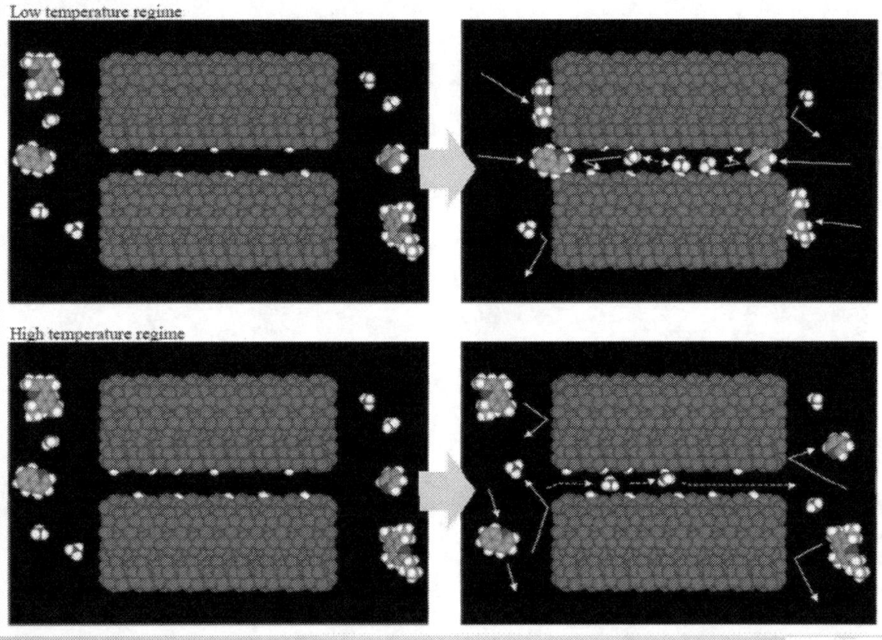

2014 AIChE Conference Page 17

AIChE Southwest Process Technology Conference

Dewaxing Challenging Paraffinic Feeds in North America

1st and 2nd Stage Dewaxing

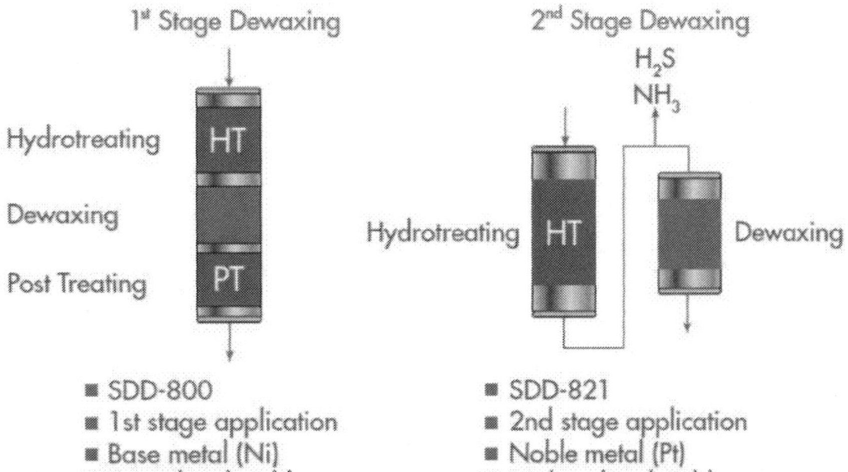

- SDD-800
- 1st stage application
- Base metal (Ni)
- Base diesel yield
- Lower CAPEX, lower cat cost
- Low H_2 consumption, no cetane boost

- SDD-821
- 2nd stage application
- Noble metal (Pt)
- Higher diesel yield
- Higher CAPEX, higher cat cost
- Higher H_2 consumption, cetane boost, hydrogenation

2014 AIChE Conference Page 18

AIChE Southwest Process Technology Conference

Dewaxing Challenging Paraffinic Feeds in North America
Selective Cracking & Isomerization Dewaxing (1)

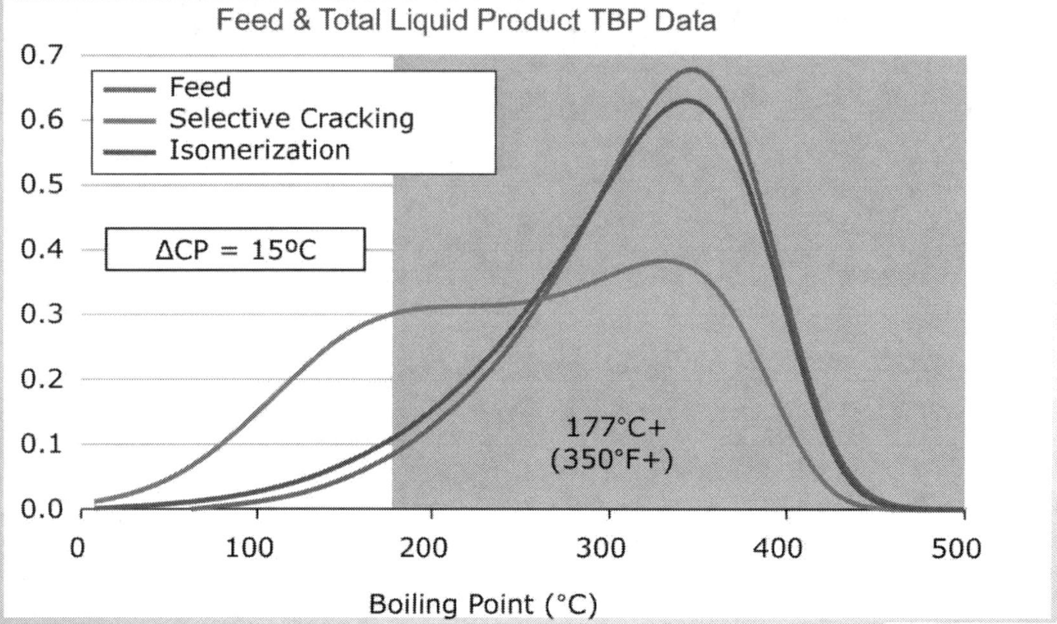

Dewaxing Challenging Paraffinic Feeds in North America
Selective Cracking & Isomerization Dewaxing (2)

	SDD-800 Selective Cracking Dewaxing	*SDD-821 Isomerization Dewaxing*
Yields (wt%)		
$C_1 - C_4$	Base	(Base - 4.1)
C_5 - 350°F	Base	(Base - 3.4)
> 350°F	Base	Base + 7.5
Cetane Index (D-4737)	Base	Base + 6
H_2 Consumption (SCFB)	Base	Base + 160*

ΔCP= 35°F * Significant contribution from ASAT

Dewaxing Challenging Paraffinic Feeds in North America

Developing Solutions for Deep Dewaxing of Highly Paraffinic Feedstocks

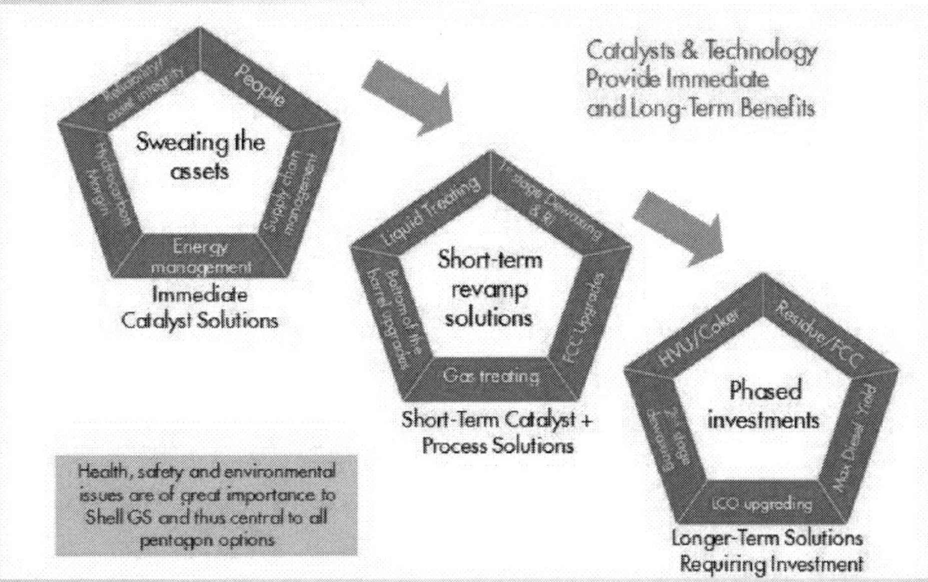

Dewaxing Challenging Paraffinic Feeds in North America

Case Study 1 & 2

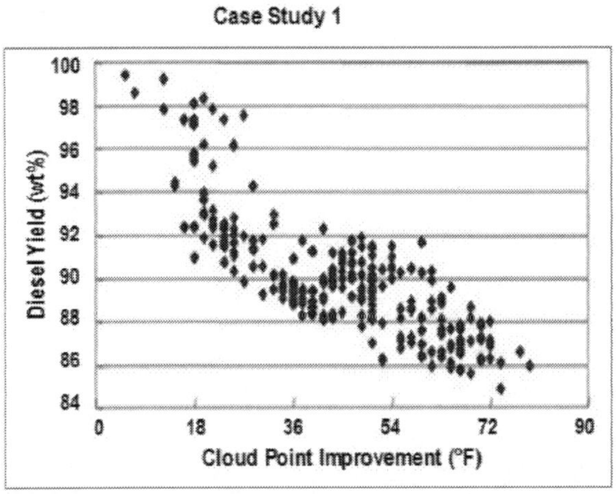

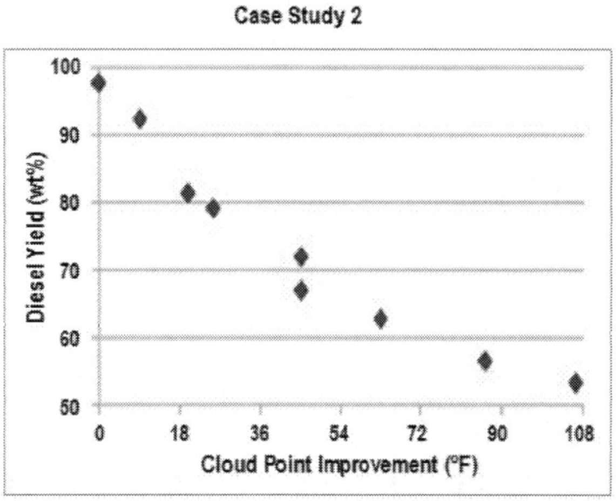

Dewaxing Challenging Paraffinic Feeds in North America

Case Study 3

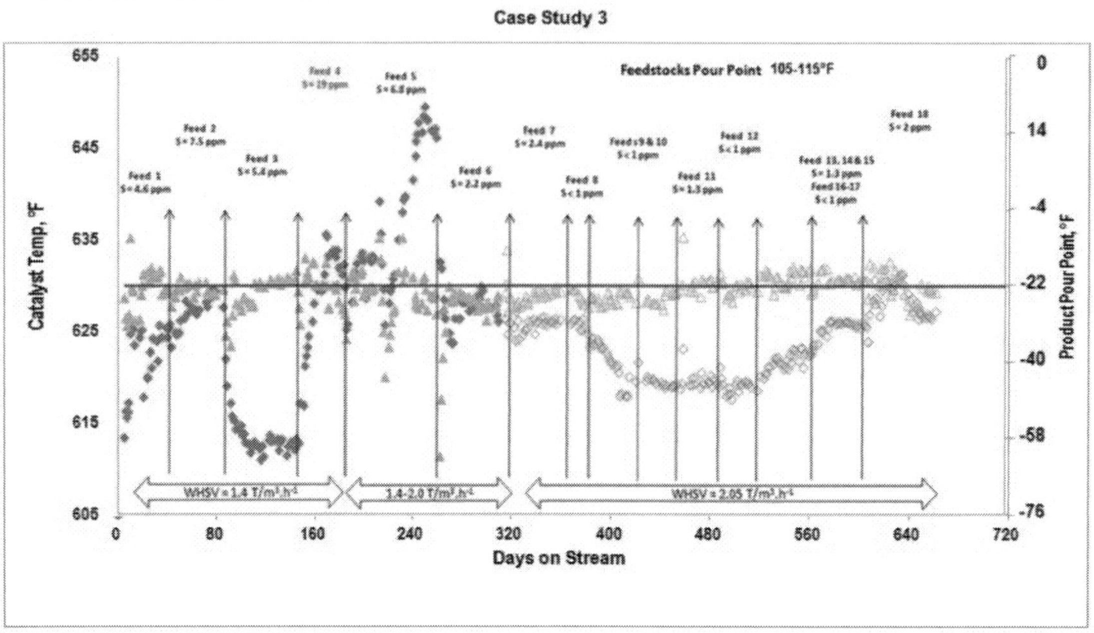

AIChE Southwest Process Technology Conference

Dewaxing Challenging Paraffinic Feeds in North America

Benefits of Catalyst Dewaxing

Catalytic Dewaxing as Cold Flow Improvement solution:

- Progressive
- Alternative
- Cost-efficient
- Profitable
- Tailored to application and feedstock
- Invaluable in development of Light Tight Oils (LTO)

AIChE Southwest Process Technology Conference

AIChE®

Southwest Process Technology Conference

Q&A

UOP
A Honeywell Company

Veronica Deak
Paula Bogdan
Patrick Whitchurch

6th AIChE Southwest Process Technology Conference
Catalysis and Reaction Engineering Session
October 9-10, 2014
Galveston, TX

1914 - 2014 | A Century of Innovation in the Oil and Gas Industry

Topics

UOP
A Honeywell Company

- **Project Background**
- **Scientific Platforms at UOP**
- **Experimental Approach**
- **Scale-up from Concept to Production**
- **Commercial Performance**
- **Summary**
- **Acknowledgements**

Continuous Technology Improvement

A Honeywell Company

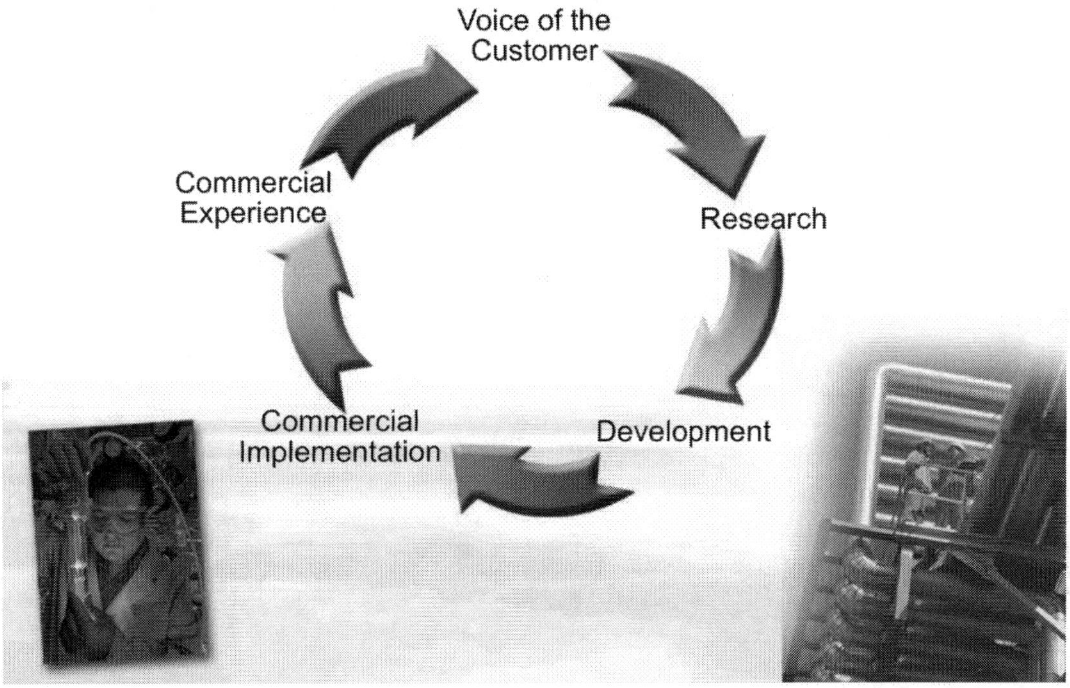

Feedback from Aromatics Complex Users

A Honeywell Company

Maximize

- **Para-Xylene Production**
- **Feed Efficiency**
- **Catalyst Life**

Minimize

- **Energy Consumption**
- **Xylene Loss**

I-500 catalyst helps commercial users address their goals

Typical Integrated UOP *p*-Xylene Complex

UOP
A Honeywell Company

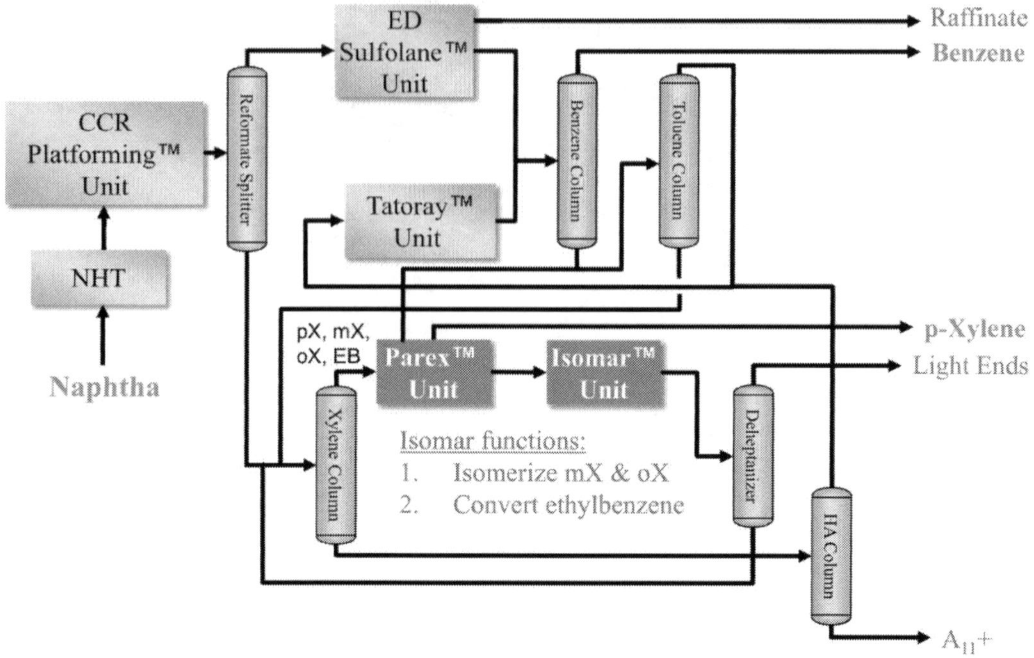

Isomar Background

UOP
A Honeywell Company

- **The Isomar process re-establishes equilibrium concentration of xylene isomers to a stream which has been depleted of one or more xylene isomers**

From a feed with mostly mX and oX

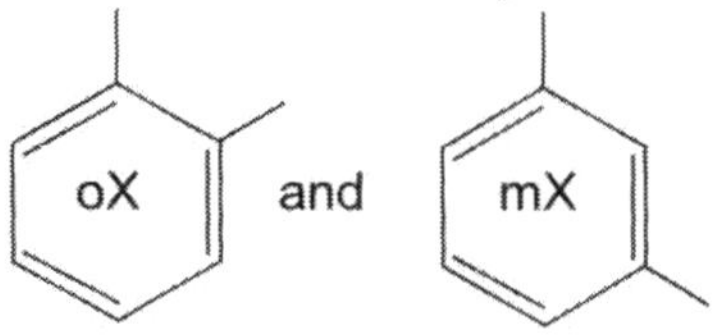

To a product with equilibrated xylene isomers

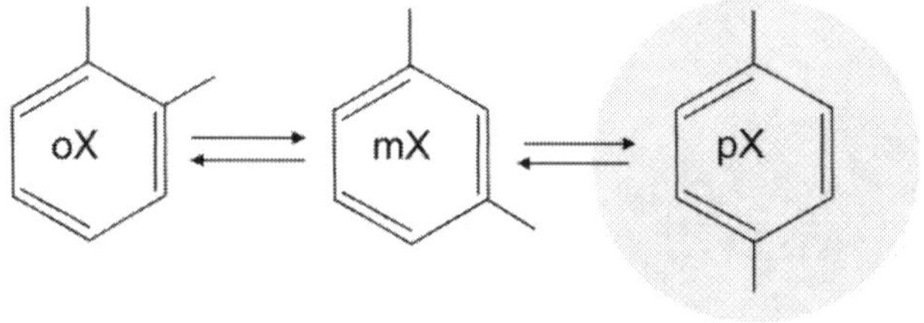

Isomar Background

A Honeywell Company

- The Isomar process re-establishes equilibrium concentration of xylene isomers to a stream which has been depleted of one or more xylene isomers
- **Customer can choose how to handle EB:**
  - *EB can be isomerized*
    - **Allows more efficient use of incoming feed**
    - **Requires larger recycle stream through Xylene Frac / Parex**

EB Isomerization

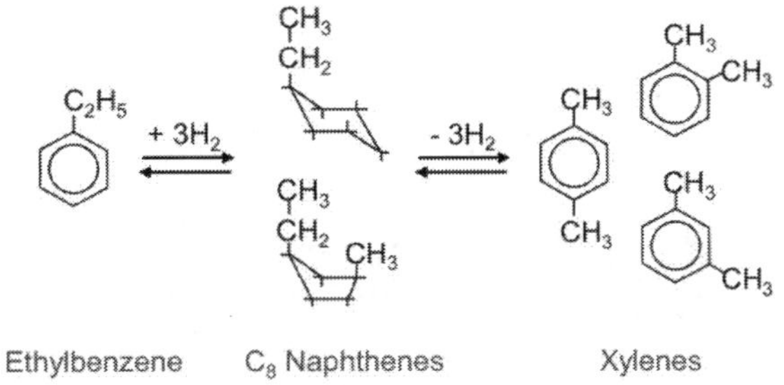

Isomar Background

A Honeywell Company

- The Isomar process re-establishes equilibrium concentration of xylene isomers to a stream which has been depleted of one or more xylene isomers
- Customer can choose how to handle EB:
  - *EB can be isomerized*
    - Allows more efficient use of incoming feed
    - Requires larger recycle stream through Xylene Frac / Parex
  - ***or EB can be dealkylated***
    - **Less recycle through Xylene Frac / Parex**
    - **Good for maximizing throughput and production**

EB Dealkylation

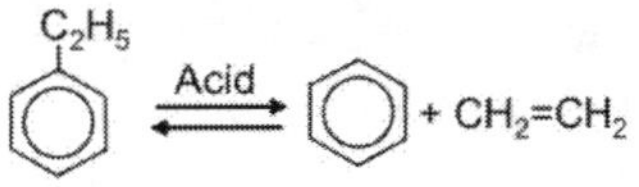

Isomar Background

UOP
A Honeywell Company

- The Isomar process re-establishes equilibrium concentration of xylene isomers to a stream which has been depleted of one or more xylene isomers
- Customer can choose how to handle EB:
  - *EB can be isomerized*
    - Allows more efficient use of incoming feed
    - Requires larger recycle stream through Xylene Frac / Parex
  - *or EB can be dealkylated*
    - **Less recycle through Xylene Frac / Parex**
    - **Good for maximizing throughput and production**

Most customers choose EB Dealkylation, leading to UOP development of I-500 catalyst

UOP Scientific Platforms
Driving Innovation

UOP
A Honeywell Company

Next Generation Process Technologies | **High Efficiency Catalytic Processes** | **Breakthrough Separation Processes**

Advanced Characterization & Modeling Capabilities

- In-situ Characterization
- Single Atom Analysis
- 3D Tomography
- DFT and Predictive Models
- Combinatorial Chemistry
- Advance Organic Structure

New Materials & Catalysis Platform

- New Zeolite Structures
- Nano-Materials
- Novel metal catalysis
- Organometallics
- Inorganic-Organic Hybrids

Pilot Plant Capabilities

- Statistically significant product quality and yields
- Predicts commercial performance

Technology Scale-Up Capabilities

UOP
A Honeywell Company

Pilot Plants
Analytical & Informatics

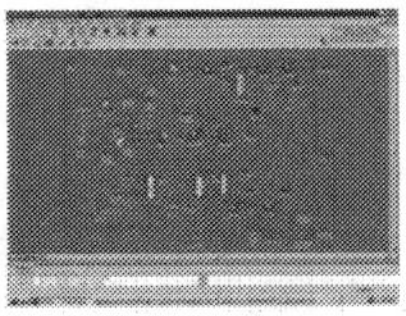
UniSim Simulation
Process Synthesis & Economics

New Processes

Computational Fluid and
Particle Dynamics

Cold Flow Modeling
Multiphase Flow

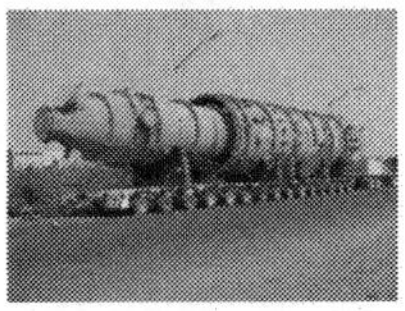
New Equipment

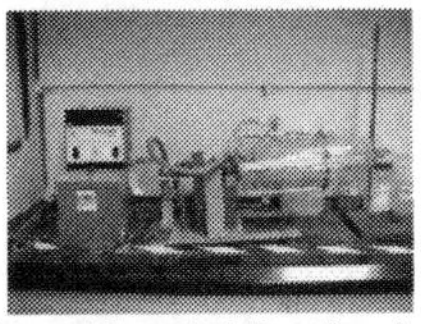
New Material Synthesis
Lab Scale Prototypes

Commercial Semi-Works Plant
Crystallization and Forming

New Catalysts
& Adsorbents

Nanotechnology Critical to Industrial Catalysis

UOP
A Honeywell Company

Control in Industrial Catalysts:

Size of Zeolite Crystals

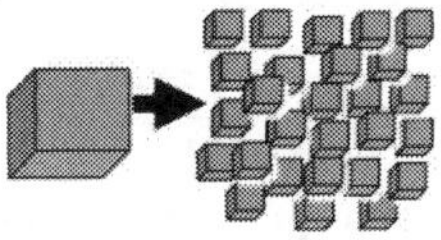

Composition of Crystals

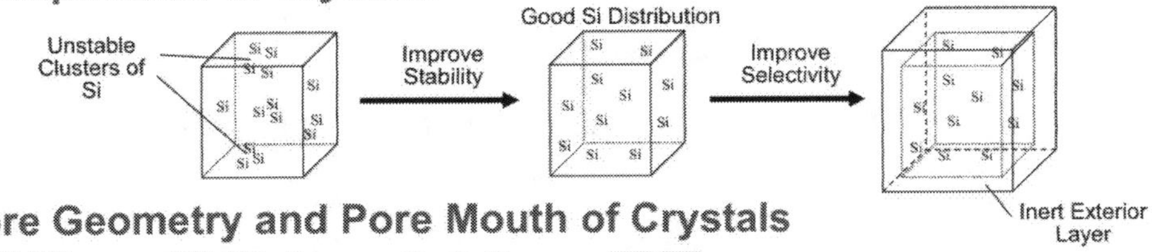

Pore Geometry and Pore Mouth of Crystals

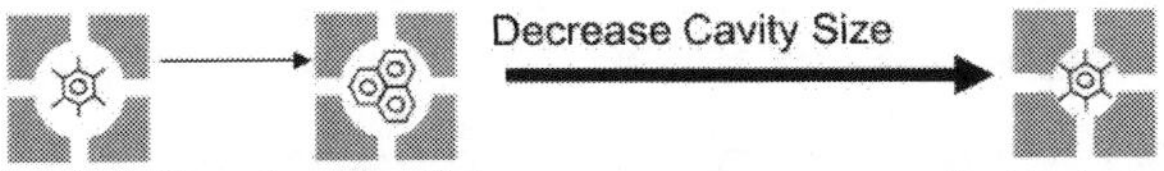

Particle Diffusional Properties

Zeolite Particle Morphology

Robust Work Processes

UOP
A Honeywell Company

- **Six Sigma/Lean**
  - **Methodology and tools to ensure development and manufacturing process capabilities**
  - **Focused on meeting customer requirements**
- **Stage Gate Process**
  - **Standardized process to ensure successful new product commercialization**

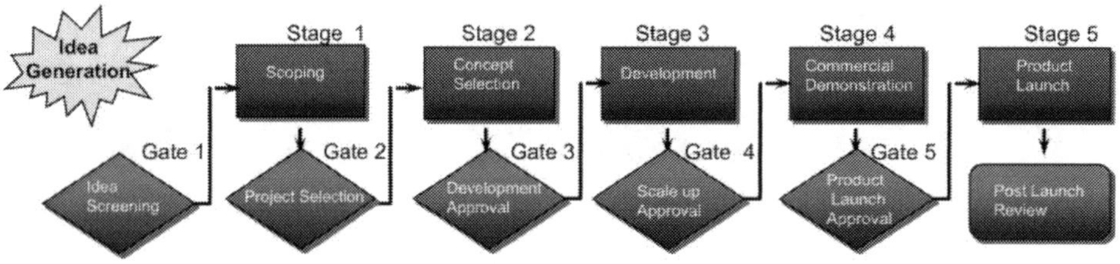

The Newest UOP Isomar Catalyst: I-500

UOP
A Honeywell Company

- I-500 delivers improved performance for every critical Isomar process variable
  - **Improved Xylene Retention**
  - **Higher pX/X Selectivity**
  - **High EB Conversion and Catalyst Stability**
  - **H2/HC as low as 1**
  - **Saleable Benzene without Extraction**
- I-500 is currently operating in 7 units, and will be loaded in 1 more unit later in 2014

I-500 Raises the bar set by UOP's previous EB Dealkylation catalysts

Experimental Approach: Desired Rxns

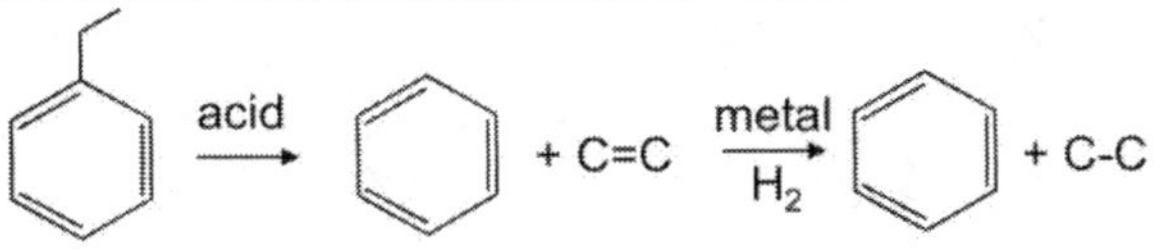

EB-Dealkylation and ethylene saturation

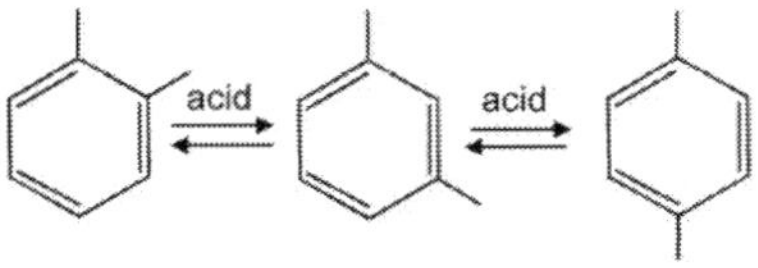

Xylene Isomerization

Explore new materials that <u>promote</u> these reactions...

Experimental Approach: Undesired Rxns

...and new materials that <u>limit</u> these reactions

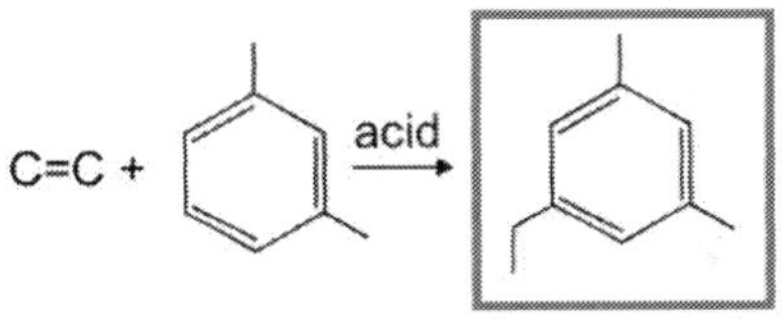

Some xylenes lost by alkylation of unsaturated ethylene

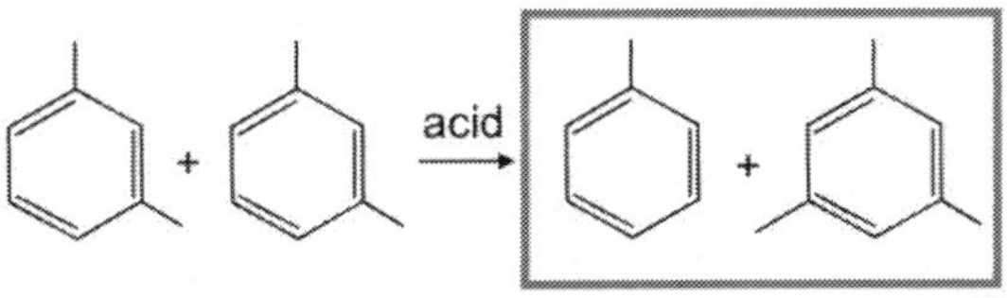

Significant xylene loss from transalkylation (many pathways)

Screening Pilot Plant Test Methodology

uop
A Honeywell Company

- Screening Pilot Plants
  - High through-put, short tests
  - Fast turnaround
  - Once-through hydrogen
  - 4 reactors in parallel
  - Whole catalyst pellets
- Feeds can be either
  - Commercially relevant feeds; or
  - Theoretical lab blended feeds
- Activity and yield results have good reproducibility
- Choose catalysts leads that are statistically significant

Scale-up Pilot Plant Test Methodology

uop
A Honeywell Company

- Catalyst leads from pilot plant screening results are selected for testing in more precise, scale-up pilot plants
- Scale-up pilot plants
  - Single Reactor
  - Recycle Gas
  - Adiabatic reactors
  - Commercial feeds
  - Published UOP methods for product analyses
  - Results match commercial performance
- Statistical Data Analysis
  - Measurement system analysis used to reduce run-to-run variation
  - Use 95% confidence intervals for comparisons and predictions
  - Regular pilot plant referencing to build confidence over time

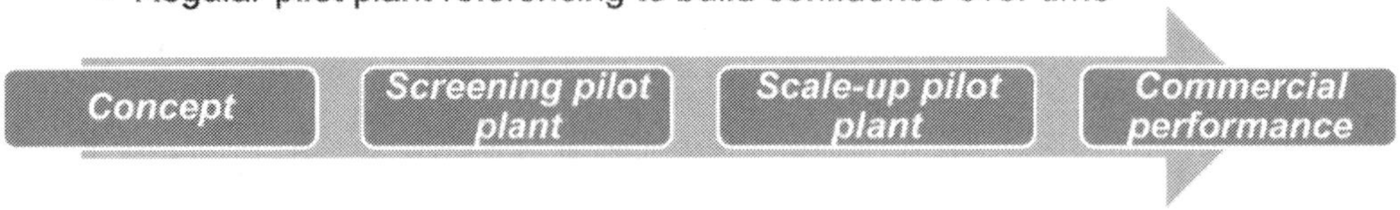

Materials Scale-Up

- In parallel with pilot plant testing scale-up, materials are scaled-up too
- Catalyst development scale equipment provides a chance to create a robust commercial catalyst recipe
  - Study mixing effects
  - Study impact of force/work at different scales
  - Determine sensitivity to raw material changes
- Semi-works Development ties concepts to production
  - Collaboration first between Research and Development
  - Then between Development and Manufacturing

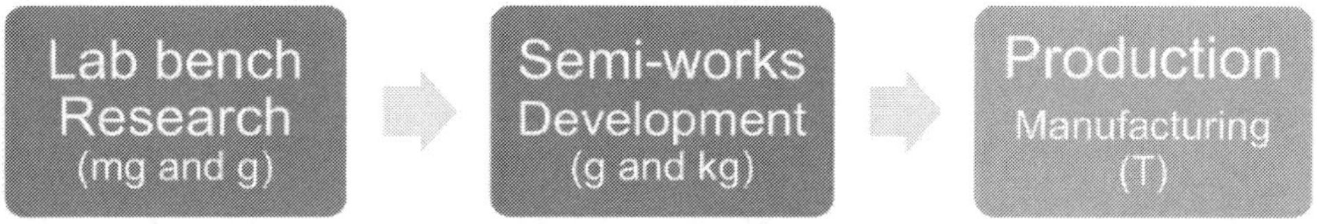

I-500 Catalyst Unit A Commercial Performance versus Pilot Plant Trends

I-500 Unit A: EB Conversion Activity

UOP A Honeywell Company

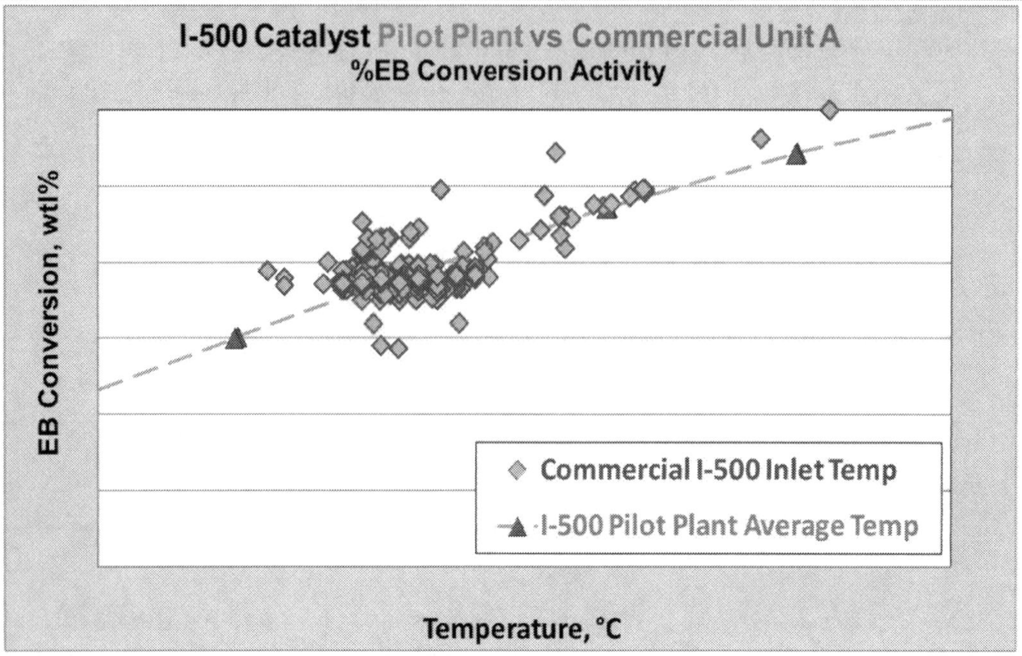

I-500 Unit A: Xylene Retention

UOP A Honeywell Company

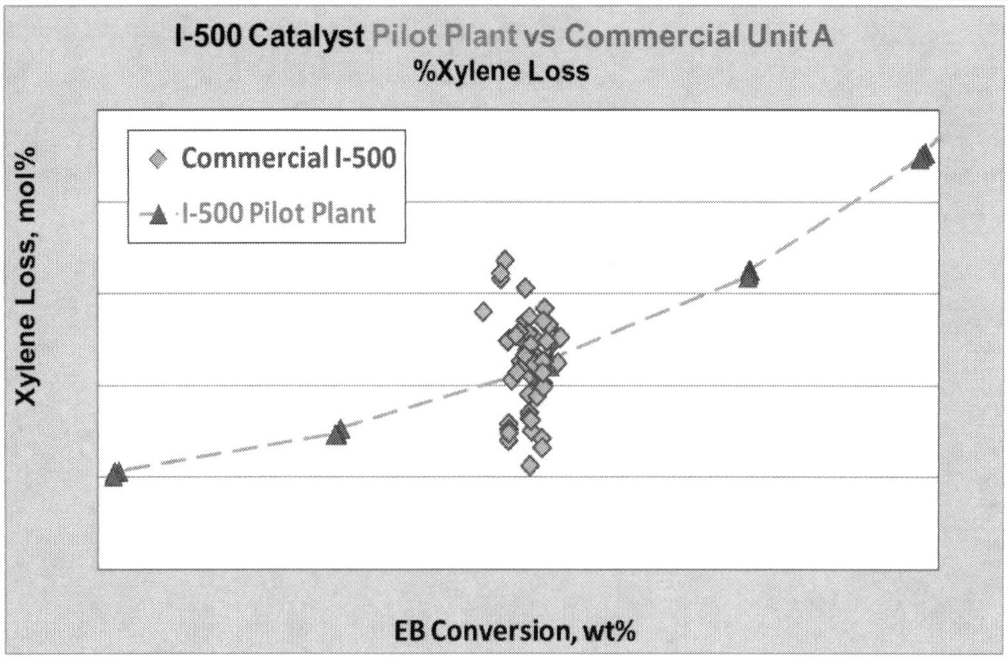

I-500 Unit A: Xylene Isomerization

UOP
A Honeywell Company

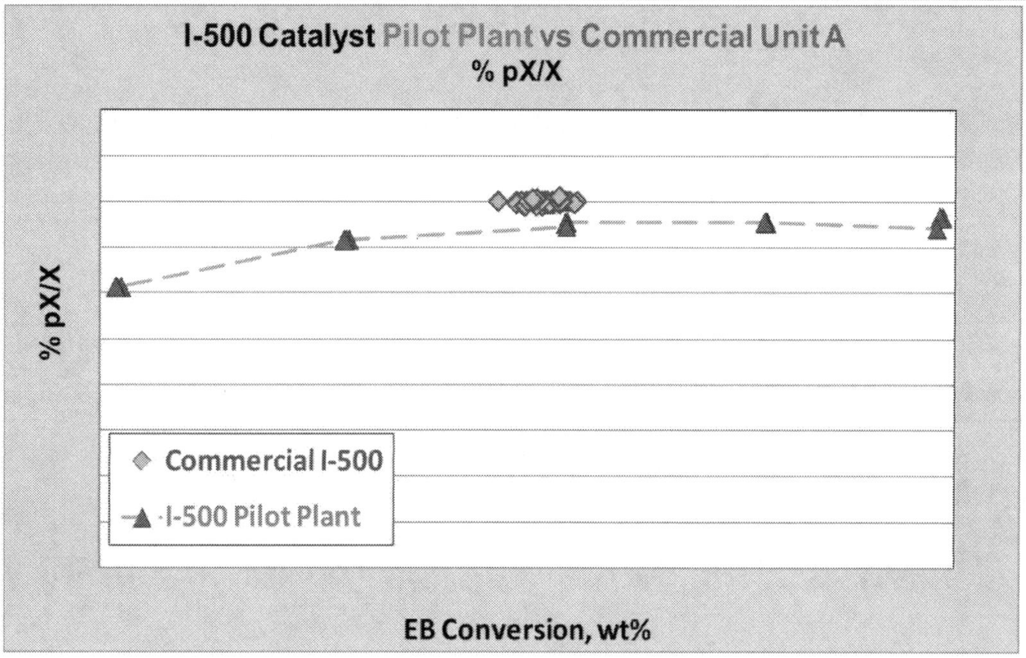

Summary

UOP
A Honeywell Company

- I-500 has helped Aromatics complex customers with:
  - Achieve higher pX production
  - Minimal xylene loss

- I-500 catalyst is meeting customer expectations in seven commercial Isomar units around the world.

- Collaboration among UOP's scientific platforms help drive innovation, checked by statistics and UOP's stage gate process.

Acknowledgements

UOP
A Honeywell Company

- Honeywell UOP, Research and Development
- Honeywell UOP, Manufacturing Product Technology
- Hard work and efforts of many technical specialists and support staff, including
  - Paula Bogdan, Patrick Whitchurch, Rich Gajeck, New Materials Group, Pilot Plant Testing Group, Characterization, and Analytical

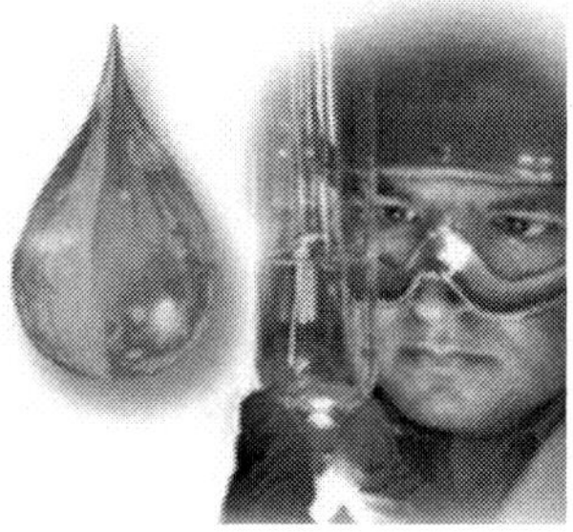

Scientific Design's Integrated Ethanol to Ethylene Oxide/ Ethylene Glycol Process

Jan Malik
Scientific Design Company, Inc.

AIChE
Southwest Process Technology Conference

Scientific Design Company

- Founded in 1946 by Dr. Ralph Landau
- Joint ownership of SABIC and Clariant
- Head Quarter at Little Ferry, NJ with branch offices at Bahrain and Shanghai
- Develops Processes and licenses technology to the worldwide petrochemical industry
- Develops, manufactures and supplies catalyst for its proprietary technologies

AIChE
Southwest Process Technology Conference

Scientific Design Company

- Sole technology provider offering EO/EG process both from petro-ethylene and Ethanol
- Leader in EO/EG technology with ~ 75% market share of non-captive plants licensed since 1980
- Participated in the licensing, design and/or construction of more than 300 major chemical plant projects in 30 countries

Southwest Process Technology Conference

Current Licensed Technologies

- **Ethylene Oxide / Ethylene Glycols**
- **Ethylene from Ethanol**
- Ethylene Glycol Ethers
- Ethanolamines
- Ethoxylation / Propoxylation
- Maleic Anhydride
- Polyether Polyols

Southwest Process Technology Conference

EO - The Value Chain

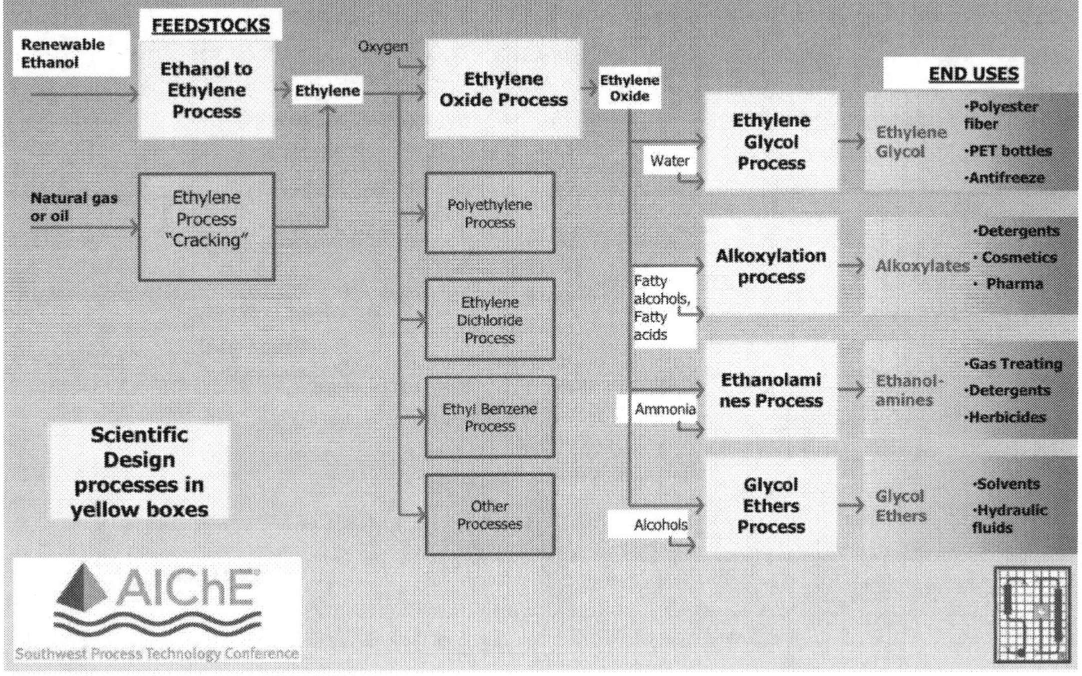

Ethanol feedstock for Ethylene production

- Ethylene is typically produced in a cracker unit, using ethane as feedstock.
- There are EO users without access to cracker-derived ethylene.
- Ethylene Oxide cannot be shipped easily over long distances.
- 1973 oil embargo coincided with peak oil in the USA, spurring sustainable feedstock technology.
- In Ethanol-to-EO technology, catalyst development is critical to process that is competitive to petrochemical Ethylene.

AIChE
Southwest Process Technology Conference

Ethanol to EO – Reaction Chemistry

The Dehydration Reaction is represented by the stoichiometry:

$$CH_3-CH_2-OH \rightleftarrows CH_2=CH_2 + H_2O$$

Endothermic reaction ΔH @ 25 ºC = –230 kcal / kg Ethanol

Forward reaction favored by low P and high T

Catalyst: Al_2O_3-based SynDol® produced by Scientific Design

Key concern for catalyst development: Low by-product formation

Southwest Process Technology Conference

Ethanol To Ethylene Process Block Diagram

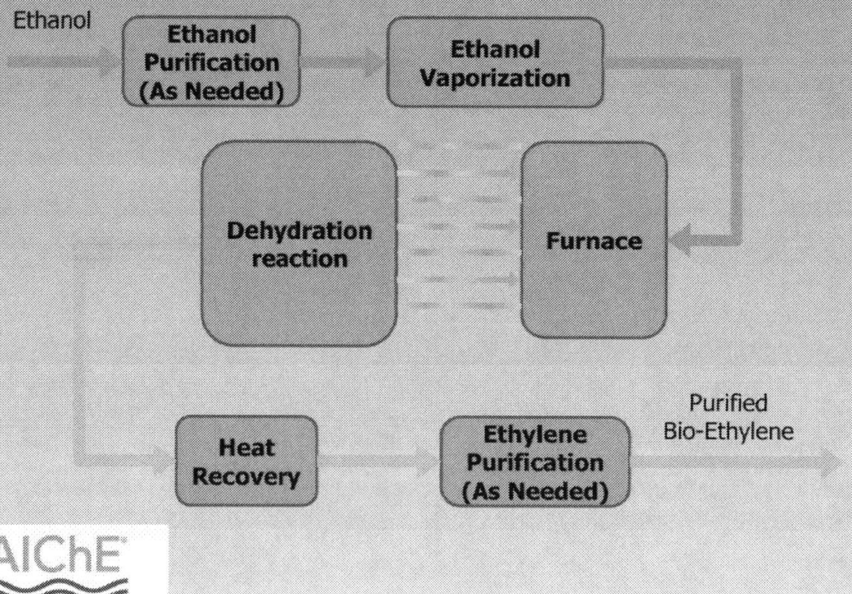

Southwest Process Technology Conference

Ethanol to EO – Reactor

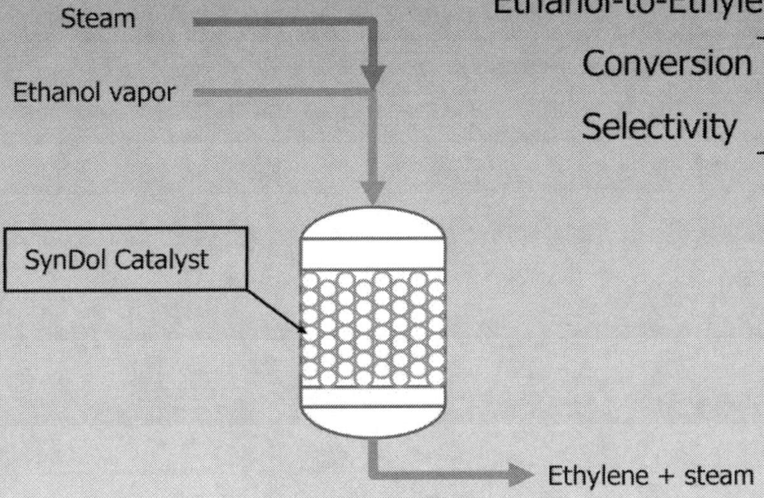

Ethanol-to-Ethylene Reaction:

Conversion / Selectivity } Essentially 100%

Southwest Process Technology Conference

Ethanol to Ethylene Technology

The Timeline

- 1960's: Development and supply of ethanol dehydration catalyst, SynDol
- 1980's: SD developed technology to produce ethylene from ethanol
- 1988: Sold the technology to Chematur AB
- 2005: Agreement with Chematur to work jointly on new E-to-E projects

Southwest Process Technology Conference

Ethanol to Ethylene Technology

The Timeline (Continued)

- 2006: 60,000 MTPY Solvay plant to be built in Brazil (Project on hold)
- 2010: Licensed three 60,000 MTPY plants to TCI Sanmar, Egypt
- 2014: First of the three TCI Sanmar plants commissioned in March'14

Southwest Process Technology Conference

Integrated Ethanol to EO/EG Process

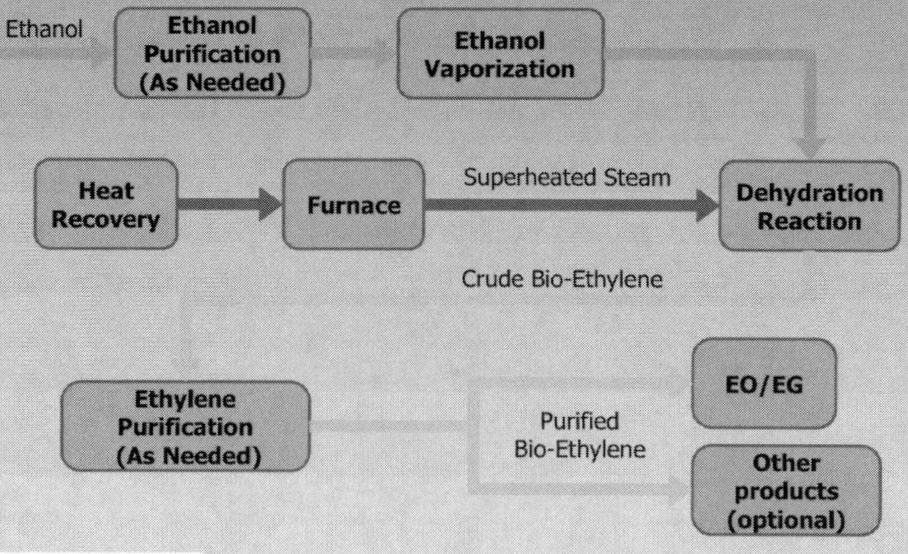

Southwest Process Technology Conference

Integrated Ethanol to EO/EG Process

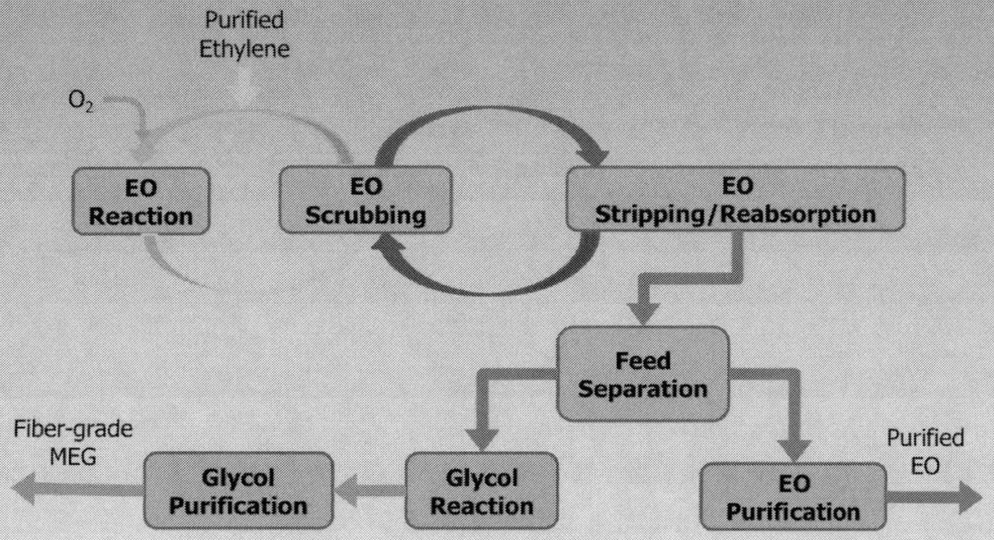

Southwest Process Technology Conference

Integrated Ethanol to EO / EG Technology

The Timeline

- 1986: Licensed World's First Fully Integrated Ethanol to EG Plant
  - Under Commercial production since 1989
  - Plant capacity expanded many times from initial design of 20,000 MTPY MEG
  - Current design capacity of 160,000 MTPY MEG
  - Supplies 'Green MEG' for the production of PlantBottle™
  - 'Green MEG' fully comparable to petro-derived MEG meeting all quality specifications and can be used in food grade applications

Southwest Process Technology Conference

EO Chemistry

Main Reaction:

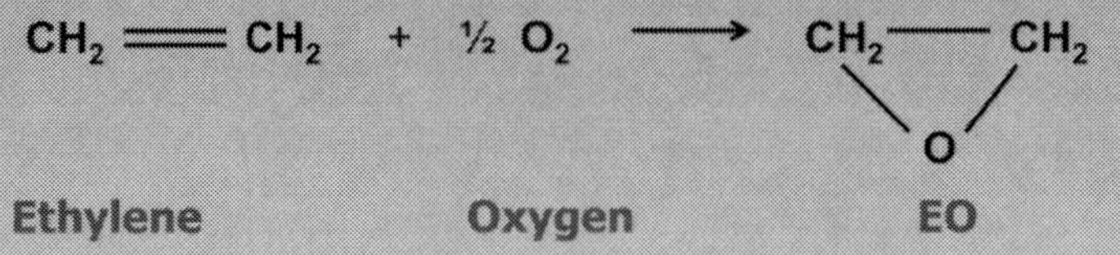

ΔH @ 25º C = −25 550 kcal/kg-mol of C_2H_4

Side Reaction:

$$CH_2{=}CH_2 + 3\ O_2 \longrightarrow 2\ CO_2 + 2\ H_2O$$

ΔH @ 25º C = −316 220 kcal/kg-mol of C_2H_4

Catalyst: Silver-based SynDox® produced by Scientific Design

Key concern for catalyst development: Selectivity

AIChE Southwest Process Technology Conference

15

Ethylene Oxide Reactor

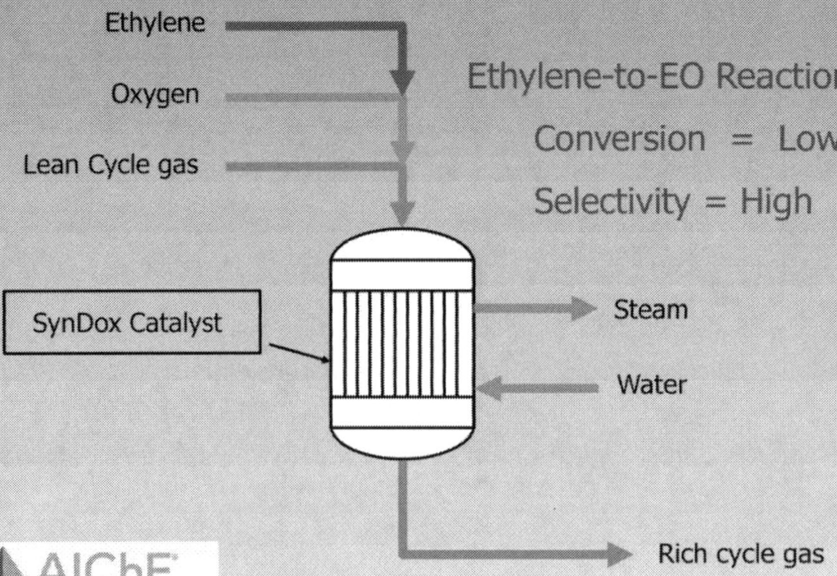

Ethylene-to-EO Reaction:

Conversion = Low

Selectivity = High

AIChE Southwest Process Technology Conference

Integrated Ethanol to EO / EG Technology

- 2007-2010: Licensed six Ethanol to EO plants in China
  - Four of them with flexi-feed option
  - Three of these six successfully commissioned in 2011 - 2013
  - Rest under different project stage with start-ups in 2014 – 2017

2011: Licensed second Ethanol to EO plant to one of the above clients

2012: Licensed three Ethanol to EG plants in China

2014 : Developing PDP for a US based Ethanol to EO plant

Southwest Process Technology Conference

Integrated Ethanol to EO / EG Technology

Highlights

- Fully integrated two stage process:

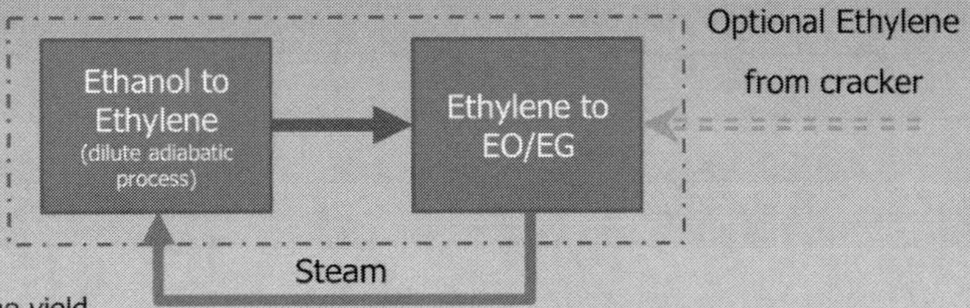

- High Ethylene yield
- No need of intermediate storage of ethylene
- Energy integration between Ethanol Dehydration and EO process
- Flexibility to switch between bio-ethylene and petro-ethylene as and when required

Southwest Process Technology Conference

Integrated Ethanol to EO / EG Process

Highlights (Continued)

- Flexibility to design for excess bio-ethylene production with a quality suitable for other downstream products such as VCM and HDPE
- Fiber-Grade MEG quality – same as with petro-ethylene
- 25 years of rich operating experience with Bio- EO/ EG
- Strong R&D support for continuous improvement in
  - Process efficiency
  - Handling of ethanol from traditional and 2nd generation raw materials

Southwest Process Technology Conference

Recent Interest in Green MEG

- Green PET Bottles (Pepsi; Coca Cola)
- Green PET film for packaging (Nestle, P&G)
- Green Anti-freeze (Toyota)
- Green Plastics for high end automobiles (Toyota)

Southwest Process Technology Conference

Green MEG Market Potential

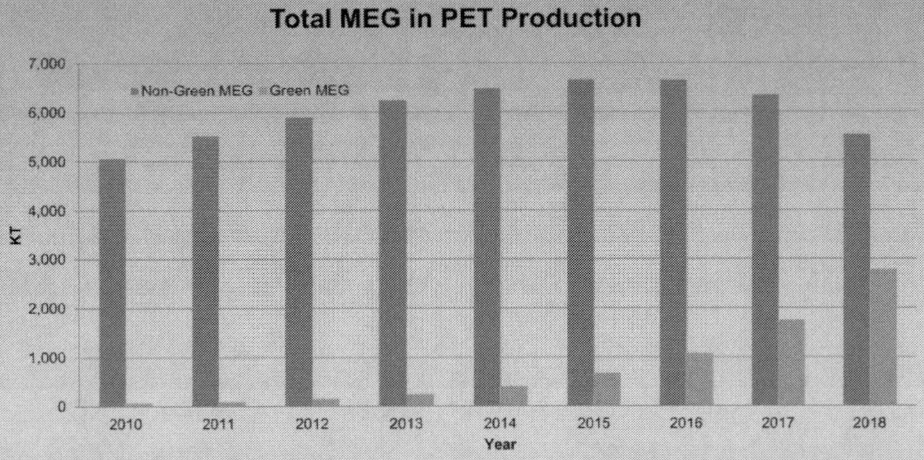

Green MEG production could be more than 30% of total MEG used in PET by 2018 or almost 3,000 KT of Green MEG

AIChE
Southwest Process Technology Conference

Typical Bio-Refinery Process Layout

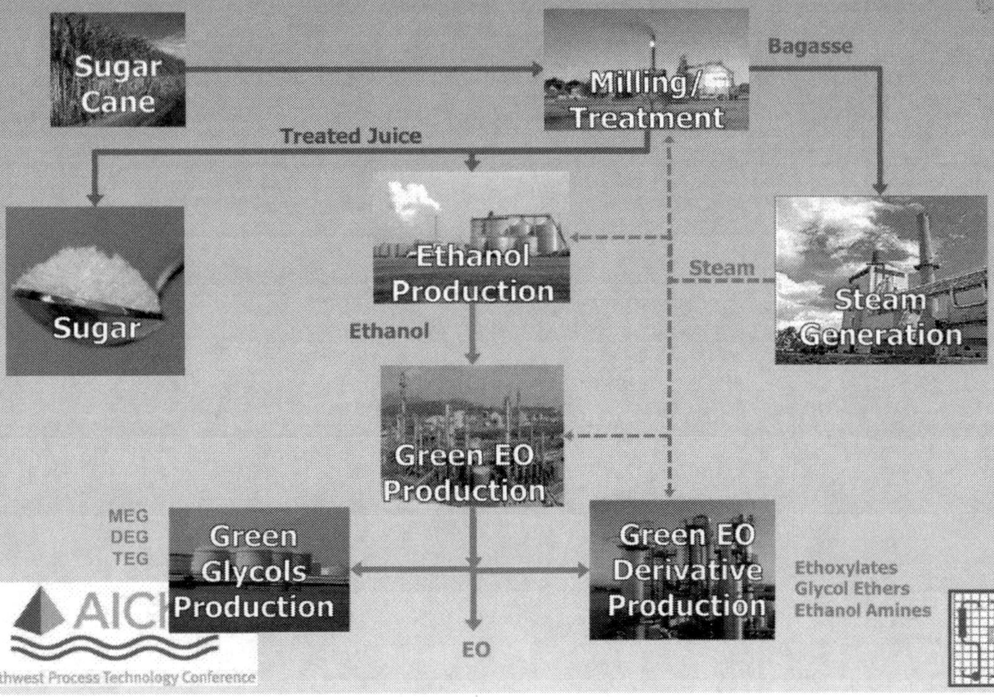

AIChE
Southwest Process Technology Conference

The Economics Of Bio-Ethylene

USAGES[1]		
Ethanol	MT/MT	1.8
MP Steam	MT/MT	3.3
Power[2]	KW/MT	300
COSTS[1]		
Ethanol[3]	$ / MT	945 - 1260
Utilities / Catalyst	$ / MT	70-100
Total	$ / MT	1015 - 1360
Ethylene Market Price	$ / MT	1050 - 1450

[1] Per MT of Ethylene [2] With Ethylene Purification [3] Ethanol cost = $525 - $700 / MT

AIChE
Southwest Process Technology Conference

Conclusion

- Addresses a niche but growing market for Green Glycol
- Viable Option for regions where petro-ethylene is expensive or not accessible
- Provides flexibility to switch between petro- & bio- ethylene based on raw material cost and market demand
- Can be a major process link for other ethylene or EO/EG based bio-products

AIChE
Southwest Process Technology Conference

THANK YOU!

AIChE
Southwest Process Technology Conference

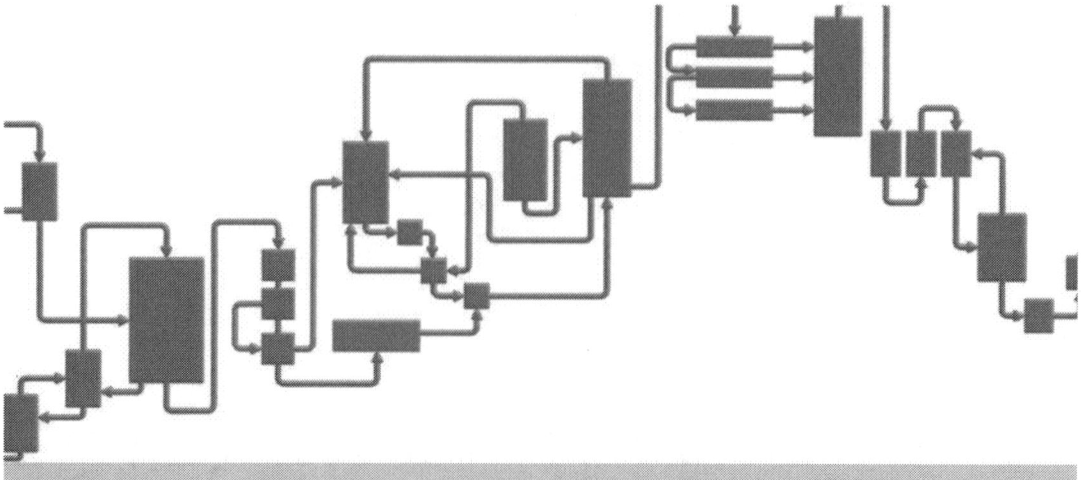

Best Practices for SOx Emissions Control

AIChE – October 2014

Jarred Drewry
Sr. Technical Services Engineer

JM
Johnson Matthey
Process Technologies

Introduction

JM
Johnson Matthey
Process Technologies

- **Many FCC units have already experienced stricter SO_2 emission regulations**
  - Primarily refiners operating close to large population centers
- **Many other refiners are now beginning to face such restrictions**
  - These refiners may benefit from what has been learned by other refiners
- **$INTERCAT_{JM}$ has crystallized this experience into a set of Best Practices for different FCC unit operations:**
  - Full combustion
  - Partial burn
  - Two stage regeneration
  - Maximum LCO operations
  - From SO_2 trimming to ultralow emissions

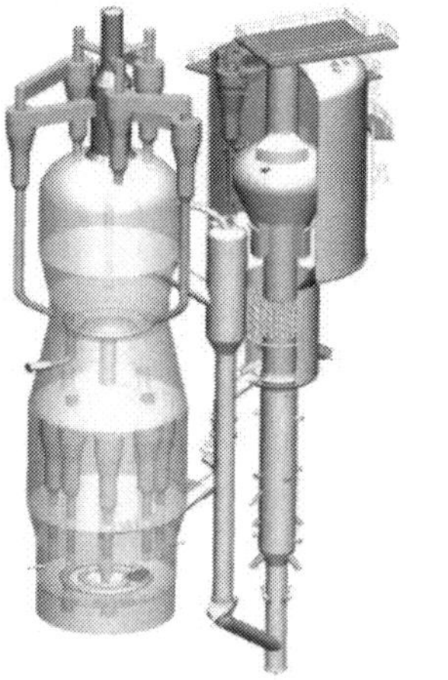

INTERCAT JM

2

Where Does SOx Come From?

Johnson Matthey
Process Technologies

Some sulfur in the feed is trapped on coke

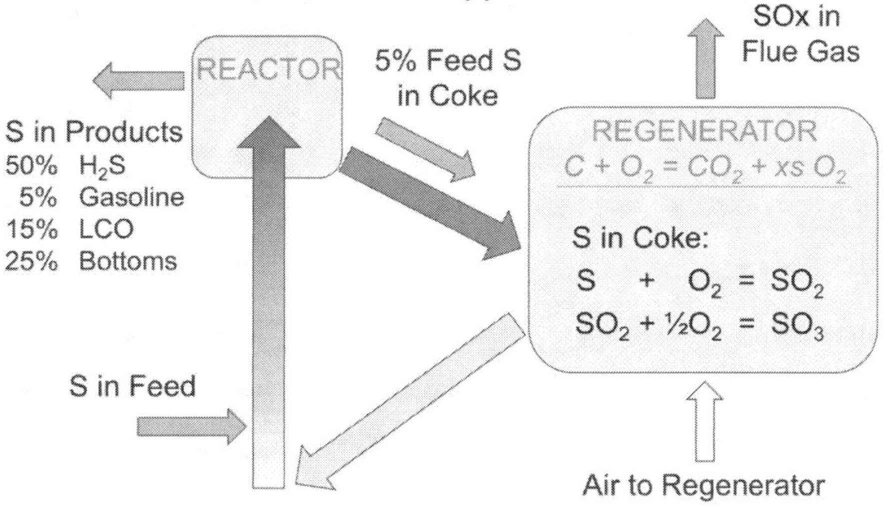

INTERCAT

3

How do SOx Additives Work?

Johnson Matthey
Process Technologies

- **Fundamental principle:** SOx additives contain MgO which bonds with SO_3
  - $SO_3 + MgO = MgSO_4$
- Equilibrium sets an approximate 90:10 SO2:SO3 ratio in complete combustion
  - Since MgO only bonds with SO_3 and the oxidation of SO_2 to SO_3 is very slow, cerium is used to promote the oxidation reaction
  - $SO_2 + ½\ O_2 = SO_3$

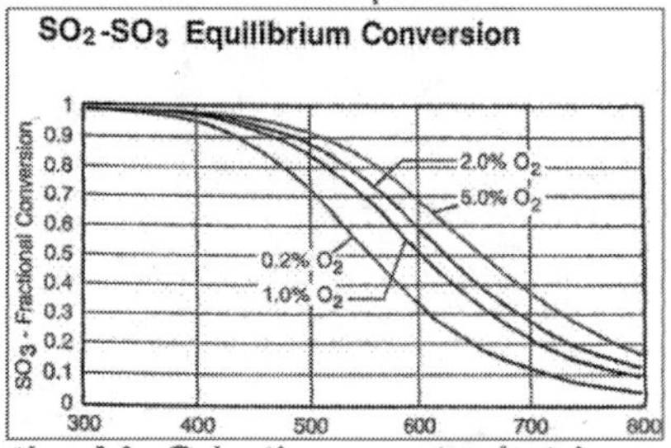

- The catalyst adsorbed SO_3 is removed from the MgO in the reactor/ stripper regenerating the additive to continue the cycle
  - Vanadium is embedded on the catalyst to assist in the sulfur release
  - $MgSO_4 + 14[H] = MgO + 4H_2S + 3H_2O$

INTERCAT

4

SOx Additive Structures

Johnson Matthey
Process Technologies

- **SO_3 adsorption occurs on the Mg structure**
  - Traditional additives
    - *Sorption structure – Spinel*
    - *Oxidation package – CeO_2 + V*
    - *Release assistant – V*
    - *Active Mg content is ~35%*
    - *Designed or full burn/gasoline mode operation*
    - *O_2 availability and additive regenerator are not limiting*
  - Second Generation
    - *Similar oxidation and release package*
    - *Sorption phase – Hydrotalcite*
    - *Higher active Mg, >55%*
  - Chemical equations shown throughout presentation refer to SOx additive as MgO
    - "Short-hand" for the Mg-OH-Al structure

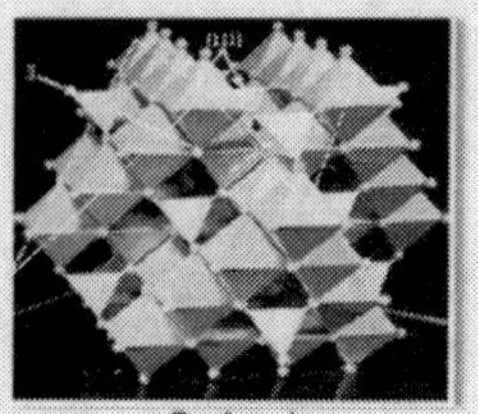

Spinel

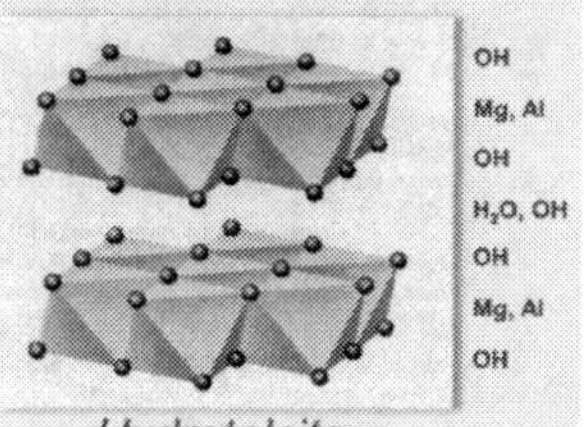

Hydrotalcite

INTERCAT

5

SOx Additive Mechanism – Full Burn

Johnson Matthey
Process Technologies

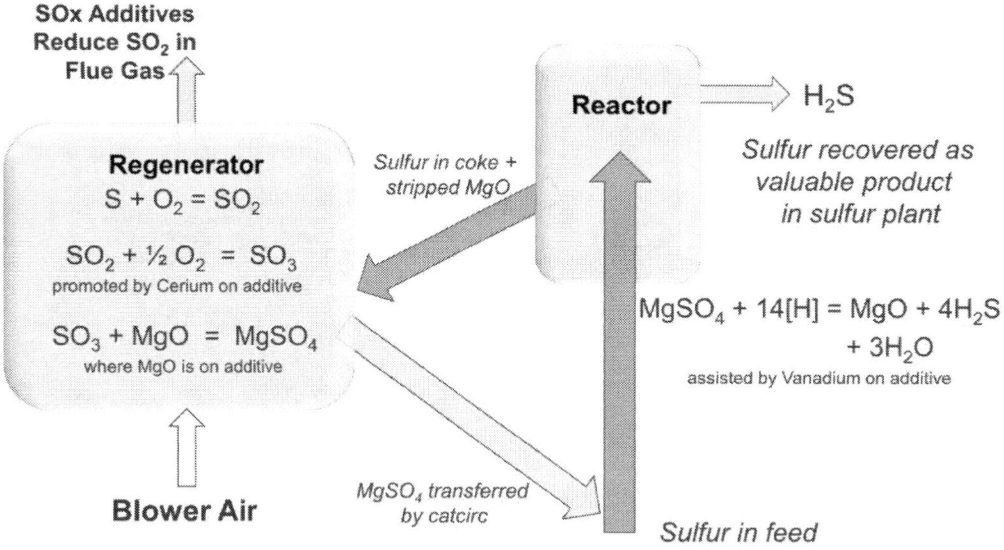

INTERCAT

6

SOx Additive Mechanism – Partial Burn

Johnson Matthey
Process Technologies

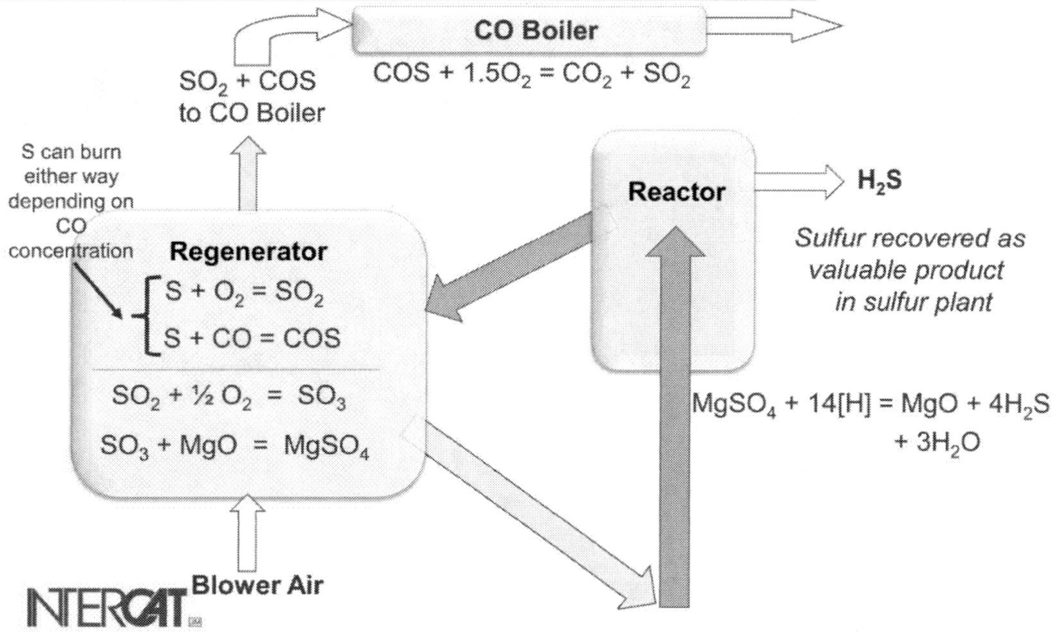

INTERCAT

7

SOx Additive Mechanism – 2 Stage Regen

Johnson Matthey
Process Technologies

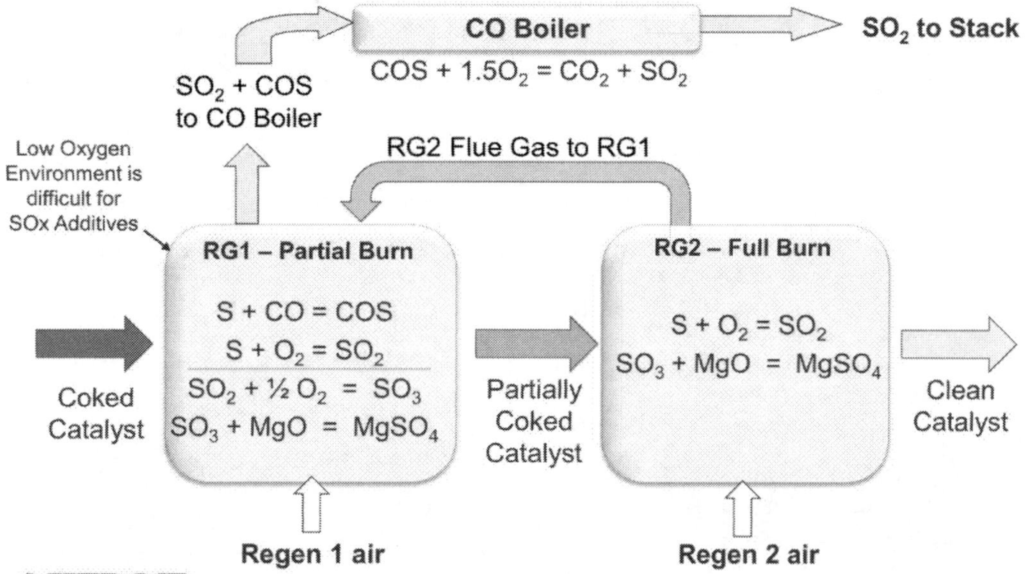

INTERCAT

8

Overview

JM
Johnson Matthey
Process Technologies

- **Staged emission controls**
  - Regulators tend to impose reductions in SO_2 emissions in stages
  - Typical stages:
    - *Mild SO_2 trimming*
    - *Moderate reductions*
    - *Ultralow emissions*
- **Additive uses in differing units**
  - Full combustion, partial combustion, & 2-stage regen
- **Variable effects**
  - Predicting and monitoring SO_2
  - Optimizing SO_2 control
- **Conclusions**

INTERCAT

9

SO_2 Trimming Operations

JM
Johnson Matthey
Process Technologies

- **Minor reductions typical of initial stage SO_2 emission controls**
  - Typically < 50% reduction
  - Zero Ce additives may be acceptable for this application
    - *Pick Up Factor (PUF) ~5 kg/kg*
    - *PUF is an additive efficiency calculation defined as mass of SO_2 removed per mass of additive used*
  - High activity, high Mg additives much more effective
    - *PUF >> 40-50*
- **Also occasionally used to reduce caustic usage with flue gas scrubbers**

Additive	SG	SG	SG	SG	SG
Charge rate, tpd	2750	2590	10170	9480	2790
Sulfur, wt%	0.62	0.62	0.36	0.31	0.78
Slurry Sulfur, wt%	1.50	1.44	0.86	0.79	2.2
Fluegas O2	1.4	1.6	1.0	1.0	<1
SO2 Uncontrolled, ppm	780	800	386	331	559
SO2 Controlled, ppm	398	336	158	126	235
SO2 Reduction, %	51	42	41	38	42
Additive Addition, kg/day	125	117	153	115	168
Additive Efficiency, kg/kg	17	15	15	15	15.3

- Data from previous generation Ce based additives
- Latest high Mg additives much more effective

INTERCAT

10

Moderate SO_2 Reductions

Johnson Matthey
Process Technologies

- **Typical of more restrictive emission standards**
  - Normally 60-90% reduction
  - Pick-up factors are typically 20-40
    - *Function of several factors including baseline SO_2 emissions, regen hydrothermal conditions, etc.*
- **High activity additives recommended to avoid dilution effects**

Additive	SSG	SSG	SSG	SSG	SSG
Charge rate, MTPD	2319	15194	2210	2120	5114
Sulfur, wt%	0.80	1.08	0.75	0.80	0.72
Slurry sulfur, wt%	2.00	1.99	2.10	1.89	1.60
Fluegas O_2	1.90	1.38	2.10	2.60	2.50
Additive addition, kg/day	92	276	170	147	125
SO_2 Uncontrolled	1189	689	1255	1061	744
SO_2 Controlled	498	241	137	238	288
SO_2 reduction, %	58	65	89	78	59
Additive efficiency, kg SO_2/kg additive	36	33	29	28	35

Latest generation high Mg based additives

INTERCAT

11

Ultralow SO_2 Emissions

Johnson Matthey
Process Technologies

- Most US FCC Units now required to meet 25ppm
- California has the most stringent emissions standards
  - 0-5 ppm SO_2
- High activity SOx reducing additives required to achieve 95-100% removal of SOx
- Additive additions are strongly a function of efficiency of air / catalyst mixing

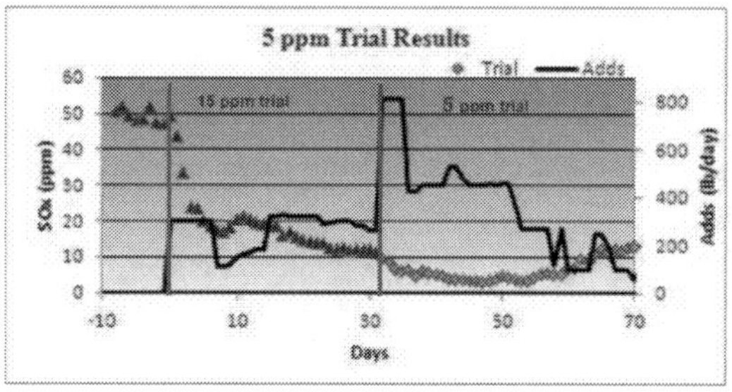

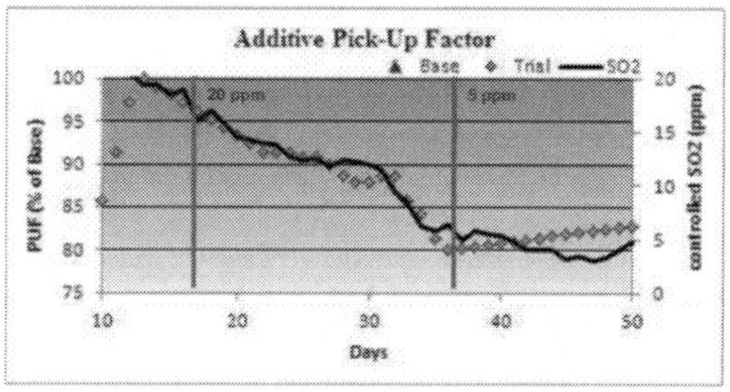

INTERCAT

12

Overview

JM Johnson Matthey
Process Technologies

- **Staged emission controls**
  - Regulators tend to impose reductions in SO_2 emissions in stages
  - Typical stages:
    - *Mild SO_2 trimming*
    - *Moderate reductions*
    - *Ultralow emissions*
- **Additive uses in differing units**
  - Full combustion, partial combustion, & 2-stage regen
- **Variable effects**
  - Predicting and monitoring SO_2
  - Optimizing SO_2 control
- **Conclusions**

INTERCAT

13

Additives in Full Combustion Operations

JM Johnson Matthey
Process Technologies

- **Addition rate increases to meet target as uncontrolled SO_2 increases**
  - No theoretical limit to percentage SO_2 reductions
  - Practical limit set by dilution effects
    - *Commercial experience sets limit at about 20%*
  - Units with higher catalyst addition rates will be able to achieve lower emissions
- **Can I Achieve 25 ppm SO_2?**

Uncontrolled Emissions (ppm)	Controlled Emissions < 25 ppm
>2000	Unlikely
1000 - 2000	Dependent upon catalyst addition rate
< 1000	Likely

Effect of Different Addition Rates

	Case 1 2 TPD Catalyst Additions		Case 2 5 TPD Catalyst Additions	
Additive Usage (kg/day)	SOx Emissions (ppm)	Additive Concentration	SOx Emissions (ppm)	Additive Concentration
0	1519	-	1519	-
500	425	20%	425	9.1%
1000	53	>20%	53	16.6%

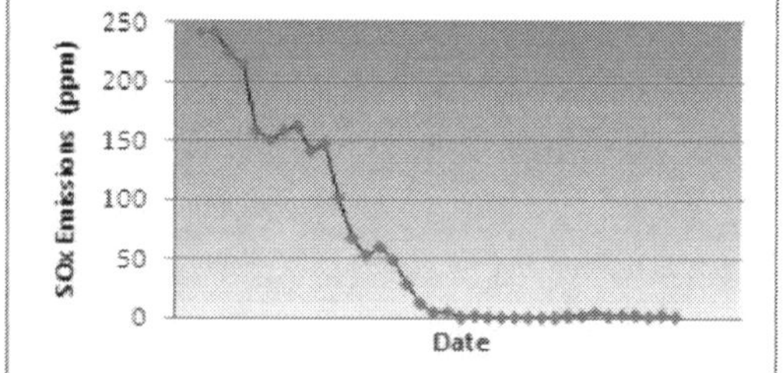

Total elimination of SOx emissions is achievable

INTERCAT

14

Additives in Partial Combustion Operations

JM Johnson Matthey
Process Technologies

- **Oxygen availability in partial burn:**
  - Some of the S will be found in the reduced form
    - *COS, H_2S*
    - *Additives do not absorb COS or SO_2*
  - Oxidation of $S \rightarrow SO_2 \rightarrow SO_3$ limiting step
  - Tailored oxidation component required for maximum additive effectiveness
- **Ratio of COS-to-SO_2 is variable**
- **Partial burn guidelines:**
  - There is a practical upper limit
    - *Monitor SO_x concentration in flue gas upstream of CO boiler thru trial*
  - Monitor regenerator CO-to-CO_2 ratio as additive concentration increases

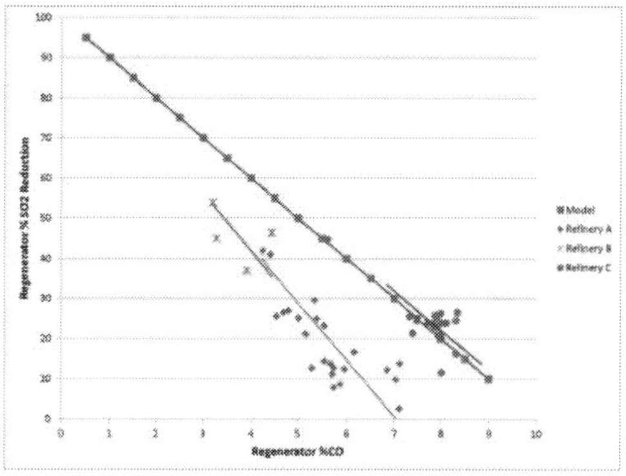

Flue gas analysis (mol/hr)	COB Inlet	COB Outlet	%S in COB inlet in oxidized form
SO_2	669	2566	-
SO_3	179	57	-
Total SOx	849	2623	32%

Sulfur Species Data from European FCC @ 7% CO

Additive Usage (kg/day)	Super SOXGETTER	Lo-SOx PB
Dates	10/12-11/2 '06	2/28-3/5 '07
Charge rate, BPD	22147	21174
Feed sulfur, wt%	0.53	0.43
Slurry sulfur, wt%	1.99	1.84
SO2 uncontrolled, ppm	175	148
SO2 controlled, ppm	125	60
Reduction, %	29	59.5
Additive efficiency (PUF)	Base	2.2XBase

Deep SOx reductions in partial burn are feasible

INTERCAT JM

15

Additives in Two Stage Regenerators

JM Johnson Matthey
Process Technologies

- **SO_2 removal in two-stage regenerators can be very challenging**
  - SO_2 removal partially retains characteristics of both operations
- **Specific SOx reducing technology has been designed for R2R units**
  - Effective additives require optimized oxidation capacity
- **INTERCAT$_{JM}$ has developed unit specific guidelines for two-stage regenerator operations**

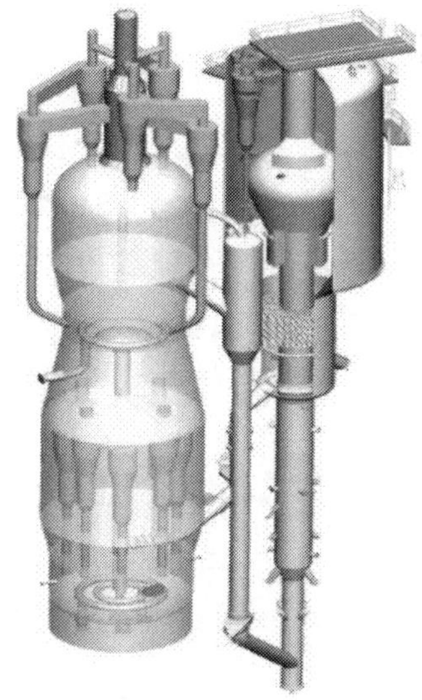

INTERCAT JM

16

Overview

Johnson Matthey
Process Technologies

- **Staged emission controls**
  - Regulators tend to impose reductions in SO_2 emissions in stages
  - Typical stages:
    - *Mild SO_2 trimming*
    - *Moderate reductions*
    - *Ultralow emissions*
- **Additive uses in differing units**
  - Full combustion, partial combustion, & 2-stage regen
- **Variable effects**
  - Predicting and monitoring SO_2
  - Optimizing SO_2 control
- **Conclusions**

INTERCAT

Predicting SO_2 Emissions

Johnson Matthey
Process Technologies

- **All sulfur in coke will burn to SOx**
  - Typically 90:10 SO_2:SO_3 in complete combustion
- **SO_2 emissions will correlate with feed sulfur in some units**
  - Amount of feed sulfur going to coke varies with feed stock
  - Feed slate changes make feedstock based correlations ineffective
- **Coke sulfur correlates well with slurry sulfur**
  - Gulf correlation relates coke & slurry sulfur
    - *Coke S = UF*(slurry S)$^{1.265}$*

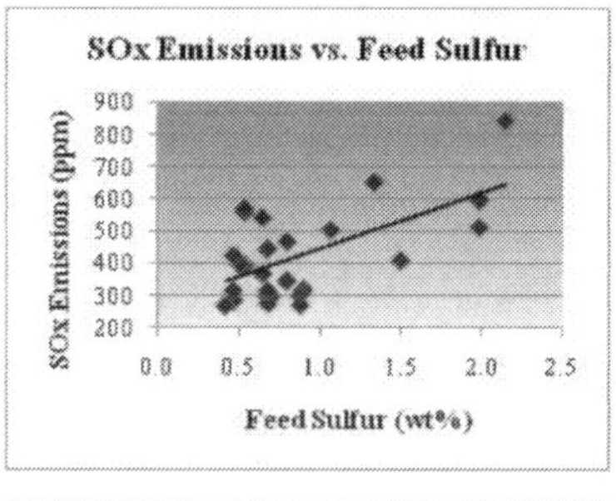

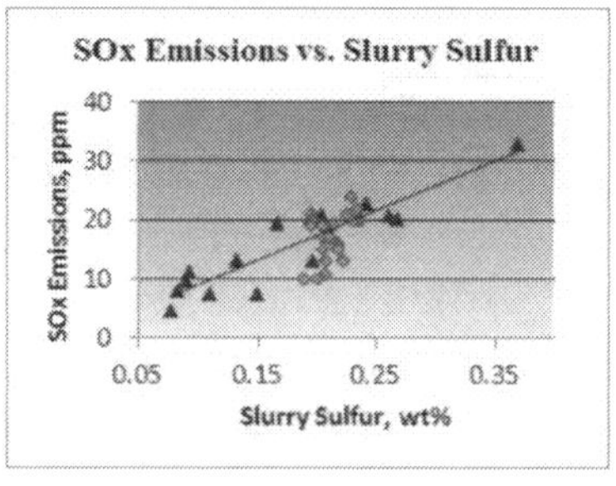

INTERCAT

Ensure Accurate Readings

Johnson Matthey
Process Technologies

- **Flue gas analyzer reliability**
  - Inaccurate readings will affect the actual SOx emissions and the heat balance calculations
  - Regular maintenance required
  - Flue gas conditioning systems most common cause of poor readings
    - *Condensate removal*
    - *Catalyst fines removal*
  - Regular calibration essential

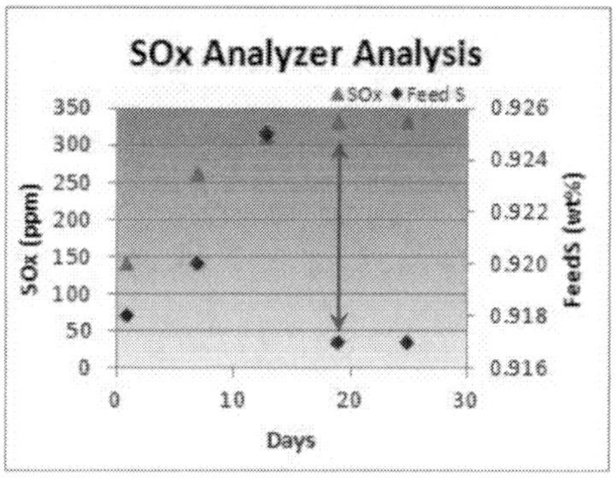

SOx analyzer gas conditioning system subject to plugging

INTERCAT

19

Controlling Operating Variables

Johnson Matthey
Process Technologies

- **Minimize Feed Sulfur Content**
  - Reduces the sulfur in FCC products including coke
  - Feed sulfur speciation impacts the fraction of sulfur distributed to coke. Aromatic sulfur species have a higher tendency to form coke.
  - Can be achieved by selectively choosing FCC feeds
- **Maximize Regenerator Pressure**
  - SO_2 oxidation and SO_3 capture reactions are favored at higher pressures
- **Maximize Catalyst Circulation Rate**
  - Increase frequency that SOx additive particles flow through the regenerator and reactor

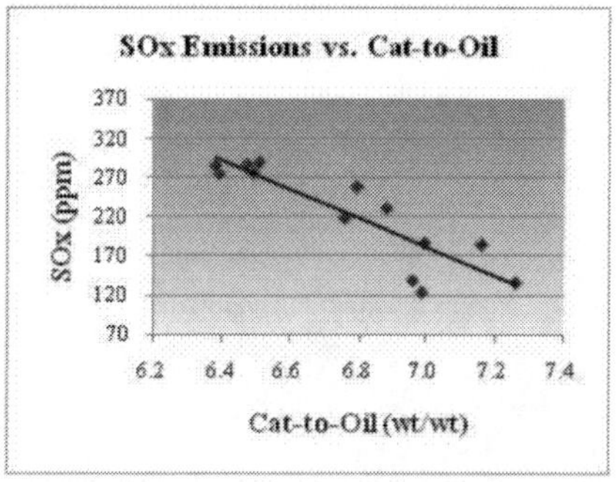

INTERCAT

20

Controlling Operating Variables

Johnson Matthey
Process Technologies

- **Monitor for Feed Iron Excursions**
  - Iron increases SO_2 emissions by acting as an "anti-SOx" additive
  - SOx excursions typically occur at ecat iron levels of over ~0.8 wt%
  - Iron absorbs H2S releasing it the regenerator
  - Iron half life and effects appears to be short
    - *Typically 2-3 days*
  - Reaction Mechanism:
    - Reactor: $FeO + H_2S = FeS + H_2O$
    - Regenerator: $FeS + 1.5\ O_2 = FeO + SO_2$

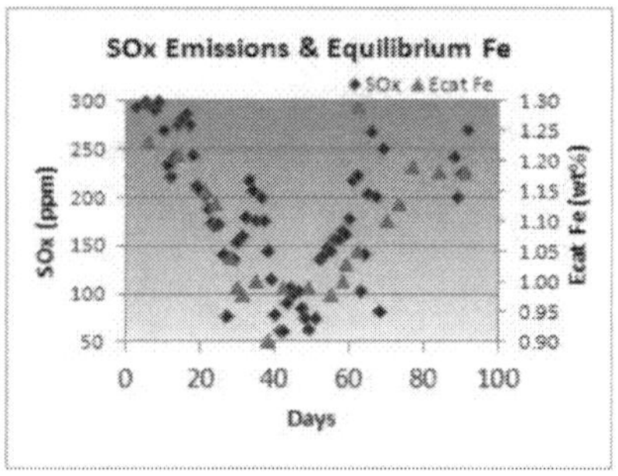

INTERCAT

21

Controlling Operating Variables

Johnson Matthey
Process Technologies

- **Increase Excess O_2 (Full Burn Units) & Minimize CO Content (Partial Burn Units)**
  - Increased oxygen raises rate of SO_2 to SO_3 oxidation
  - A special additive designed with enhanced oxidation functionality should be used in partial burn units

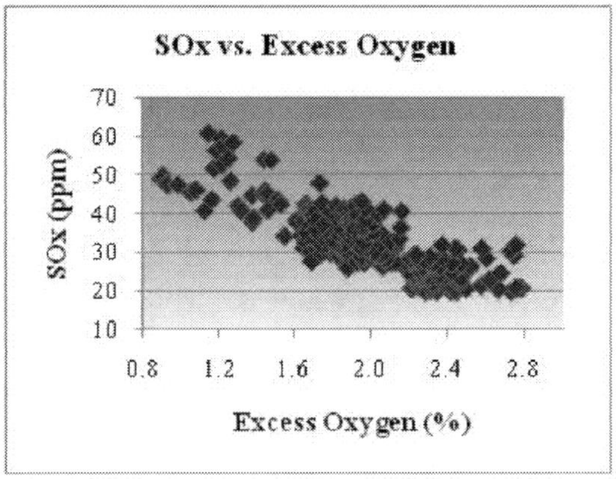

INTERCAT

22

Controlling Operating Variables

JM Johnson Matthey
Process Technologies

- **Maximize Riser Outlet Temperature**
  - Regeneration of the SOx additive occurs in the reactor / riser
  - Regeneration reaction can be rate limiting at low riser outlet temperatures, noticeably at <950°F
  - There are SOx additives designed specifically for low riser temperature operation such as SUPER SOXGETTER-II-DM
    - *Helps achieve deep SOx reductions in maximum LCO operations*

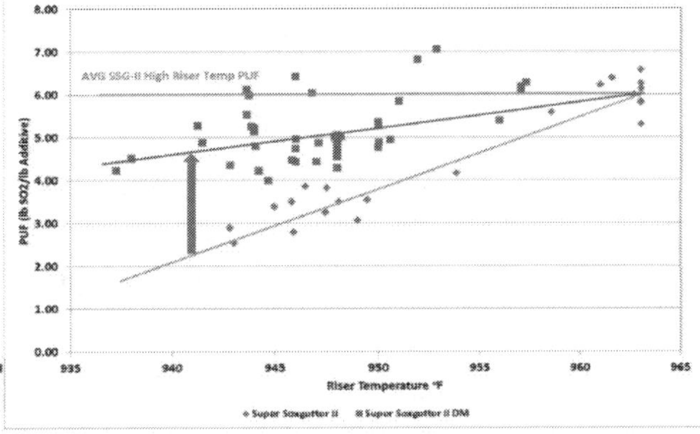

Sulfur release mechanism:

$MgSO_4 + 4H_2 \rightarrow MgO + 4H_2S + 3H_2O$

INTERCAT

23

Controlling Operating Variables

JM Johnson Matthey
Process Technologies

- **Other Considerations**
  - Use of CO promoter also promotes SO_2 to SO_3 helping additive effectiveness promotion
  - High catalyst losses
  - Large regenerator inventories
  - Poor stripper efficiencies
  - High Fe on purchased e-cat

INTERCAT

24

Hardware Problems

Johnson Matthey
Process Technologies

- **Minimize Regenerator Maldistribution**
  - Regenerator air and catalyst mixing is primarily controlled by hardware design
  - Maldistribution may result in both oxygen rich and lean zones leading to reduced SOx removal
- **Commercial example:**
  - Fluegas SO_2 reading = 248 ppm
  - Substantial over dosing required to drop below 83 ppm SO2

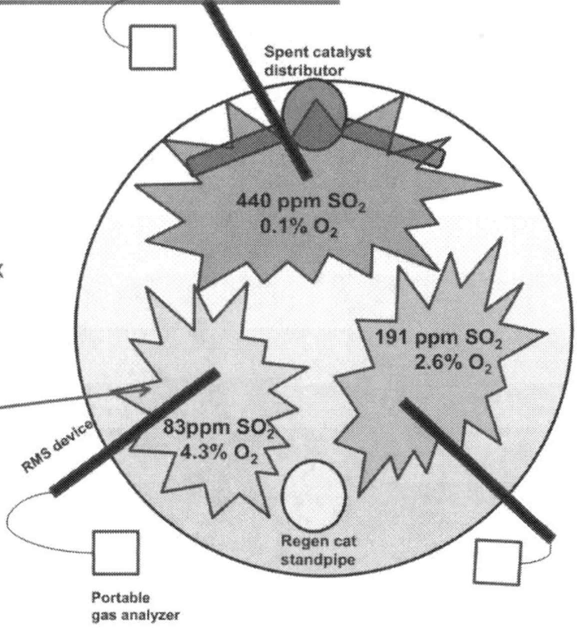

INTERCAT

25

Improving Air / Catalyst Distribution

Johnson Matthey
Process Technologies

- **Several regenerator designs are highly effective at catalyst / air distribution**
  - These designs tend to operate at very low NOx emissions levels
  - These designs are capable of achieving very low SOx emissions with modest SOx additive injections
- **Specialized CFD* technology can be used to assist refiners improve regenerator mixing**
  - Example on right demonstrates an exceptionally efficient distributor
  - "Christmas tree" design containing 17 arms

*BARRACUDA software provided by cpfd COMPUTATIONAL PARTICLE FLUID DYNAMICS

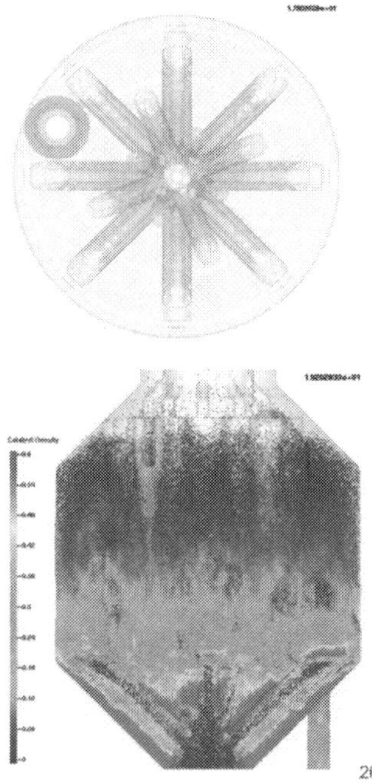

INTERCAT

26

Overview

Johnson Matthey
Process Technologies

- **Staged emission controls**
  - Regulators tend to impose reductions in SO_2 emissions in stages
  - Typical stages:
    - *Mild SO_2 trimming*
    - *Moderate reductions*
    - *Ultralow emissions*
- **Additive uses in differing units**
  - Full combustion, partial combustion, & 2-stage regen
- **Variable effects**
  - Predicting and monitoring SO_2
  - Optimizing SO_2 control
- **Conclusions**

INTERCAT

27

Conclusions

Johnson Matthey
Process Technologies

- **SOx additives are commonly used to reduce FCC SOx emissions**
  - Trimming, moderate, & ultralow SOx
- **SOx additives are a flexible & economic solution for differing FCC types**
  - Reductions over 95% common in full combustion
  - Reductions of up to 70% common in partial burn operations
- **SOx additive use can be optimized by simple operating condition and hardware changes**
  - Process variables
  - Improved air distribution

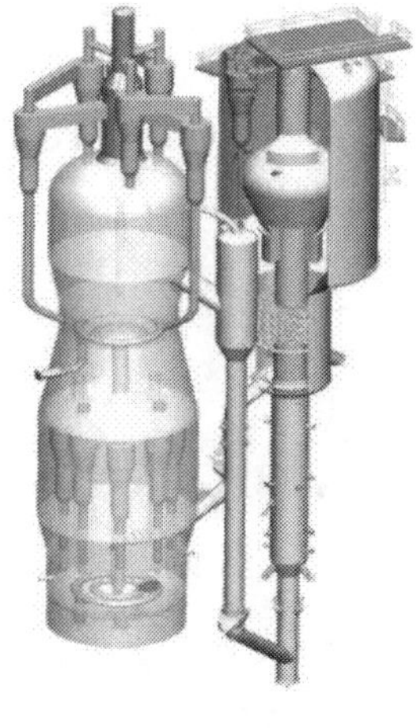

INTERCAT

28

Identifying Energy Efficiency Opportunities in the Process

Alan Rossiter,
Rossiter & Associates
alan@rossiters.org

Rossiter & Associates
Energy Efficiency by Design

Why We Don't Look at the Process…

- Adverse Impact on Yield, Quality or Throughout.
- Lack of Expertise.
- Safety Concerns.

Why We Should Look at the Process...

- We are in the Process Industries!
- Often Large Energy Savings.
- Outside of heat integration (pinch analysis, etc.), we lack good methods for identifying improvement. Or do we?

The Solution: PFD Reviews

General Approach

- Structured Brainstorming
- Use Existing (or new) Process Flow Diagrams
- Mark up with:
  - temperatures
  - pressures
  - flow rates
  - heat loads

The Team

- Facilitator
- Senior Plant Operator
- Maintenance Supervisor
- Control Engineer
- Process Engineer
- Energy Efficiency Specialist

The Procedure

- Review main streams, equipment and systems
- Identify inefficiencies
- Brainstorm improvements
- Replicate and build on ideas from other sites
- Document opportunities
- Conduct preliminary ranking

Later

- Screen options (costs, benefits, risks, etc.)

Example 1:
Keep the hot streams hot

What's Wrong With This Picture?

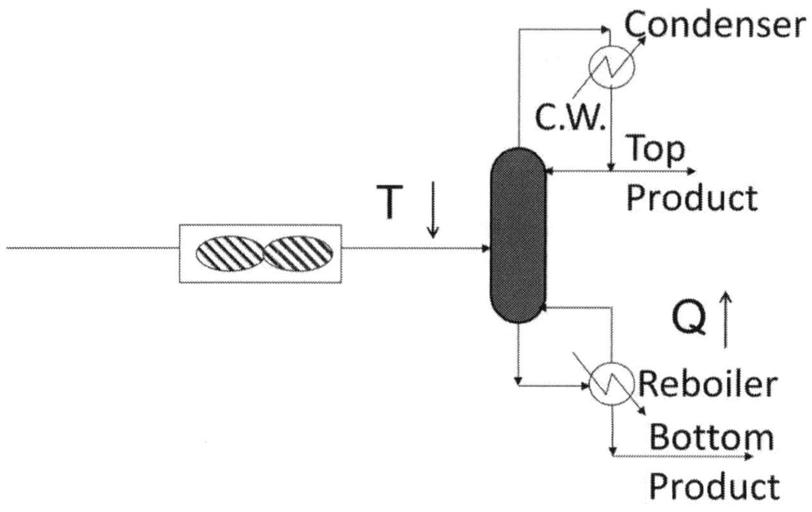

A Simple Partial Fix...

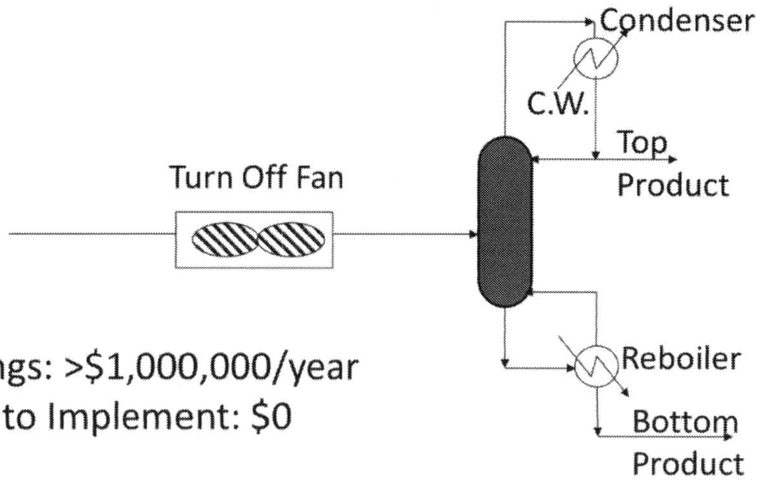

Savings: >$1,000,000/year
Cost to Implement: $0

A Further Correction...

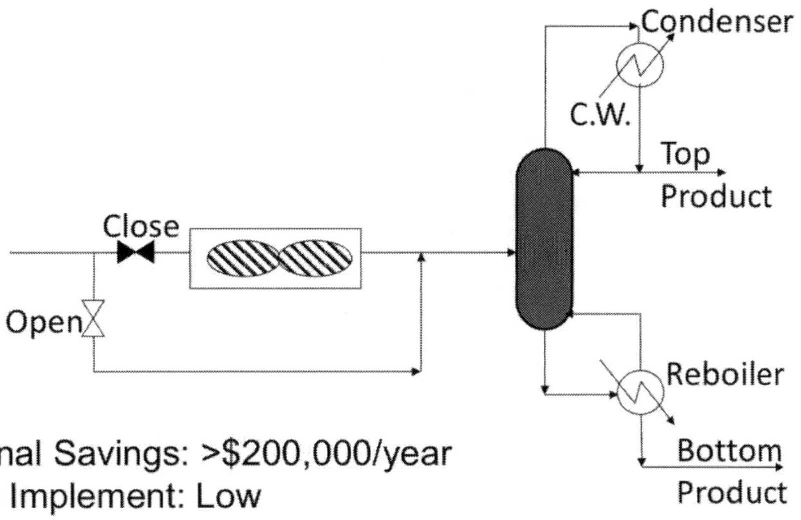

Additional Savings: >$200,000/year
Cost to Implement: Low

Example 2:
If you've got it, use it!

Why Bypass?

Operators opened the bypass because the pumparound duty became too high in the winter. They did not close it in the summer.

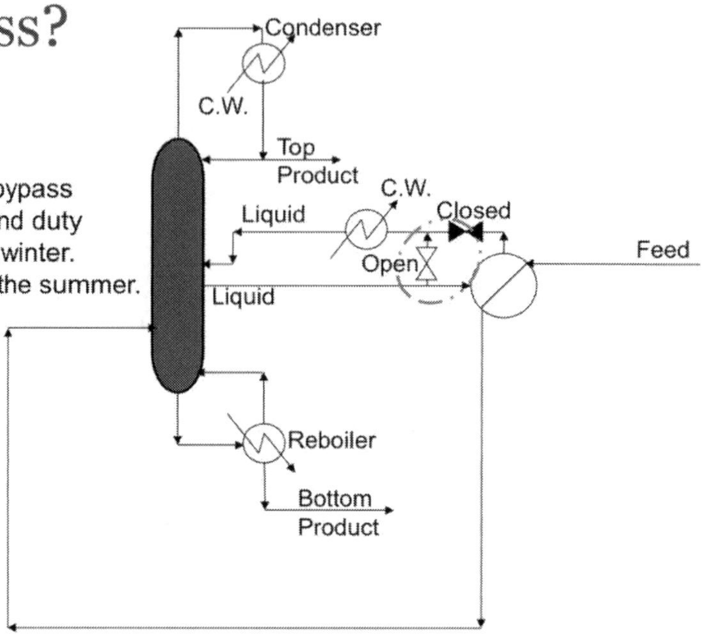

A Simple Fix...

Close the bypass.
Monitor bypass valve.
Only open when authorized.

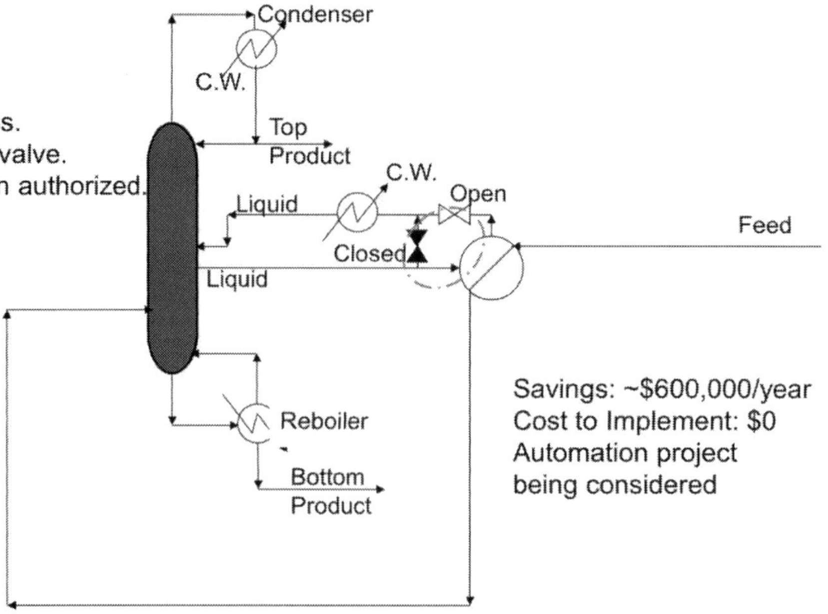

Savings: ~$600,000/year
Cost to Implement: $0
Automation project being considered

Example 3:
Don't Pump More Than You Need

What's Wrong Here?

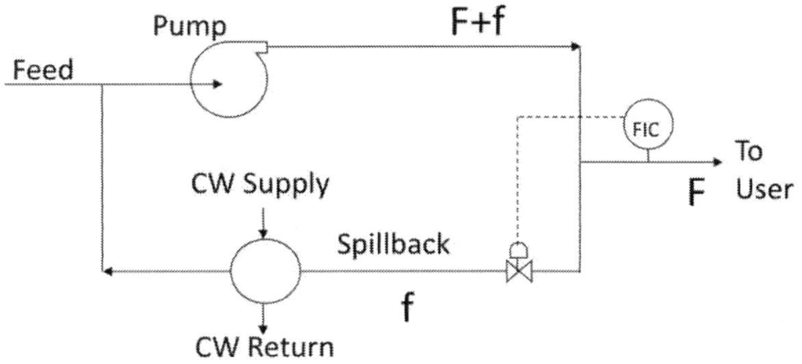

Inefficient Flow Control

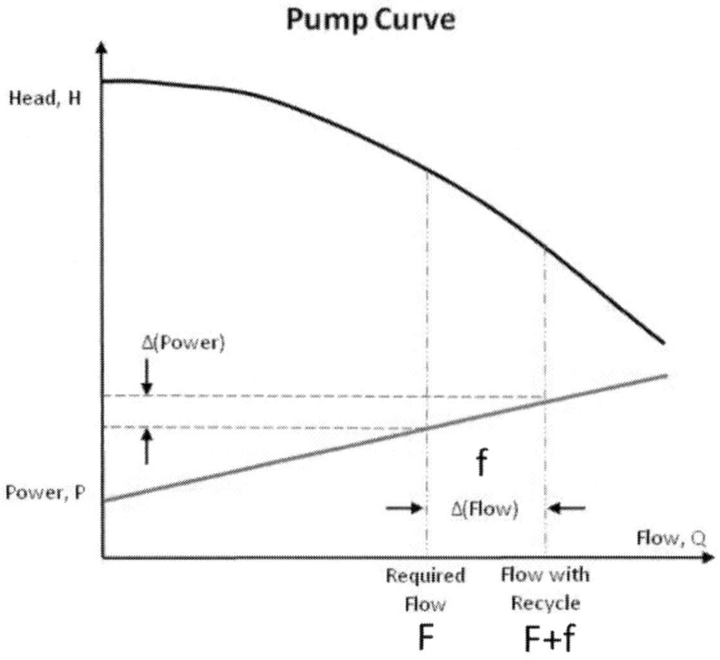

Take Advantage of the Pump Curve!

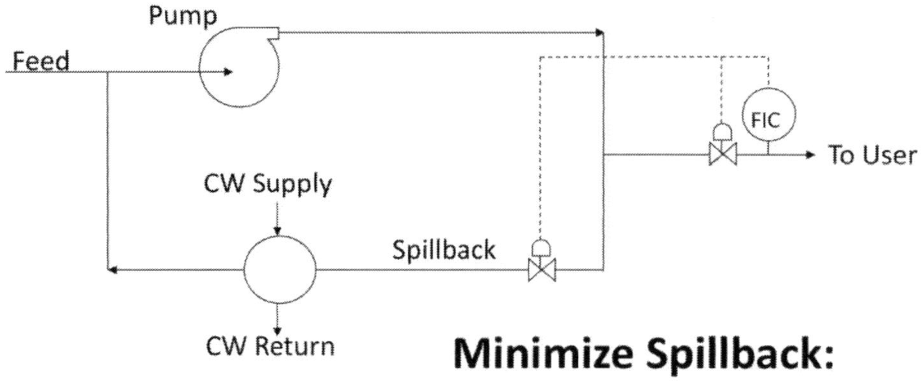

Minimize Spillback: Saves $150,000/year

Comparing Pump Control Options

Bypass Control:

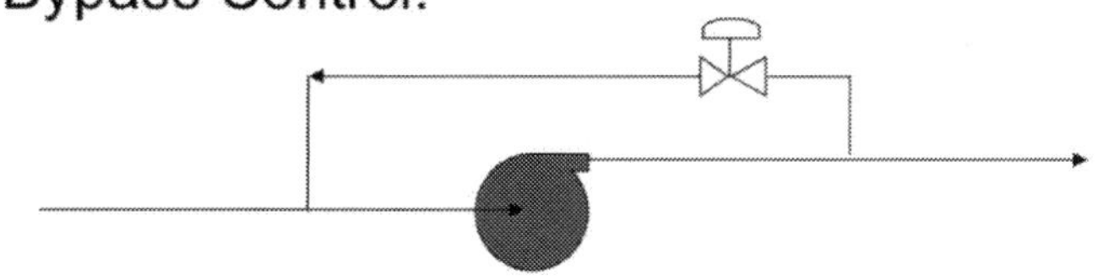

Throttle Control:

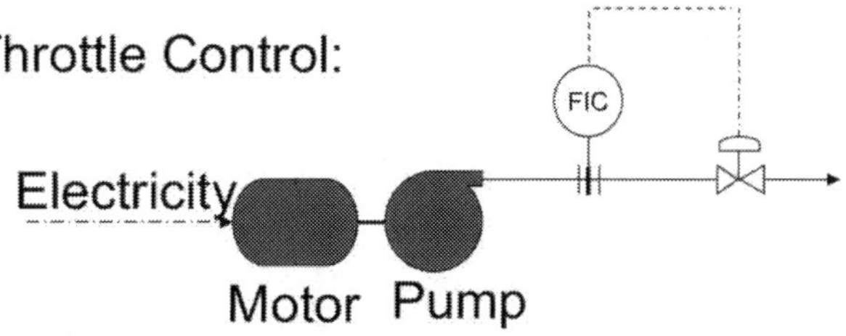

Another Option

Speed Control:

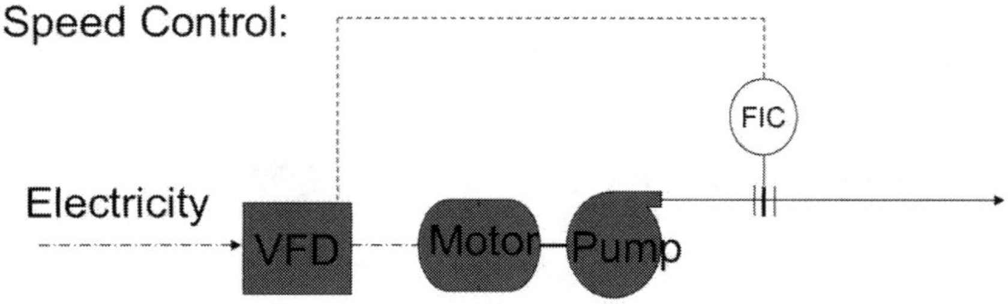

VFD = Variable Frequency Drive

Power Savings with VFDs

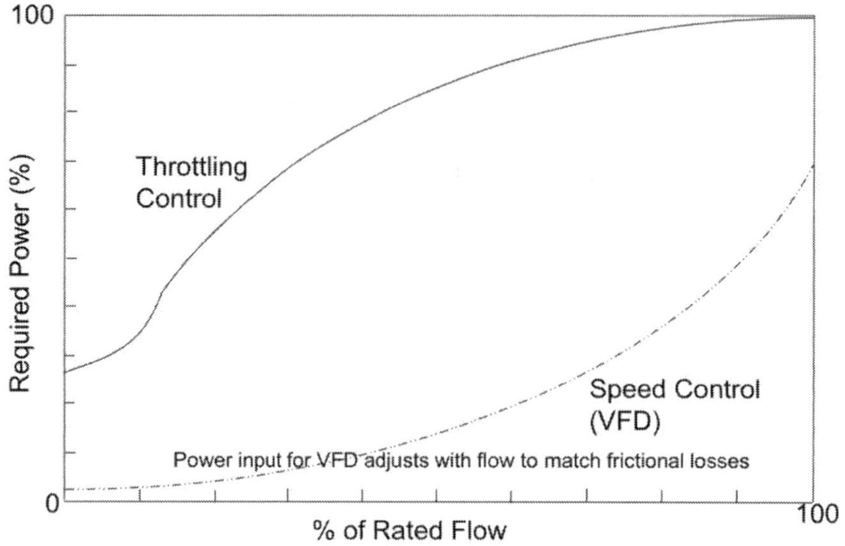

Additional Benefits of PFD Reviews

- Knowledge sharing
- Staff can present suggestions
- Build on ideas
- Address concerns
- Motivate for success

New Damper Design Saves Energy in Fired Heaters

Garg, A., Furnace Improvements Services

Originally presented on: 10/9/2014 10:00:00 - 10:30:00

Damper operation is one of the most important control element in the fired heater operation. Current damper designs are inadequate and are not able to effectively control the draft in the heaters. In most heaters, dampers are left fully open. A new energy efficient damper design, which will save energy especially during turn down operation, will be described. Case studies demonstrating benefits of better damper design will be presented.

Energy Systems Model
Development of a Discrete-Event Simulation Tool to Facilitate Capital Decisions

Intan Hamdan, Rene Diaz, Scott Bury
Thursday, October 9, 2014

Dow Energy Business

- Dow Energy provides advantaged power and utilities, creates extraordinary value by leveraging Dow's energy capabilities, and is recognized as a global leader in sustainable production and consumption of energy.
- Dow Energy business manages and provides power, steam and other utilities to some of Dow's largest sites located around the world
- At certain sites, 100% of the steam requirements are generated in-house
- In several locations, Dow is a net exporter of power to the external grid

Dow Energy's Mission: Deliver safe, reliable and advantaged power and utilities

1

Energy Supply Reliability is Essential

Disruptions to the steam and power supply would impact site operations. Especially for sites that are 100% reliant upon in-house steam generation

Factors that Impact Energy Reliability

- ***Variability in Generation Capacity***
  - Energy plants typically contain gas and steam turbines, boilers, synchronous condensers, and transformers
  - All assets have potential to experience unplanned shutdowns and scheduled maintenance
- ***Variability in Demand***
  - Variability in steam and power demand
- **It is important to ensure that steam and power requirements are met despite inherent variability in the supply and demand**

2

Key Operational Decisions

When facing upstream unreliability, businesses minimize impact to customers by (1) increasing inventory and (2) having longer lead times

Operational Decisions to Be Made

- Determine effective and robust operating strategies of available assets
- Manage assets in the long-term to ensure reliability of steam and power supply despite inherent variability in supply and demand of these utilities

3

Motivation for Simulation Tool

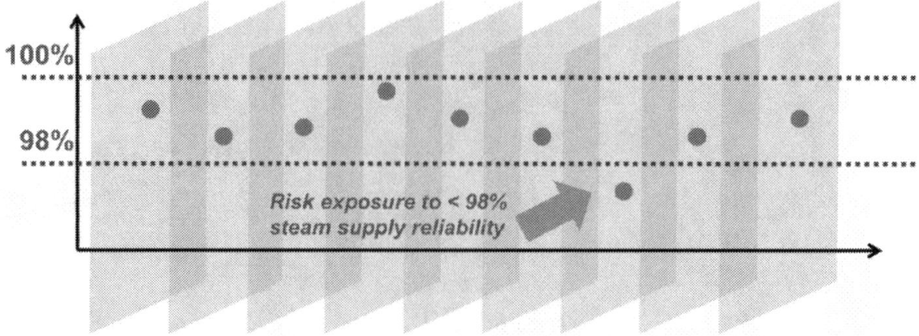

- ***Monte Carlo Simulation:***
  - Provides reliability projections of steam supply based on planned and unplanned events (*e.g.*, takes into account impact of asset reliability)
  - Enables quantification of risk and facilitates decision-making

Decision-Making Through Modeling

Objective

- **Enable** the evaluation of steam and power supply reliability under various conditions through computationally inexpensive simulations
- **Support** decision-making by quantifying risk of various scenarios and facilitate capital expansion/reductions decisions
- **Facilitate** decisions regarding long-term asset management and capital investments
- **Provide** summary comparisons for differences between operating scenarios in terms of overall margin loss/gain potential
- **Develop** a tool that can be leveraged to other sites

Generic Energy Systems Schematic

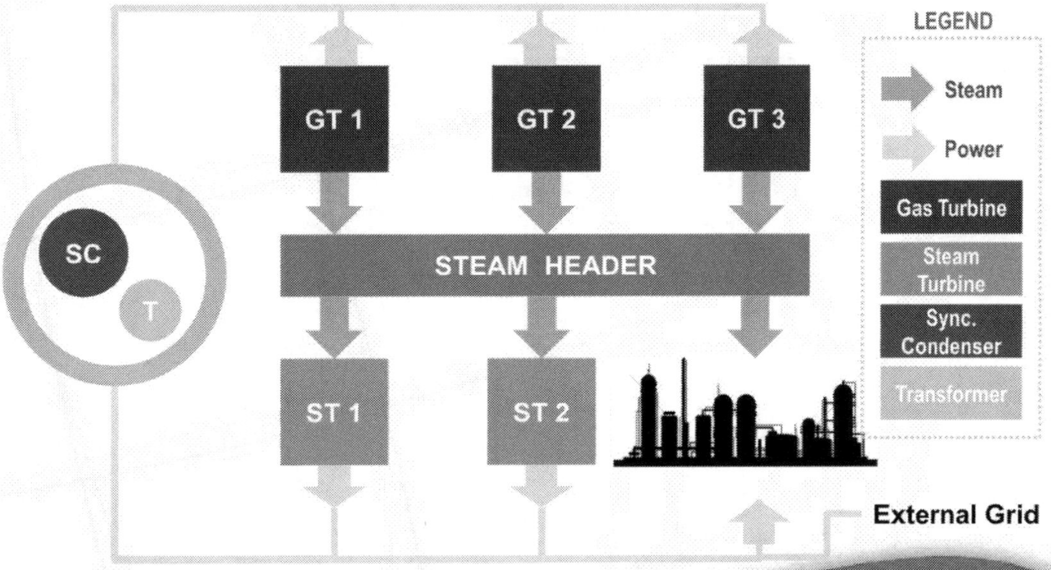

4

Modeling Energy Systems
Variability in Supply

- We consider two main reasons that equipment are not operational: (1) unplanned shutdowns, and (2) scheduled maintenance.

Time	Status
1/13/14 23:00	Off
1/16/14 09:00	On
4/09/14 08:00	Off
4/20/14 07:30	On

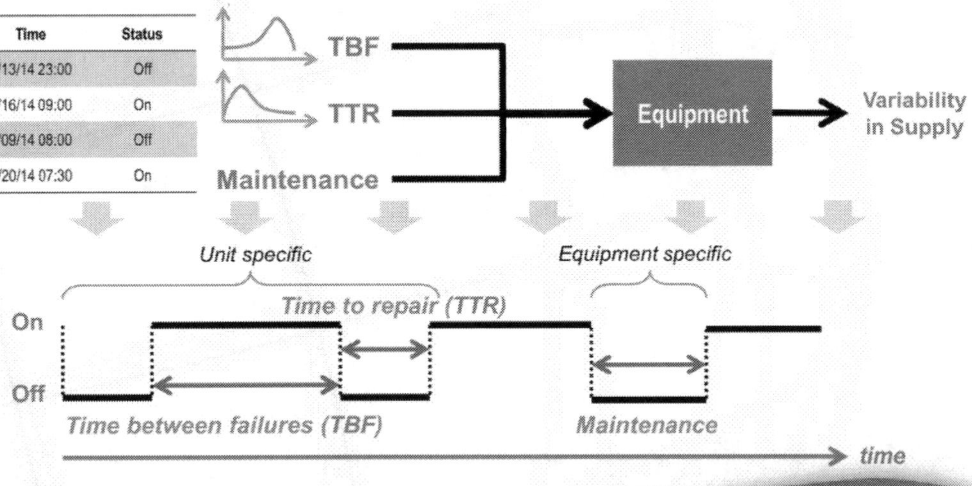

6

Modeling Energy Systems
Variability in Demand

- Utility demands are incorporated as deterministic schedules in order to capture the periodicity inherent to the demand

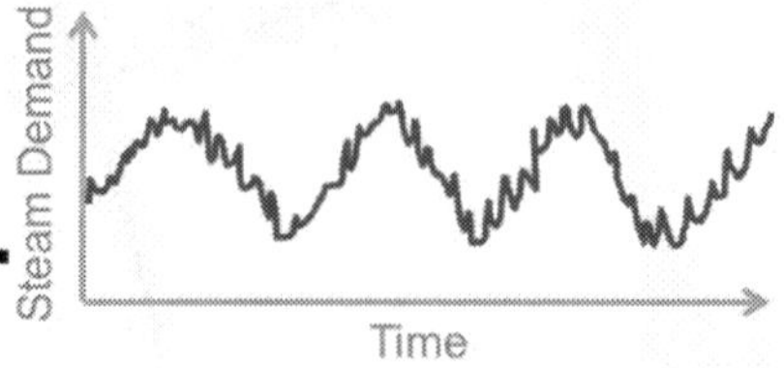

Variability Introduced By:
- Product demand seasonality
- Seasonality
- Temperature fluctuations
- Power export to the grid

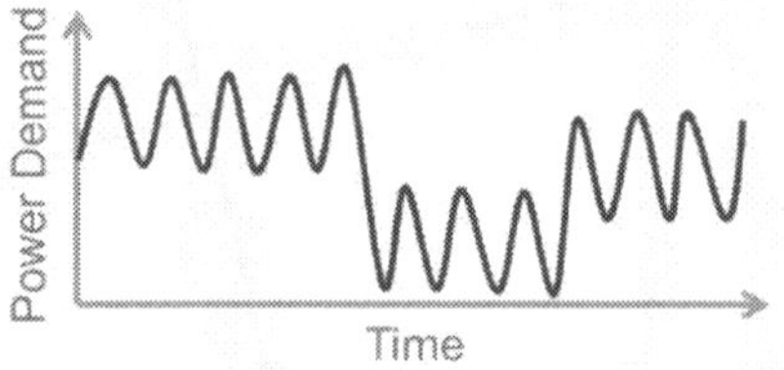

Variability Introduced By:
- Power export to the grid
- Power requirement fluctuates throughout the day
- Power requirement fluctuates over the year

7

Modeling Energy Systems
Demand Dictates Supply

- Steam and power are operated as *a pull system* in which demand dictates the rates at which gas and steam turbines are operated to produce steam.

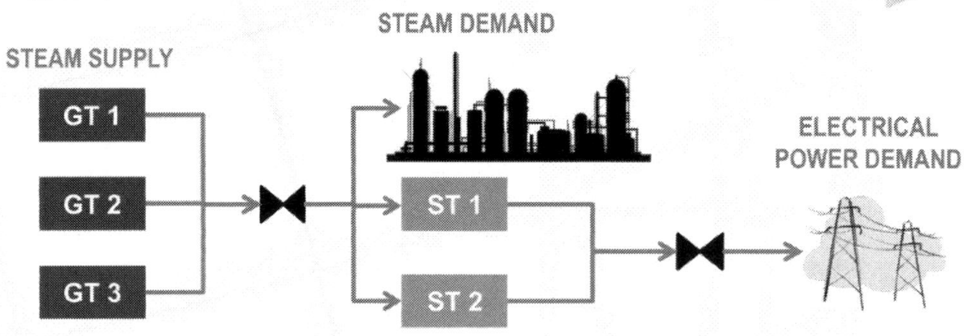

- The objective is to ensure that steam is neither excessively nor insufficiently produced to meet demands

8

Modeling Energy Systems
Curtailment Plan

- During periods of insufficient steam production, steam is prioritized over power production.

CURTAILMENT PLAN

Reduce condensing loads of operational steam turbines to no lower than their minimum capacities

↓

Shutdown steam turbines in sequence of steam turbine priority

↓

Curtail steam to customers (from lowest to highest priority customers)

Curtailing electrical power production in order to maintain steam supply to the site

Energy plant is unable to meet steam demands; lower-priority customers are curtailed

9

Evaluation Process

SCENARIOS
- *Add/remove assets from the simulation*
- *Increase the steam demand*
- *Adjust maintenance schedule of assets*
- *Update asset reliability*

1 → 2 **User Interface** (extend SIM) → 3 **Model** (extend SIM) → 4 **Data Analysis** (Tableau ®, Excel) → 5

Modeling approach separates data from model structure and **enables** analyses of various scenarios

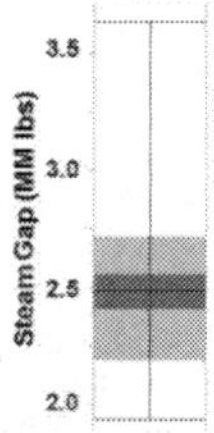

6 ← *Meaningful interpretation of results in context of decision*

Simulations are repeated XX times to incorporate the variability inherent to the system

10

Hypothetical Case Study
Should We Decommission an Asset?

- **Scenario** : An energy asset requires extensive repairs to bring back online. Should we decommission it instead?
- **Question to Be Answered**: Will the site still be able to maintain the reliability of its steam and power supply without the asset under consideration?

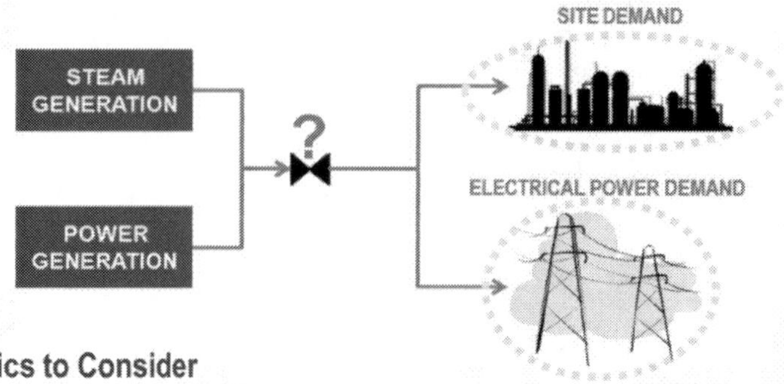

- **Key Metrics to Consider**
  - Unmet steam demand → $$ loss due to unmet steam demand
  - Unmet power demand → $$ loss due to unmet power demand

11

Hypothetical Case Study
Facilitating Decision-Making

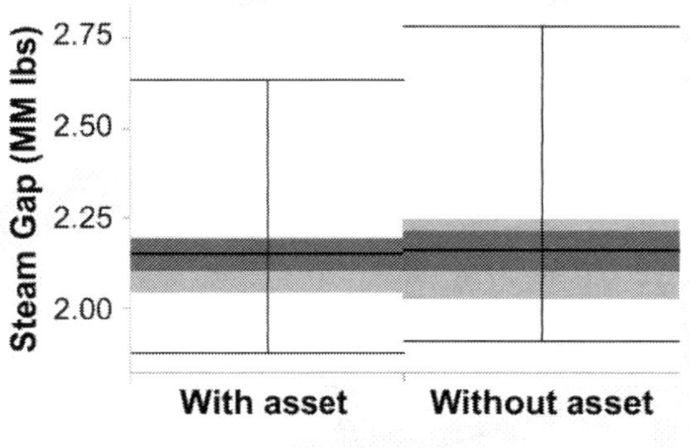

- Overlapping 95% confidence intervals indicate that there is no statistically significant difference between the two cases
- Wider confidence intervals and interquartile regions of the "without asset" indicates greater variability and uncertainty
- The min and max values are higher for the "without asset" case – there is a risk that we obtain steam gaps higher than "with asset"

- **Conclusion**: There is no statistically significant difference in reliability of supply, with or without asset in question. Though there is greater variability in "without asset", the risk of deviating greatly from the "with asset" case is small
- **Decision:** Decommission energy asset.

12

Summary

- Our modeling approach involved the development of a discrete event simulation-based tool used to facilitate operating decisions by quantitatively characterizing a site's exposure to risk
- The tool provided a quick, computationally inexpensive method of evaluating various scenarios for the purpose of long-term asset management
- The methodology can be leveraged to many different systems, keeping in mind the following points:
  - The decision to be made must be clearly defined to enable correct scenario evaluations
  - Model must sufficiently capture the details that are to be evaluated in the various scenarios
  - Metrics chosen must be relevant to the decision being made; metrics are a function of the system being modeled and the decision being made
  - Meaningful interpretations of the results must be made, within the context of the decision

13

Thank You

Questions?

Flare Minimization during Ethylene Plant Turnaround Operations via Dynamic Simulation and Optimization

Xu, Q., Lamar University

Originally presented on: 10/9/2014 11:30:00 - 12:00:00

Flaring emissions during CPI (chemical process industry) plant startups, shutdowns, and upsets causes negative environmental and social impacts and also results in tremendous material and energy losses to plants. For instance, an ethylene plant with 1.2 billion pounds of ethylene production per year may flare five million pounds of ethylene during one typical startup, which generates 15.4 million pounds of CO2, 7.5 K lbs of NOX, 40 K lbs of CO, 100 K lbs of highly reactive VOCs. Obviously, minimizing flaring and CO2 emissions bring significant benefits to environmental, societal, and CPI industrial sustainability.

Unfortunately, current practices in flare minimization almost exclusively depend on industrial experience and "end-of-the-pipe" control strategies. For instance, the installation of flare gas recovery units (FGRU) can capture flare gas for recycle and reuse. However, it is less desirable since the "waste gas" is already generated. Besides, the capital expenditure and operating cost of the FGRU is considerable. Thus, current practices in flare minimization are not enough for CPI today because of increasingly strict environmental regulations and economic competitions.

This presentation introduces systematic methodologies that have been developed at Lamar University for industrial flare minimization via plant-wide dynamic simulation and optimization. Since off-specification streams are inevitable during plant turnaround operations, to significantly reduce flaring emission, they must be either recycled to the upstream process for online reuse, or stored somewhere temporarily for future reprocessing, when the plant manufacturing returns to stable operation. Thus, the off-spec products will be able to be reused instead of being flared. This can be achieved through the identification of viable operation strategies through plant-wide dynamic simulation and optimization. The dynamic simulation/optimization provide an insight into process dynamic behaviors, which is crucial for a plant to minimize the flaring; meanwhile, maintain the operating safety. The effectiveness of the developed methodologies will be demonstrated by multiple case studies involving plant start-up, shutdown, and process upset from different ethylene processes.

Dow Oil & Gas

Modeling Acid Gas and Mercaptan Removal with Hybrid Solvents

Diego Ortiz Vega, John R. Dowdle, Diego Cristancho, Ajay Badhwar

6th AIChE Southwest Process Technology Conference, October 2014

Dow.com

Agenda

- Dow Oil, Gas & Mining solutions
- Motivation
- Development of simulator capabilities
- Foundations: Thermodynamics, transport and kinetics
- Plant data comparison
- Conclusions

Dow Oil, Gas and Mining

Solutions for:

e — exploration	p — production	eor — enhanced oil recovery	t — transmission	g — gas processing	r — refining
• Drilling • Cementing • Stimulation	• Phase separation • Asset integrity • Flow assurance	• Gas injection • Chemical flooding • Thermal recovery applications	• Offshore flow assurance • Gas recompression • Gas dehydration • Gas compressor engines • Gas pipeline line heaters	• Onshore and offshore gas processing • LNG • Gasification • Acid gas enrichment/sulfur plants • Ammonia synthesis	• Acid gas removal • LPG treating • Isomerization and reformer regeneration • Heat transfer
Our brands:					
• EMBARK™ • CELLOSIZE™ • ANTISOL™ • POLYOX™	• DEMTROL™ • ROMAX™ • ACCENT™ • AQUCAR™	ELEVATE™ ROMAX™ DOW FILMTEC™ DOW IntegraPac™	• NEPTUNE™ • HYPERLAST™ • NORKOOL™ • DESITHERM™	• UCARSOL™ • SELEXOL™ • NORKOOL DESITHERM™ • DOWTHERM™ • SYLTHERM† • SulFerox® • UCARSORB™	• AMINE MANAGEMENT℠ Program • UCARSOL™ • SELEXOL™ • UCARSEP™ • ISOFORM™

Key Offerings
Amine Sweetening

UCARSOL™ Solvents
Multiple formulations used for gas treating; each provides key benefits, depending on customer specifications

Selective H_2S – preferentially absorbing H_2S while allowing CO_2 to slip
- UCARSOL™ HS-101 for gas treating
- UCARSOL™ HS-102 for acid gas enrichment
- UCARSOL™ HS-103 for tail gas treating
- UCARSOL™ HS-115 for liquid hydrocarbon treating

Tunable CO_2 removal – removal of CO_2 to a desired specification
- UCARSOL™ AP-802, AP-804, AP-806 – removal of CO_2 to pipeline specification
- UCARSOL™ AP-810, AP-814 – removal of CO_2 to cryogenic

Mercaptan and organic sulfur removal – capturing organic sulfur compounds requires unique solutions
- UCARSOL™ LE Series– removal of organic sulfur and meeting total sulfur specification

Refinery, low-energy solvents – where removal of bulk components is desired, but minimizing energy is key
- UCARSOL™ LE-713, LE-777, LE-801 – maximize energy savings

SELEXOL™ Solvent
Removal of acid gas, organic sulfur and hydrocarbons when there is a high partial pressure of target components

Why Hybrid Solvents in Gas Processing?

Motivation

- Need to reduce sulfur emissions to meet regulatory limits. Mercaptans contribute to these emissions.
- Mercaptans removal from gas is very challenging using conventional treating solvents.
- Additional treatments are required to remove mercaptans and meet total sulfur specifications.

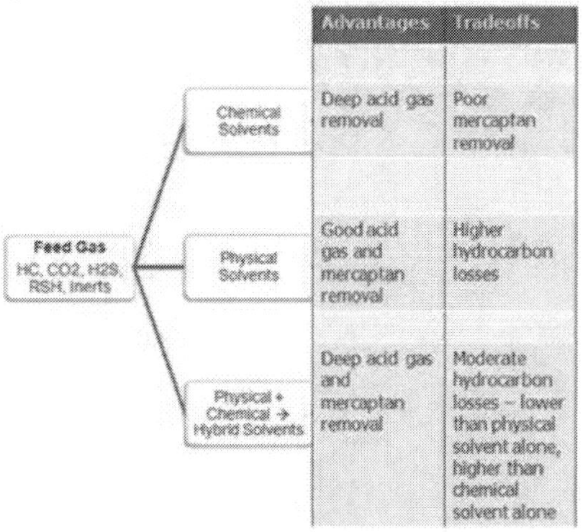

DOW – For External Release 5

Why Hybrid Solvents in Gas Processing?

Motivation

- Need to reduce sulfur emissions to meet regulatory limits. Mercaptans contribute to these emissions.
- Mercaptans removal from gas is very challenging using conventional treating solvents.
- Additional treatments are required to remove mercaptans and meet total sulfur specifications.

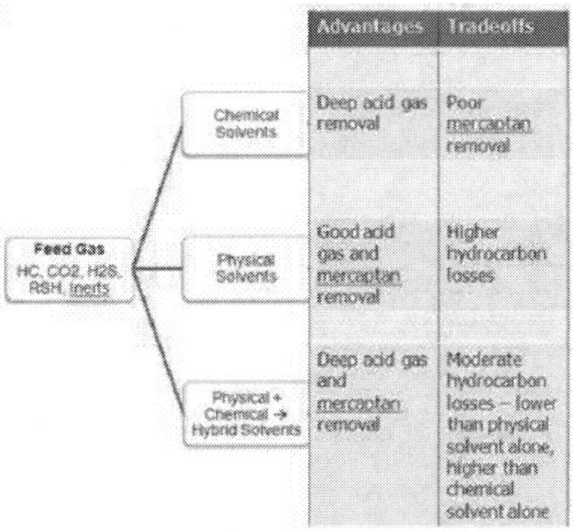

DOW – For External Release 6

Why Hybrid Solvents in Gas Processing?

Advantages

- Simultaneous absorption of acid gases and mercaptans.
- Lower regeneration energy demand for given circulation.
- Formulation targeted to decrease hydrocarbon co-absorption.
- Potential for significant operating cost savings.

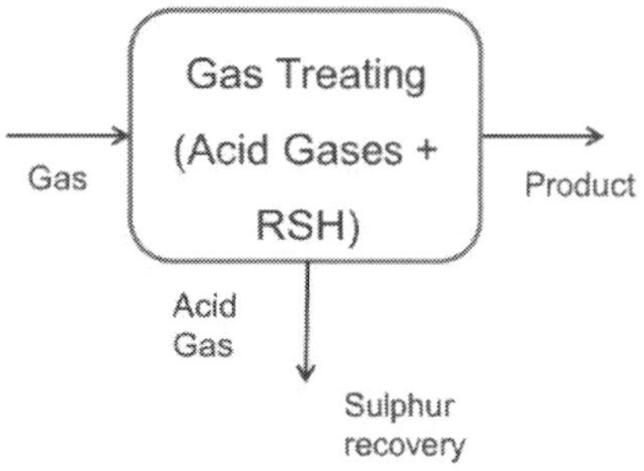

Challenge: Accurate equipment sizing and process design require a high fidelity simulation tool for Hybrids Solvents → Difficult.

Solution: Develop a property package that includes thermodynamic, transport properties, kinetics and mass transfer models for the major components in hybrid solvents applications (hydrocarbons, H2S, CO2, Amine, water, mercaptans and a physical solvent)

Development of Simulator Capabilities

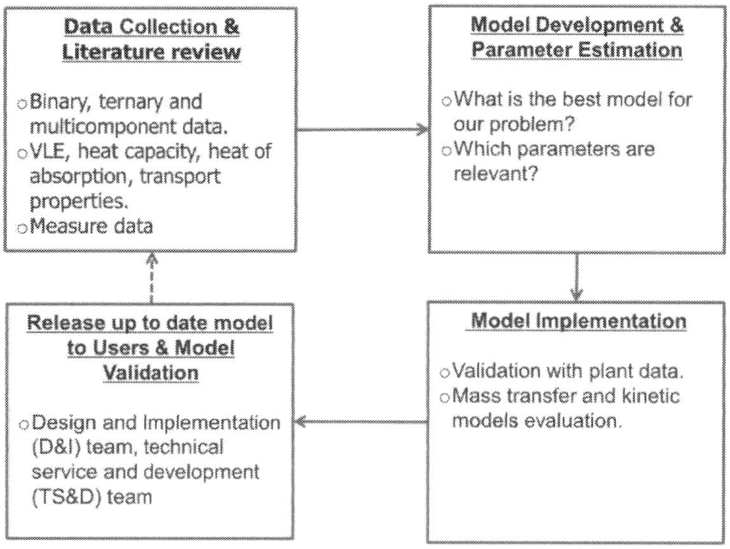

Model development: Thermodynamic Framework

Vapor phase described by any Fugacity model

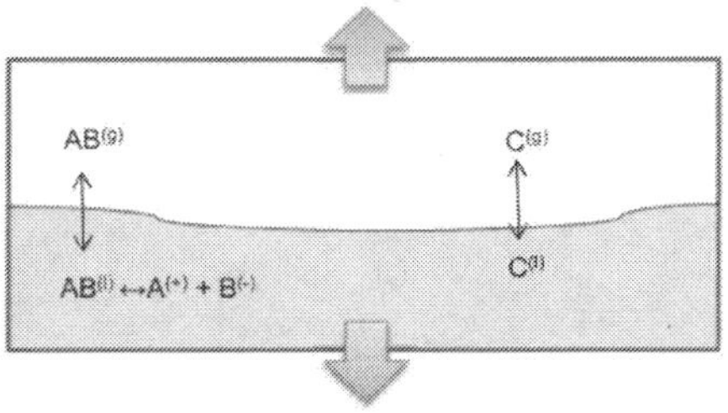

- Chemical equilibrium present in liquid phase → Physical and Chemical solubility for Acid Gases.
- Challenge: Electrolyte Systems → Speciation, Electroneutrality, dielectric constant of solvents.
- Activity coefficient model has to be able to describe short range interactions, long range interactions as well as the effect of mixed solvents.

$$G^{*,\mathrm{ex}} = G^{*,\mathrm{SR}} + G^{*,\mathrm{LR}} + \Delta G^{*,\mathrm{Born}}.$$

Excess Gibbs Energy Model

Model development: Mercaptans approach

- Mercaptans exhibit properties of both acid gases and hydrocarbons.
- Mercaptans are less acidic than CO_2 and H_2S, this reduces their solubility in liquid phase with loading.

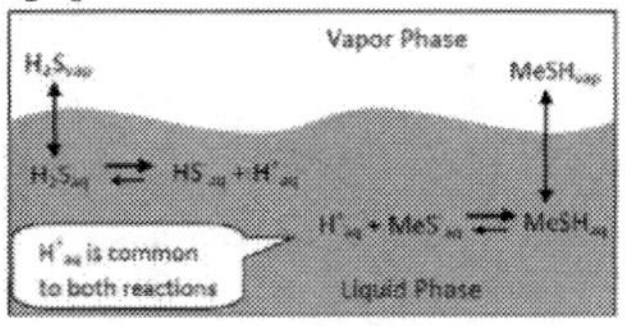

"Mercaptans Removal from Gases by Absorption into Amines and Caustic " by Ralph Weiland, Nate Hatcher, and Clay Jones *(Optimized Gas Treating)*

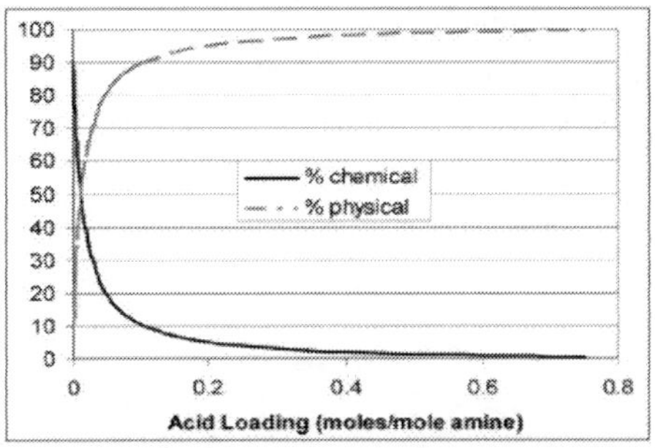

Degree of physical and chemical solubility of methyl mercaptan in 50% MDEA at 40 ° C with various acid loadings. Bedell *Ind. Eng. Chem. Res.* ***2007, 46, 3729-3733***

Physical Solubility vs. Chemical Solubility

- At low loadings, the chemical solubility dominates the total solubility. As the acid loadings in the solution increase, chemical solubility is substantially reduced and physical solubility dominates.
- Hybrid solvents allow for the same lean amine reactions plus sustained greater physical solubility at the higher loadings.

Model implementation

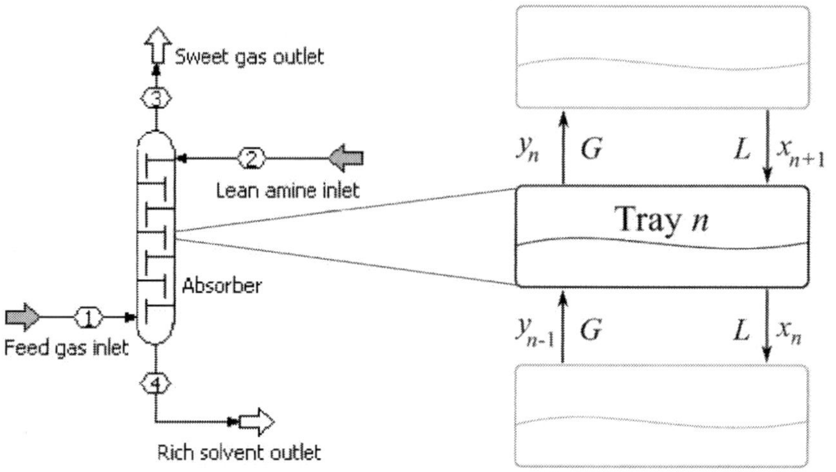

Thermodynamics + Mass Transfer + Kinetics

Model development: Mass Transfer framework

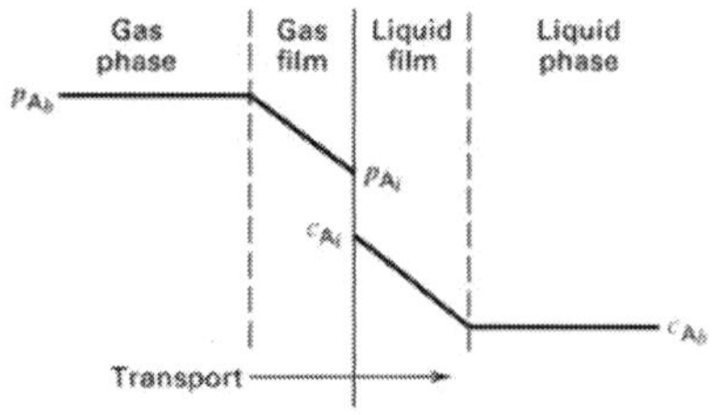

A simplified representation of the molar flow of a species across the vapor liquid interface:

$$N = K_{OV} a_e \left(c_g - c_l / m \right)$$

Thermodynamic driving force

Product of overall mass transfer coefficient and effective gas-liquid contact area

The overall mass transfer coefficient is a function of:

- Vapor film coefficient (k_g)
- Liquid film coefficient (k_l)
- Chemical enhancement factor (E_a)
- Equilibrium partition coefficient ($m = c_l^*/c_g^*$)

$$K_{ov} = \frac{1}{\frac{1}{k_g} + \frac{1}{m k_l E_a}}$$

Model development: Mass Transfer framework

Molar Flow

$$N = K_{OV} a_e \left(c_g - c_l / m\right)$$

Overall Mass Transfer Coefficient

$$K_{ov} = \frac{1}{\frac{1}{k_g} + \frac{1}{mk_l E_a}}$$

Some particular trends seen in our applications.

CO_2
- $E_a \sim 1$
- $1/k_g$ Low
- $1/mk_lE_a$ High → Controlling

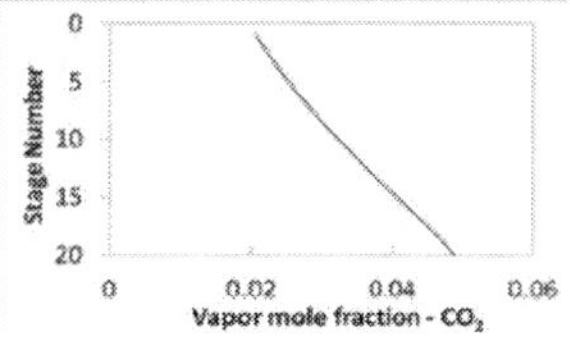

H_2S
- $E_a >>1$
- $1/k_g$ Controls at the top
- $1/mk_lE_a$

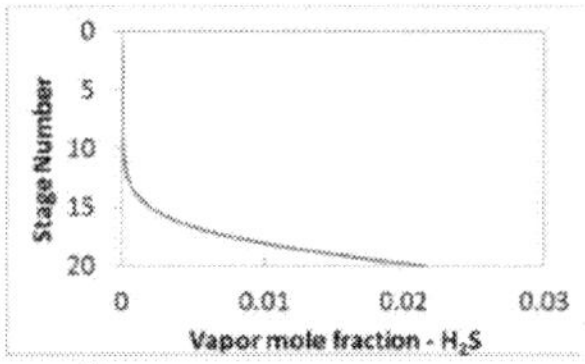

CH_4S
- $E_a \sim 1$
- $1/k_g$ High → Controlling
- $1/mk_lE_a$ Low

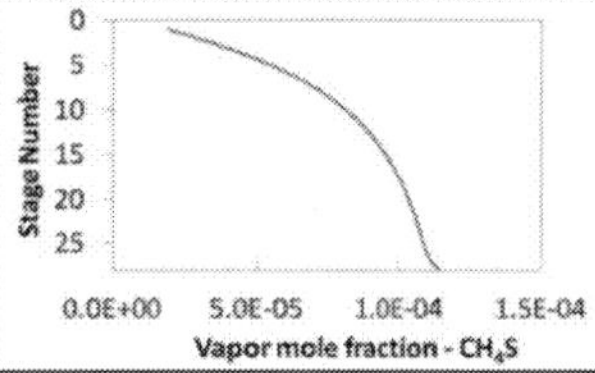

Model Implementation: Plant Data Comparison

Component	Plant - molar	Predicted – molar
CH4	87.4%	88.5%
CO2	2.6%	2.6%
H2S	8.5 ppm	8.37 ppm
CH4S	13 ppm	19 ppm
Other	10%	8.9%

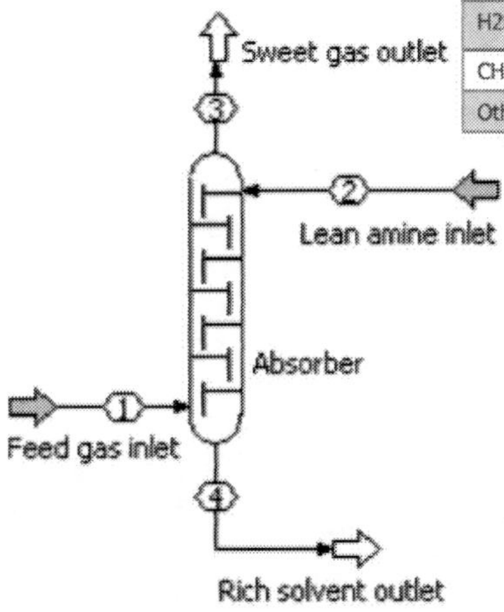

Component	Plant - molar
CH4	82.2%
CO2	5.6%
H2S	3.9%
CH4S	121 ppm
Other	8.3%

Model reproduces plant data.

Conclusions

- Dow's hybrid solvents represent an excellent alternative for Acid Gas and Mercaptan removal.
- Modeling hybrid systems is a complex task. Our simulator for Hybrid Solvents was developed to account for the effect of the mixed solvents: Thermodynamics, transport properties, mass transfer and kinetics.
- Our simulator for Hybrid Solvents matches plant data.

Thank You

Questions?

Model development: Adding components

Pure component properties → Binary Data → Ternary Data → Multicomponent Data

- Pure component properties: Molecular and ionic species
- Binary Data: binary parameters
- Ternary Data: Tune binary parameters if needed
- Multicomponent Data: Electrolyte-electrolyte binary parameters if needed

- Thermodynamic data: dielectric constant, Gibbs energy and enthalpy of formation, VLE, LLE, heat capacity, heat of absorption, density.
- Transport data: Viscosity, surface tension, conductivity, diffusivities.
- Trends: Excess properties for solvents blends, salting-in and salting-out curves for loaded solutions, extrapolation capabilities.

Effect of Organics Solvents on Equilibrium and kinetics

- Solvents of lower dielectric constants reduce the basic strength of the alkanolamines → lower pKa
- From a Brønsted plot, a lower basic strength implies a lower reaction rate of the alkanolamine with an acid gas (e.g. CO2). Although, for accurate determination of reaction rate constants, kinetic measurements have to be considered.

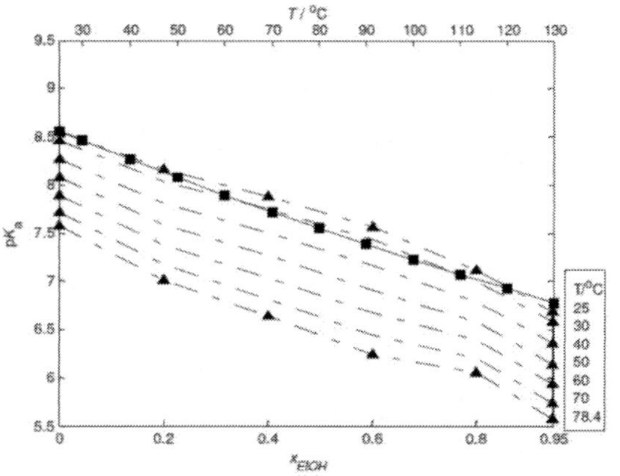

Comparison of the change of the basic strength at infinite dilution of MDEA in ■pure water, ▲aqueous ethanol (lower axis).

Versteeg et al. Energy Procedia 4 (2011) 187–194

Fundamental Properties and Simulation Results

The Importance of Having the Foundation Correct

Loren Wilson Shell International E&P

Abstract

- Most simulations are done using models based on equations of state. The quality of the parameters (Tc, Pc, and acentric factor) used by most equations of state can lead to success or failure of the model.
- Two case studies will be presented where errors in the fundamental data resulted in significant errors in the model.
- Based on a compilation of case studies by Henry Z. Kister in *Distillation Design*, about one out of six problems in separations are caused by fundamental issues with the quality of the model; while the other five out of six are caused by mistakes during design, fabrication, installation, or operation.
- Fixing the model beforehand is usually cheaper than retrofitting the column.

The Principle of Corresponding States

- Proposed by van der Waals in the 1870's.
- This principle states that the PVT behaviour (compressibility) of compounds in the fluid state are the same at the same properly scaled temperature and pressure.
- The scaling parameters are the critical temperature and pressure (Tc and Pc).
- Basis of most equations of state used in simulation and modelling today (Redlich-Kwong family, Peng-Robinson family, etc.).
- Notable exceptions: Some predictive cubic equations of state get A and B from group contributions, skipping Tc and Pc entirely.
- SAFT family equations of state do not use critical constants or acentric factor for input

Compressibility Factor Chart

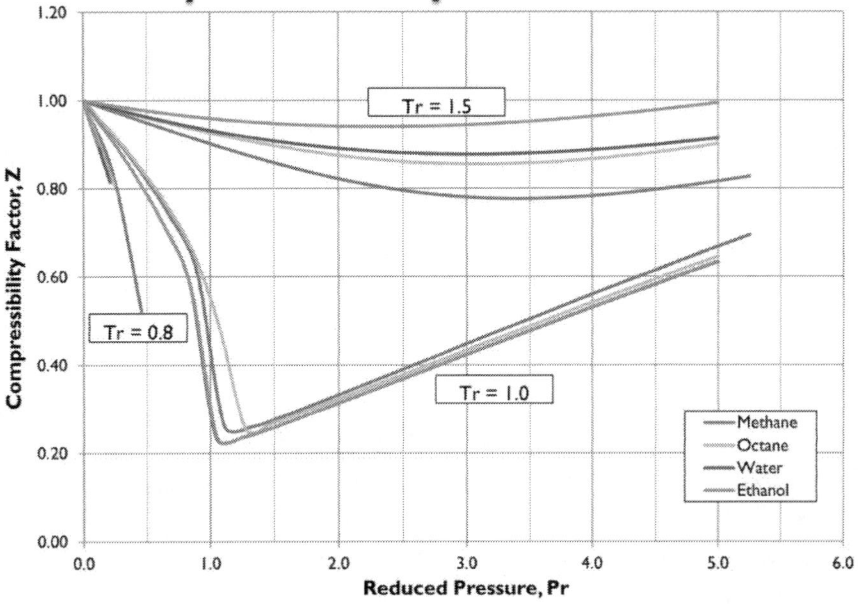

Tc and Pc

- The critical point is the highest temperature and pressure that a pure material can coexist as liquid and vapor
- Varies widely by compound
  - Helium: 5.2 K and 2.3 bar
  - n-Octane: 568.7 K and 24.9 bar
  - Water: 647.1 K and 220.6 bar
  - Mercury: 1735 K and 1608 bar

How Tc and Pc fit in

Original Peng-Robinson Equation of State

$$P = \frac{RT}{v-b} - \frac{a\alpha}{v(v-b)+b(v-b)}$$

$$a = 0.45724\frac{R^2T_c^2}{P_c}$$

$$b = 0.07780\frac{RT_c}{P_c}$$

$$\alpha = \left\{1 + (0.37464 + 1.54226\omega - 0.26992\omega^2)(1 - T_r^{1/2})\right\}^2$$

$$\omega = -\log_{10}(P_{vap} / P_c) - 1 \ where\ P_{vap}\ is\ calculated\ at\ Tr = 0.7$$

A Bit of History

- The first reported experiments into the critical point were by Charles Cagniard de la Tour in 1822.
  - Investigating acoustics in fluids at high temperature and pressure.
  - Used a cannon barrel as a pressure vessel.
  - A flint ball inside the barrel was rolled from end to end and he listened for the sound it made when entering the liquid.
  - Above a given temperature for a fluid, there was no more sound of the ball entering the liquid phase.
  - He deduced that above that temperature, there was only one fluid phase.
- In 1869, Thomas Andrews studied the phase behaviour of carbon dioxide, discovering that above 30.92°C, CO2 could not be liquefied at any pressure.
  - The accepted value today is 31.06°C.
  - He used the phrase "critical point" to describe this phenomenon.
- Critical points of many stable compounds have since been measured.
- However, most compounds are either not stable at the critical temperature or too toxic or corrosive to work with.

Current Experimental Efforts

- Can be measured in any PVT apparatus used by the petroleum industry, but much easier to observe if the vessel is rocked.
- LTP in Oldenburg, Germany measures Tc and Pc in a traditional visual pressure vessel.
- Dr. Amyn Teja at Georgia Tech developed a visual flow method in 1989 by pumping the fluid through a heated line and glass tube, and using a regulating valve on the output to control the pressure.
- The author while at Wiltec Research improved this method using a small pressure vessel with windows that could be rocked inside an oven. The fluid was pumped from one pump to another, with a residence time as short as 15 seconds
- A sub-second method was developed by Dr. Grant Wilson at Wiltec by pumping the fluid through a coil in a bath and observing the phase change from liquid to vapor using thermocouples.
- Another sub-second method has been developed by Dr. Eugene Nikitin at Ekaterinburg, Russia. The critical point is detected by the lack of phase change of the fluid surrounding a small platinum wire that is heated electrically. He has been able to measured the critical temperature and pressure of alkanes up to C60.

Very Close to the Critical Point

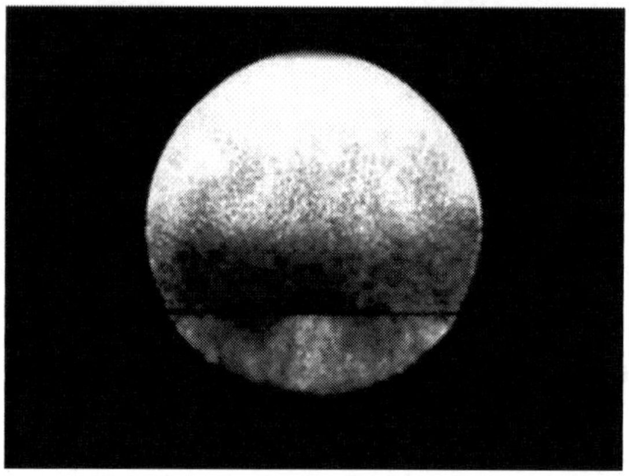

Compilations

- API Research Project 44
  - Work done by Thermodynamic Research Center (TRC), then at NBS, to collect and evaluate physical properties for hydrocarbons for the refining industry
  - TRC moved to Carnegie-Mellon, then Texas A&M, now back to NIST
- Dortmund Data Bank
  - Started in the 1970s at the University of Dortmund to enable them to develop parameters for the UNIFAC group contribution activity coefficient model.
  - Now distributed by DDBST Gmbh, Oldenburg.
- Design Institute for Physical Properties (DIPPR®) Project 801
  - Organized under AIChE in 1980
  - Consortium of companies to collect and evaluate physical property data, and to sponsor joint research
  - Project 851 started in 1985 to measure critical points
  - 150 critical points measured to date
- NIST now sponsors TRC, with ongoing efforts to collect and evaluate physical property data

Quality of the Measured Data

- In the DIPPR® Project 801 database, we have approximately 2200 compounds

	Uncertainty					
	<0.2%	<1%	<3%	<5%	<10%	Total
Critical Temperature	218	255	161	21	3	658
Critical Pressure	16	123	246	97	11	493

Predictive Methods

- Usually based on group contributions.
- Lydersen's method (1955) has been widely used.
- Many more in the literature.
- Predictive ability depends on the class of compound and how much data were available to develop the groups.
- In the first case study, we will compare the results of a simulation using Tc, Pc, and ω predicted by three methods versus using measured values of Tc and Pc.

Groups for Lydersen's Method

Group	G (T.)	G (P.)	G (V.)	Group	G (T.)	G (P.)	G (V.)
-CH3,-CH2-	0.02	0.227	55	>CH	0.012	0.21	51
-C<	-	0.210	41	=CH2,#CH	0.018	0.198	45
=C<,=C=	-	0.198	36	=C-H,#C-	0.005	0.153	36
-CH2-(Ring)	0.013	0.184	44.5	>CH-(Ring)	0.012	0.192	46
>C<(Ring)	-0.007	0.154	31	=CH-,=C<, =C=(Ring)	0.011	0.154	37
-F	0.018	0.224	18	-Cl	0.017	0.32	49
-Br	0.01	0.5	70	-I	0.012	0.83	95
-OH	0.082	0.06	18	-OH(Aromat)	0.031	-0.02	3
-O-	0.021	0.16	20	-O-(Ring)	0.014	0.12	8
>C=O	0.04	0.29	60	>C=O(Ring)	0.033	0.2	50
HC=O-	0.048	0.33	73	-COOH	0.085	0.4	80
-COO-	0.047	0.47	80	0	0.031	0.095	28
>NH	0.031	0.135	37	>NH(Ring)	0.024	0.09	27
>N	0.014	0.17	42	>N-(Ring)	0.007	0.13	32
-CN	0.06	0.36	80	0	0.055	0.42	78
-SH,-S-	0.015	0.27	55	-S-(Ring)	0.008	0.24	45
=S	0.003	0.24	47	>Si<	0.03	0.54	-
-B<	0.030	-	-				

Lydersen's Method

- $Tc = Tb/(0.567 + \Sigma G_{ti} - (\Sigma G_{ti})^2$
- $Pc = Mwt/(0.34 + \Sigma G_{pi})^2$
- $Vc = 40 + \Sigma G_{vi}$

Case Study #1 – Gas Treating

- Methyl diethanolamine (MDEA) is used to remove CO2 and H2S from natural gas streams.
- Calculate solvent loss to sweet gas and to rejected acid gas.

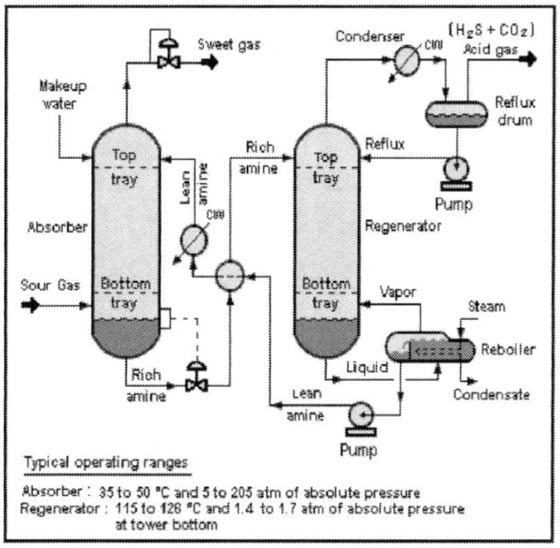

The Model

- Absorber condition: 40°C and 100 bar.
- Regenerator condition: 120°C and 2 bar.
- Literature values of T_c, P_c, and ω used for methane and water.
- T_c, P_c, and ω of MDEA predicted by three methods: Lydersen (1955), Nannoolal (2007), Wilson-Jasperson (1996).
- In all three predictive methods, we use the literature value of the normal boiling point of MDEA as input.
- $\omega = -\log_{10}(P_{vap}/P_c) - 1$, where P_{vap} is the vapor pressure calculated at Tr = 0.7
- Vapor pressure predicted from Reidel's method using the normal boiling point from the literature and the predicted critical temperature and pressure.
- Peng-Robinson equation of state used to calculate the K-value of MDEA in the absorber and regenerator.

Input Parameters

Method	Tb, K	Tc, K	Pc, bar	VP at Tr = 0.7	Acentric Factor
Lydersen	520.7	678.2	38.8	0.21	1.264
Nannoolal	520.7	772.2	41.9	1.57	0.426
Wilson-Jasperson 1st order	520.7	728.1	41.5	0.75	0.744
Experimental	520.7	741.9	41.6	0.98	0.628

Method	Tc, K	Meas - Predicted	% Dev.	Pc, bar	Meas - Predicted	% Dev.
Lydersen	678.2	-63.7	-8.6%	38.8	2.8	6.8%
Nannoolal	772.2	30.3	4.1%	41.9	-0.3	-0.7%
Wilson-Jasperson 1st order	728.1	-13.8	-1.9%	41.5	0.1	0.3%
Experimental	741.9			41.6		

Results

Method	Absorber		Regenerator	
	K-value	K/K Exp	K-value	K/K Exp
Lydersen	2.4E-05	0.02	1.4E-03	0.26
Nannoolal-TC	6.9E-03	5.16	9.1E-03	1.67
Wilson-Jasperson 1st order	5.4E-04	0.41	4.1E-03	0.75
Experimental Tc, Pc, ω	1.3E-03	1.00	5.5E-03	1.00

Why the large differences?

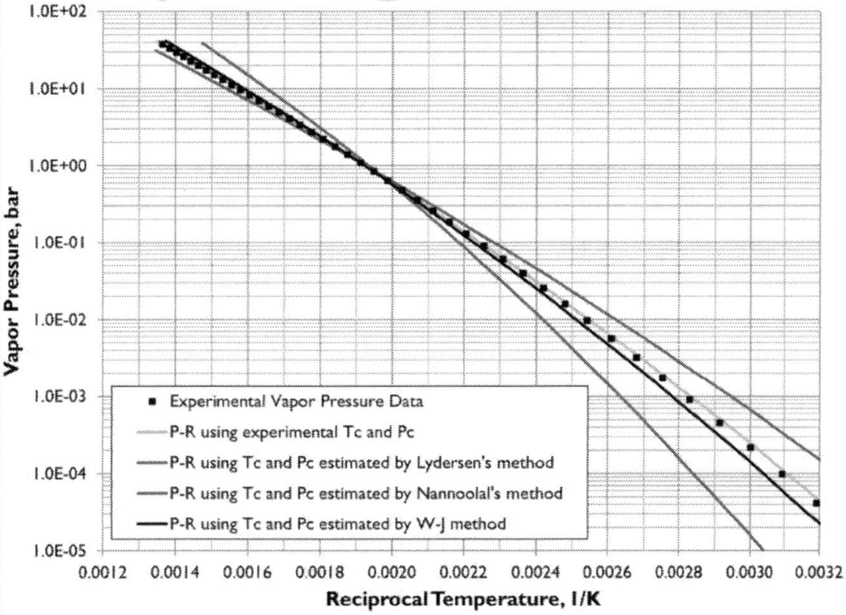

Implications

- Depending on the set of parameters, you will either under or over-predict solvent losses.
- This is an issue if you license a process to other companies, or if you want to compare performance with competing processes.

Case Study #2

- Process change to new solvent to increase recovery of the feedstock.
- Previously unknown reaction between feed and new solvent produced a new by-product that rendered the material unfit to sell.
- In the meantime, millions of pounds of product were manufactured.
- Urgent need to distill the by-product out .
- Simulation showed the by-product could be removed into the overhead, with the pure product out the bottom.
- Reality: the by-product was actually less volatile than the product (light key was actually the heavy key).
- It took an afternoon for the model to be fixed, and the simulation to be realistic.
- Then K-values were measured to improve the model.

How to avoid bad parameters

Easier

- Look at your parameters. My in-house database was not as up-to-date as I expected. It had the estimated values of Tc and Pc for MDEA instead of the measured data which we have access to.
- Commercial simulator companies do a good job of keeping current.
- Use an experimental value from a reliable database if available.
- Use an estimated value that an experienced practitioner has reviewed, either from a database or by someone in your company.
- For new compounds, you may have to estimate the value.
- Try methods of Nannoolal and Wilson-Jasperson to start with, especially if you have some vapor pressure data, W-J if you don't have any data.
- Compare a predicted vapor pressure with a measured value in the range of interest.
- Compare the estimated Tc and Pc with experimental values in the same family.
- Convince your manager to pay for measurements.

Harder

Questions?

Citations

1. Henry Z. Kister, Distillation Design, McGraw-Hill, 1992.
2. Ding-Yu Peng, Donald B. Robinson, " A New Two-Constant Equation of State," *Ind. Eng. Chem., Fundam., 15,* No. 1, p. 59-64, 1976.
3. Charles Cagniard de la Tour, "Exposé de quelques résultats obtenu par l'action combinée de la chaleur et de la compression sur certains liquides, tels que l'eau, l'alcool, l'éther sulfurique et l'essence de pétrole rectifiée" (Presentation of some results obtained by the combined action of heat and compression on certain liquids, such as water, alcohol, sulfuric ether [i.e., diethyl ether], and distilled petroleum spirit), *Annales de chimie et de physique, 21* : 127-132, 1822.
4. Andrews, Thomas, The Bakerian Lecture: On the Continuity of the Gaseous and Liquid States of Matter, *Phil. Trans. R. Soc. Lond.* 1869 159, doi: 10.1098/rstl.1869.0021.
5. Rosenthal, D. J. and Teja, A. S., "The Critical Properties of n-Alkanes Using a Low Residence Time Flow Apparatus," *AIChE Journal*, 35, 1829-1834 (1989).

Kinetic Modeling and Optimization of Polymerization Reactions

Balakrishna, S., Optience Corporation

Originally presented on: 10/9/2014 11:00:00 - 11:30:00

Specialty polymers often involve polycondensation mechanisms. Our goal is to identify the kinetics and to develop optimization strategies for achieving desired polymer properties while minimizing by-product formation. The tasks typically involved in the analysis of acid-alcohol based polycondensation reactions are:

- What are the kinetics of the polycondensation reactions?
- What is the optimal alcohol to acid ratios to ensure desired molecular weights?
- What is the optimum pressure profile and pressure transition points to ensure desired polymer properties?

In this work, we illustrate our approach to answer these questions. Polycondensation systems are usually characterized by reversible reactions that require the removal of water or alcohol, usually by pressure reduction, to drive the growth in molecular weights and polymer properties. Furthermore, by-product formation reactions need to be minimized to the extent possible. Accurate kinetic model development is a crucial first step in order to optimize these systems.

For a single phase acid-alcohol polymerization, equi-molar ratios are ideal to achieve high molecular weights. However, for multiphase systems, this is not the case, since the alcohol is distributed between the phases, and therefore, some excess of alcohol is needed. The optimum alcohol excess depends on reactor conditions such as the temperature and pressure profiles, which presents an interesting dynamic optimization problem.

We will present our approach to develop the kinetic model and strategies for reactor optimization. In solving the problem for a variety of projects, we have developed a systematic tool that delivers substantial productivity in the processes of estimation, simulation and reaction optimization. Project execution is improved by having a common framework for experimental data storage, kinetic estimation and reactor optimization. Results for the polycondensation kinetic estimation and optimization will be shown along with explanation of the generalized infrastructure we have built to analyze reaction systems.

Tridiagonal Solutions
compute | innovate | develop

Enhancing polymerization reactor performance via better micro-mixing

A modeling study using PDF methods

Mothivel Mummudi
Principal Engineer
Tridiagonal Solutions Inc.

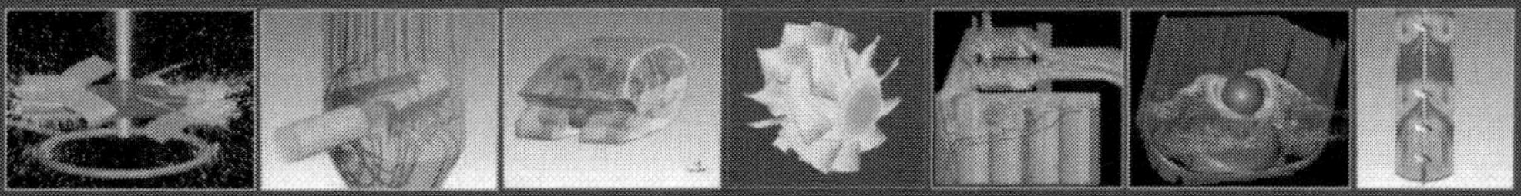

www.tridiagonal.com
info@tridiagonal.com

Turbulence and Mixing

Tridiagonal Solutions
compute | innovate | develop

- Turbulence is practically everywhere
  - Fluid turbulence is a naturally occuring phenomenon
  - Characterised by random/stochastic variations in velocity, pressure, species concentrations etc.
  - Industrial flows are highly turbulent
- Characterised by relative magnitudes of diffusion and convection

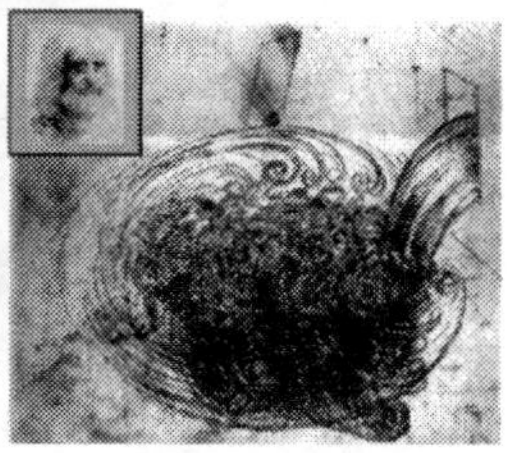

- Turbulence enhances mixing
  - Stretching and folding
  - More contact between fluid elements
- Mixing occurs at different "scales"
  - Large-scale mixing - Macromixing
  - Small-scale/molecular mixing – micromixing

Scales of Turbulent Mixing

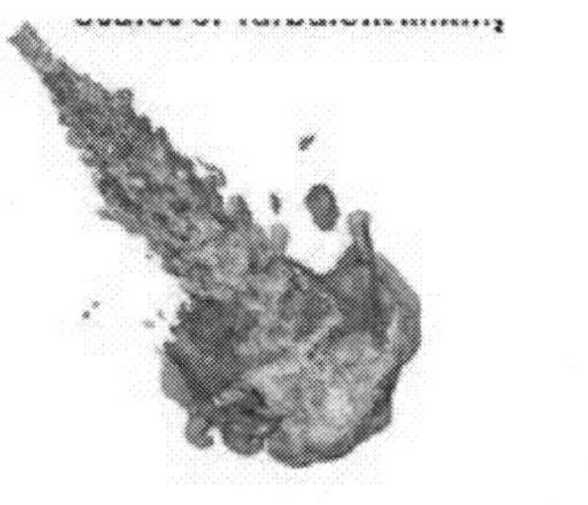

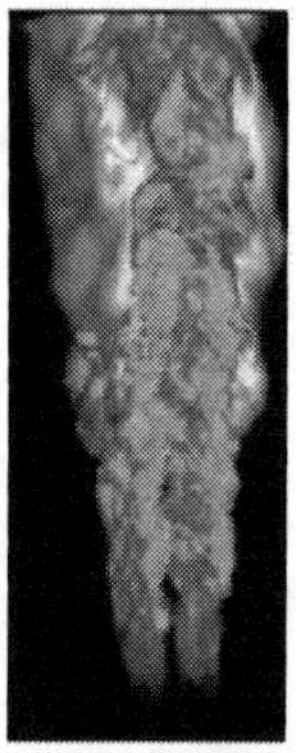

Source:www.olcf.ornl.gov

- Chemical reactions are molecular processes
  - Species interact at molecular levels
  - Good micromixing is essential for enhanced reaction

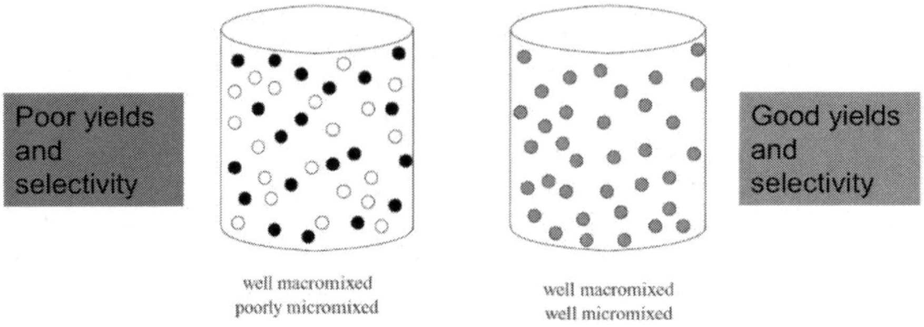

- Residence time distribution cannot provide micromixing information
- More sophisticated models than RTD-based ones are needed

Source: www.bakker.org

Multi-Environment Micromixing Models

Tridiagonal Solutions
compute | innovate | develop

- A two-environment symmetric (2E) model for micromixing
  - Each environment corresponds to a set of scalar compositions
  - Micromixing can be described in terms of "fluid-exchange" between the environments in composition space
  - Interaction-by-exchange-with-the-mean (IEM) model used to describe the "fluid-exchange"

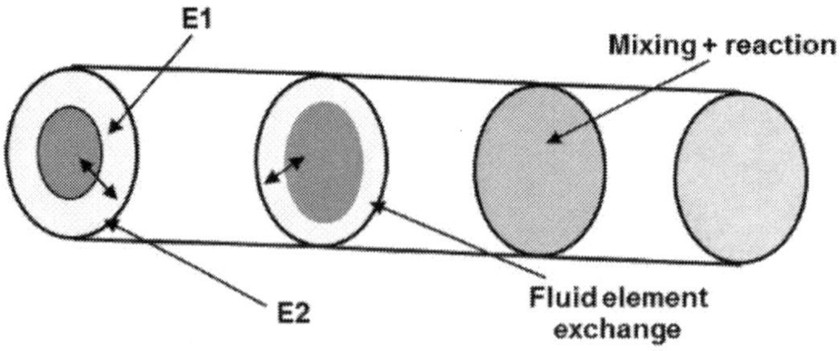

Modeling Mixing & Reactions

Tridiagonal Solutions
compute | innovate | develop

A 1-D mathematical representation of a reacting flow system

Convection/Dispersion — Diffusion — Source Terms

$$\frac{\partial \langle \phi_\alpha \rangle}{\partial t} + \boxed{\langle U_j \rangle \frac{\partial \langle \phi_\alpha \rangle}{\partial x_j} = -\frac{\partial}{\partial x_i} \langle u_i' \langle \phi_\alpha \rangle \rangle} + \boxed{D_{\phi_\alpha} \frac{\partial^2 \langle \phi_\alpha \rangle}{\partial x_i \partial x_j}} + \boxed{\langle S_{\phi_\alpha}(\phi) \rangle}$$

"Closure Problem" $\boxed{\langle S_{\phi_\alpha}(\phi) \rangle \neq S_{\phi_\alpha}(\langle \phi \rangle)}$

- Convection and dispersion – Macromixing
- Molecular diffusion – micromixing
- Numerical solution
  - Length scale resolution necessary to describe mixing across scales
  - Resolution is a strong function of Reynolds number

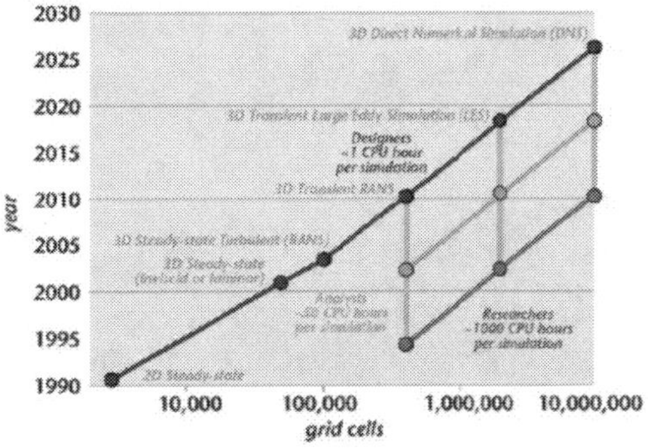

Source: www.bakker.org

Utility Of Probabilistic Methods

Tridiagonal Solutions
compute | innovate | develop

- No closure strategies required for reaction terms
- Involves using probability density function (pdf) of species composition
- Use a finite mode representation for better computational efficiency

$$\underbrace{\frac{\partial f_\phi}{\partial t} + \langle U_i \rangle \frac{\partial f_\phi}{\partial x_i}}_{\text{Macromixing}} = \underbrace{\boxed{\Gamma_T \frac{\partial^2 f_\phi}{\partial x_i \partial x_i}}}_{\text{Mesomixing}} - \frac{\partial}{\partial \psi_\alpha} \underbrace{\left[\left\langle \Gamma_\alpha \frac{\partial^2 \phi_\alpha}{\partial x_i \partial x_i} \mid \psi \right\rangle f_\phi \right]}_{\text{Micromixing}} + \underbrace{\boxed{S(\psi) f_\phi}}_{\text{Source terms}}$$

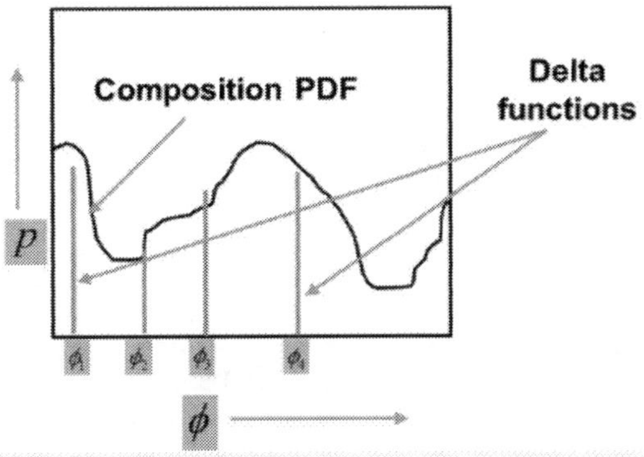

Finite Mode Concept

Source: www.bakker.org

The "Connection"

The modes of the composition PDF ⇔ Environments in an MEM model !

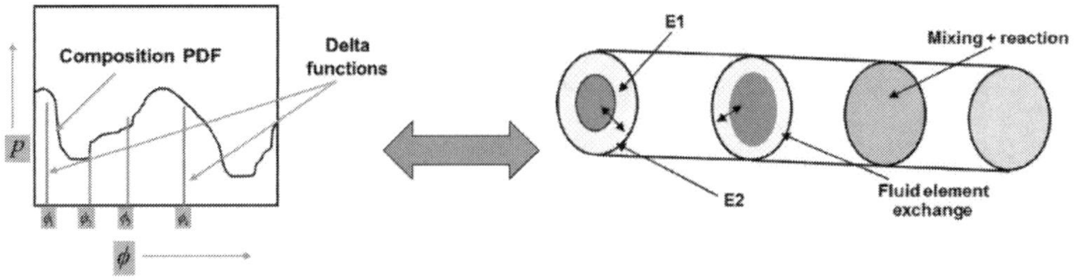

Micromixing between the "environments" drives the shape evolution of the FMPDF

Described by changing location of the different "modes"

Industrial Tubular LDPE Processes

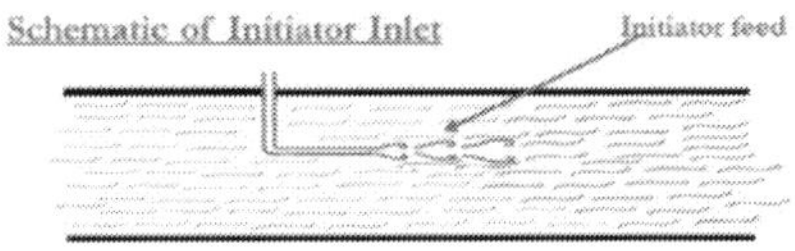

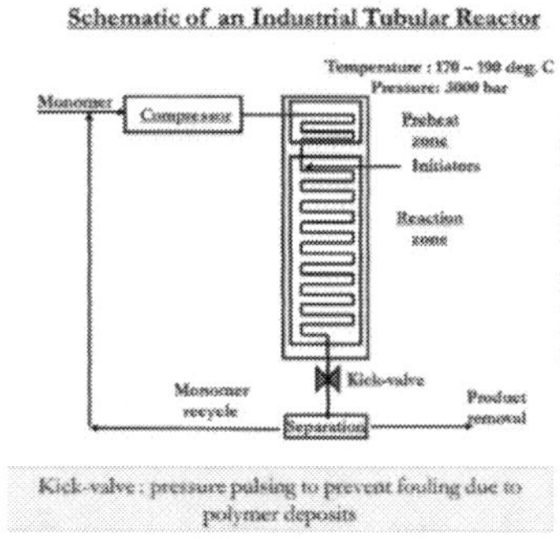

Product quality : Molecular Weight Distribution (MWD) is strongly influenced by operating conditions

Ethylene decomposition or "decomp" → runaway

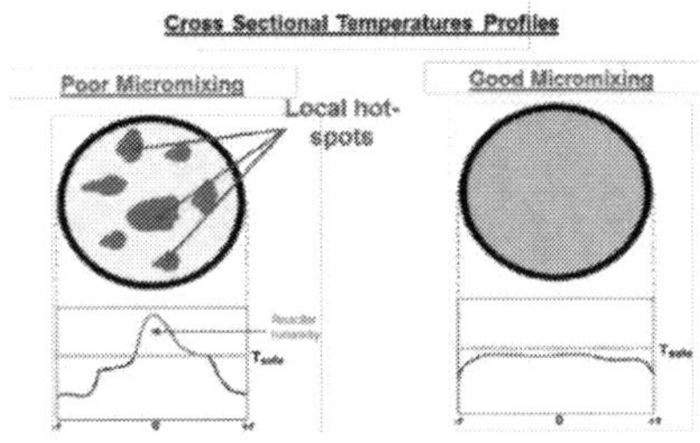

- Reliable prediction of conversion, yield, temperature and MWD from process conditions
- Good description of dynamical phenomena in the reactor : micromixing, reactions, heat-transfer and kick-valve operation
- Ability to detect process conditions which favor decomps

Reaction Chemistry

Tridiagonal Solutions
compute | innovate | develop

Free-radical Polymerization Reactions

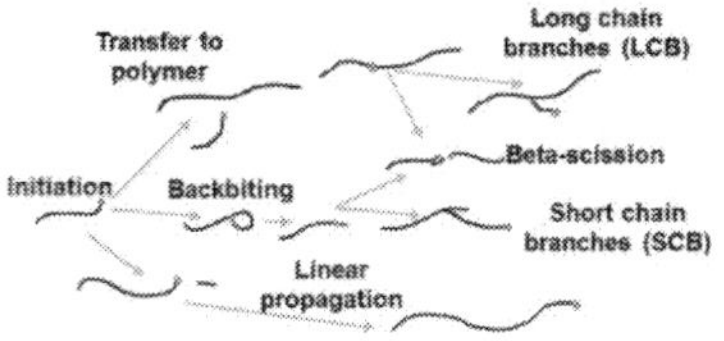

- Source terms for
  - Classical chemical species
  - Pseudo species
    - Moments of MWD (0,1,2)
    - Double bonds, SCB, LCB
  - Temperature
- Higher-order MWD moments – Hulburt and Katz closure

Ethylene Decomposition Reactions

- Ethylene decomposition free-radical mechanism
  - Initiation, propagation, termination
- Rates increase rapidly with temperature. Highly exothermic
- Large volumes of CH_4,C_2H_2 and C_2H_6 released – rapid pressure buildup
- Kinetic equations for C, CH_4,C_2H_2 and C_2H_6

Combined Kinetic Model

- Total of 28 species including 12 free-radical species
- Quasi-steady state approximation (QSSA) for the 12 free-radical species concentrations.
- Rate equations for 16 species – classical, pseudo and temperature

Model Prediction Vs Plant Data

Tridiagonal Solutions
compute | innovate | develop

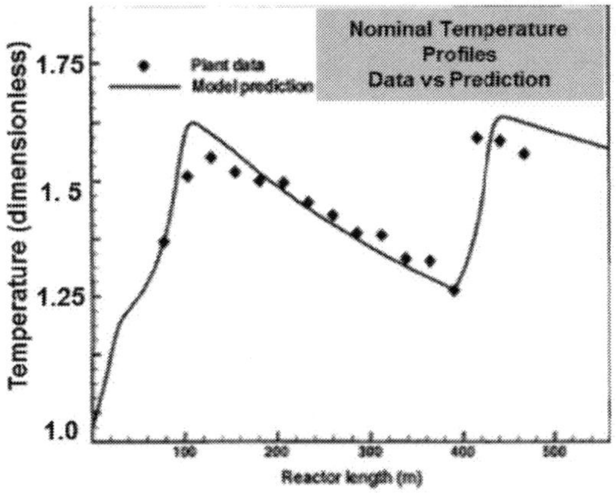

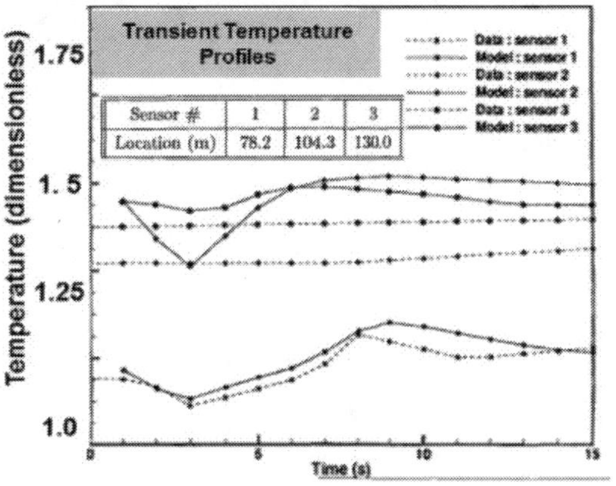

- Model predictions compared reasonably well with plant data for both nominal steady state as well as transient profiles of temperature

Effect Of Micromixing

Tridiagonal Solutions
compute | innovate | develop

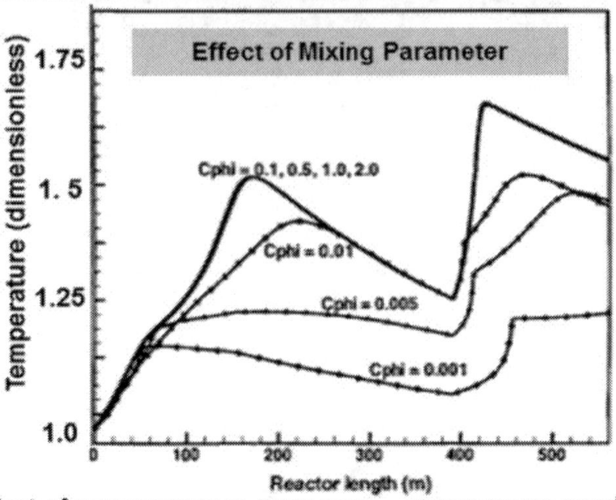

- The effect of micromixing is clear here – temperature profiles show that the extent of reaction is a strong function of degree of micromixing
- Lower C_ϕ implies poor micromixing and higher C_ϕ implies good micromixing

Detection of "Decomp" Events

Tridiagonal Solutions
compute | innovate | develop

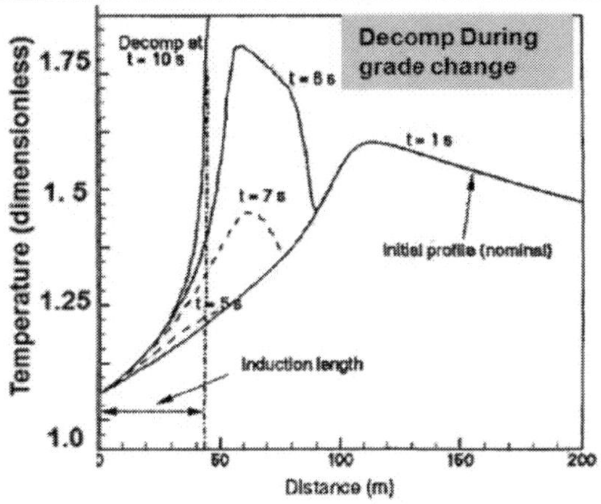

- Some types of polymer product grade changes are performed by increasing initiator concentrations
- Poor micromixing during such grade changes can lead to the onset of decomposition
- The induction period during which the system is "building up towards decomp" is determined by degree of mixing and the underlying chemistry

Key Process Parameters

Tridiagonal Solutions
compute | innovate | develop

- Sensitivity analysis (SA) helps in understanding the key physical variables that drive reactor behavior/dynamics
- Statistical techniques are typically used to perform SA

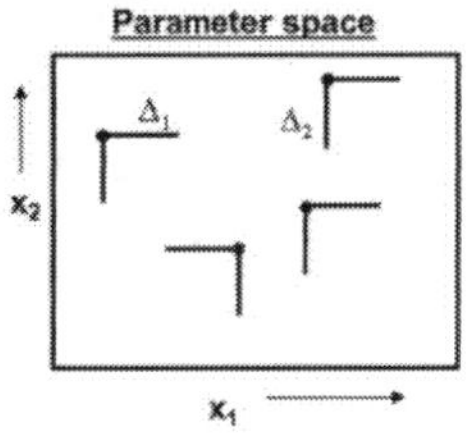

Parameters that have a significant effect on ΔT_{max}.

Parameter	Individual effects	Interactive effects
P_{atm}		√
T_{in}	√	
$TBPIN$		√
MFR	√	

Parameters that have a significant effect on $\Delta T_{max}/L$.

Parameter	Individual effects	Interactive effects
P_{atm}		√
T_{in}	√	
$TBPP$		√
$TBPIN$		√
CTA		√
MFR		√
$TCW0$	√	

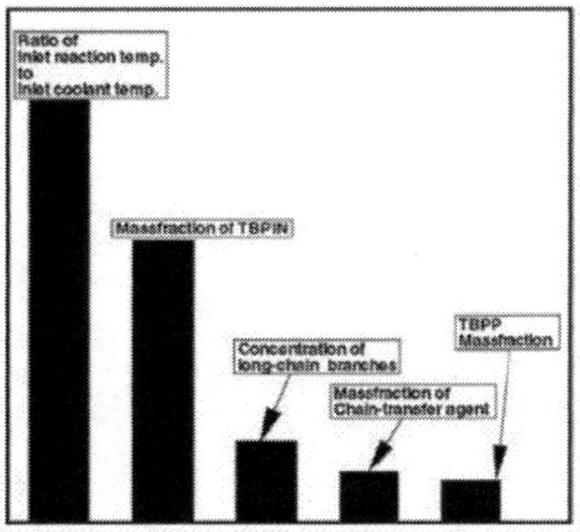

Summary

Tridiagonal Solutions
compute | innovate | develop

- Micromixing in turbulent reactive flows can be effectively treated using the probability density function (PDF) approach.
- Using this approach the dynamic behavior of LDPE reactors can be described adequately

6th AIChE Southwest Process Technology Conference

Moody Gardens Hotel and Convention Center • Galveston, TX

October 9-10, 2014

Comprehensive Optimization Methodology for Stimulation Design of Unconventional Gas Reservoirs

Srimoyee Bhattacharya, Ph.D.

Reservoir Engineer, Shell

October 9th, 2014

Bhattacharya S. Comprehensive Optimization Methodology for Stimulation Design of UGRs.

Outline

- Introduction
- Challenges
- Fracture Design
- Well Planning Optimization
- Conclusions

Bhattacharya S.
Comprehensive Optimization Methodology for Stimulation Design of UGRs.

- Introduction
- Challenges
- Methodology
- Fracture Design Module
- Production Estimation Module
- Objective Function
- Case Study
- Conclusions

Unconventional Gas Resources

AIChE Southwest Process Technology Conference

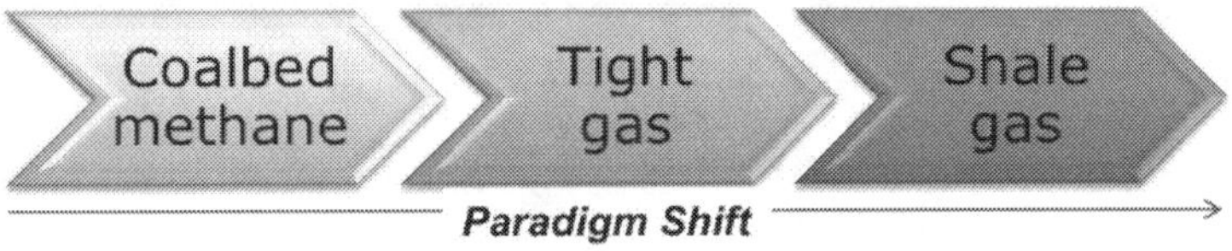

Figure 5. U.S. wet natural gas reserves, production, and imports, 2001-12

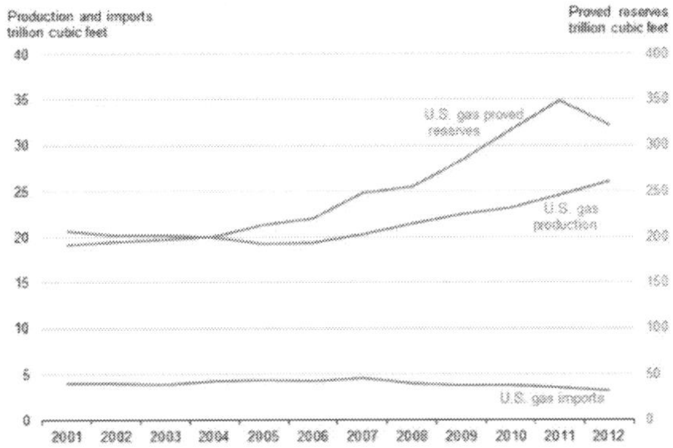

Sources: U.S. Energy Information Administration, Form EIA-23L, "Annual Survey of Domestic Oil and Gas Reserves" and U.S. Department of Energy, Office of Fossil Energy, "Natural Gas Imports and Exports."

From 2011 to 2012 –

- 34% drop in natural gas price resulted in 7.5% drop in proved reserves.
- Still, production increased by 6%.

(EIA, 2014, U.S. Crude Oil and Natural Gas Proved Reserves 2012)

3

Bhattacharya S.
Comprehensive Optimization Methodology for Stimulation Design of UGRs.

- Introduction
- Challenges
- Methodology
- Fracture Design Module
- Production Estimation Module
- Objective Function
- Case Study
- Conclusions

Natural gas resources

AIChE Southwest Process Technology Conference

Figure MT-42. Total natural gas production, consumption, and imports in the Reference case, 1990-2040 (trillion cubic feet)

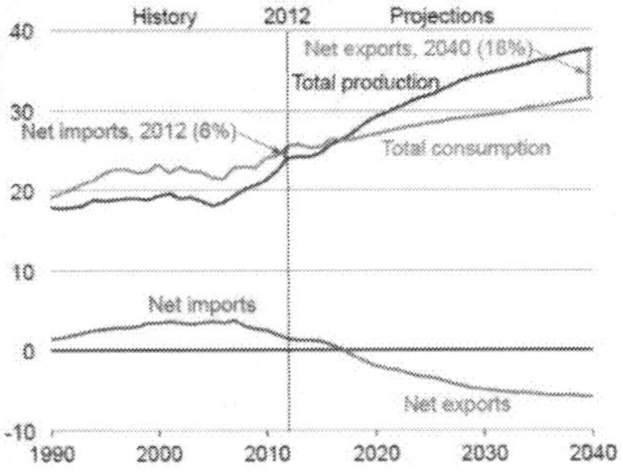

- Transition from net importer of 1.5 Tcf natural gas in 2012 to a net exporter of 5.8 Tcf in 2040.
- Net exporter of natural gas before 2020.

Figure MT-44. U.S. natural gas production by source in the Reference case, 1990-2040 (trillion cubic feet)

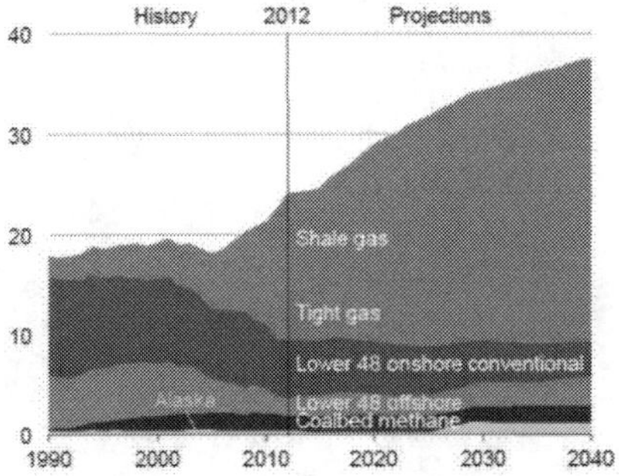

- Largest contributor: Shale gas.
- Growing from 9.7 Tcf in 2012 to 19.8 Tcf in 2040.

(EIA, 2014, Annual Energy Outlook)

4

Bhattacharya S.
Comprehensive Optimization Methodology for Stimulation Design of UGRs.

AIChE Southwest Process Technology Conference

Challenges

Introduction
Challenges
Methodology
Fracture Design Module
Production Estimation Module
Objective Function
Case Study
Conclusions

- No two shales are alike
- Extremely low permeability formations
- Complex geology
- Lack of historical production data
- Costly completion strategies

5

Bhattacharya S.
Comprehensive Optimization Methodology for Stimulation Design of UGRs.

AIChE Southwest Process Technology Conference

Optimal Well Planning

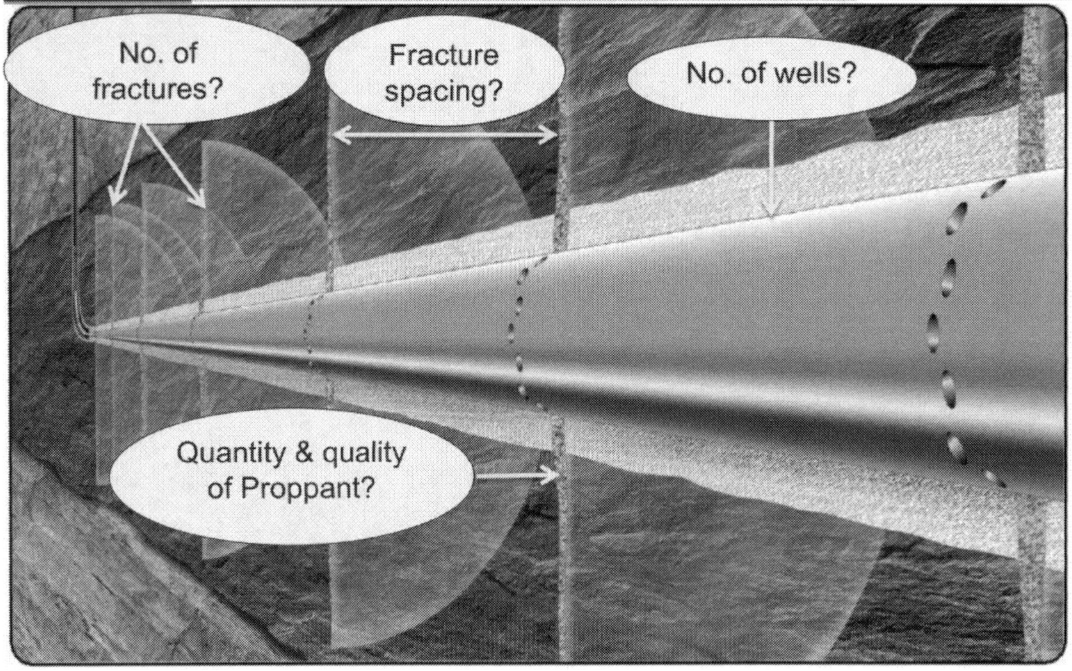

Bhattacharya S.
Comprehensive Optimization Methodology for Stimulation Design of UGRs.

Existing Approaches

AIChE Southwest Process Technology Conference

Introduction
Challenges
Methodology
Fracture Design Module
Production Estimation Module
Objective Function
Case Study
Conclusions

- Historical databases of stimulation jobs
- Ad-hoc / Trial & error methods
- Parametric sensitivity analysis using simulators
  - (Porcu et. al., 2008,2009; Bagherian et al., 2010; Meyer et al., 2010)

Calls for systematic computer-aided methodology!

7

Bhattacharya S.
Comprehensive Optimization Methodology for Stimulation Design of UGRs.

Optimization Methodology

AIChE Southwest Process Technology Conference

Introduction
Challenges
Methodology
Fracture Design Module
Production Estimation Module
Objective Function
Case Study
Conclusions

Input Variables:-
Reservoir/ Well Properties
Fluid properties
Proppant type

Design Variables:-
No. of wells, n_w
No. of fractures, n_f
Mass of proppant, M_p (lbs)

Fracture Design Module:
Unified Fracture Design, PKN, P3D
x_f, w_f, h_f, p_{treat}

Production Estimation Module:
Q(t)

Objective Function
NPV

Optimization Routine

Best Design

(Previous version: Bhattacharya and Nikolaou, 2011)

8

Bhattacharya S. Comprehensive Optimization Methodology for Stimulation Design of UGRs.

Fracture Design Flowchart

AIChE Southwest Process Technology Conference

- Introduction
- Challenges
- Methodology
- Fracture Design Module
- Production Estimation Module
- Objective Function
- Case Study
- Conclusions

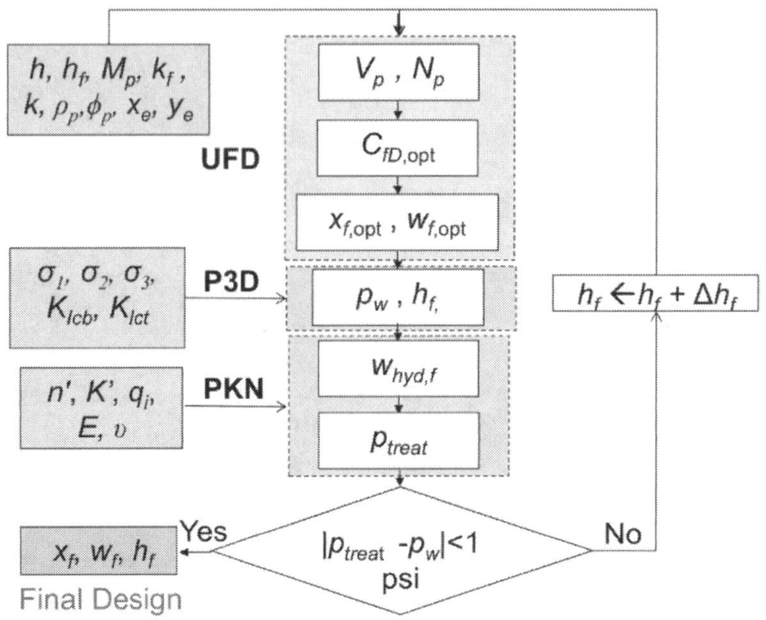

9

Bhattacharya S. Comprehensive Optimization Methodology for Stimulation Design of UGRs.

Unified Fracture Design (UFD)

(Economides and Valkó, 2002)

AIChE Southwest Process Technology Conference

- Introduction
- Challenges
- Methodology
- Fracture Design Module
- Production Estimation Module
- Objective Function
- Case Study
- Conclusions

Inputs: h, h_f, M_p, k_f, k, ρ_p, ϕ_p, x_e, y_e

$$C_{fD} = \frac{k_f w_f}{k x_f} \qquad I_x = \frac{2x_f}{x_e} \qquad y_{eD} = \frac{y_e}{x_e}$$

$$V_p = \frac{M_p\left(h/h_f\right)}{\rho_p\left(1-\phi_p\right)} \qquad N_p = \frac{2k_f V_p}{k h x_e y_e} = \frac{I_x^2 C_{fD}}{y_{eD}}$$

$$J_{D,\max} = \frac{0.6 y_{eD}^{\ 0.7} + 14 y_{eD}^{\ 1.6} N_p + 28 y_{eD}^{\ 2.7} N_p^{\ 2}}{1 + 35 y_{eD}^{\ 1.75} N_p + 14 y_{eD}^{\ 3.7} N_p^{\ 2}}$$

$$C_{fD,\text{opt}} = 1.6\sqrt{\frac{\left(1+0.03N_p^{\ 2}\right)\left(1+1.3N_p^{\ 2}\left(y_{eD}^{\ -1.2}\right)\right)}{1+0.1N_p^{\ 2}\left(y_{eD}^{\ -3.2}\right)}}$$

$$x_{f,\text{opt}} = \sqrt{\frac{k_f V_f}{C_{fD,\text{opt}} k h}} \qquad w_{f,\text{opt}} = \sqrt{\frac{C_{fD,\text{opt}} k V_f}{k_f h}}$$

Output: $x_{f,\text{opt}}$, $w_{f,\text{opt}}$

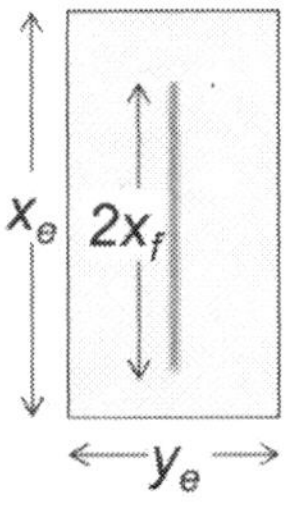

(Bhattacharya, Nikolaou and Economides, 2012)

10

Bhattacharya S.
Comprehensive Optimization Methodology for Stimulation Design of UGRs.

AIChE Southwest Process Technology Conference

Results: UFD

- Introduction
- Challenges
- Methodology
- Fracture Design Module
- Production Estimation Module
- Objective Function
- Case Study
- Conclusions

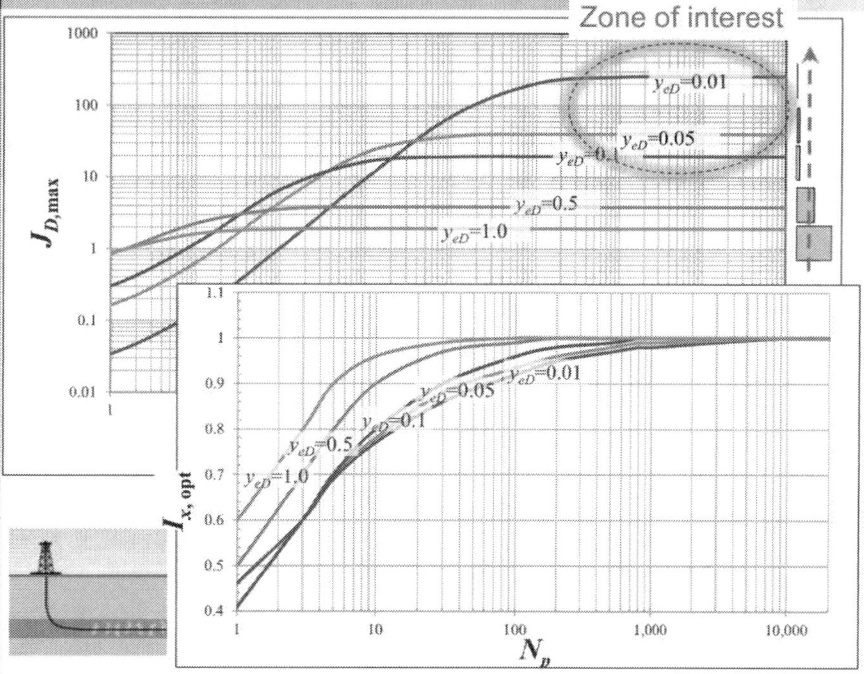

11

Bhattacharya S.
Comprehensive Optimization Methodology for Stimulation Design of UGRs.

AIChE Southwest Process Technology Conference

P3D

- Introduction
- Challenges
- Methodology
- Fracture Design Module
- Production Estimation Module
- Objective Function
- Case Study
- Conclusions

σ_1, σ_2, σ_3, K_{Icb}, K_{Ict} →

$$\sqrt{\frac{\pi}{a}}\left(\frac{K_{Icb}+K_{Ict}}{2}\right)=(\sigma_2-\sigma_1)\sin^{-1}\left(\frac{b_2}{a}\right)+(\sigma_3-\sigma_1)\sin^{-1}\left(\frac{b_3}{a}\right)-(\sigma_2+\sigma_3-2p_w)\frac{\pi}{2}$$

$$\sqrt{\pi}\left(\frac{K_{Icb}-K_{Ict}}{2}\right)=(\sigma_2-\sigma_1)\sqrt{a^2-b_2^2}-(\sigma_3-\sigma_1)\sqrt{a^2-b_3^2}$$

$$b_2+b_3=h$$

→ p_w, h_f

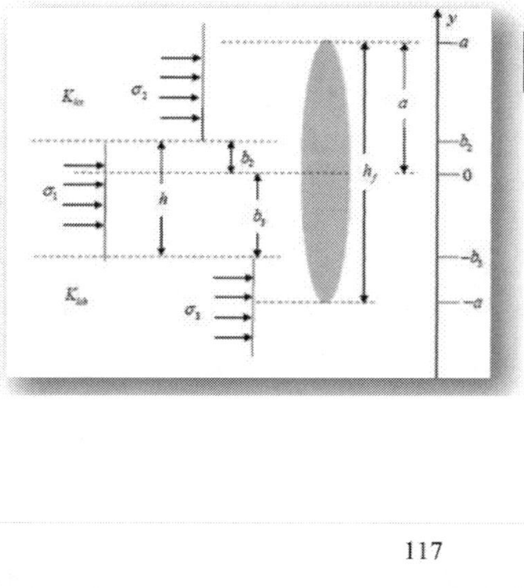

(Economides and Nolte, 2000)

12

Bhattacharya S.
Comprehensive Optimization Methodology for Stimulation Design of UGRs.

PKN

- Introduction
- Challenges
- Methodology
- Fracture Design Module
- Production Estimation Module
- Objective Function
- Case Study
- Conclusions

UFD: x_f
P3D: h_f
n', K', q_i, E

$$w_{hyd,f} = 9.15^{\frac{1}{2n'+2}} 3.98^{\frac{n'}{2n'+2}} \left[\frac{1+2.14n'}{n'}\right]^{\frac{n'}{2n'+2}} K'^{\frac{1}{2n'+2}} \left(\left(\frac{q_i}{2}\right)^{n'} \frac{h_f^{1-n'} x_f}{E'}\right)^{\frac{1}{2n+2}}$$

$$p_{net} = \frac{E}{\left(1-\upsilon^2\right) 2h_f} w_{hyd,f}$$

p_{treat}

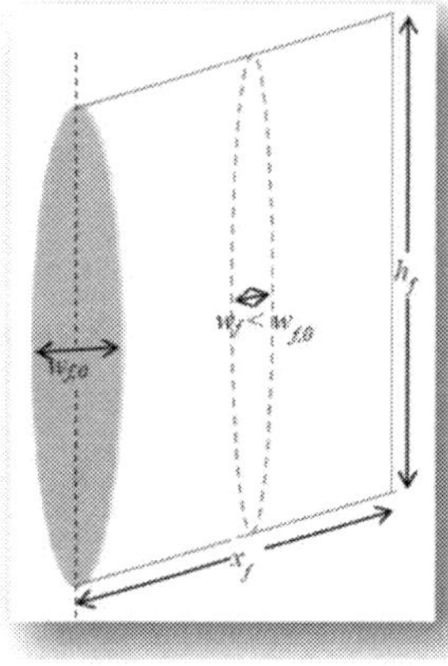

(Economides and Nolte, 2000)

13

Bhattacharya S.
Comprehensive Optimization Methodology for Stimulation Design of UGRs.

Fracture Height Variation

- Introduction
- Challenges
- Methodology
- Fracture Design Module
- Production Estimation Module
- Objective Function
- Case Study
- Conclusions

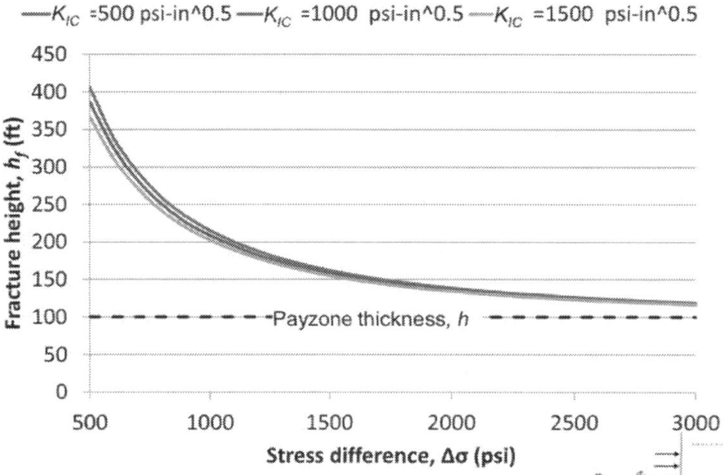

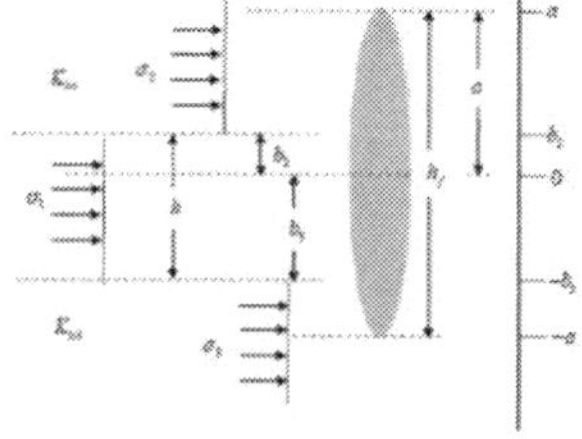

14

Bhattacharya S.
Comprehensive Optimization Methodology for Stimulation Design of UGRs.

Production Estimation Models

AIChE Southwest Process Technology Conference

Introduction

Challenges

Methodology

Fracture Design Module

Production Estimation Module

Objective Function

Case Study

Conclusions

Two kinds of model:

- Numerical
  - accurate modeling of near-fracture flow behavior
  - good tools for in depth analysis
  - computational intensive for optimization purpose
- Analytical
  - quick and easy to use
  - allows variation in n_f and n_w easily

Grid

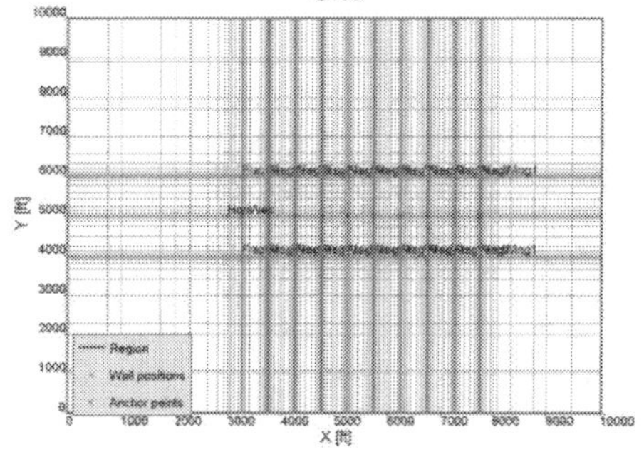

15

Bhattacharya S.
Comprehensive Optimization Methodology for Stimulation Design of UGRs.

Analytical Model Assumptions

AIChE Southwest Process Technology Conference

Introduction

Challenges

Methodology

Fracture Design Module

Production Estimation Module

Objective Function

Case Study

Conclusions

- Single phase gas flow model
- Homogeneous formation; Uniform thickness
- Finite conductivity fractures
- Identical and equidistant vertical fractures
- Horizontal well produces through multiple transverse fractures interaction
- Contribution outside SRV is ignored
- Non-Darcy effect:

$$k_{f,e} = \frac{k_{f,n}}{1+N_{\mathrm{Re}}}$$

(Gidley,1991)

Gas phase pressure - Layer (Fracs_study)

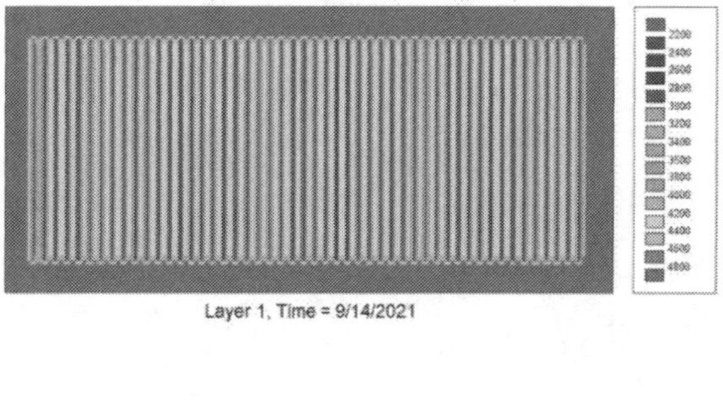

Layer 1, Time = 9/14/2021

16

Analytical Production Model

- **Transient regime**

Matrix:- $$k_m \frac{\partial^2 p_m}{\partial y^2} = (\phi\mu c_t)_m \frac{\partial p_m}{\partial t}$$

Fracture:- $$k_f \frac{\partial^2 p_f}{\partial x^2} - \mu \bar{q}_m(p_f, t) = (\phi\mu c_t)_f \frac{\partial p_f}{dt}$$

Solution → $$\bar{q}_D = \frac{C_{fD}}{\pi}\frac{u}{s}\tanh(u)$$

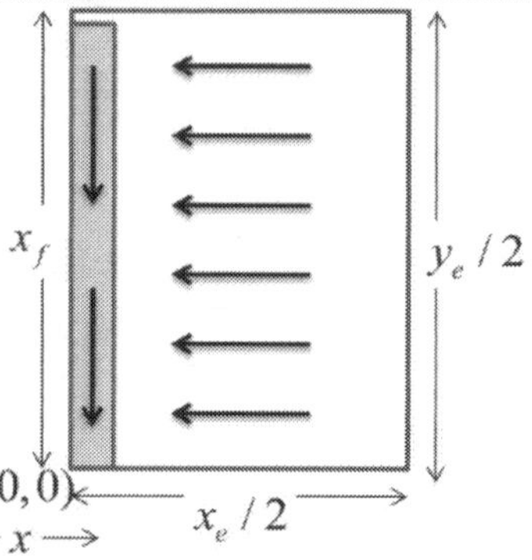

where $$u^2 \approx sC_1 + \sqrt{s}\frac{2}{C_{fD}}$$ $$q = n_f \frac{kh(m(p_i) - m(p_{wf}))}{1424T} q_D$$

- **Boundary-dominated regime** (after fracture interference)

$$t_{eplf} = \frac{948\pi\phi\mu c_t z_e^2}{4k}$$ $$q_{pss-Linear} = q_{pssi} \times \frac{m(p_r) - m(p_{wf})}{m(p_{ri})_{pss} - m(p_{wf})_{pss}}$$

(Ehlig-Economides,1992) (Williams-Kovacs and Clarkson , 2011)

17

Bhattacharya S.
Comprehensive Optimization Methodology for Stimulation Design of UGRs.

Comparison With Numerical Simulator

Introduction
Challenges
Methodology
Fracture Design Module
Production Estimation Module
Objective Function
Case Study
Conclusions

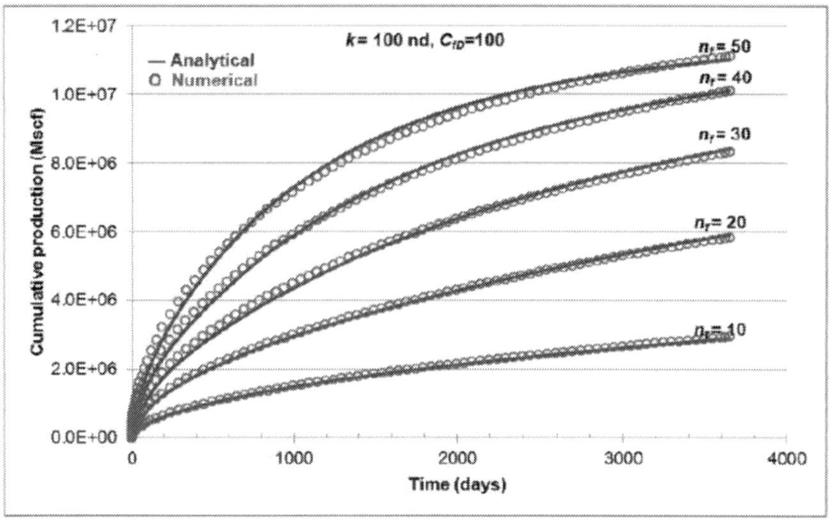

18

Bhattacharya S. Comprehensive Optimization Methodology for Stimulation Design of UGRs.

Objective Function

- Introduction
- Challenges
- Methodology
- Fracture Design Module
- Production Estimation Module
- Objective Function
- Case Study
- Conclusions

- Maximize NPV ($)

$$NPV = \sum_{ny=1}^{N_{yr}} \frac{R_{ny}}{(1+i_r)^{ny}} - I$$

$$R_{ny} = G_{p,ny} \times SP \times (1-f_r) \times (1-f_o) \times (1-f_t)$$

$$I = n_f (P_{ff} \times V_{ff} + P_{pr} \times M_p) + FC + DC$$

19

Bhattacharya S. Comprehensive Optimization Methodology for Stimulation Design of UGRs.

Case Study

- Introduction
- Challenges
- Methodology
- Fracture Design Module
- Production Estimation Module
- Objective Function
- Case Study
- Conclusions

Reservoir properties

Drainage Area	640 acres	Maximum horizontal stress	5000 psi
Reservoir dimension (side length)	5280 ft	wellbore radius	0.35 ft
Depth	10,000 ft	Flowing bottomhole pressure	1500 psi
Well length	5000 ft	Tubing inside diameter	3 in.
Pay thickness	100 ft	Closure Stress	6000 psi
Porosity	10%	Young's Modulus	5.0E6 psi
Permeability	100 nd.,.10 µd	Poisson's Ratio	0.2
Initial reservoir pressure	5000 psi	Leak-off coefficient	2.5e-4 $m/s^{0.5}$
Reservoir Temperature	200 F	Spurt loss coefficient	0
Gas gravity	0.85	Burst strength of the tube	13,000 psi
Gas saturation	80%	Pump horse power	14,000 psi
Water compressibility	3.0E-6 psi^{-1}	Proppant Type	20/40 Westprop
Pore compressibility	8.6E-6 psi^{-1}	Specific gravity	2.66
Skin factor	0	Fluid Consistency Index(n')	0.45
Minimum horizontal stress σ	6000 psi	Fluid Performance Index(K') lbf s^n_dash/ft^2)	0.07

Economic data

Proppant cost	$1 /lb
Fracturing Fluid cost	$0.4 /gal
Facilities cost	$100,000
Gas price	$6/Mscf
Discount rate	10%
Number of Operating years	10
Horizontal well drilling& completion cost	$1,000,000
Opex	20%
Royalties	20%
Taxes	30%

20

Bhattacharya S. Comprehensive Optimization Methodology for Stimulation Design of UGRs.

AIChE Southwest Process Technology Conference

Optimum Design: n_f and M_p

Introduction | Challenges | Methodology | Fracture Design Module | Production Estimation Module | Objective Function | Case Study | Conclusions

Design Variables	k=100 nd	k=500 nd	k=1 µd	k=5 µd	k=10 µd
$n_{f,opt}$	38	24	18	11	8
$M_{p,opt}$(lbs)	633,000	660,000	666,000	699,000	683,000
Results per fracture					
N_p	167,396	21,966	8,301	1,060	376
C_{fD}	5,289	1,018	506	103	50
t_i (min)	50	53	53	56	55
p_{treat} (psi)	6,563	6,563	6,564	6,564	6,566
I_x	0.89	0.92	0.93	0.94	0.95
h_f (ft)	212	213	213	214	214
w_f (in.)	0.07	0.07	0.07	0.08	0.07
V_i (gal)	63,585	66,678	67,347	70,404	69,243
Results for whole system					
G_i (Bcf)	62.32	62.32	62.32	62.32	62.32
G_p (Bcf)	23.01	27.07	27.69	29.11	29.8
C_{treat} (MM USD)	26.12	17.58	13.57	9.1	6.79
Rev (MM USD)	38.33	47.94	49.53	54.91	56.09
NPV (MM USD)	12.21	30.36	35.96	45.81	49.3

21

Bhattacharya S. Comprehensive Optimization Methodology for Stimulation Design of UGRs.

AIChE Southwest Process Technology Conference

Optimum Design: n_w, n_f and M_p

Introduction | Challenges | Methodology | Fracture Design Module | Production Estimation Module | Objective Function | Case Study | Conclusions

Design Variables	k=100 nd	k=500 nd	k=1 µd	k=5 µd	k=10 µd
$n_{w,opt}$	3	5	9	13	17
$n_{f,opt}$	41	35	23	17	7
$M_{p,opt}$(lbs)	160,000	94,000	50,000	39,000	22,000
Results per fracture					
N_p	182,901	34,122	11,925	2,098	330
C_{fD}	15,667	5,098	4,640	1,543	840
η_e	0.81	0.84	0.87	0.9	0.9
p_{treat} (psi)	6,491	6,453	6,406	6,376	6,354
I_x	0.90	0.95	0.98	0.99	0.99
h_f (ft)	159.02	142.49	128.27	121.41	117.2
w_f (in.)	0.07	0.08	0.08	0.09	0.07
V_i (gal)	14,435	8,170	4,180	3,170	1,783
Results for whole system					
G_i (Bcf)	62.32	62.32	62.32	62.32	62.32
G_p (Bcf)	24.54	30.16	34.16	40.15	55.55
C_{treat} (MM USD)	23.69	22.52	20.6	23.2	21.4
Rev (MM USD)	41.25	57.15	62.91	78.91	92.82
NPV (MM USD)	17.57	34.64	42.3	55.66	71.41

NPV increased with more wells

22

Bhattacharya S.
Comprehensive Optimization Methodology for Stimulation Design of UGRs.

SA For n_f

- Introduction
- Challenges
- Methodology
- Fracture Design Module
- Production Estimation Module
- Objective Function
- Case Study
- Conclusions

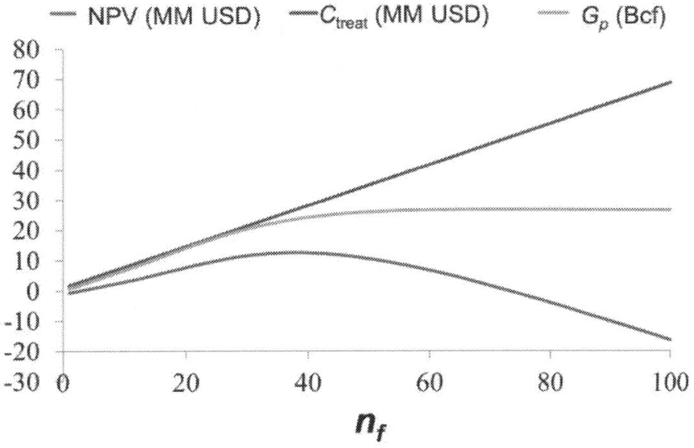

23

Bhattacharya S.
Comprehensive Optimization Methodology for Stimulation Design of UGRs.

SA For M_p

- Introduction
- Challenges
- Methodology
- Fracture Design Module
- Production Estimation Module
- Objective Function
- Case Study
- Conclusions

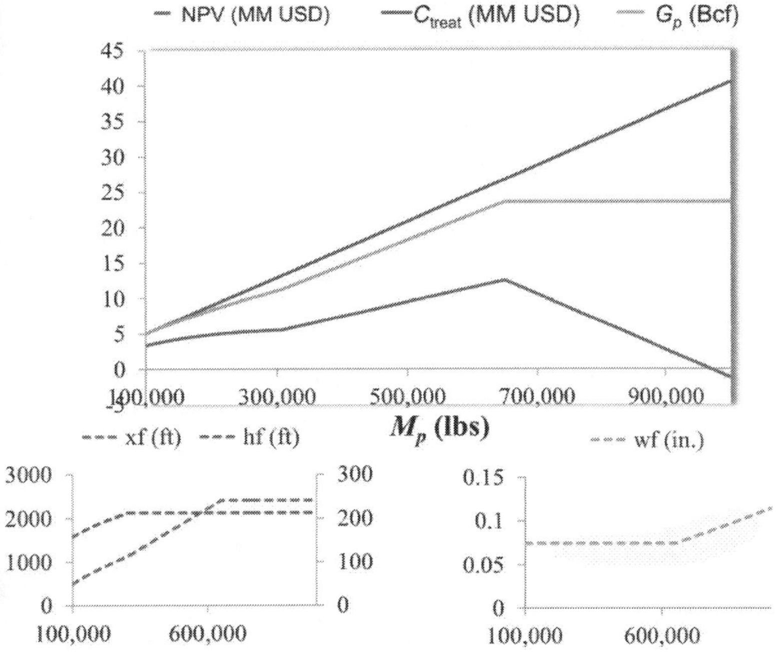

24

Bhattacharya S.
Comprehensive Optimization Methodology for Stimulation Design of UGRs.

Introduction
Challenges
Methodology
Fracture Design Module
Production Estimation Module
Objective Function
Case Study
Conclusions

Conclusions

- **Design**
  - Empirical equations for optimal x_f and w_f.
  - Iterative approach for h_f and p_{treat}.
- **Optimization**
  - Comprehensive optimization methodology reduces complexity in decision making for planning of horizontal wells.

25

Bhattacharya S.
Comprehensive Optimization Methodology for Stimulation Design of UGRs.

Introduction
Challenges
Methodology
Fracture Design Module
Production Estimation Module
Objective Function
Case Study
Conclusions

Acknowledgement

- Advisors
  - Michael Nikolaou, University of Houston
  - Late Michael J. Economides , University of Houston

- Funding
  - RPSEA
  - Landmark, Halliburton

- Collaborators
  - Texas A&M University
  - Anadarko Petroleum Corporation
  - Lawrence Berkeley National Laboratory

26

Bhattacharya S.
Comprehensive Optimization Methodology for Stimulation Design of UGRs.

Questions??

AIChE
Southwest Process Technology Conference

Introduction

Challenges

Methodology

Fracture Design Module

Production Estimation Module

Objective Function

Case Study

Conclusions

27

Thank You!

Thermodynamic Modeling of the Phase Behavior of Asphaltenes in Crude Oils

Vargas, F., Rice University

Originally presented on: 10/9/2014 14:00:00 - 14:30:00

In this talk, the advances in modeling the complex behavior of asphaltenes in crude oil systems will be reviewed. The effects of asphaltene poly-dispersity on the phase stability and thermodynamic properties of crude oils will be discussed from experimental as well as modeling perspectives.

Erosion in Under-Developed Multiphase Flows

Vedapuri, D., Tridiagonal Solutions Inc.

Originally presented on: 10/9/2014 14:30:00 - 15:00:00

The presence of flow control components such as control valves and chokes along with elbows and Tees in close proximity to one another can lead to under-developed multiphase flows. In gas-producing fields, under-developed wet gas flows can exhibit localized high velocity regions and high liquid entrainment rates. If sand is present, such flows can potentially accelerate the erosion rates. In this work, the erosion of a carbon steel elbow downstream of an orifice plate in low liquid loading multiphase flows is discussed. Experiments were conducted in a 2-inch diameter multiphase flow loop. Erosion in the range of 0.04 to 0.14 mils/lb of sand was observed.

Computational Study of Water Wetting Regimes Established in Horizontal Oil-Water Flows: Flow Regime Maps and Corrosion

Srinivasan, V., ANSYS

Originally presented on: 10/9/2014 15:30:00 - 16:00:00

Internal corrosion in oil-water pipelines is generally associated with the water phase being in contact with the metal surface at the bottom of the pipe. Prediction of phase distribution, especially the information whether oil or water wets the pipe wall is an important factor, which can affect the corrosion mitigation strategy and increase confidence in the integrity of the pipeline. There are large knowledge gaps in this area of research and consequently only empirical criteria are used in the industry.In the present work, detailed numerical study for turbulent oil-water multiphase flow is carried out in a horizontal carbon steel pipeline to predict different water wetting regimes as well as corrosion kinetics of fluid-metal interaction. Computational Fluid Dynamics simulations include Eulerian multiphase models with appropriate sub-models for interfacial forces, turbulent interaction and population balance equations. Population balance techniques identify and follow the dispersed (water) phase dynamics such as droplet breakage and coalescence in a detailed fashion. In addition, concentration equation for ferrous ion ($Fe2+$) is modeled to capture corrosion kinetics. The current study focuses on flow pattern computations for water cut of 20% and lower. Numerical simulations captured various water wetting regimes inclusive of intermittent, fully wetting, dispersed and semi-dispersed water-in-oil behavior. Results were in very good agreement with the experimental observation. Calculations clearly predicted that for given superficial water velocity, elevating the superficial oil velocity leads to transition of regimes from stable water wetting to intermittent water wetting and subsequently stable oil wetting scenario. Probing into the flow profile of the two-phase flows under different wetting regimes revealed the existence of severely under-developed system during a water-oil intermittent wetting regime. For a complete wetting regime, flow profile fluctuations were considerably reduced. However, fully developed oil-water flow conditions weren't achieved. Phasic fraction and velocity distribution at different flow cross-section reveal the flow modulations as a function of the turbulence inherent to the two-phase system. Corrosion computations clearly indicate a substantial generation of $Fe2+$ ion under stable water wetting regime while the concentration and corrosion levels deteriorate with intermittent water wetting, an effect well characterized in flow loop tests.

CFD modeling of water-in-crude oil emulsions

Nathanael Inkson - CD-adapco

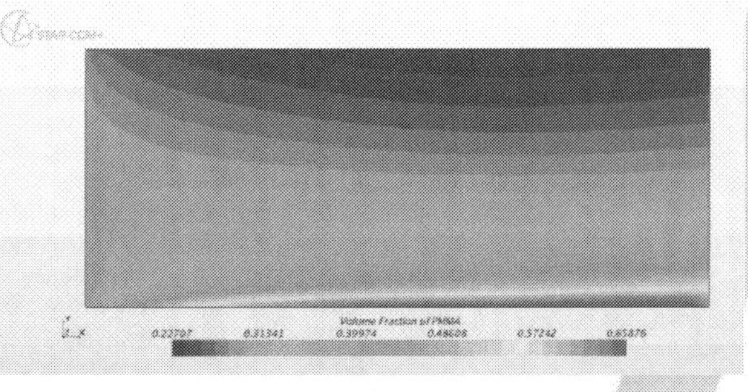

CD-adapco

Non-Newtonian fluids in STAR-CCM+ CFD software

Single phase fluids
- Generalized Newtonian Models
- Carreau-Yasuda, Cross Model
  - Shear-thinning and shear thickening viscosity
- Herschel-Bulkley Model (power law)
  - Bingham fluids, yield stress
- Laminar or turbulent

Multiphase flow
- One continuous phase can have non-Newtonian behaviour (as above)
- Volume of fluid method or Eulerian multiphase
- Eulerian multiphase has modified drag laws
- New relative (mixture) viscosity models for suspensions and emulsions

Viscoelastic fluids
- Finite element solver in development
- Viscoelastic models consider the full stress tensor, beyond just viscosity
- Time dependent effects, elasticity, extrudate swell, relaxation, cooling
- Extensional rheology, extension strain-hardening
- Multiphase injection moulding

CD-adapco

Under-prediction of emulsion pressure drop

The "problem" of emulsion rheology

A recent collaboration with academics from European universities and oil companies industries, indicated to us that our multiphase code under-predicted pressure drop in water-in-oil emulsions in a series of flow loops experiments, due to the increase of the mixture viscosity with volume fraction of the dispersed phase.

A study of the literature of emulsion rheology didn't lead to A clear modelling approach, however the research on suspensions turned out to be very rewarding

When these suspension models were applied to some emulsion data we were Able to get very good agreement when the emulsion is behaving like a suspension of particles

CD-adapco

Emulsions and suspensions

Emulsions and suspensions are multiphase complex fluids

The rheological behaviour is a function of the volume fraction of the dispersed phase and the microstructure

Stabilised emulsions share many properties of suspensions

- As the volume fraction of the dispersed phase is increased the effect of maximum packing leads to divergent viscosity due to formation of gel "network" and flocs

- for suspensions maximum packing is reached and a solid-like material with a yield stress develops,

For emulsions a phase inversion occurs

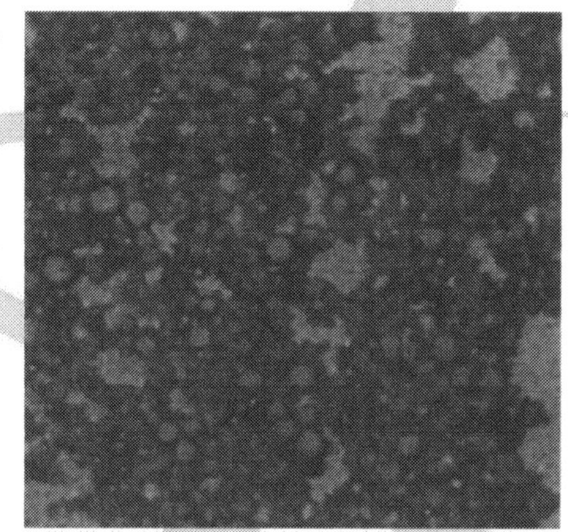

CD-adapco

Emulsions and suspensions

In STAR-CCM+ 9.02 we have implemented two well known relative viscosity models from the literature.
The new models allow the correct prediction of pressure drop for stabilised water-in-oil emulsions in pipe flow.

1) Krieger-Dougherty model
   Models divergent relative viscosity, zero normal stresses

2) Morris and Boulay model
   Models divergent relative viscosity, negative normal stresses

CD-adapco

Suspensions – literature review

Relative viscosity models

The relative viscosity is defined as

$$\eta_r = \frac{\eta}{\mu_c}$$

where η is the mixture viscosity and μ_c is the continuous phase (Newtonian) viscosity.

The mixture viscosity >> Newtonian viscosities of the other phases

Maximum packing

An important aspect of the physics of suspensions is that a maximum packing of the particles occurs at a critical volume fraction.
The relative shear and normal viscosities both diverge to infinity at maximum packing.

There are other relative viscosity models developed such as the Pal and Rhodes models for emulsions

CD-adapco

Suspensions – relative viscosity

Relative viscosity models implemented into STAR-CCM+

1) Krieger-Dougherty model

$$\eta_r = \left(1 - \frac{\phi}{\phi_m}\right)^{-[\eta]\phi_m}$$

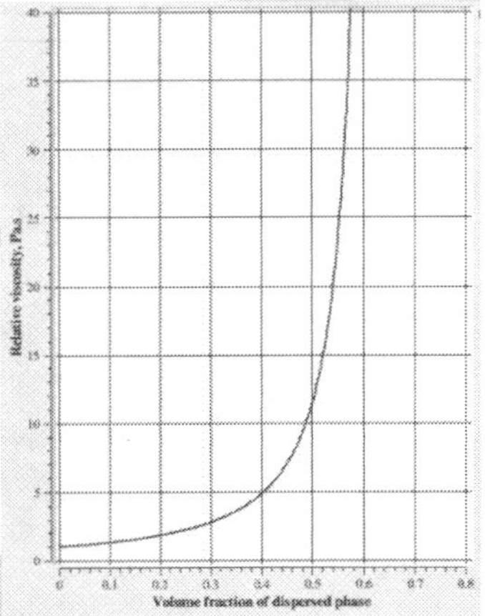

where $[\eta]$ is the intrinsic viscosity. $[\eta] = 2.5$ for spherical particles and ϕ_m is the maximum critical packing fraction for which ($\phi_m = 0.645$) for hard spheres.

CD-adapco

I. M. Krieger and T. J. Dougherty. A mechanism for non-newtonian flow in suspensions of rigid spheres. *Transactions of the Society of Rheology*, 3:137–152, 1959.

Suspensions – relative viscosity models

2) Morris and Boulay model

This model introduces normal stress components that represent the pressure of the particles due to collisions. The model has been verified using rheology experiments that measure normal stress via the Weissenberg effect

Total particle stress is:

$$\boldsymbol{\Sigma}_p = -\mu_c \eta_n(\phi)\dot{\gamma}\boldsymbol{Q} + 2\mu_c \eta_r(\phi)\boldsymbol{D},$$

web.mit.edu

where ϕ is the particle volume fraction, μ_c is the continuous liquid viscosity $\boldsymbol{Q}$ is given by

$$\boldsymbol{Q} = \begin{pmatrix} \lambda_1 & 0 & 0 \\ 0 & \lambda_2 & 0 \\ 0 & 0 & \lambda_3 \end{pmatrix}$$

$\lambda_1 = 1.0$, $\lambda_2 = 0.8$ and $\lambda_3 = 0.5$ are the anisotropy parameters.

$$\eta_r(\phi) = 1 + 2.5\phi_{max}\left(1 - \frac{\phi}{\phi_{max}}\right)^{-1} + K_s\left(\frac{\phi}{\phi_{max}}\right)^2\left(1 - \frac{\phi}{\phi_{max}}\right)^{-2},$$

and the normal stress by

$$\eta_n(\phi) = K_n\left(\frac{\phi}{\phi_{max}}\right)^2\left(1 - \frac{\phi}{\phi_{max}}\right)^{-2}.$$

CD-adapco

J. F. Morris and F Boulay. Curvilinear flows of noncolloidal suspensions: The role of normal stresses. *J. Rheol.*, 43:1213, 1999.

Suspensions – Eulerian multiphase

Two-fluid formulation

The momentum equation for phase *k* is

$$\frac{\partial}{\partial t}(\alpha_k \rho_k \boldsymbol{v}_k) + \boldsymbol{\nabla}.(\alpha_k \rho_k \boldsymbol{v}_k \boldsymbol{v}_k) =$$

$$-\alpha_k \boldsymbol{\nabla} p + \alpha_k \rho_k \boldsymbol{g} + \boldsymbol{\nabla}.(\alpha_k(\boldsymbol{\tau}_k + \boldsymbol{\tau}_k^t)) + \boldsymbol{M}_k + \boldsymbol{F}_k + \boldsymbol{S}_k + \Sigma(m_{kj}\boldsymbol{v}_j - m_{jk}\boldsymbol{v}_k)$$

From the suspension balance model, the Morris and Boulay total stress is split between phases
Drag similar to Schiller-Naumann model

Total continuous phase stress $\boldsymbol{\tau}_{c,total} = \boldsymbol{\tau}_{c,Newtonian} + \boldsymbol{\tau}_{c,extra}$

$$\boldsymbol{\tau}_{c,Newtonian} = 2\mu_c \alpha_c \boldsymbol{D}_c, \qquad \boldsymbol{\tau}_{c,extra} = 2\alpha_c \mu_c (\eta_r(\alpha_d) - 1)\boldsymbol{D}_c.$$

Particle extra stress $\boldsymbol{\tau}_{d,extra} = 2\alpha_d \mu_c (\eta_r(\alpha_d) - 1)\boldsymbol{D}_d - \mu_c \eta_n(\alpha_d)\dot{\gamma}_d \boldsymbol{Q}.$

CD-adapco

Suspensions – Eulerian multiphase

Drag formulation

The drag force is

$$\boldsymbol{F}_D = A_D|\Delta \boldsymbol{v}|.$$

where $|\Delta \boldsymbol{v}|$ is the slip between phases and the linearised drag, A_D is given by

$$A_D = \frac{a'''}{8}\rho_c |\Delta \boldsymbol{v}| C_D.$$

where a''' is the symmetric interaction area density and is defined as

$$a''' = \frac{6\alpha_c \alpha_d}{d}$$

where d is the droplet diameter. The drag coefficient, C_D is a Schiller-Naumann drag for particles

$$C_D = \frac{24}{Re}(1 + Re^{0.681}).$$

We modify the drag to tend to infinity as maximum packing is approached to prevent slip between the phases due to the network structure of particles.

CD-adapco

Suspensions – pipe flow modelling

Modelling results of Hampton et. al NMR suspensions in pipe experiments

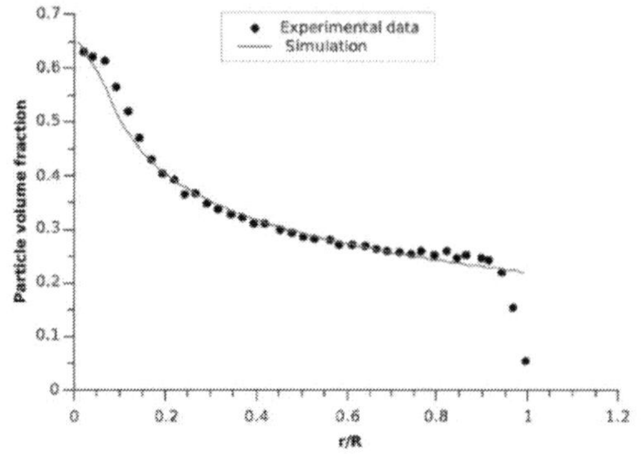

Kn = 3.2

R. E. Hampton, A. A. Mammoli, A. L. Graham, N. Tetlow, and S. A. Altobelli. Migration of particles undergoing pressure-driven flow in a circular conduit. *Journal of Rheology*, 41:621, 1997.

CD-adapco

At volume fraction of particles = 0.3

A good agreement with experimental data is achieved.
The Morris and Boulay relative viscosity model is capable of causing particle migration towards the centre of the pipe.

When the normal stresses are turned off then no particle migration occurs.

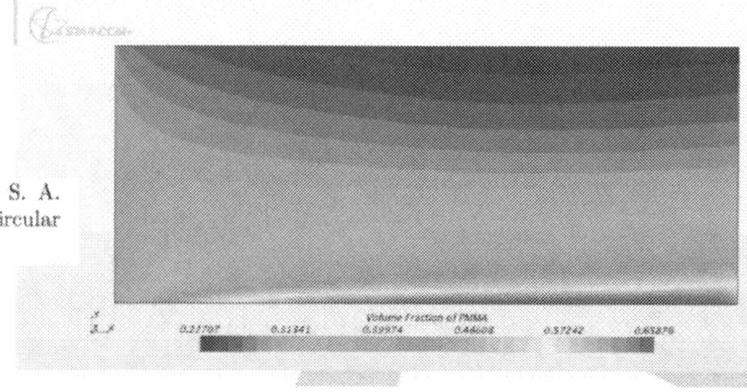

Emulsions Pressure Drop in STAR-CCM+

Modelling the pressure drop in pipes, (experimental data from Dr. Jose Plasencia)

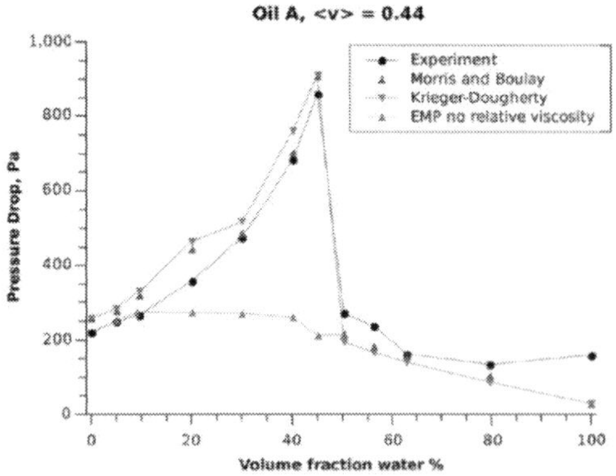

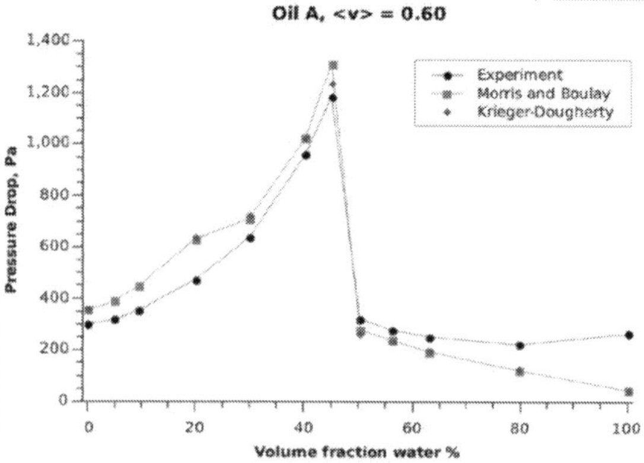

Crude oil A and seawater emulsion in horizontal pipe of diameter 2.21 cm,
At velocities of 0.44 and 0.60 m/s (laminar regime).

CD-adapco

"Pipe flow of water-in-crude oil emulsions: Effective viscosity, inversion point and droplet size distribution"
Jose Plasencia, Bjørnar Pettersen and Ole Jørgen Nydala,
Journal of Petroleum Science and Engineering Volume 101, January 2013, Pages 35–43

Emulsions Pressure Drop in STAR-CCM+

Model parameters

Table 1: Krieger-Dougherty and Morris-Boulay Model parameters.

Dispersed phase	ϕ_m^{KD}	ϕ_m^{MB}	K_n^{MB}	ϕ_i
particle suspension	-	0.7	3.2	-
Oil A	1.05	0.71	0.75	0.45
Oil B	0.88	0.7	0.75	0.57

Oil A

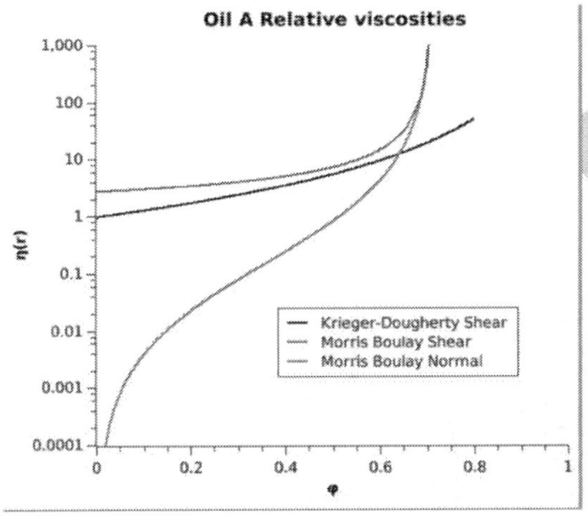

Relative viscosities are not exactly the same between models but close.

Unknown role of normal relative viscosity on pressure drop.

Multiphase Rheology

Rheology experiments show even more complexity than relative viscosity models presented

Relative viscosity models are purely a function of volume fraction $\eta_r(\phi)$

However the shear strain-rate can also be important

Can we find $\eta_r(\phi, \dot{\gamma})$

CD-adapco

Emulsions: New rheology model

CD-adapco

Rheology shows more complexity than the simple relative viscosity models, shear thinning and thickening present as well as a yield stress

Experimental data: Lee, Lee and Park: Journal of colloid and interface science, 185, 297–305 (1997)

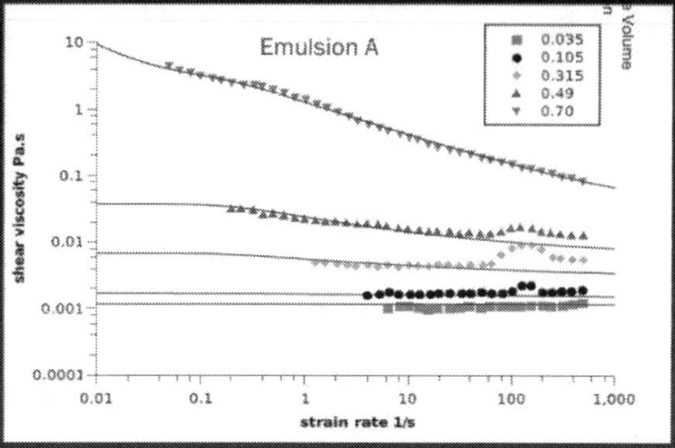

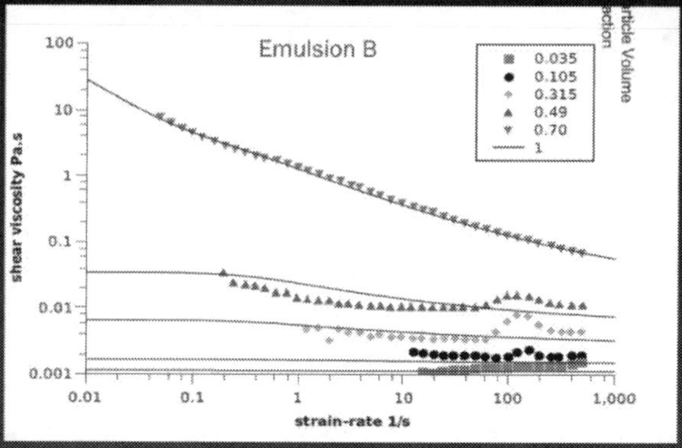

Line: model curve fit
Points: test data

Yield stress appears at Volume Fraction > critical packing = 64% for spheres, and effects can be observed in the curves for volume fraction = 0.7.

The bump in viscosity at high shear-rates is due to jamming, which is neglected in the model

Suspensions: New rheology model

CD-adapco

Heymann et. al. Suspension data: log shear stress vs. Log Strain rate

Figure 1: Log shear stress against log strain-rate for a $5\mu m$ suspension at particle volume fractions of $\phi = 0.126, 0.259, 0.352, 0.400, 0.449, 0.474$ and 0.499.

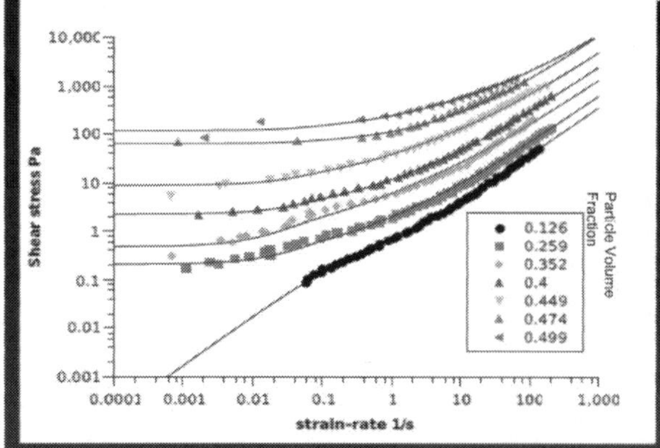

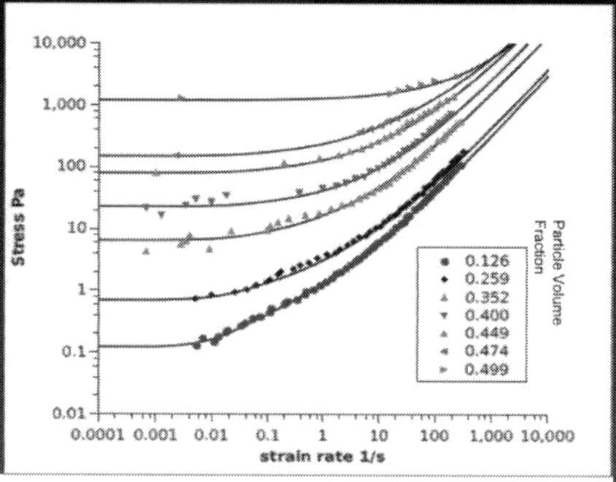

Line: model curve fit
Points: test data

Figure 2: Log shear stress against log strain-rate for a $3\mu m$ suspension at particle volume fractions of $\phi = 0.126, 0.259, 0.352, 0.400, 0.449, 0.474$ and 0.499.

Stress-strain rate curves show the yield stress plateau on the left side of the graphs showing how it increases with dispersed volume fraction. All the curves were fitted with a new model that will be implemented into STAR-CCM+.

Suspensions: New rheology models

CD-adapco

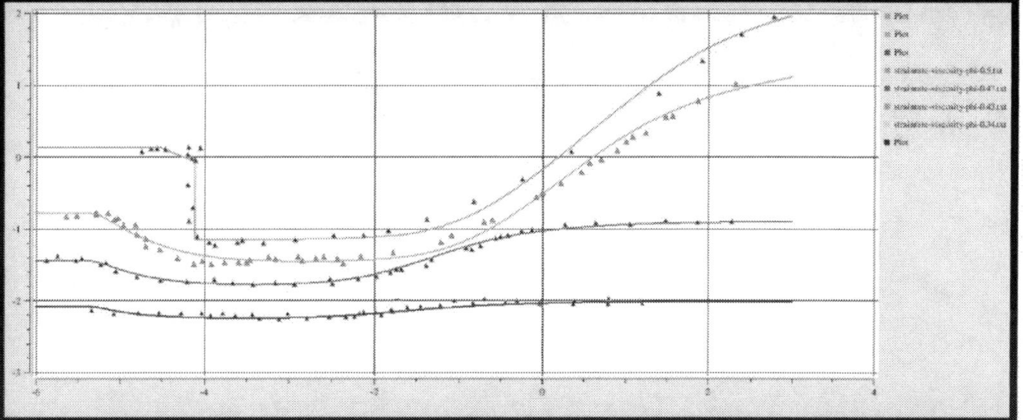

Log viscosity vs. Log shear strain-rate. Lines are model curve fit, Points: test data

Experiments: H. M. Laun, Die Angewandte Makromolekulare Chemie 123/124 335-359 (1984)

2 mode Carreau-Yasuda model fits charged latex sphere suspension shear strain-thinning and thickening rheology at different volume fractions

Applications of models

Emulsion rheology model:
- Oil-in-water and water-in-oil emulsions
- Paints
- Foods

Suspension rheology model:
- Sand in water suspensions
- Hydrates
- Drilling muds
- Slurries
- Pastes
- Pharmaceuticals
- Fillers
- Colloids

CD-adapco

Conclusions

Relative viscosity models have proved to be useful for modeling
Both suspensions of particles and emulsion systems

The success of the emulsion model is due to the droplets behaving as particles, i.e
they have small droplet sizes,
asphaltene coverage may be important

CD-adapco

Future work

Extend suspension and emulsion rheology model
- Shear-thinning and thickening
- Extensional yield stress (Osmotic pressure)

Extension of emulsion modelling when emulsions have larger drops

Coalescence and breakup
- S gamma model,
- Musig

Viscoelasticity at high volume fractions

CD-adapco

Crude Vacuum Tower Wash Bed Optimization Based on Tomography Gamma-Scans

Pless, L., Tracerco
Cantley, G.,

Originally presented on: 10/9/2014 13:30:00 - 14:00:00

The crude unit of today's modern refinery is where it all starts. Good, clean fractionation from the Crude Atmospheric and Crude Vacuum Towers sets the tone for all the downstream operations.

The Wash Bed in the Crude Vacuum Tower is one of those evil necessities. When it operates well it does not seem to garner much attention but when things go wrong all hell can break loose downstream. Refinery operators are constantly trying to minimize wash oil loss versus keeping the wash bed from coking.

Packed columns are widely used in distillation, absorption and stripping processes. It is well known that liquid maldistribution can severely reduce column efficiency and operability. For packed towers a grid-scan, 3 or 4 equal-distant scans would be done to diagnose macro problems of severe liquid maldistribution. But more subtle patterns of maldistribution can be better diagnosed by a more comprehensive method.

The most prevalent use of tomography is still in medical applications, for instance the Computerized Tomography (CT) device that is used for non-destructive examination of the human body. Gamma-ray tomography takes advantage of the penetrating properties of high-energy gamma radiation to generate a cross sectional density profile through a packed bed.

This paper presents a case study describing gamma-ray computer-aided tomography (CAT-Scan) to monitor the density distribution profile through a Vacuum Column Wash Bed enabling this refiner to correlate operation data to the inevitable wash bed coking and to optimize tower assets.

How Sensitive is Your Treating Plant to Operating Conditions

Weiland, R., Optimized Gas Treating, Inc.
Santos, J., 2INEOS GAS/SPEC Technology Group
Portillo, J., Sulzer
Summers, D.,
Pilling, M.,
Duss, M., Sulzer
Schultes, M., RASCHIG GMBH
Praderio, A.,
Originally presented on: 10/9/2014 14:00:00 - 14:30:00

One of the most neglected areas in gas treating is the proper selection of tower internals in both absorbers and regenerators. Apart from hydraulic considerations, the mass transfer performance of tower internals unfortunately has tended to be almost completely ignored. When parts-per-million specifications must be met on product gases, selecting the wrong packing size or bed depth can result in a failed design. For many years, packing has had a bad reputation in absorption and distillation at high pressure. Part of the reason is inattention to proper distributor design. Another is the persistent and still unresolved difficulty in translating ideal stages to actual packed bed depths and the selection of a particular commercial packing. The effect in the contactor's diameter and height of different random packing sizes and types is compared to structured packing and two types of commercially proven valve trays for an LNG project under study that initially will process 745 KNCMH of a gas containing 11.5% of CO2.

Improving Tower Reliability

Nieuwoudt, I., Koch-Glitsch

Originally presented on: 10/9/2014 14:30:00 - 15:00:00

The overall performance of a tower is not only a function of the separation performance and the hydraulic capacity. Due to the high cost of unplanned shutdowns the mechanical integrity and run lengths are extremely important in assessing the overall economics. In the case of trays, fouling resistance and mechanical integrity (i.e. upset resistance) are important in assessing reliability. The choice of valves, tray construction and support mechanisms are important in improving mechanical integrity and run lengths. Examples of such tray installations will be presented. In packed towers fouling resistance has a significant impact on the run length of the tower. Fouling resistance in a packed tower could be a function of the liquid distribution, vapor distribution and the packing itself. Ensuring good liquid and vapor distribution, and choosing a packing with high fouling resistance can be used to increase the run length of towers prone to fouling. Examples of such packed tower installations will be presented.

CH2MHILL.

Atmospheric Preflash Options

Andrew W. Sloley – CH2M HILL, Bellingham, WA

Prepared for Presentation at the AIChE 2014 6th Southwest Process Technology Conference
Galveston, TX, 9-10 October, 2014

Atmospheric Preflash Definition

- A drum or tower upstream of the crude unit fired heater
- Downstream of the desalter
- Vaporizes lighter material and water
- Many different configurations

CH2MHILL.

Configuration Options

Configuration Options

- Preflash drum
  - High-pressure preflash
  - Low-pressure preflash
- Preflash tower
  - One product
  - Two product
  - Three product
  - Feed preheat only
  - Reboiled
  - Steam stripped
  - Separate overhead system
  - Shared overhead system with atmospheric tower
  - Side-tower or complex overhead configuration

CH2MHILL.

Unit Survey – 50 Units

Configuration	Units
No preflash	25
Preflash drum	10
Preflash tower	15

Drum Type	10 Units
HP Drum	8
LP Drum	2
Tower connection	2
FZ connection	8

Tower Type	15 Units
Separate	11
Shared overhead	2
Complex	2

Tower Heat	15 Units
Feed – Exch only	8
Feed – Fired	2
Reboiled	3
Steam Stripped	1
Fd Fired + Reboiled	1

CH2MHILL.

Basis

Basis Areas

- Crude type selection
- Overall distillate yields
  - API gravity
- Specific yield pattern
- Dil-bit mixtures
- Light ends
- Performance parameters
- Overhead
  - LN-HN split performance or
  - FRN-Kero split performance or
  - Temperature constraints
- Middle products
  - Constant separation or
  - Constant yield or
  - Constant heat removal
- Lowest side-draw
  - AGO or diesel as lowest product
- Feed enthalpy
  - Constant cut-point or
  - Constant heater outlet or
  - Constant duty input

CH2MHILL.

Basis Chosen

Crude	°API Gravity	Specific Gravity
BCF 24	23.3	0.9141
Tia Juana Lt	31.6	0.8676
Bakken	42.1	0.8151
Gippsland	49.7	0.7810

- Feed enthalpy
  - Constant heater outlet
- Other
  - Heater outlet vaporization
- Overhead
  - Temperature constraints
  - Reduces corrosion and makes overhead heat integration practical
  - ~70% of units have heat recovery from overhead
- Upper products
  - LN and HN products
- Middle products
  - Constant separation or
  - Constant heat removal
- Lowest side-draw
  - AGO or diesel as lowest product

CH2MHILL.

Why Use a Preflash

Why Use a Preflash

- Increases temperature before vaporization
  - Allows more heat input upstream of heater pass-balancing valves
  - Light-ends vaporization
  - Water vaporization
- Reduces pressure drop in system
  - Reduced crude rate through exchangers
  - Reduced vapor formation in atmospheric heater
- Bypasses lower section of atmospheric column (sometimes)
  - Increases capacity by reducing vapor load in bottom of column
- Bypasses atmospheric tower completely (sometimes)
  - Preflash towers may add overhead systems

CH2MHILL.

The Importance of Water

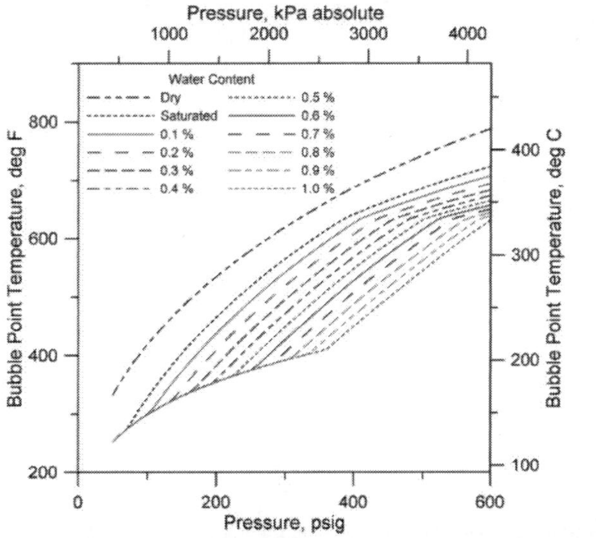

Tia Juana Light

Comparing Crudes

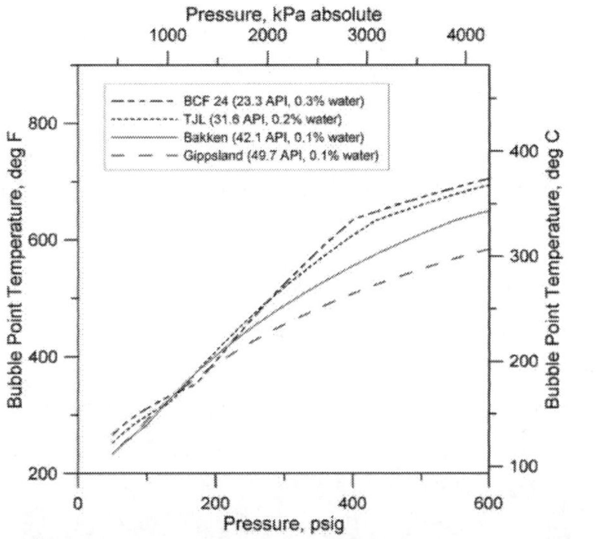

Atmospheric Unit Comparisons

	Units	BCF 24	TJL	Bakken	Gippsland
LNap	Bbl/day	11,980	22,330	35,900	54,580
HNap		2,140	3,870	7,890	2,700
Kero		10,010	14,740	15,460	17,010
Dies		9,860	9,220	11,070	7,160
AGO		7,390	9,980	12,380	11,250
ARC		58,740	39,920	17,350	7,360
Cond L	Mbtu/hr	96	102	104	142
Cond H		22	32	53	123
Nap PA		14	36	76	17
Kero PA		7	36	40	58
Dies PA		50	42	53	0
Cut Pt	°F	714	747	797	*777

* Heater outlet temperature limited by minimum wetting rate.

CH2MHILL.

Preflash Yield and Capacity Effects

- Depends upon constraints
- For a heater outlet temperature limited system
  - Adding a preflash drops the atmospheric tower cutpoint (lower distillate yield)
- For a heater duty limited system
  - Adding a preflash increases the atmospheric tower cutpoint (higher distillate yield)
- For a heat removal limited system
  - Reduces heat removal capability from the atmospheric tower
- For a heat input limited system
  - Increases heat recovery possible into the crude feed
- Many, many complex interactions possible with different equipment constraints
- Every case needs to be examined

CH2MHILL.

Previous Work

- Limited fundamental analysis published
- Heat integration with a vacuum unit makes preflash options more attractive
  - Ji and Bagajewicz 2002
  - Author's analysis agrees with this (unpublished)
- Energy recovery economics highly dependent on energy prices
  - Updating energy economics from Feintuch et al. 1985
  - Most complete published evaluation
  - Capital costs too low

Basis	US $ / Mbtu	Energy Source	No Preflash	Preflash
			Payback Months	
Nymex 2014 08	3.82	Natural Gas	126	80
Europe 2014 08	7.27	Natural Gas	66	42
Japan 2014 06	15.23	LPG	28	20

CH2MHILL.

15

Preflash Drums

16

Preflash Drums

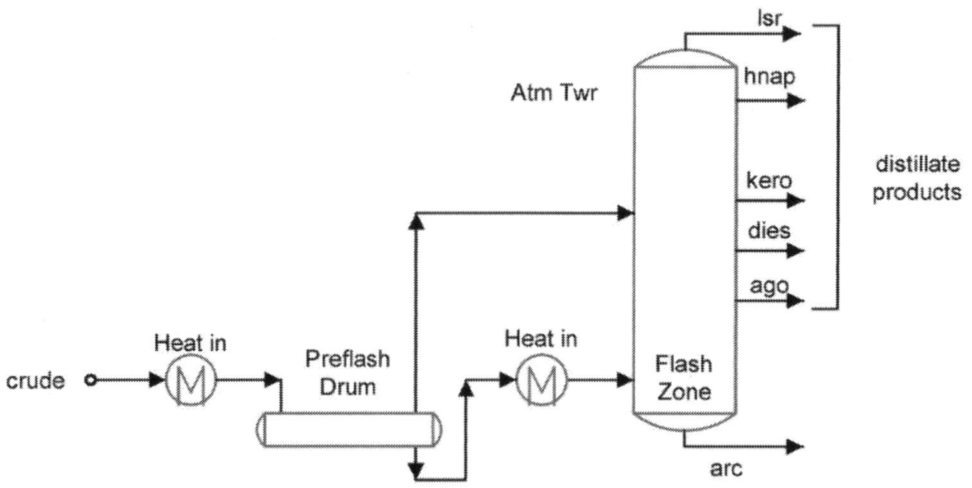

High-Pressure Preflash Drums

- Pressure controlled independently of atmospheric tower
- Pressure controller in vapor line from preflash drum to atmospheric tower
- Low vapor volumes
- Lower likelihood of foam carryover
- Majority of preflash drums
- Constraints relaxed
  - Liquid hydraulics downstream of preflash
  - Vaporization upstream of fired heater
  - Pressure drop in fired heater
  - Vapor load in atmospheric tower – depends upon connection

Low-Pressure Preflash Drums

- Pressure floats on atmospheric tower pressure
- Direct vapor line from preflash drum to atmospheric tower
- High vapor volumes
- Higher likelihood of foam carryover
- Constraints relaxed
  - Liquid hydraulics downstream of preflash
  - Vaporization upstream of fired heater
  - Pressure drop in fired heater
  - Vapor load in atmospheric tower – depends upon connection

Drum Connection to Atmospheric Tower

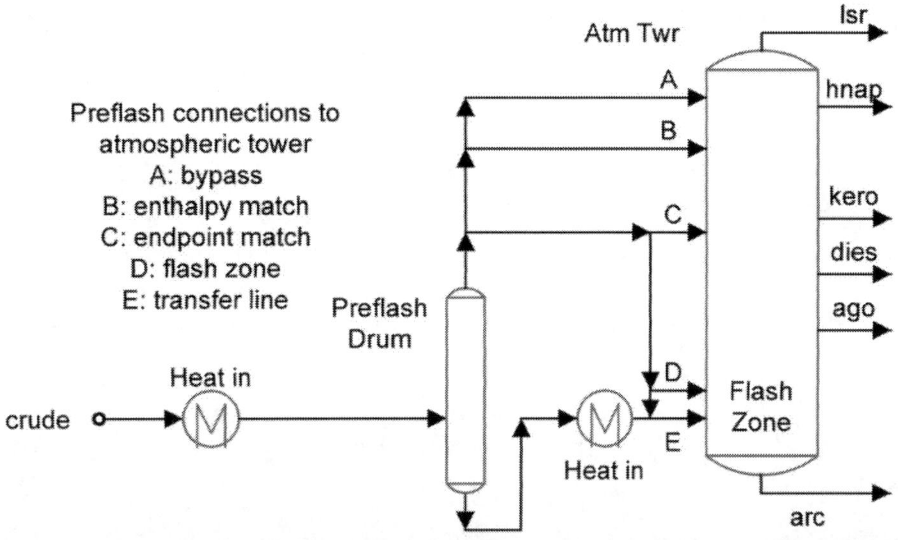

Drum Connection to Atmospheric Tower

- Bypass
  - Moves vapor around a section of the tower
  - Hydraulic debottleneck
- Enthalpy
  - Most thermodynamically efficient
- Endpoint
  - Match endpoint of vapor phases for best yields
- Flash zone (segregated)
  - Reduces risk from foam carryover
  - Any residue in the foam goes into feed area
- Transfer line (or mixed flash zone)
  - Eliminates nozzle on the tower
  - Same as flash zone with heater outlet mixing with the preflash vapor

CH2MHILL.

Comparison – LP Preflash – TJL Crude – Constant Sep

	Units	None	A	B	C	D	E
LNap	Bbl/day	22,330	22,590	22,360	22,430	22,360	22,340
HNap		3,870	3,800	3,870	3,890	3,790	3,790
Kero		14,740	12,810	12,770	14,050	14,860	14,990
Dies		9,220	10,260	10,500	8,840	7,680	7,830
AGO		9,980	6,920	6,890	7,170	7,670	7,310
ARC		39,920	43,670	43,670	43,670	43,700	43,800
Cond L	Mbtu/hr	102	96	103	101	104	104
Cond H		32	26	32	30	32	32
Nap PA		36	39	30	55	28	61
Kero PA		36	12	8	0	56	5
Dies PA		42	41	42	30	0	12
Cut Pt	°F	747	706	706	706	706	705

CH2MHILL.

Conclusions

- Bypass
  - If a bypass configuration is necessary, no other choice
- Enthalpy
  - Most thermodynamically efficient – large yield cost
- Endpoint
  - Best overall yields
- Flash zone (segregated)
  - Most common option
  - Reduces risk of foam degrading product quality
  - Significant yield and heat recovery costs
- Transfer line (or mixed flash zone)
  - Avoid

CH2MHILL.

Why is the Flash Zone Connection Used So Often?

- Preflash services are highly foaming
- Preflash drums and towers need *large* cross-section and holdup
- Many units undersized
- Undersized preflash leads to entrainment
- Black products to atmospheric column
- Black kerosene, diesel, ago common
  - Very expensive, shuts unit down
- Flash zone and transfer line connection removes this risk or solves problem
- Critical issue in preflash design and integration

CH2MHILL.

Preflash Towers

Preflash Towers – Conceptual Two-Product Example

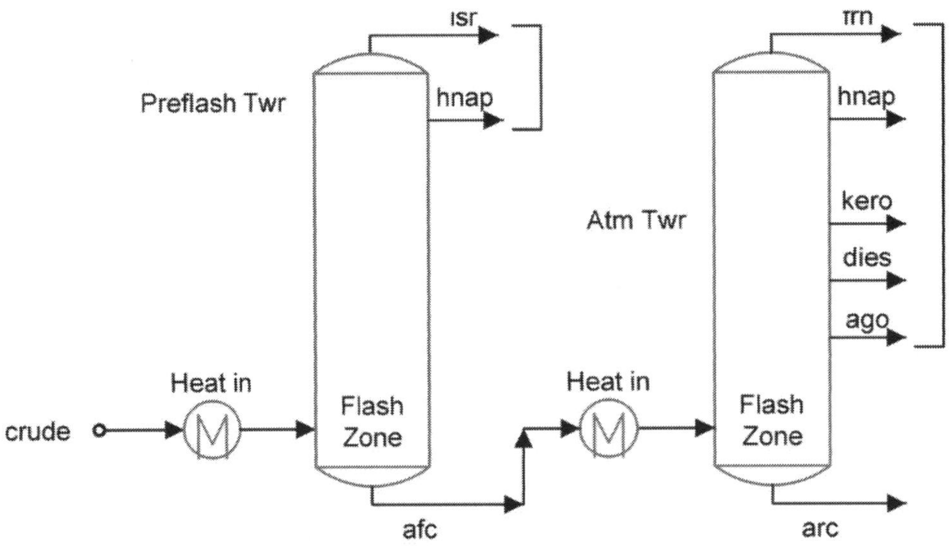

CH2MHILL.

Preflash Towers

- Advantage – large change in constraints
  - Liquid hydraulics downstream of preflash
  - Vaporization upstream of fired heater
  - Pressure drop in fired heater
  - Large capacity increases possible
  - Increased naphtha fractionation for T90 control possible
- Disadvantage
  - Expensive
  - Overhead systems

CH2MHILL.

Preflash Towers – Options

- Products
  - One product
  - Two product
  - Three product
- Atmospheric tower connection
  - None (independent)
  - Vapor off-gas to atmospheric overhead
  - Shared overhead system
  - Complex (linked) configurations
- Heat input
  - Exchangers only
  - Fired heater
- Stripping section
  - None
  - Reboiled
  - Steam stripped

CH2MHILL.

Preflash Towers – Evaluated

- Products
  - One product
  - Two product
- Atmospheric tower connection
  - None (independent)
  - Shared overhead system
- Stripping section
  - None
- All four crudes
- *But remaining work includes*
  - *Final shared overhead*
  - *Reboiled stripping*
  - *Steam stripping*

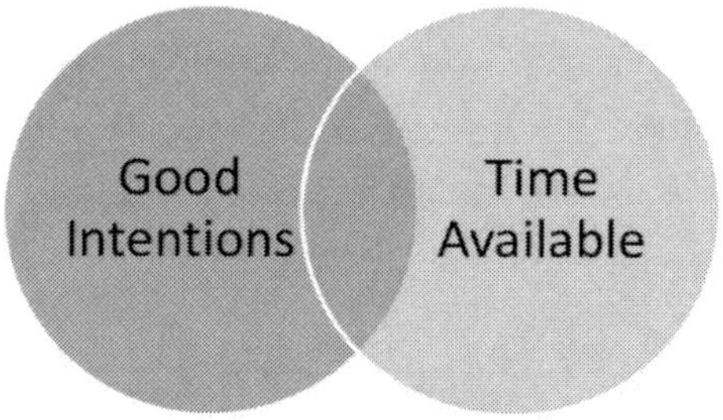

- *Some complex results on the shared overhead system configurations*

Two-Product Preflash Tower Comparisons

	Units	BCF 24	TJL	Bakken	Gippsland
LNap	Bbl/day	11,220	20,670	32,890	51,260
HNap		3,030	6,800	12,140	6,580
Kero		8,850	12,990	13,420	14,650
Dies		9,510	9,360	10,040	9,130
AGO		4,820	6,110	7,430	9,740
ARC		62,640	45,140	24,150	8,700
Cond L	Mbtu/hr	62	57	46	37
Cond H		8	8	8	10
Nap PA		13	23	92	15
Kero PA		13	27	0	41
Dies PA		41	37	13	33
Cut Pt	°F	671	690	706	761

Observations

- Shift from light naphtha and kerosene to heavy naphtha
- HN-rich Bakken drives duty to HNPA to get separation between HN and kersosene
- Specifics of distillation curves count
- HN yield increase also driven by constraints used
  - LN-HN split imposed on preflash tower and atmospheric tower individually
  - Running split to mixed-LN stream and mixed-HN stream from both units likely to change results
- Constraints used lead to different answers
  - Some results differ from previous work based on constant yields and heat balance but varying separation

CH2MHILL.

Summary

Factors That Favor a Preflash

- Capacity more important than yield
- Capacity more important than energy recovery
- Heater duty limits instead of outlet temperature limits
- Vaporization in preheat train constraint
- Pressure limitations in preheat train (equipment capability)
- Energy very expensive

Preflash Drum versus Preflash Towers

- ***Preflash drums*** favored when
- Atmospheric column limitations in lower section
- Plot plan constraints

- ***Preflash towers*** favored when
- Atmospheric column has significant constraints
- Atmospheric overhead system has significant constraints

- Specific plant constraints make every plant unique
- Crude slate has a significant effect
- Carefully examine preflash options
- Other options not included here possible – many configurations

Factors That Favor a Preflash Tower

- Multiple users for assays
- Level of detail required depends upon user
- Results normally reported as a TBP distillation
- Properties measured on cuts depend upon the boiling range of the cut
- Watch for typical properties
- If the property moves in unexpected conditions, find out why

CH2MHILL.

Questions ?

CH2MHILL.

Atmospheric Preflash Options

Andrew W. Sloley – CH2M HILL, Bellingham, WA

Prepared for Presentation at the AIChE 2014 6th Southwest Process Technology Conference
Galveston, TX, 9-10 October, 2014

CH2MHILL.

Distribution or
Printed Presentation
Ends on Previous Slide

Appendix –
Shared and Complex Configuration Examples

Pemex Tula – Shared Overhead Reflux
Based on O&GJ, 1999 June 21: 41-46

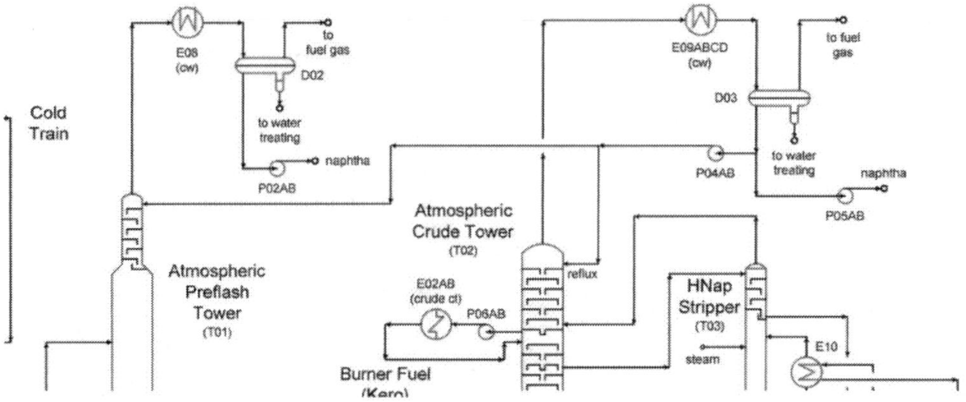

CH2MHILL.

Refineria Panama Bahia – Complex
Based on www.ventech.com

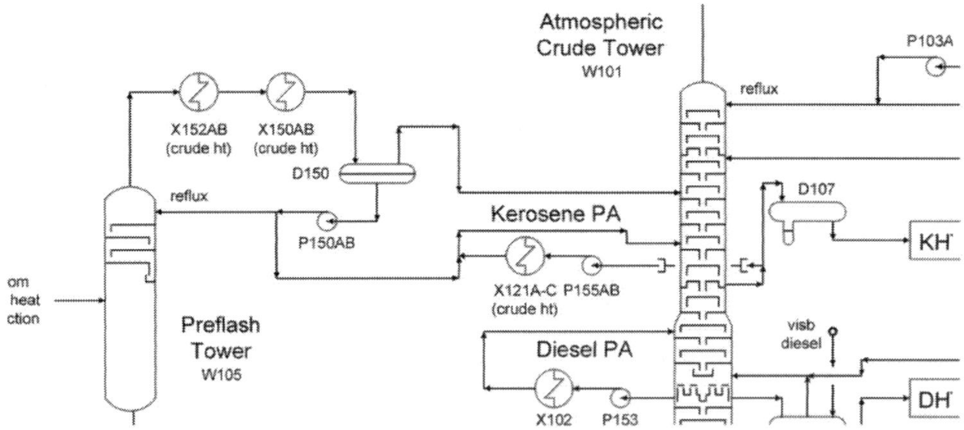

CH2MHILL.

Complex – Prefractionator

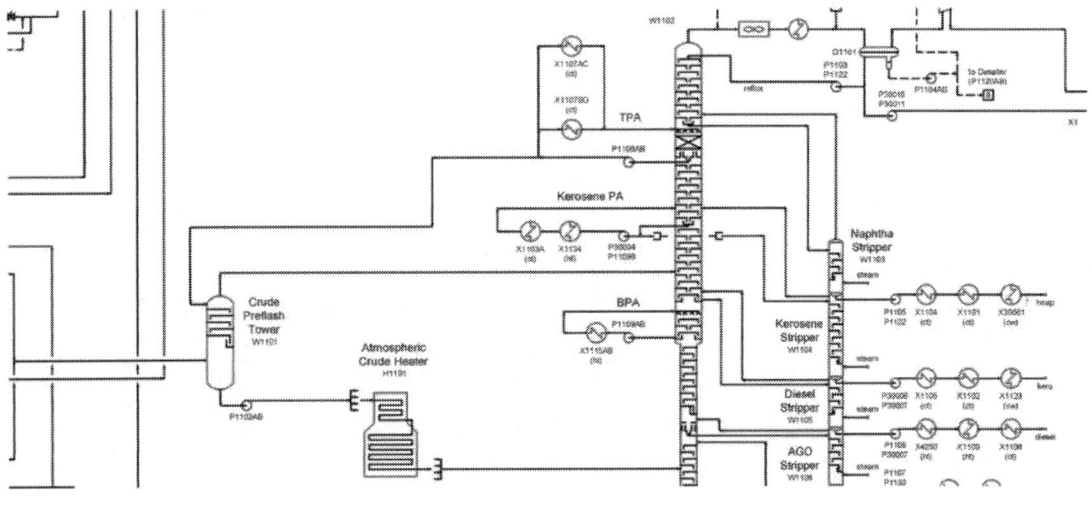

CH2MHILL.

RASCHIG

Superior Performance by Design

PRODUCTS & SERVICES | DISTRIBUTOR TESTING | APPLICATIONS | DOWNLOADS | ABOUT US | CONTACT US | HOME

Packing Trays & Column Internals

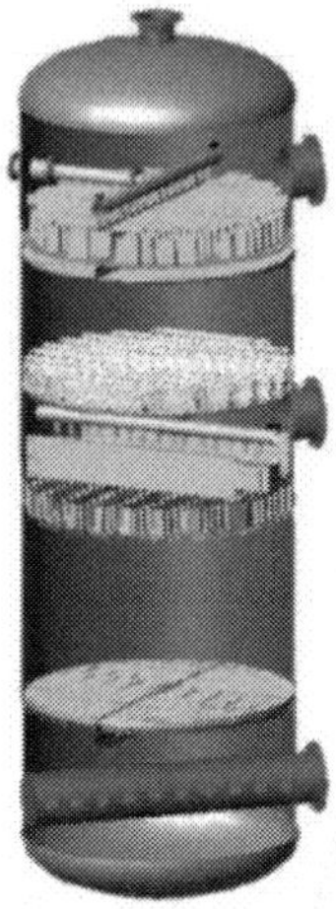

RASCHIG

Optimization of Column Performance

„What Subject to look at“

Michael Schultes
Raschig GmbH

RASCHIG

Superior Performance by Design

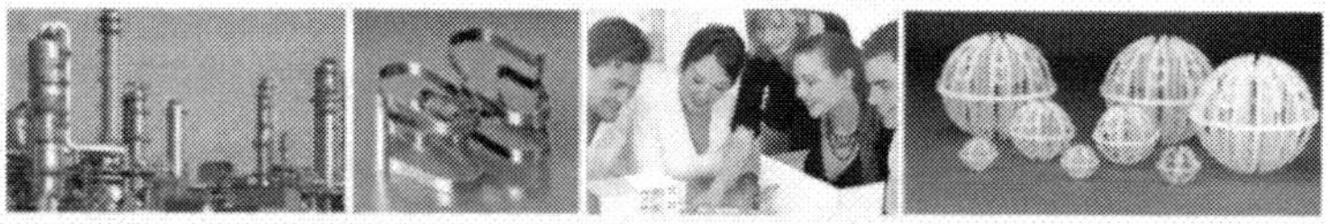

PRODUCTS & SERVICES | DISTRIBUTOR TESTING | APPLICATIONS | DOWNLOADS | ABOUT US | CONTACT US | HOME

Packing Trays & Column Internals

Disclamer

IMPORTANT: The following supercedes Buyer's Documents. SELLER MAKES NO REPRESENTATION OR WARRANTY, EXPRESS OR IMPLIED , INCLUDING OF MERCHANTABILITY OR FITNESS FOR A PARTICULAR PURPOSE. No statements herein are to be construed as inducements to infringe any relevant patent. Under no circumstances shall Seller be liable for incidental, consequential or indirect damages for alleged negligence, breach of warranty, strict liability, tort or contrace arising in connection with the product(s). Buyer's sole remedy and Seller's sole liability for any claims shall be Buyer's purchase price. Data and results are based on controlled or lab work and must be confirmed by Buyer by testing for intended conditions of use. The product(s) has not been tested for, and is therefore not recommended for, uses for which prolonged contact with mucous membranes, abraded skin, or blood is intended; or for uses for which implantation within the human body is intended.

Content

- Raschig Super-Ring® tested with different tower internals
- Performance of liquid distributor at flashing feed locations
- Lessens learned from industrial applications

What is the Impact of Tower Internals on Packing Performance?

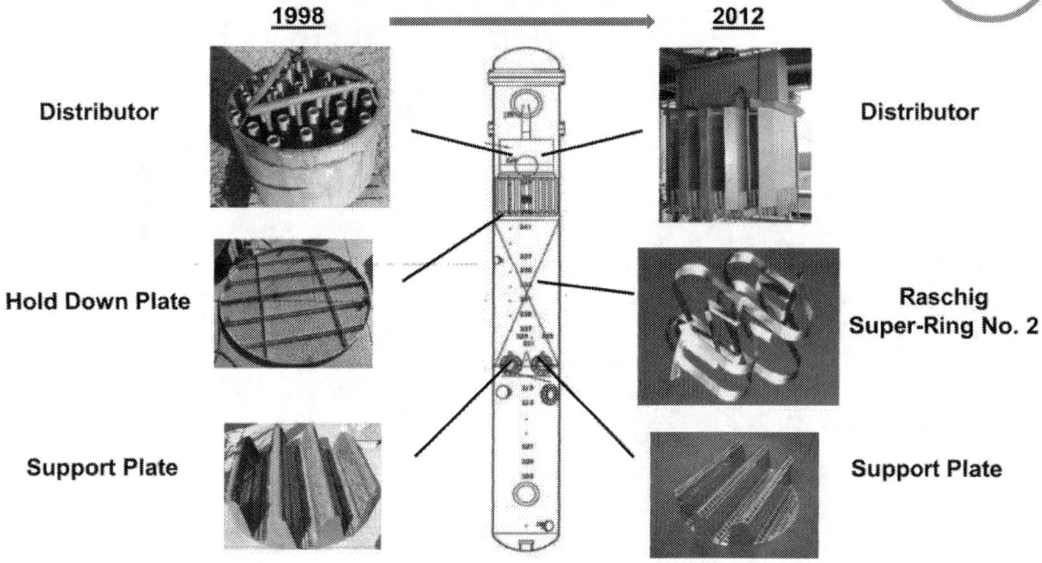

FRI Test Column

FRI-Test: 1998 TDP-Distributor

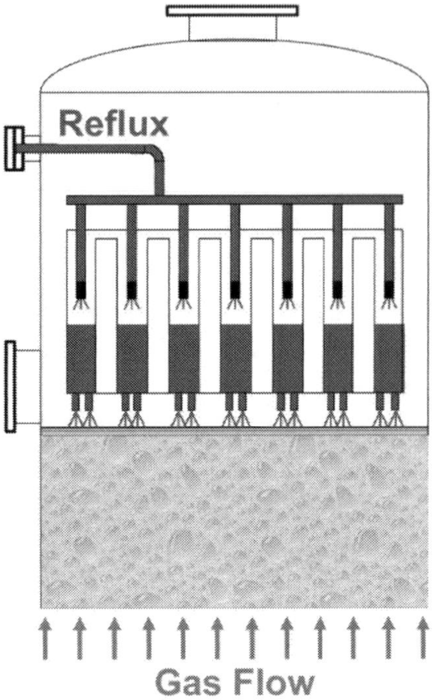

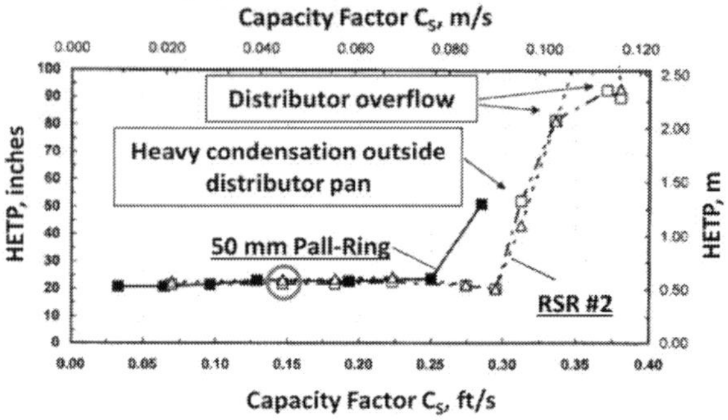

System: Cyclohexane/N-heptane, p = 1.62 bar = 24 psia

FRI-Test: 1998 TDP-Distributor

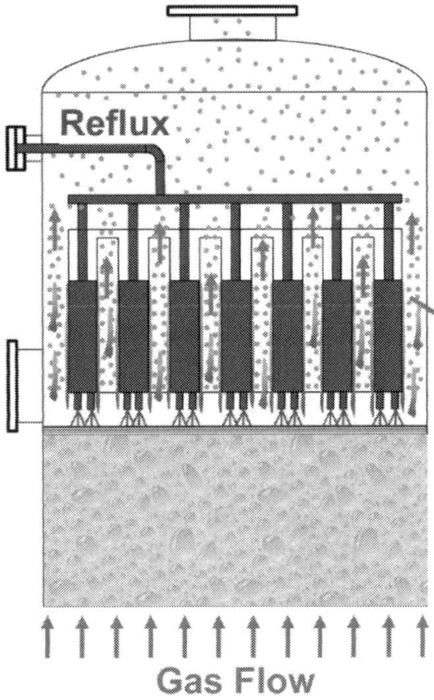

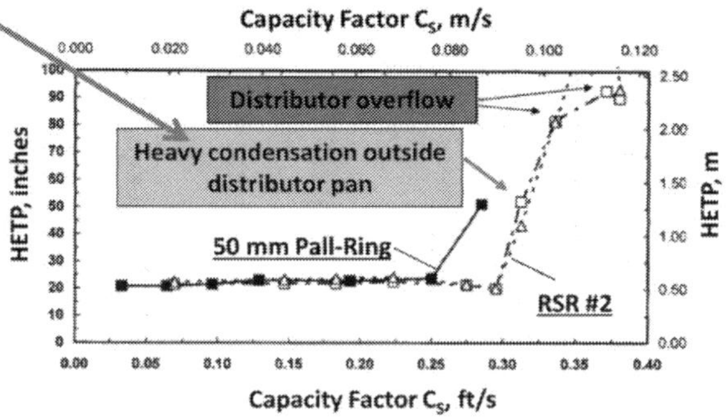

System: Cyclohexane/N-heptane, p = 1.62 bar = 24 psia

FRI-Test: 2012 DT-S-Distributor

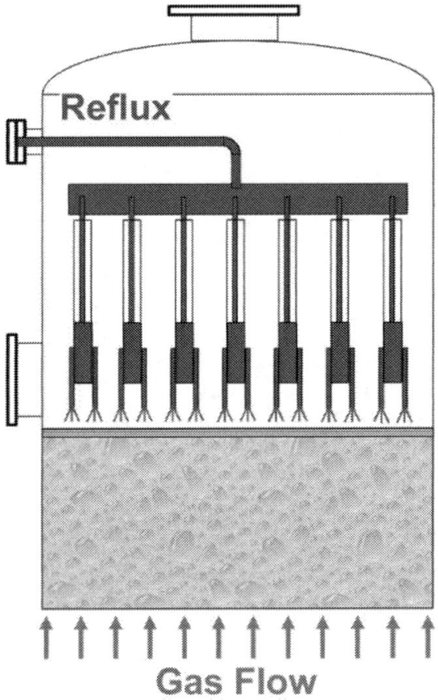

ε = 50 %

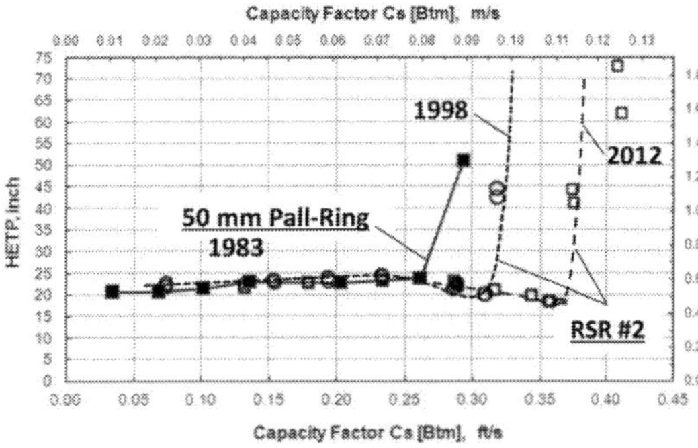

System: Cyclohexane/N-heptane, p = 1.62 bar = 24 psia

Raschig Super-Ring #2
FRI Test Results 1998, 2012

System: iso-butane/n-butane
p = 165 psia = 11.4 bar

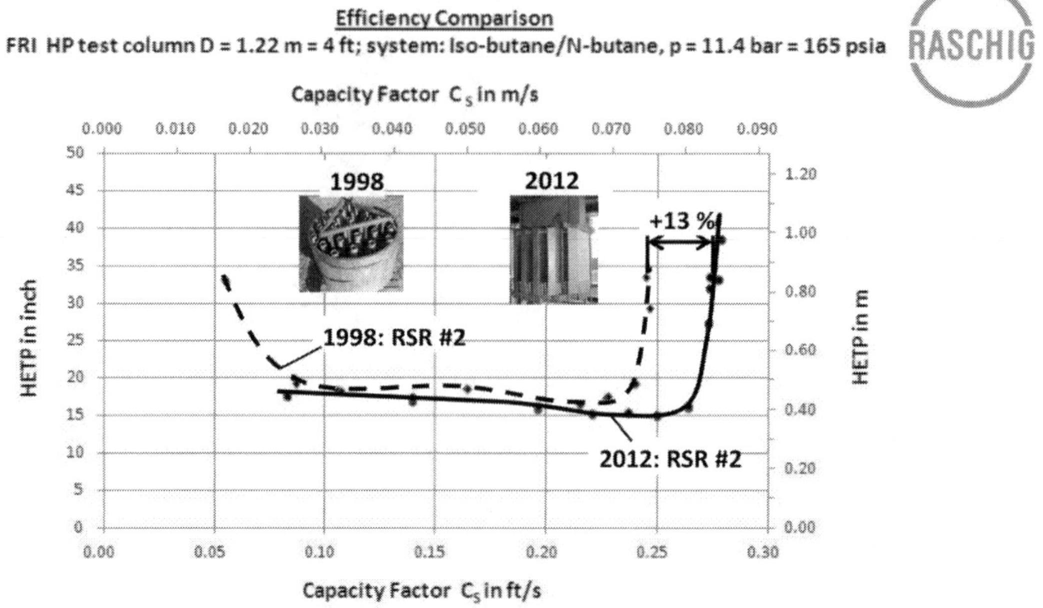
Efficiency Comparison
FRI HP test column D = 1.22 m = 4 ft; system: Iso-butane/N-butane, p = 11.4 bar = 165 psia
RASCHIG
Capacity Factor C_S in m/s
1998
2012
+13 %
1998: RSR #2
2012: RSR #2
HETP in inch
HETP in m
Capacity Factor C_S in ft/s

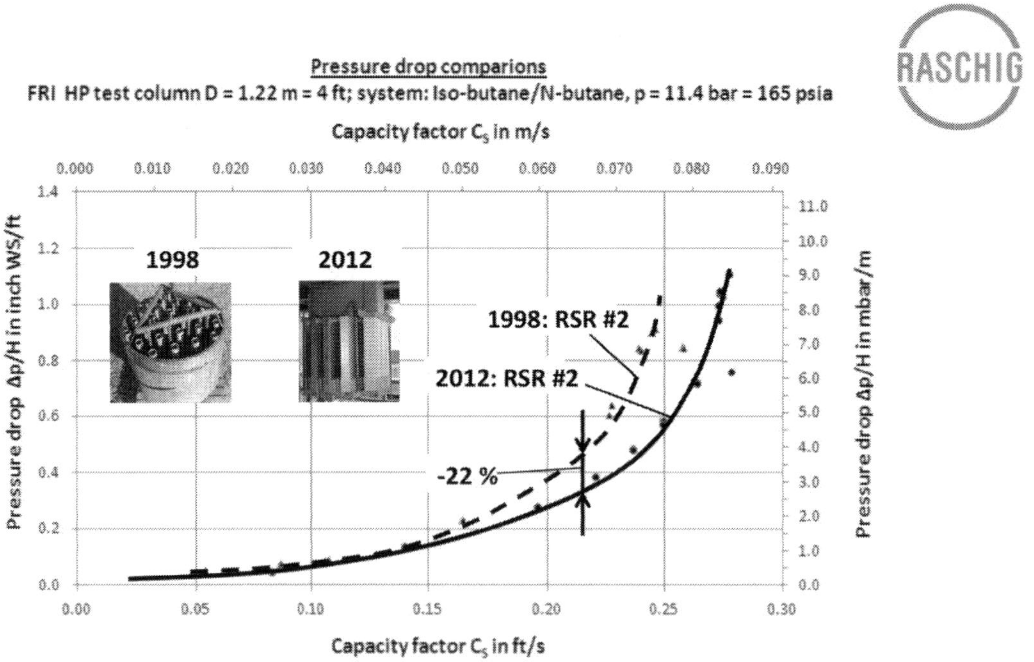
RASCHIG
Pressure drop comparions
FRI HP test column D = 1.22 m = 4 ft; system: Iso-butane/N-butane, p = 11.4 bar = 165 psia
Capacity factor C_S in m/s
1998
2012
1998: RSR #2
2012: RSR #2
-22 %
Pressure drop Δp/H in inch WS/ft
Pressure drop Δp/H in mbar/m
Capacity factor C_S in ft/s

Raschig Super-Ring #2
FRI Test Results 1998, 2012

System: cyclohexane/n-heptane
p = 24 psia = 1.62 bar

Efficiency Comparison

FRI HP test column D = 1.22 m = 4 ft; system: Cyclohexane/N-heptane, p = 1.62 bar = 24 psia

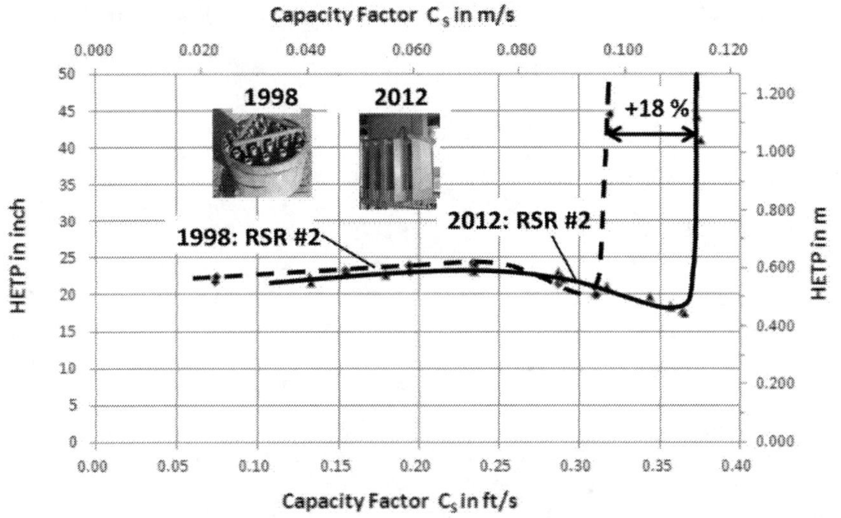

Pressure drop comparions

FRI HP test column D = 1.22 m = 4 ft; system: Cyclohexane/N-heptane, p = 1.62 bar = 24 psia

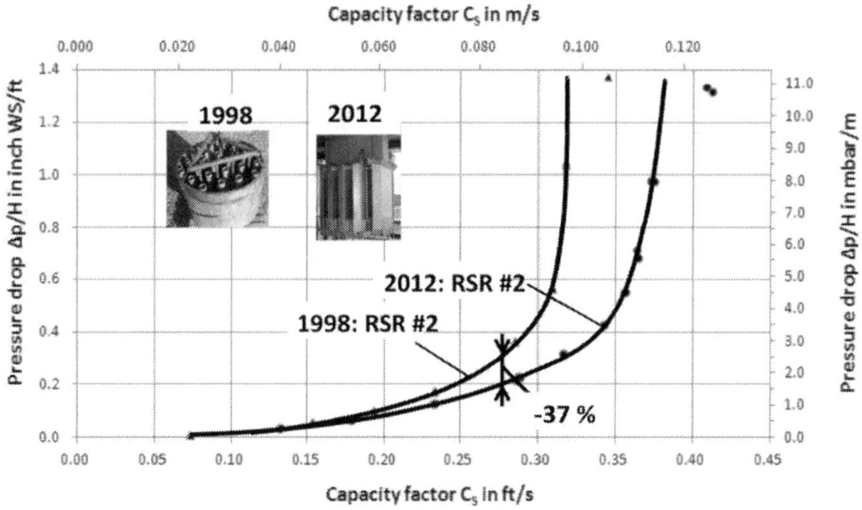

High Performance Random Packing

Raschig Super-Ring

has proven it's

Superior Performance

1. Result:

The proper selection of tower internals are as important as the correct selection of packing type

iso-/n-butane	65 psia 11.4 bar	+ 13 %	- 22 %

Efficiency (HETP): up to 14 % efficiency improvement

Raschig Test Facility

D_{max} = 12 m = 40 ft; L_{max} = 4400 m^3/h = 20000 gpm

Liquid distributor with feed pipe

Question:
How does a distributor operate at a flashing feed location?

Industrial Case: Flashing feeds

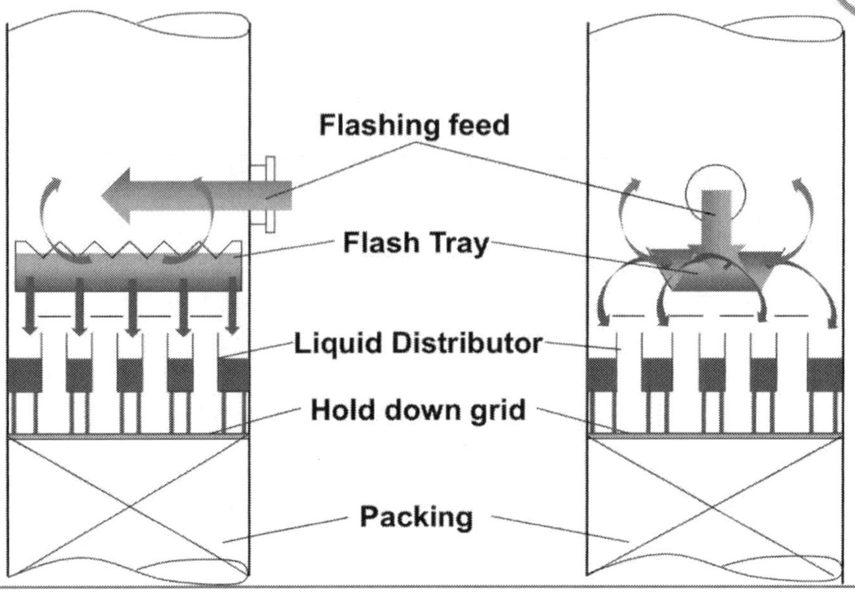

Industrial Case: Flashing feeds

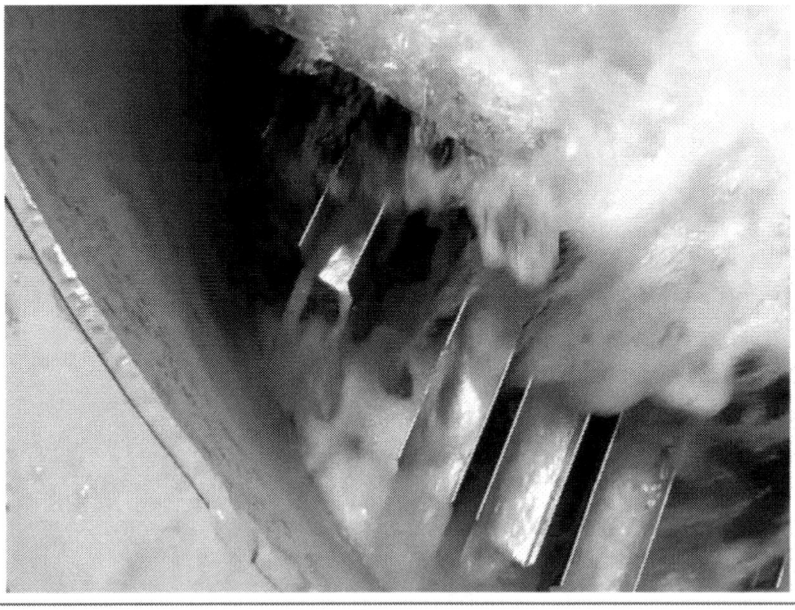

Industrial Case: Flashing feeds

Industrial Case: Flashing feeds

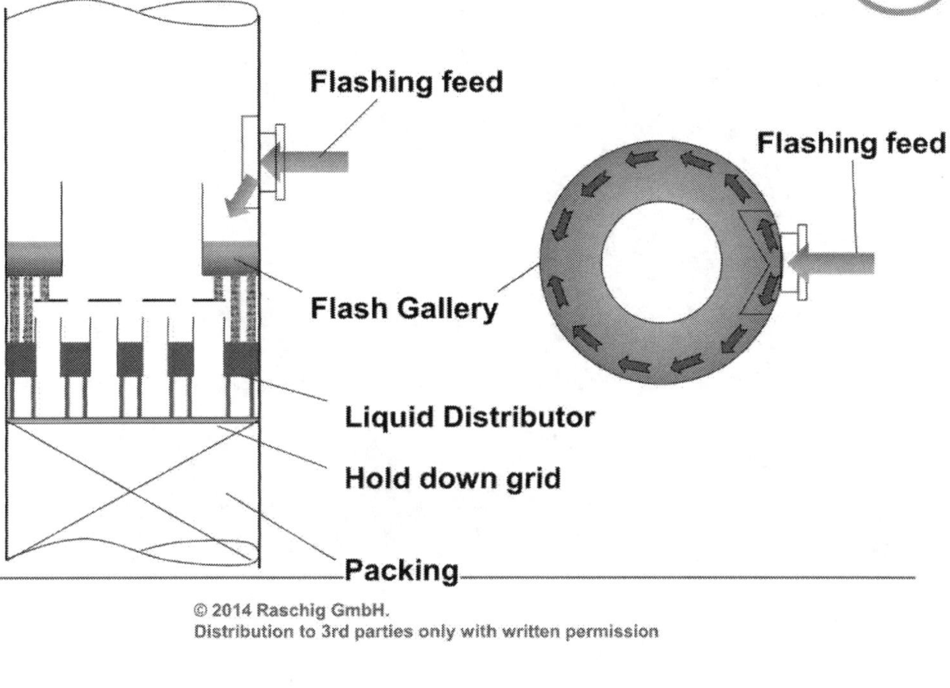

Flash-Gallery and Liquid Redistributor

Flash-Gallery and Liquid Redistributor

Industrial Case: Flashing feeds

D = 5 m = 16 ft; L_{max} = 2100 m^3/h = 9200 gpm

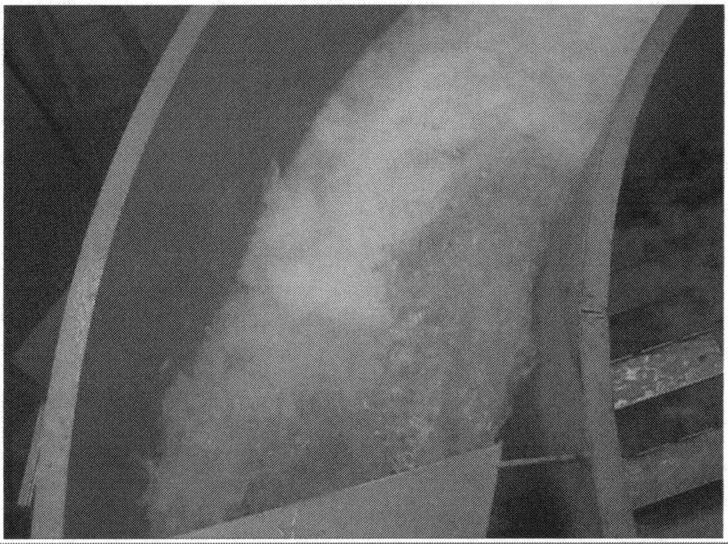

Flashing Feeds

D = 7.0m = 23 ft

Liquid distributor test facility

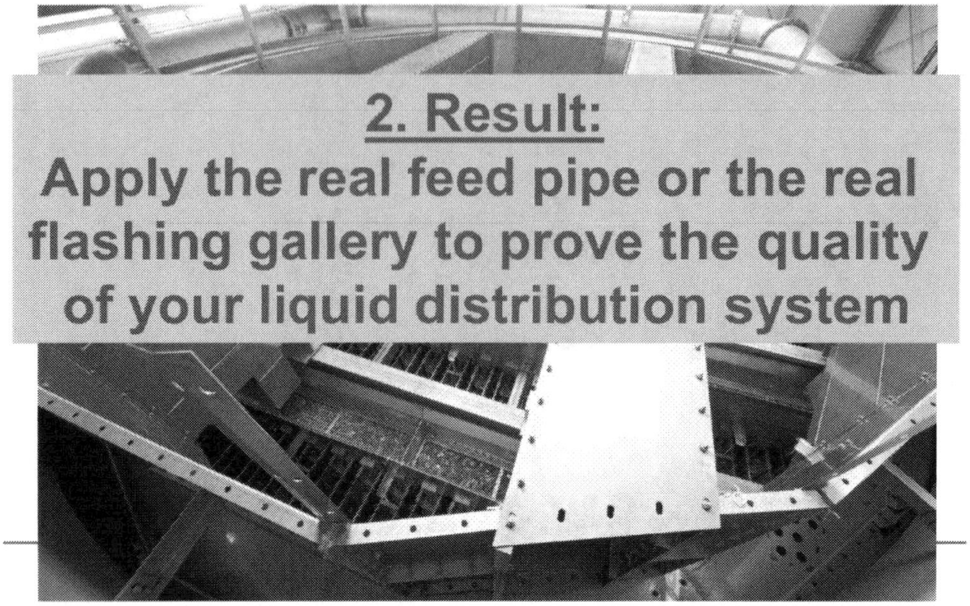

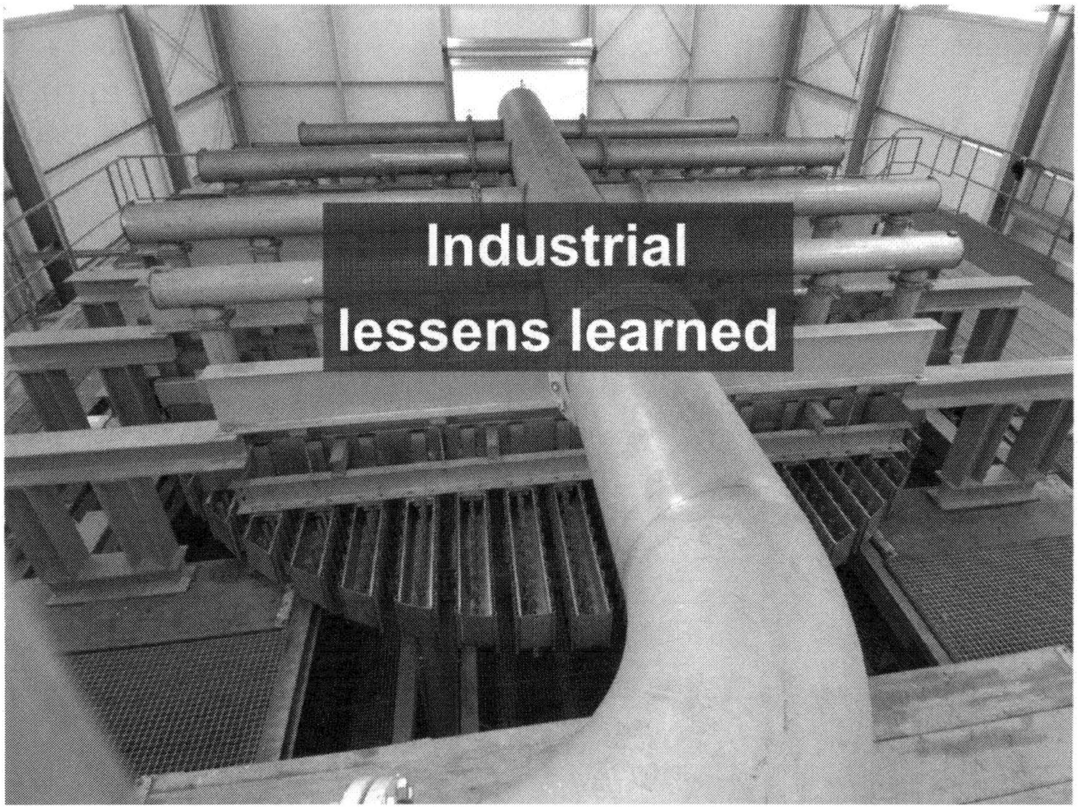

Column malfunction

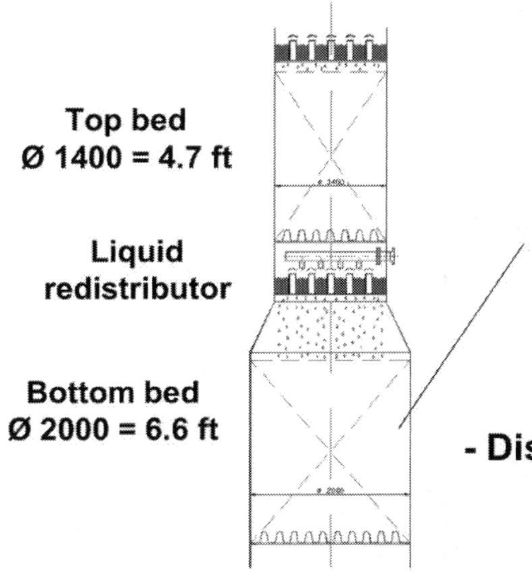

1st example:

Revamp of bottom bed from ceramic saddles to high capacity metal packings

- Distributor mispositioned !

Column malfunction

2nd Example

Customer ordered X m3 of modern plastic random packing

3 months later customer called Raschig:
! COLUMN TOTALLY FAILS !

Verification over the phone:

- Ceramic packing had been revamped
- Column operates under dirty service
- Relatively high liquid rate was used
- Trough type distributor with V-notches was in use
- Raschig asked for a photo of distributor

Column malfunction

2nd example

D = 2000 mm
= 6.6 ft
3 troughs
with v-notches

Modern type
plastic packing:
Ralu-Flow No. 2

Column malfunction

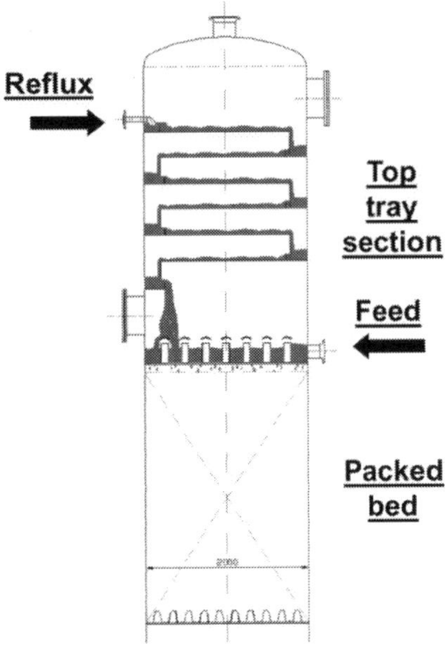

3rd example:

Transition from trayed to packed section and feed line in between

- No space between feed nozzle and distributor deck
- No feed pipe
- No Parting box
- No transition channel

Column malfunction

RASCHIG

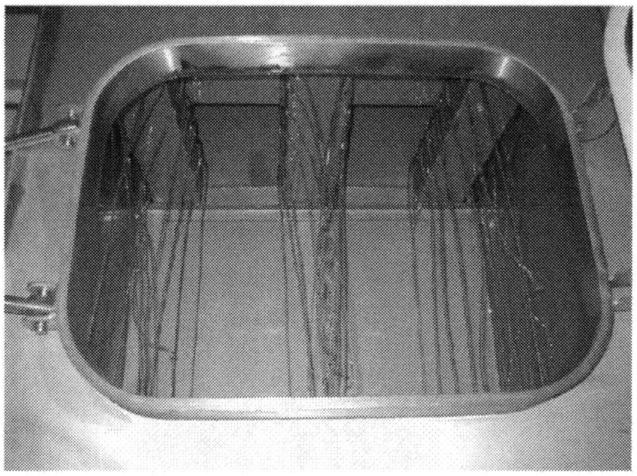

4th example:

Distribution of viscous media (oil) via side notches

- No guide channels
- Long distance to packed bed

Column Revamp: Installation/Supervision

RASCHIG

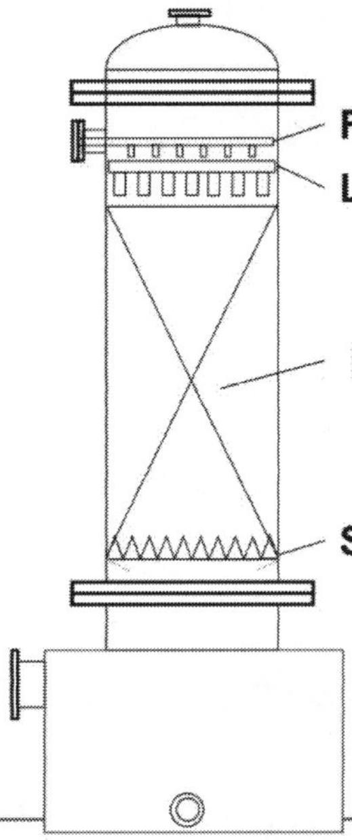

5th example:

Customer ordered a Raschig inspector last minute

Column Revamp: Installation/Supervision

5th example:

Customer ordered a Raschig inspector last minute

- Turned feed pipe

Column Revamp: Installation/Supervision

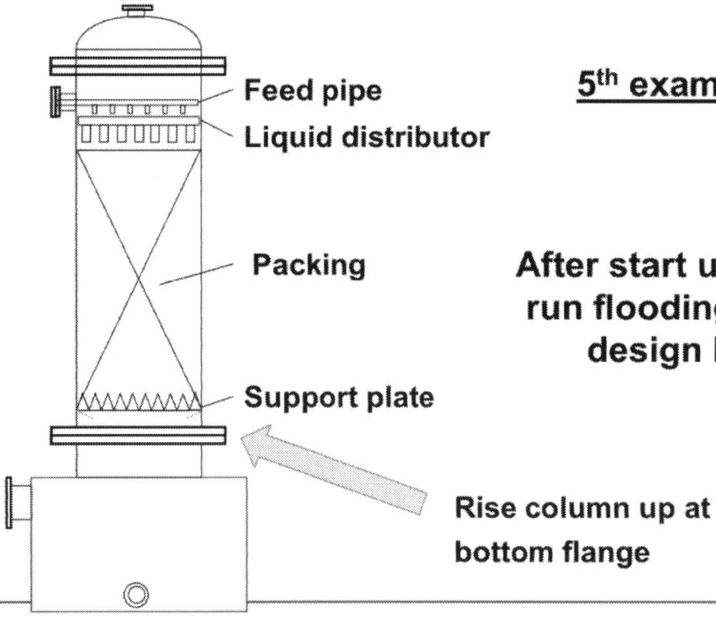

5th example:

After start up column run flooding at 50 % design loads

Column Revamp: Installation/Supervision

RASCHIG

5th example:

- **Conical liquid collector remained during installation**

Column Installation/Supervision

RASCHIG

6th example:

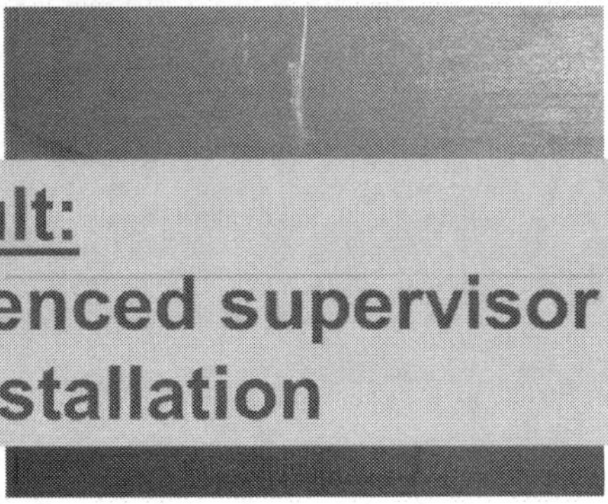

3. Result:
Ensure that an experienced supervisor witness the installation

? Support Ring levelness ?

Thank you for listening

Implementation Issues With Regard to Process Safety Programs

Dr. M. Sam Mannan, PE, CSP, DHC
Regents Professor and Director
Holder of T. Michael O'Connor Chair I
Mary Kay O'Connor Process Safety Center
Artie McFerrin Department of Chemical Engineering
Texas A&M University System
College Station, Texas 77843-3122, USA
(979) 862-3985, mannan@tamu.edu
URL: http://process-safety.tamu.edu

(6th AIChE Southwest Process Technology Conference, Moody Gardens Hotel and Convention Center, Galveston, Texas, October 9-10, 2014)

TEES | MARY KAY O'CONNOR PROCESS SAFETY CENTER
TEXAS A&M ENGINEERING EXPERIMENT STATION

Outline

- Introduction and Background
- Process Safety Management (PSM) Program
- Implementation Challenges for the 14 Elements of PSM
- Concluding Remarks

TEES | MARY KAY O'CONNOR PROCESS SAFETY CENTER
TEXAS A&M ENGINEERING EXPERIMENT STATION

Reasons for Increased Awareness of and Concern About Hazardous Chemical Releases

- Highly Publicized Accidents
  - Bhopal, India - 3,000+ fatalities
  - Mexico City, Mexico - 542 fatalities
  - Three-Mile Island (Nuclear)
  - Chernobyl (Nuclear)
  - Phillips Petroleum (Pasadena, TX) - 23 fatalities; >$750 Million
  - Arco, Channelview, TX - 17 fatalities; ~$90Million
  - Shell Refinery (Norco, LA) - 7 fatalities; >$300Million
  - Seveso, Italy
  - Flixborough, UK - 28fatalities
  - Piper Alpha - 165 fatalities
  - Texas City, US - 15 fatalities
  - Macondo Gulf Disaster - 11 fatalities
  - Buncefield, UK
  - West, Texas - 15 fatalities
- Increasing Number and Size of Facilities
- Government and Industry Regulations and Recommended Practices

TEES | MARY KAY O'CONNOR PROCESS SAFETY CENTER
TEXAS A&M ENGINEERING EXPERIMENT STATION

Some Questions/Observations

- Are we getting better of worse?
- Is there data that provides any information or conclusions?
- Is PSM the right program? What are the implementation challenges and pitfalls?
- If performance is not getting better, what should we do?

TEES | MARY KAY O'CONNOR PROCESS SAFETY CENTER
TEXAS A&M ENGINEERING EXPERIMENT STATION

PSM Program: What Does it Include?

The 14 Elements

- Employee Participation
- Process Hazards Analysis
- Training
- Pre-startup Safety Review
- Hot Work Permit
- Incident Investigation
- Compliance Audit
- Process Safety Information
- Operating Procedures
- Contractors
- Mechanical Integrity
- Management of Change
- Emergency Planning
- Trade Secrets

TEES | MARY KAY O'CONNOR PROCESS SAFETY CENTER
TEXAS A&M ENGINEERING EXPERIMENT STATION

Employee Participation – paragraph (c)

- Written plan of action
- Consult with employees on the conduct and development of process safety programs
- Provide access to PHAs and all other information developed for the process safety programs

Objective: Accountability

TEES | MARY KAY O'CONNOR PROCESS SAFETY CENTER
TEXAS A&M ENGINEERING EXPERIMENT STATION

Employee Participation – Implementation Challenges

- Provide Education and Awareness Information
  - Demonstrate the need and value of PSM Programs
- Have employees help develop and implement procedures for input and access to information
- Get employee participation as team members in PHAs and other process safety activities
- Allow easy access to PHA reports and involve employees in completion of the action items
- Communicate PHA findings to affected employees and involve them in determining what additional training is needed

TEES | MARY KAY O'CONNOR PROCESS SAFETY CENTER
TEXAS A&M ENGINEERING EXPERIMENT STATION

Process Safety Information – paragraph (d)

Chemicals

- Toxicity Information
- Permissible Exposure Limits
- Physical Data
- Reactivity Data
- Corrosivity Data
- Thermal and Chemical Stability Data
- Hazardous Effects of Inadvertent Mixing

Technology

- Block Diagram
- Process Flow Diagram
- Process Chemistry
- Maximum Inventory
- Safe Upper and Lower Limits
- Consequences of Deviation

Equipment

- Materials of Construction
- P&ID
- Electrical Classification
- Relief Systems Design and Design Basis
- Ventilation System Design
- Design Codes Employed
- Material and Energy Balances
- Safety Systems

Objective: Knowledge and Control of the Operation

TEES | MARY KAY O'CONNOR PROCESS SAFETY CENTER
TEXAS A&M ENGINEERING EXPERIMENT STATION

Process Safety Information –
(Implementation Challenges)

- **Availability of all process safety information**
  - **Required before conducting PHAs**
  - **Understand the hazards of the processes**
- **Document management system**
- **Access and update procedures**
- **Equipment information**
  - **Detailed information**
  - **Document that equipment complies with recognized codes, standards and good practices**

MARY KAY O'CONNOR
PROCESS SAFETY CENTER
TEXAS A&M ENGINEERING EXPERIMENT STATION

Process Hazards Analysis – paragraph (e)

A PHA Should be Performed at Least Every Five Years to Minimize the Risk (Consequence and Probability) of a Dangerous Substance Release

Methodologies:

- **What If**
- **Checklist**
- **What If/Checklist**
- **FMEA** (Failure Modes and Effects Analysis)
- **FTA** (Fault Tree Analysis)
- **HAZOP** (Hazard and Operability)

Objective: Hazards Identification and Control of Hazards

MARY KAY O'CONNOR
PROCESS SAFETY CENTER
TEXAS A&M ENGINEERING EXPERIMENT STATION

Process Hazards Analysis –
(Implementation Challenges)

- Prioritization of PHAs and schedule
- Develop PSI before conducting PHAs
- Use appropriate methodology
- Consideration of past incidents
- Composition of teams and appropriate reports
- Facility siting and human factors
- PHA recommendations and follow up actions
- Schedule for implementation of PHA findings
- Updating/revalidating of PHAs
- Keep PHAs for the life of the process

TEES | MARY KAY O'CONNOR PROCESS SAFETY CENTER
TEXAS A&M ENGINEERING EXPERIMENT STATION

Attributes of a Good PHA Facilitator

- Preparation
- Good Meetings vs. Bad Meetings
- Seeks Consensus
- Facilitates vs. Dictates
- Knowledgeable in Methodology
- Works well with People/Respected

TEES | MARY KAY O'CONNOR PROCESS SAFETY CENTER
TEXAS A&M ENGINEERING EXPERIMENT STATION

PHA Report Contents

- Executive summary
- Introduction and scope of work (includes process description)
- Summary of significant hazards identified
- Action item list/Recommendations
- Accidental release scenarios considered (consequence analysis)
- PHA methodology used and worksheets
- Documentation reviewed

MARY KAY O'CONNOR PROCESS SAFETY CENTER
TEXAS A&M ENGINEERING EXPERIMENT STATION

Operating Procedures – paragraph (f)

Written Operating Procedures Which Provide Clear Instructions These Procedures Must Address Each of the Following:

- **Operating Phases**
  - Initial Startup
  - Temporary Operations
  - Normal Shutdown or emergency shutdown
  - Normal Operation
  - Emergency Operations
  - Startup after Turnaround
- **Operating Limits**
  - Consequences of Deviation
  - Steps to Correct Deviations
- **Safety and Health Considerations**
  - Hazards of the Chemicals
  - Control Measures
- **Safety Systems and their Functions**

Objective: Knowledge and Control of the Operation

MARY KAY O'CONNOR PROCESS SAFETY CENTER
TEXAS A&M ENGINEERING EXPERIMENT STATION

Operating Procedures - Implementation Challenges

- Availability of current operating procedures to:
  - Operations personnel
  - PHA team
    - PHA findings may require modification of operating procedures
- Operating limits, safety systems
  - Determination of control system required
- Document control and update of procedures
  - Annually certify current and accurate
  - Management of Change

MARY KAY O'CONNOR PROCESS SAFETY CENTER
TEXAS A&M ENGINEERING EXPERIMENT STATION

Training - paragraph (g)

Training Must be Provided to all Personnel Responsible for Operating the Facility

Training Should Include:

- Initial Training
  - Before operating the process
- Refresher and Supplemental Training
  - At least every 3 years
- Communication of Change
  - Training, if required
- Contractor Training
- Verification of Understanding

Objective: Knowledge and Control of the Operation

MARY KAY O'CONNOR PROCESS SAFETY CENTER
TEXAS A&M ENGINEERING EXPERIMENT STATION

Training – Implementation Challenges

- Availability of training programs and procedures
- PHA findings may require modification of training program and procedures
- Training triggered by management-of-change
- Employee Feedback may require modification of training program and procedures
- Document all training

- Refresher training and documentation
  - Get/document employee input as to frequency and content

MARY KAY O'CONNOR PROCESS SAFETY CENTER
TEXAS A&M ENGINEERING EXPERIMENT STATION

Contractors (Host Employer Responsibilities) - paragraph (h)

Host employer must:

- Consider Safety Records in Selecting Contractors
- Inform Contractors of Potential Process Hazards
- Explain the Facility's Emergency Action Plan
- Control Entrance, Presence and Exit in Covered Process Areas
- Evaluate Contractor Safety Performance
- Maintain Injury/Illness Log for Contractors Working in Process Areas

Objective: Knowledge and Control of the Operation

MARY KAY O'CONNOR PROCESS SAFETY CENTER
TEXAS A&M ENGINEERING EXPERIMENT STATION

Contractors (Contractor Employer Responsibilities) - paragraph (h)

Contractor employer must:

- **Train Their Employees in Safe Work Practices and Document That Training**
- **Assure That Employees Know About Potential Process Hazards and Emergency Action Plan**
- **Assure That Employees Follow Safety Rules of Facility**
- **Advise Employer of Hazards Caused by Contract Work or Hazards Found by Contractor**

Objective: Knowledge and Control of the Operation

TEES | MARY KAY O'CONNOR PROCESS SAFETY CENTER
TEXAS A&M ENGINEERING EXPERIMENT STATION

Contractors – (Implementation Challenges)

- **Pre-Approval of Contractors**
- **Training Discrepancy**
- **Continued Monitoring of Contractors**
- **Contractor Employees' Lack of Awareness Regarding Responsibility to Report Hazards**
- **Floating/temporary Contractor Employees**
- **Sub-contracting Issues**

TEES | MARY KAY O'CONNOR PROCESS SAFETY CENTER
TEXAS A&M ENGINEERING EXPERIMENT STATION

Pre-Startup Safety Review – paragraph (i)

Pre-Startup Safety Review for New and Modified Facilities

- **Construction is in Accordance with Design Specifications**
- **Safety, Operating, and Emergency Procedures**
- **PHA Recommendations**
- **Training of Employees**

Objective: Hazard Identification and Control of Hazards

MARY KAY O'CONNOR PROCESS SAFETY CENTER
TEXAS A&M ENGINEERING EXPERIMENT STATION

Pre-Startup Safety Review – Implementation Challenges

- **Checklist Approach**
- **Design and Construction Issues**
- **Procedures, Training, PHA Recommendations**
- **Procedures for Non-Essential Personnel**
- **SIMOPS**

MARY KAY O'CONNOR PROCESS SAFETY CENTER
TEXAS A&M ENGINEERING EXPERIMENT STATION

Mechanical Integrity – paragraph (j)

Establish and Implement Written Procedures to Maintain the Mechanical Integrity of:

- **Pressure Vessels and Storage Tanks**
- **Piping Systems**
- **Relief and Vent Systems**
- **ESD Systems**
- **Controls, Alarms, and Interlocks**

Objective: Hazard Identification and Control of Hazards

TEES | MARY KAY O'CONNOR PROCESS SAFETY CENTER
TEXAS A&M ENGINEERING EXPERIMENT STATION

Mechanical Integrity – Implementation Challenges

- **Written Program**
- **Covered Items**
- **Procedures Specific to Equipment**
- **Training**
- **Inspection Frequency**
- **Deferred Maintenance or Inspection**
- **Repetitive Maintenance**

TEES | MARY KAY O'CONNOR PROCESS SAFETY CENTER
TEXAS A&M ENGINEERING EXPERIMENT STATION

Safe Work Practices – paragraph (k)

- Hot Work Permit
- Fire Prevention and Protection Requirements
- Safe Conduct of Operating, Maintenance Modification Activities
- Control of Materials and Substances
- Control of Access to Process Areas

Objective: Knowledge and Control of the Operation

TEES | MARY KAY O'CONNOR PROCESS SAFETY CENTER
TEXAS A&M ENGINEERING EXPERIMENT STATION

Safe Work Practices – (Implementation Challenges)

- Move object to a safe place if possible
- Is there an alternate to hot work?
- Remove fire hazard from hot work area
- Protect immovable fire hazards
- Protect remote fire hazards
- Prevent flammables from entering hot work zone
- Provide fire extinguishing capabilities and fire watch
- Test atmosphere to confirm no detectable flammables
- Secure approvals before issuing permit
- Must be kept on file until work is completed
- Shift change issues

TEES | MARY KAY O'CONNOR PROCESS SAFETY CENTER
TEXAS A&M ENGINEERING EXPERIMENT STATION

Management of Change – paragraph (l)

Establish and Implement Written Management of Change Procedures Which Address:

- Technical Basis
- Impact on Safety and Health
- Modifications to Operating Procedures
- Necessary Time Period for the Change
- Authorization Requirements

Objective: Hazard Identification and Control of Hazards

TEES | MARY KAY O'CONNOR PROCESS SAFETY CENTER
TEXAS A&M ENGINEERING EXPERIMENT STATION

Management of Change – Implementation Challenges

- Change Definition and Recognition of Change
- Relationship of Change Definition with Design Specifications
- Management of Organizational Changes
- Managing Temporary Changes
- Process Safety Information Must be Updated, as appropriate
- Impact on Procedures
- Impact on Training

TEES | MARY KAY O'CONNOR PROCESS SAFETY CENTER
TEXAS A&M ENGINEERING EXPERIMENT STATION

Incident Investigation - paragraph (m)

Incidents That Result In, or Could Reasonably Have Resulted in a Major Incident Must be Investigated

Incident Investigation Report:

- **Date of Incident**
- **Date Investigation Began**
- **Description**
- **Factors**
- **Recommendations**

Objective: Accidents and Learning from Accidents

MARY KAY O'CONNOR PROCESS SAFETY CENTER
TEXAS A&M ENGINEERING EXPERIMENT STATION

Incident Investigation – (Implementation Challenges)

- **Incident recognition and reporting**
- **Incident investigation procedures**
  - **Causes—Contributing and Root**
- **Address findings and recommendations**
  - **Track to completion**
- **Lessons Learned system**
  - **Review findings and corrective actions with all affected personnel, including contract employees, if appropriate**
  - **Learn from other's incidents**

MARY KAY O'CONNOR PROCESS SAFETY CENTER
TEXAS A&M ENGINEERING EXPERIMENT STATION

Incident = LNG Vaporizer Shell Rupture

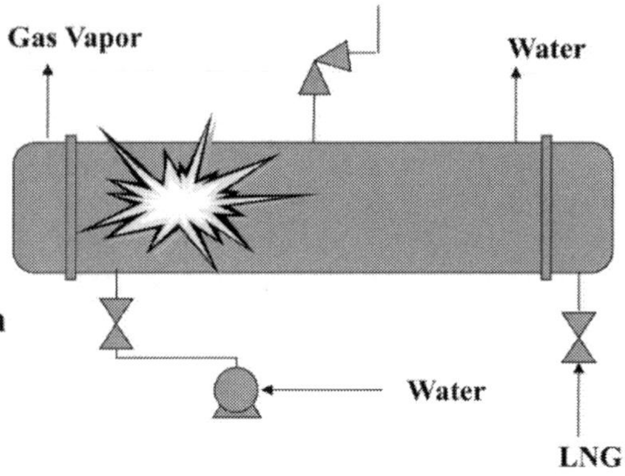

Conduct an incident investigation using a Fault Tree analysis.

Top event = Shell Rupture

Now start asking Why?

MARY KAY O'CONNOR PROCESS SAFETY CENTER
TEXAS A&M ENGINEERING EXPERIMENT STATION

Determining Causes

- **Immediate Cause**
  - Usually the last thing to happen before the incident
- **Contributing or Enabling Cause(s)**
  - Those things that helped cause the incident, but by themselves did not cause it
- **Root Cause(s)**
  - Usually the ***one*** thing that if corrected would prevent ***this*** incident from recurring. (There may be/usually is more than one root cause)
- **System Failure**
  - The management system fault that allowed all of the other causes to exist. Correction prevents this incident as well as other similar events and dissimilar events allowed by that same management system fault

MARY KAY O'CONNOR PROCESS SAFETY CENTER
TEXAS A&M ENGINEERING EXPERIMENT STATION

Emergency Planning and Response – paragraph (n)

- Establish and Implement an Emergency Action Plan in Accordance with 29 CFR 1910.38(A)

Objective: Accidents and Learning from Accidents

MARY KAY O'CONNOR PROCESS SAFETY CENTER
TEXAS A&M ENGINEERING EXPERIMENT STATION

Emergency Planning and Response – Implementation Challenges

- Escape procedures and route assignments
- Procedures and protection systems for employees who remain to operate critical equipment
- Procedures to account for employees after evacuation
- Rescue and medical duties
- Means of reporting fires and other emergencies
- Names and job titles of persons who can be contacted for further information or explanation of duties under the plan

MARY KAY O'CONNOR PROCESS SAFETY CENTER
TEXAS A&M ENGINEERING EXPERIMENT STATION

Compliance Audits - paragraph (o)

- Conduct Audit at least Every Three Years
  - Certify evaluation of compliance
  - Verify procedures and practices are adequate and being followed
  - By self, by company representatives, by a consultant
- Audit Team
  - At least one person knowledgeable in the process
- Report of Findings
- Determine Appropriate Response to Each Finding
  - Correct Deficiencies
  - Document actions
  - Track to completion
- Retain Two Most Recent Audit Reports

Objective: Compliance Audits

TEES | MARY KAY O'CONNOR PROCESS SAFETY CENTER
TEXAS A&M ENGINEERING EXPERIMENT STATION

Compliance Audits – Implementation Challenges

- Develop policies, protocol, and guidelines
- Audit team composition and training
- Appropriate reports
- Audits findings and follow up
- Schedule for implementation of improvement opportunities
  - Industry standard
  - Documentation of deviation

TEES | MARY KAY O'CONNOR PROCESS SAFETY CENTER
TEXAS A&M ENGINEERING EXPERIMENT STATION

Trade Secrets - paragraph (p)

- **Employers shall make all information necessary to comply with this rule available to those persons who are:**
  - **Responsible for compiling the PSI**
  - **Assisting in the development of PHAs**
  - **Responsible for developing operating procedures**
  - **Involved in accident investigations, emergency planning and response**
  - **Involved in compliance audits**
- **The employer may require persons receiving this information to enter into a confidentiality agreement**
- **Subject to certain rules, employees and their designated representatives shall have access to trade secret information contained within the PHA and other documents**

TEES | MARY KAY O'CONNOR PROCESS SAFETY CENTER
TEXAS A&M ENGINEERING EXPERIMENT STATION

Wish List for Progress in Process Safety

- **National surveillance system for process safety incidents**
- **Rulemaking processes should be streamlined and based on good science and risk-benefit analyses**
- **Streamline enforcement of existing regulations by governmental agencies should be improved**
- **Universities must play a role in continuing to integrate process safety into the academic and research programs in the engineering disciplines as well as in other disciplines (*e.g.*, sciences, geosciences, management, business)**
- **Strong, visible and effective leadership is a cornerstone of good safety performance**
- **Strong culture with regard to process safety.**
- **Human factors and lack of adequate attention to human factors engineering in plant design and related operations continues to be a major factor and reason behind a vast number of incidents**
- **Competency in process safety programs and activities related to process safety continues to be another major factor impeding the progress in process safety performance**

TEES | MARY KAY O'CONNOR PROCESS SAFETY CENTER
TEXAS A&M ENGINEERING EXPERIMENT STATION

Process Safety Competency Development

- **Everyone in the corporation whos**e actions/inactions may impact the process
- Chairman, CEO, Board of Directors
- **Executive Management**
- **Senior Managers**
- **Process Safety Managers/Coordinators**
- **Specialists**
  - Process Hazard Analysis Participants
  - Process Hazard Analysis Managers
  - Process Safety Management Auditors
  - Mechanical Integrity Specialists
  - Incident Investigation Specialists
  - Contractor Management Specialists
- Operations and Maintenance Supervisors
- Operations and Maintenance Workers

TEES | MARY KAY O'CONNOR PROCESS SAFETY CENTER
TEXAS A&M ENGINEERING EXPERIMENT STATION

Concluding Remarks

- **To Develop Process Safety Competency a Sound Process Safety Management Program should be implemented**
- **The fourteen elements discussed in this talk holds great promise in reducing risk and increasing the productivity**
- **Companies cannot be sustainable without successful safety and risk management programs**

TEES | MARY KAY O'CONNOR PROCESS SAFETY CENTER
TEXAS A&M ENGINEERING EXPERIMENT STATION

THANK YOU

Dr. M. Sam Mannan, PE, CSP, DHC
Regents Professor and Director
Holder of T. Michael O'Connor Chair I
Mary Kay O'Connor Process Safety Center
Artie McFerrin Department of Chemical Engineering
Texas A&M University System
College Station, Texas 77843-3122, USA
(o) 979-862-3985; mannan@tamu.edu
URL: http://process-safety.tamu.edu

4

TEES | MARY KAY O'CONNOR PROCESS SAFETY CENTER
TEXAS A&M ENGINEERING EXPERIMENT STATION

Why Are The New People Making The Same Mistakes That The Old-Timers Made 25 Years Ago? **Managing Organizational Change**

ART DOWELL
STAN ANDERSON (IN MEMORIAM)

The Problem -- History repeats itself

- The same process safety incidents occur over and over
- The same occupational safety incidents occur over and over

- Why?
- "Organizations have no memory" – Trevor Kletz

The Memory Issue

- The operating and maintenance personnel
  - Are promoted
  - Retire
  - Take vacation
  - Are absent
  - OR
  - Job duties are reallocated.

New Employee Example

- In one unit, within the first 6 months, almost every operator had an OII or first aid from exposure to an organic acid.
- After that, no experienced operators had the exposures even though they were continuing to do the same tasks.
- Why?

New Employee Example

- Before exposure:
  - No one understood that exposure to the organic acid was as bad as they had been told,
  - They treated the job tasks in a cavalier manner, and
  - All the training was on the job.
- After exposure,
  - They understood the severity of the consequences and the reasons underlying the procedures to avoid exposure.

New Employee Example

- The procedures were appropriate, but new folks believed short cuts were OK.
- The facility began a new operator training program using water and air in the pumps and tanks of an out-of-service unit.
  - Trainees made mistakes and took short-cuts, but they were exposed only to water.
  - The trainers explained the severity of a chemical exposure in similar situations.

New Employee Example

- Result – number of injuries in the first 6 months declined.

- Follow-up –
  - The training unit was repurposed for another product.
  - Due to a economic downturn at the time, the hands-on water and air training was discontinued.
  - Anecdotally, we understand the initial injury rate went up.

Promotion

- Replacement must be able to do what incumbent did, OR
- Duties must be parceled out
- ALL duties must be identified and
- Training required must be made available and taken
  - **before** the change is made or
  - **before** the new person undertakes new duties.

Promotion Example

- Day foreman retires. Shift Foreman is promoted to Day Foreman. An operator is promoted to Shift Foreman.
- Previous Shift Foreman had managed the water quality in the cooling tower. New Foreman had no knowledge of what this all meant, and did not manage the cooling tower water.
- Legionella bacteria developed in the cooling tower basin. Result was a major public relations disaster, even though no cases of Legionella were identified.

Reorganization

- Similar to above, except more extensive.
- ALL duties of ALL organizations being reorganized must be identified.
  - Are there job descriptions for the jobs being eliminated?
  - Do they cover all aspects of the jobs?
- New system must cover all.
  - It is a massive task to identify exhaustively all tasks that are done by a worker; it is even more daunting to identify those associated with a department.
  - Extensive and intensive debriefings must be done.

Reorganization

- The more departments are being reorganized, the more work there is in identifying all the tasks that must be either:
  - Not done.
    - Can they be safely eliminated? It takes a meeting of knowledgeable people to know.
  - Done by others who already know how to do them.
    - Who and where are they? Contractors? Other Departments? How do we know they know?

Reorganization

- Parceled out to others, who then must be trained.
  - Training requirements may be extensive.
  - What training is adequate? Who will do it? How can we know they can do the training?
  - Will the incumbents be expected to train their successors? Will time be provided to allow training to be done while their work is still needed?
  - Do the persons who will be expected to do these new tasks currently have full-time jobs?
  - Will the addition of these tasks to the existing work load be such that the person REALISTICALLY will be able to do them all well?

Reorganization Example

- An extensive reorganization was done...
- Process safety audit found that **several relief valves had not been inspected and tested** as required by the company's mechanical integrity program.
- Auditors found that all the personnel who might have had the responsibility in the past to ensure the inspections and tests were done were sure that they didn't have that responsibility in their current jobs.
- The auditors could not find anyone who believed the inspections and tests were part of their current jobs. The task had been lost in the reorganization.

Reorganization Example

- Everyone thought that someone else had the task.
- Everyone was sure that he/she did not have the task
- Reality: **NO ONE** had the task.

Temporary Absence Long Term

- Long term (Vacation, Illness)
  - How will the replacement be done?
    - From a pool of equally qualified individuals?
    - Temporary help?
    - Adding to the work load of others?
  - Are the duties to be performed known?
    - Is there a job description?
    - Is it correct and up to date?
    - Is the updating of job descriptions part of the management of change process?
  - Management must clearly communicate to all involved what each individual is being asked to do, and how frequently.

Long Term Example

- A pumper had the task to stick a 15 million pound storage tank weekly to check on the inventory, in addition to the single level gauge on the tank.
- This pumper went on vacation, and nobody was designated to stick the tank during this time.
- After a week of production and barge receipts with no inventory confirmation, the tank **overflowed**, creating a potentially serious exposure hazard, a difficult cleanup, and potentially serious environmental consequences.

Temporary Absence Short Term

- Short term (e.g., Meeting, Trip into town, Doctor's Appointment)
  - Are there critical tasks being performed during this absence (e.g., startup, shutdown)?
  - Have minimum coverage standards for these tasks been established (e.g., number of people, experience level)
  - Will these tasks be adversely affected by this absence?
  - Has there been opportunity to plan for this absence (e.g., calling in a replacement, deferring critical tasks)?

Short Term Example

- One area, comprised of multiple trains producing a toxic intermediate, established a policy
  - Always starting up a train on evenings, because there were fewer people in the plant at that time.
  - Ensuring that the skill level of operations personnel on shift was adequate.
- Another area normally would hold over a shift, thereby providing double staffing, for training and assurance. All managers were cognizant of the personnel on shift during startup, and provided coverage by foremen as needed.

What Is Needed?

- The right critical ***knowledge*** and ***skills*** available
  - On site
  - At all the times
  - When they are needed
- A ***Management System*** to accomplish the above bullets

A Management System

- Document tasks, skills, knowledge
- Formal management of personnel change system
- Formal management of personnel change system

- Assignments to manage the Management System

1. Document

- Knowledge – what does the technician (operator or mechanic) need to know?
- Skills – What skills does the technician need to do the task safely, reliability, and with good quality?
- Tasks – What are **ALL** the tasks that need to be done?

- What is the best way to do the task? Everyone does the task in the best way!

2. Formal Management of Personnel Change System

- Identifies the knowledge, skills, and tasks,
- Develops an action plan to ensure the new person
  - learns the knowledge,
  - demonstrates the skills, and
  - understands his/her tasks and responsibilities, and
- Develops a plan to provide the knowledge, skills, and tasks safely while the new person is in training. ← Transition Plan

3. System to Manage Temporary Absences

- To recognize when critical knowledge and skills are missing, and
- To halt critical operations when key people are missing, or
- To provide alternate resources to ensure safe operations.

Managing the Management System

- Who is documenting the changes to tasks, skills, and knowledge?
- Who is responsible for scheduling and conducting the formal management of change processes?
- Who is responsible to ensure that the necessary training is completed on schedule?
- Who audits the management system?

Audits

- Remember that we must inspect and test the equipment and procedures for our independent protection layers that prevent serious consequences in the operating facility.
  - To repair or replace parts that cannot do the job.
- Likewise, it is critical to "inspect and test" the Management System for Organization Change to ensure it is providing personnel with the right skills and knowledge to do the right tasks.

More Reading

Guidelines for Managing Process Safety Risks During Organizational Change

CCPS

ISBN: 978-1-118-37909-7

264 pages

April 2013

Conclusion

- Managing personnel change is essential to safe operations.
- Letting things "drop through the cracks" in a multimillion dollar plant producing flammable or toxic materials is not acceptable in today's environment (or yesterday's, for that matter).
- Facilities that do not have a viable program for managing personnel changes are ripe for disaster.

AIChE®
Southwest Process Technology Conference
Track 1:
Process Safety
Afternoon
Thursday, October 9, 2014
AIChE
ETA
1

PROCESS SAFETY REGULATION
CAN ANY STYLE BE CONSIDERED SUPERIOR?
Dr. Steven D. Emerson
Emerson Technical Analysis, LLC
Corpus Christi, TX, Las Vegas, NV
AIChE
ETA
2

Safety Case vs. PSM

*"It has been reported that financial losses from refinery incidents in the US are **three times** that of industry counterparts within the 'EU cluster'"*

Rafael Moure-Eraso, Chairman US CSB

October 2013 Mary Kay O'Connor International Process Safety Symposium

AIChE

ETA

3

Can "Safety Case" regulation be that superior?

- **Wow!**
- Approach
  - Learn more about "Safety Case"
  - Find underpinnings of the claims
  - Examine cited references
  - Ask those people who developed the basis of the claims

AIChE

ETA

4

SAFETY CASE

Lord Cullen (1990)

"...a formal safety assessment of major hazards...to demonstrate that the potential major hazards of the installation and risks to personnel thereon have been identified and appropriate controls provided."

5

SAFETY CASE

Lord Cullen (1990) continued

"...operators should be required to demonstrate this to the regulatory body"

"It is inappropriate and impracticable for the regulatory body to undertake the detailed auditing of operator's compliance ...the regulatory body should be required to review operator's audits on a selective basis and carry out such further audits...and verify that the output of the system is satisfactory."

6

Other commentators…

Andrew Hopkins, Australian National University, Canberra

"…a proactive, risk-based performance approach specific to individual facilities…"

Five basic functions:

1. A risk- or hazard-management framework
2. A requirement to make the case to the regulator
3. A competent and independent regulator
4. Workforce involvement
5. A general duty of care imposed on the operator

7

Other commentators…

Nancy Leveson, MIT

"…a structured argument for why the system is safe…"

Two types of regulation:

1. Prescriptive (e.g.., FAA)
2. Performance-based or goal-setting

8

Sequence of References

CSB Report 2014 – not accepted by Board

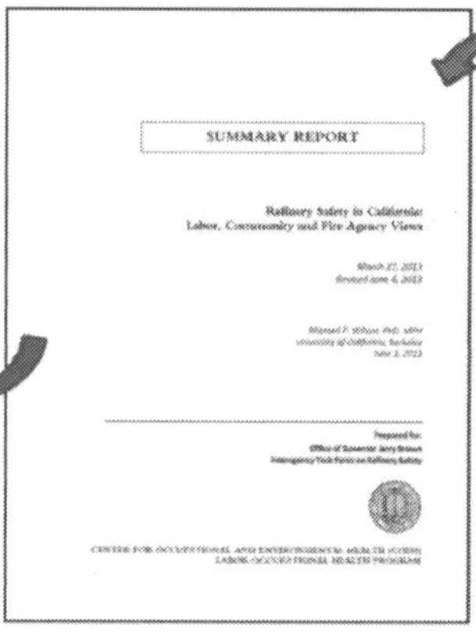

Swiss Re 2012 and earlier presentations

2013 UC Berkeley Report to Gov.

9

Telephone conversation with Ernst Zirngast, Swiss Re Technical Expert – August 2014

- Swiss Re data not intended for public
- Based on Willis Advisors worldwide loss data – available for a fee
- Country groupings *don't* allow conclusions about Safety Case vs. other approaches

ETA

10

Swiss Re Report – Country Groupings

"ANGLO" – USA (PSM), Canada (similar to PSM), UK (Safety Case), Australia (Safety Case)

"EU" – Europe, Singapore, S. Korea, Japan, Saudi Arabia, Gulf States, Egypt

"Russia" – Russia, former Soviet Union, Eastern Europe

"Other" – Latin America, N. Africa, other Middle East, rest of Asia

ETA

11

Swiss Re Results

GROUP	# Refineries	Total Capacity (MMBPD)	# Losses 1985-2006	Loss Burden 1985-2006 (MM$ BI & PD)
ANGLO	170	22	417	**$980**
EU	161	27	267	**$350**
RUSSIA	144	17	23	$15
OTHER	177	22	259	$260
			TOTAL 1985-2006	$1,605*

12

Ok then, what other sources might exist?

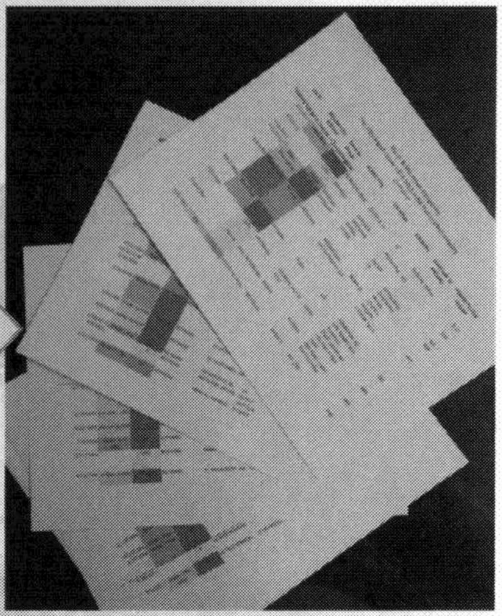

ETA

13

13 of 20 Largest Losses - since 1992

Includes:

Williams Olefins, Hurricane Ike, Kuwait
BP Macondo
Conoco-Phillips Ekofisk, France AZF, Australia Gas Plant
Algeria LNG, Argentina, India Offshore, Petrobras P-36...

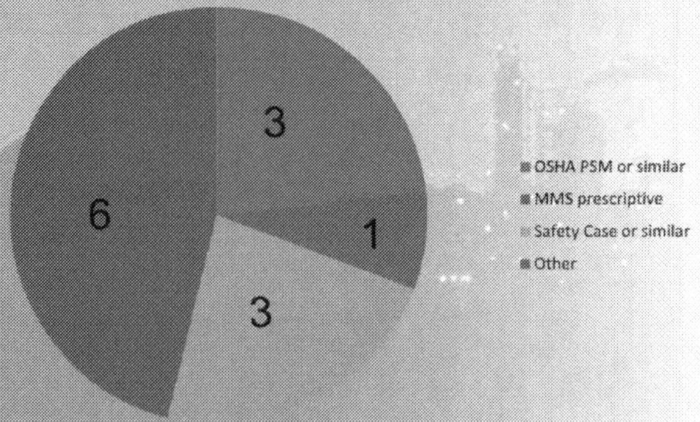

ETA

14

REFINING – since 1992

26 of 100 includes:

- **3 California, 3 Texas, 2 Mississippi, 3 Canada, Kuwait**
- **Singapore, Pembroke UK,**
- **2 Italy, 2 Japan, France, Argentina, Venezuela, Thailand, Turkey…**

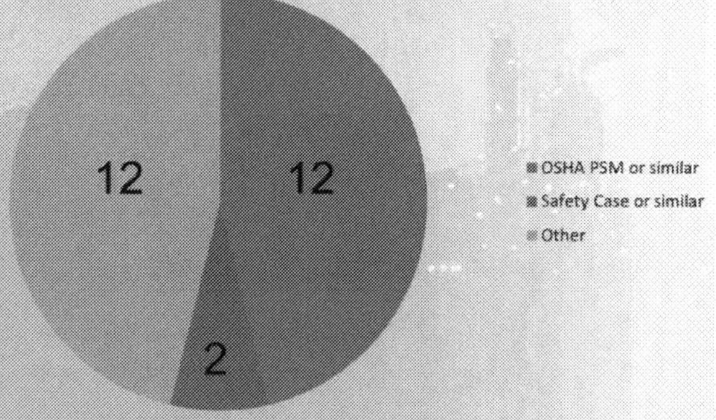

*Note: M&M 24 incidents 1985-2006 $6,310MM

AIChE

ETA

15

UPSTREAM OFFSHORE since 1992

15 of 100 includes:

- **4 GOMR including BP Macondo**
- **Ekofisk, FPSO loose off UK, Australia, leaking risers off Norway**
- **India, 2 Angola, 2 Venezuela, Egypt…**

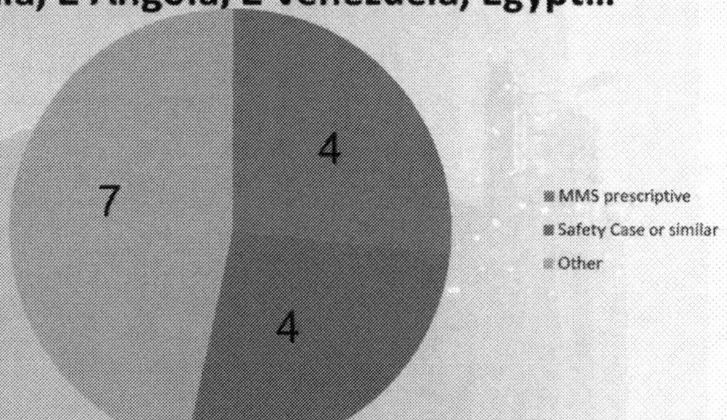

AIChE

ETA

16

CONCLUSION

- Data shortcomings
  - Very small numbers
  - Swiss Re mixed the groupings
  - Not normalized to bpd, employment, exposure, etc.
  - Not all data reported (e.g., Russian data)
- *Data do not allow firm conclusions as to Safety Case vs. PSM*

ETA

17

SUGGESTIONS

There is no magic answer.

Hard work and scrupulous attention is demanded.

Exemplars exist

Nuclear submarines
(see Leveson)

USNRC

ETA

18

Look at the "Black-Box" BEFORE the Crash....

The Use of "Black-Box" Reviews in PHA Scenario Development and Validation Against Vessel Limits

Date: 22 September 2014

Prepared by: Ralph E. King, PE

Taking on your toughest technical challenges.

STRESS ENGINEERING SERVICES INC.
an employee-owned company

Agenda

- Introduction and Background
- FAA – "Black-Box" Data Analysis
- State of our Industry
  - Our Challenge
- Process Hazards Management
  - Potential Use of "Black-Box" Analysis
- Automated Data Historian Searches
- Examples
- Conclusions

STRESS ENGINEERING SERVICES INC. an employee-owned company

Airline Accident Rate - 2009

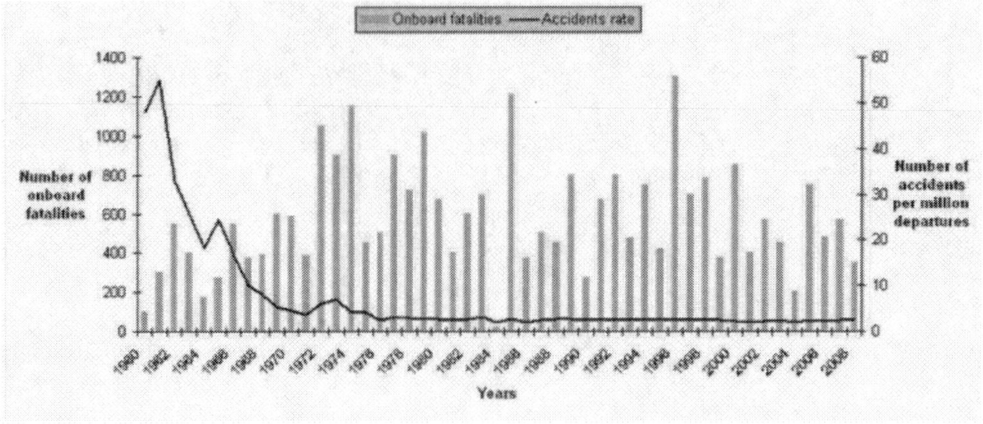

http://www.1001crash.com/index-page-statistique-lg-2.html

Airline Growth - 2009

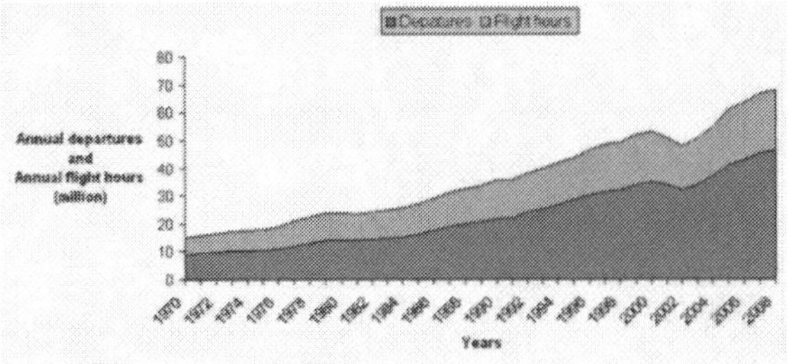

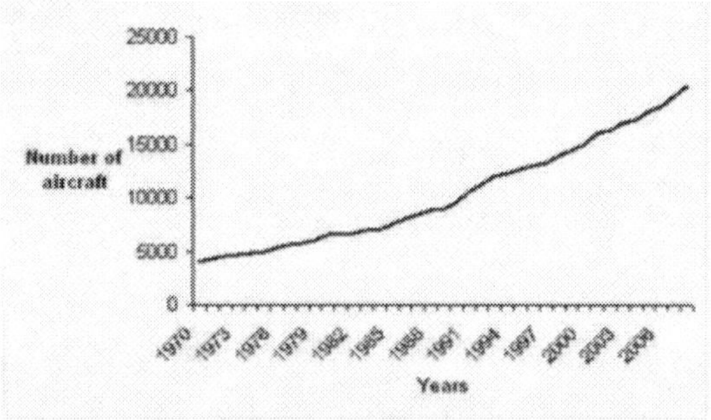

http://www.1001crash.com/index-page-statistique-lg-2.html

The Challenge

- At a constant incident rate
- At an increasing air-traffic volume
- Crash rates WILL go UP!
- How do you keep the same or reduce crash rates in the face of increasing air-traffic?
  - Reduce the incident rate!
- BUT HOW?

STRESS ENGINEERING SERVICES INC. an employee-owned company

FAA – "Black-Box" Data Analysis

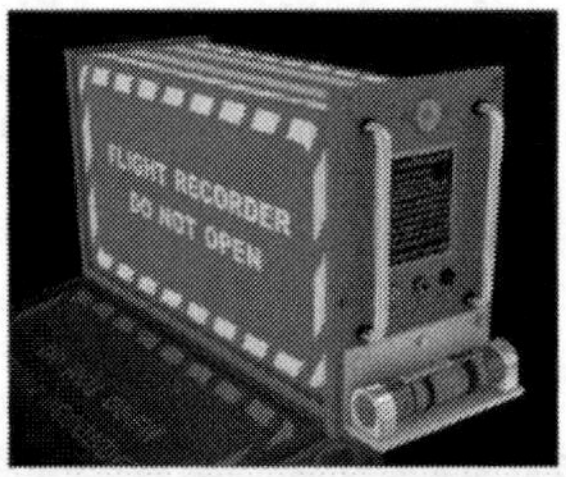

- Concept – don't wait for the plane crash to start identifying safety issues (Started in 2009)
- Routinely Pull "Black-Box" Data
- Execute data analysis (bin-analysis) to identify potential safety issues
- Integrate all flight information from multiple carriers to identify "outliers" to the normal operation
- Goal is to reduce "near-hit" events

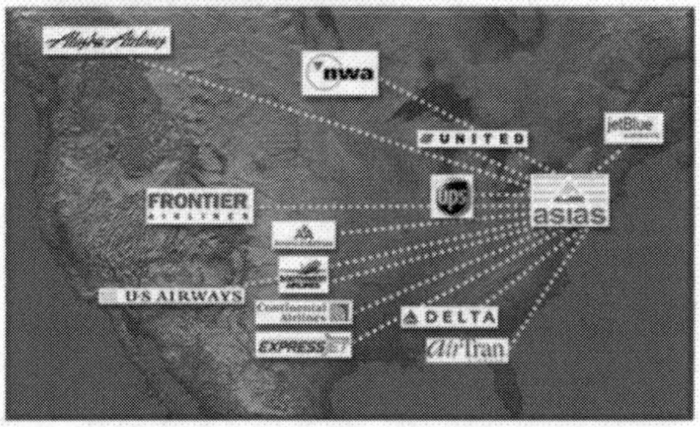

Aviation Safety Information Analysis and Sharing (ASIAS—pronounced "a-sigh-us") initiative.
http://www.mitre.org/news/digest/aviation/04_09/data.html
http://web.mit.edu/newsoffice/2011/black-box-analysis-0912.html

STRESS ENGINEERING SERVICES INC. an employee-owned company

Number Of Accidents Per Year / Number of Accidents With Greater than 99 Fatalities

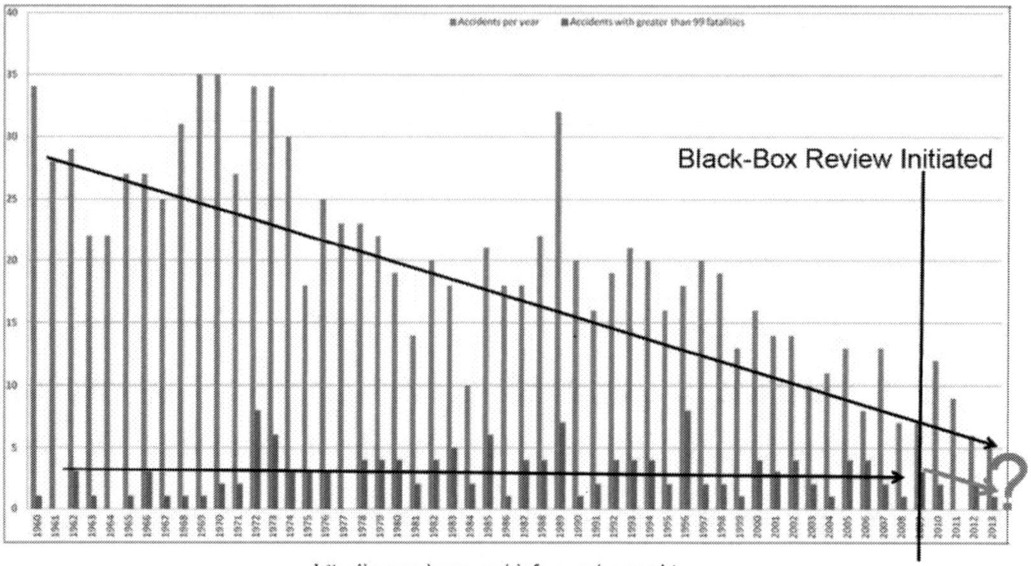

Spills and Release Performance

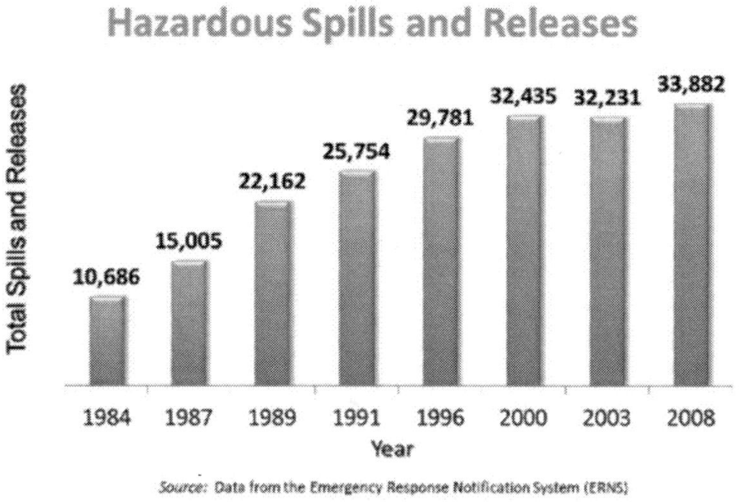

Growth in US Production

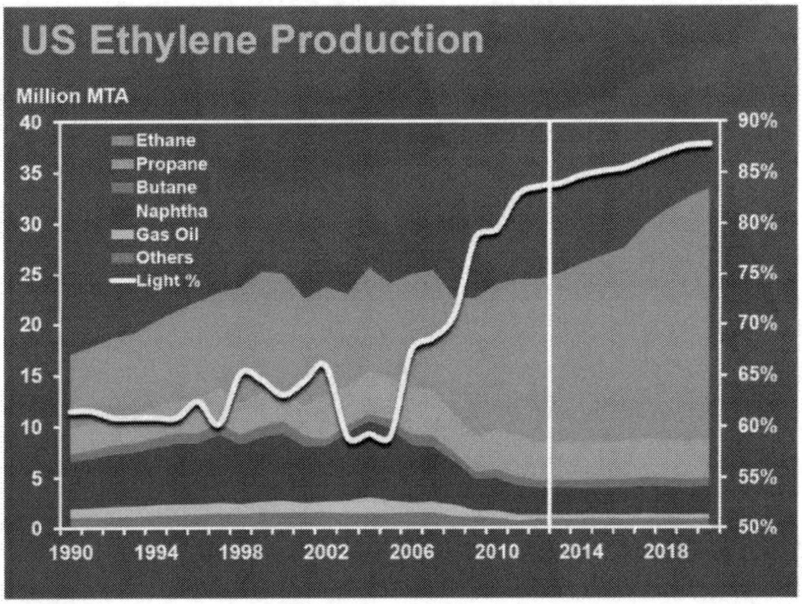

STRESS ENGINEERING SERVICES INC. an employee-owned company

State of the Industry – Global Renaissance, Challenges and Opportunities, by R. Heinen, IHS
Presented at Saudi Aramco Global Reliability Forum 2013

Our Challenge

- In the face of increasing Chemical, Refinery and Energy demand and growth;
- How do we lower the total incident rate?
- Perhaps we take a page from the book of the airline industry?
- **Pull the "Black-Box" data BEFORE the Crash**

STRESS ENGINEERING SERVICES INC. an employee-owned company

Process Hazards "Black-Box"

- Sites are running a detailed Monte Carlo simulation of their processes everyday…it's called the plant.
- DCS data historian is recording the results of this real world simulation; often many times a second
- Only the most obvious and clear hazards are alarmed
  - Single initiating event hazard
- Multiple enabling events or initiating events are recorded **if** they can be identified
- Often "at-risk" events go undetected and pass without incident due to an interruption in the chain of events
  - "Window" of vulnerability where enabling events or actions occur
  - BUT final "at-risk" action did NOT occur – Just Luck?
- Key is to routinely query historical process data for deviation of key parameters

STRESS ENGINEERING SERVICES INC. an employee-owned company

11

PSM Questions?

- Process Hazards Analysis (New/Revalidation)
  - For each "Guide-Word", have any scenarios occurred?
  - How close to the "Safety Critical" Limit have conditions reached?
  - What were the chain of events that resulted in challenging the "Guide-Word"?
- LOPA Validation
  - How often does a scenario occur? (Initiating event frequency)
  - How often are the protective layers challenged? (Mitigation effectiveness)
  - How far up the chain of events has a scenario progressed? (Layers of Protection)
  - Have the protective layers functioned as designed? (IPL Validation)

STRESS ENGINEERING SERVICES INC. an employee-owned company

12

PSM Questions? (cont'd)

- Incident Investigation
  - What were the chain of events that resulted in the excursion?
  - Did conditions reach or nearly reach the top-event?
  - During the excursion; did other sections of the process have parameters approach process safety limits?
- Process Monitoring
  - Have process conditions exceeded "normal" conditions?
  - Were operations at target?
  - Exceedance of mechanical integrity operating limit?

STRESS ENGINEERING SERVICES INC. an employee-owned company 13

PSM Questions? (cont'd)

- New Fitness-For-Service (FFS) Limit
  - Pressure Limit – permanent de-rate or temporary limitation until repair?
  - Temperature Limits?
    - High Temp. reduction due to metallurgical or thermally driven fracture?
    - Minimum Allowable Temperature (MAT) – due to API-579 / ASME FFS-1 Part 3 Analysis?
      - Potential auto-refrigeration
      - Potential metallurgical embrittlement mechanism (e.g. Hydrogen)
  - Temperature Rate of Change Limit?
    - Thermal gradient stresses or deformation

STRESS ENGINEERING SERVICES INC. an employee-owned company 14

"Black-Box" Data Review

- Examples of when to execute "Black-Box" query review
  - New FFS limit
    - How often did parameter get to the FFS limit?
    - How often did the parameter get to within 90% of the limit?
      - Indicates mitigated initiating event frequency
      - Indicates how far up the chain of events did events progress
      - How difficult it may be to limit the process to the FFS limit
  - Preparation work prior to PHA revalidation
    - Effectiveness of mitigation (LOPA validation)
    - Verification of initiating event frequency
    - Indication of the chain of events to cause parameter deviation

STRESS ENGINEERING SERVICES INC. an employee-owned company

15

"Black-Box" Data Review (cont'd)

- Examples of when to execute "Black-Box" query review (cont.)
  - After a process upset or incident
    - Helpful in determining state of the process, chain of events and potentially the root cause of the incident
    - MUST look outside the area of incident focus (root cause) to determine if other potential hazards were present or "near-hits" during the event
  - After major step change to the process
    - Shutdown
    - Startup

STRESS ENGINEERING SERVICES INC. an employee-owned company

16

“Black-Box” Data Dump

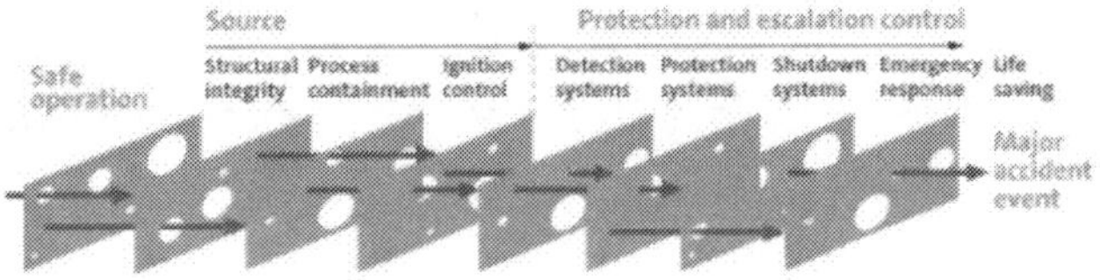

- Execute query (via spreadsheet) for:
  - Parameter conditions outside (above or below) or nearly outside (90% of) system limits
    - Pressure, temperature, flow, composition limits, etc.
    - Identifies “events” where data indicates “at-risk” deviation
      - Known initiating events or
      - Key enabling events
- Utilize root-cause analysis methods in determining the cause for the excursion including the chain of events to the parameter deviation
- How close to the top-event did excursion reach?

STRESS ENGINEERING SERVICES INC. an employee-owned company

http://www.bg-group.com/sustainability11/ManagementSystems/Pages/HSSE.aspx

17

“Black-Box” Review

- Data Historian Search
  - Locate historical results where:
    - Parameter limit was approached (e.g. minimum allowable temperature – MAT)
    - Companion parameter limit was approached (e.g. low-pressure excursion of LPG will result in low-temperature)
    - Unit or major equipment trips or upsets
  - Used to determine sequence of events:
    - Enabler and initiator to excursion (e.g. auto-refrigeration)
    - Excursion recovery
      - Control actions
      - Operations actions
  - How close to “top-event” did chain-of-events reach?
- Technology to automate data query is available

STRESS ENGINEERING SERVICES INC. an employee-owned company

18

"Black-Box" Review (cont'd)

- Data Historian Search.
  - Generate a query of the historian using unit instrument tags
  - Evaluate process variables for:
    - Approach to limits
    - Exceedance of limits
    - Sequence of events
    - Root-cause for excursion

  NOTE: Requires discipline to sift through data to locate and understand scenario – root cause analysis methods.
- Utilized as part of Layers of Protection Analysis (LOPA) to establish unmitigated event frequency.

STRESS ENGINEERING SERVICES INC. an employee-owned company

19

Data Compression

- Know how data is actually stored on your historian(s).
- Data compression (typically a box-car algorithm) is common on "plant-wide" historians located on IT networks.
  - Most of the time this approach is "good enough" for finding abnormal events BUT,
  - Where sequence of events and associated timing of events is important, higher resolution (or at least more frequent recording of uncompressed data points) may be available on a "unit" historian located on the process control network.
- As a minimum, if sequence of rapidly occurring changes is important, "actual" data points preferably from uncompressed data should be reviewed to ensure that trends created from compressed data are not masking the sequence.

STRESS ENGINEERING SERVICES INC. an employee-owned company

20

Automated Data Historian Search

1. Launch the Envoy-Stress Form

After installing the add-in launch the form by going to Envoy-Stress → Find Auto Refrigeration

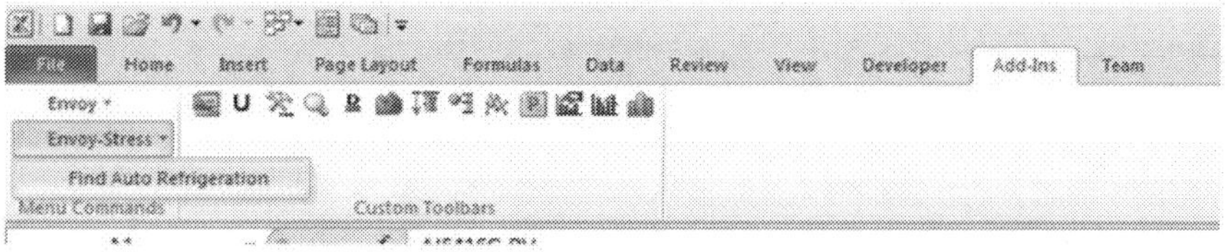

http://www.envoydevelopment.com/

STRESS ENGINEERING SERVICES INC. an employee-owned company

21

Automated Data Historian Search

Fill in the Form

Follow the directions on the form to fill it in. Ref edits are used for the tag list, limit dates, start date, and end date. It is helpful to have the information for the ref edit already on a sheet. The user may specify via the start and end dates if the application will look forwards or backwards in time. The limit dates should be configured by low limit then high limit for each tag.

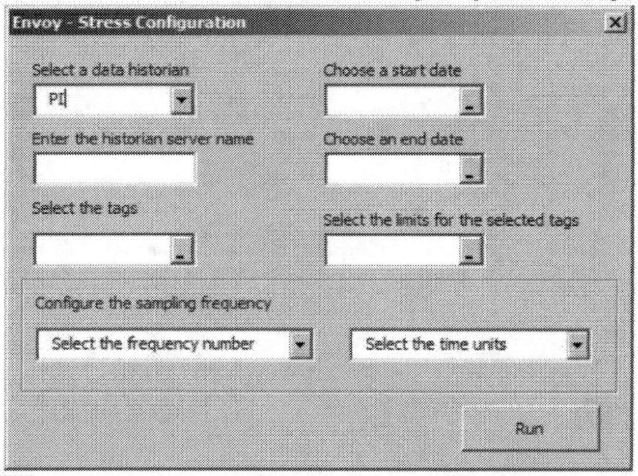

STRESS ENGINEERING SERVICES INC. an employee-owned company http://www.envoydevelopment.com/

22

Automated Data Historian Search

Form configuration with example:

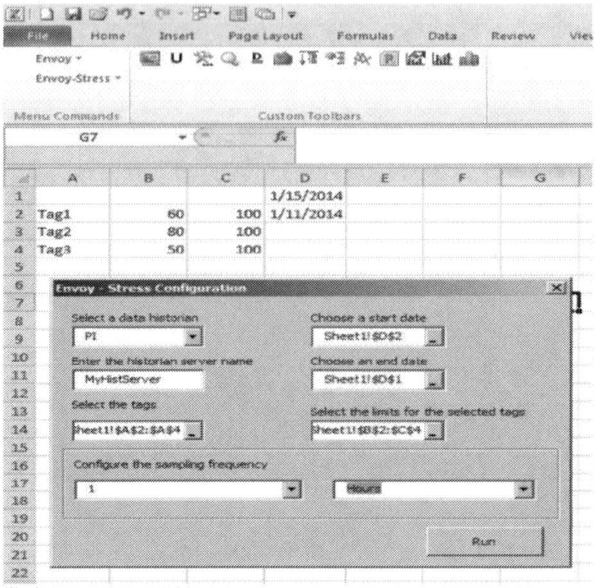

Automated Data Historian Search

Find times exceeding the given limits

Click the Run button to find times when auto refrigeration occurred as determined by the limits. The findings will be output on a new sheet.

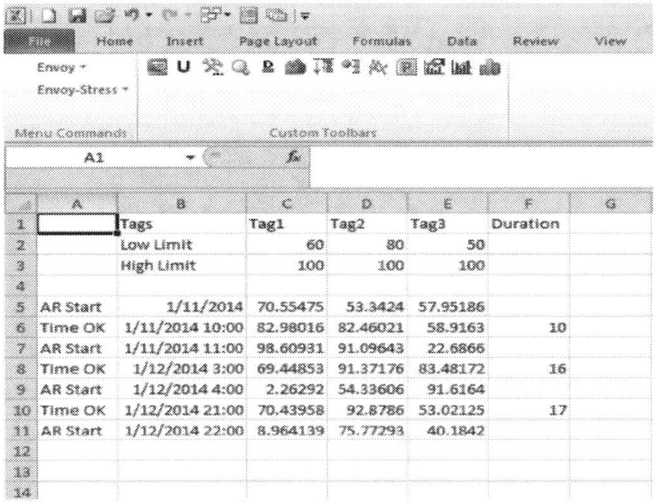

	A	B	C	D	E	F
1		Tags	Tag1	Tag2	Tag3	Duration
2		Low Limit	60	80	50	
3		High Limit	100	100	100	
4						
5	AR Start	1/11/2014	70.55475	53.3424	57.95186	
6	Time OK	1/11/2014 10:00	82.98016	82.46021	58.9163	10
7	AR Start	1/11/2014 11:00	98.60931	91.09643	22.6866	
8	Time OK	1/12/2014 3:00	69.44853	91.37176	83.48172	16
9	AR Start	1/12/2014 4:00	2.26292	54.33606	91.6164	
10	Time OK	1/12/2014 21:00	70.43958	92.8786	53.02125	17
11	AR Start	1/12/2014 22:00	8.964139	75.77293	40.1842	

Process Simulation

- **For identified scenarios often process simulation is required:**
  - To accurately indicate the process temperature/pressure.
  - To provide heat transfer rate and heat capacity – API-579/ASME FFS-1 Level 3 analysis.
  - Cases are often non-normal so compositions are often markedly different than standard operating conditions.
- **Can provide critical insight into understanding "black-box" identified excursions.**

Process Simulation

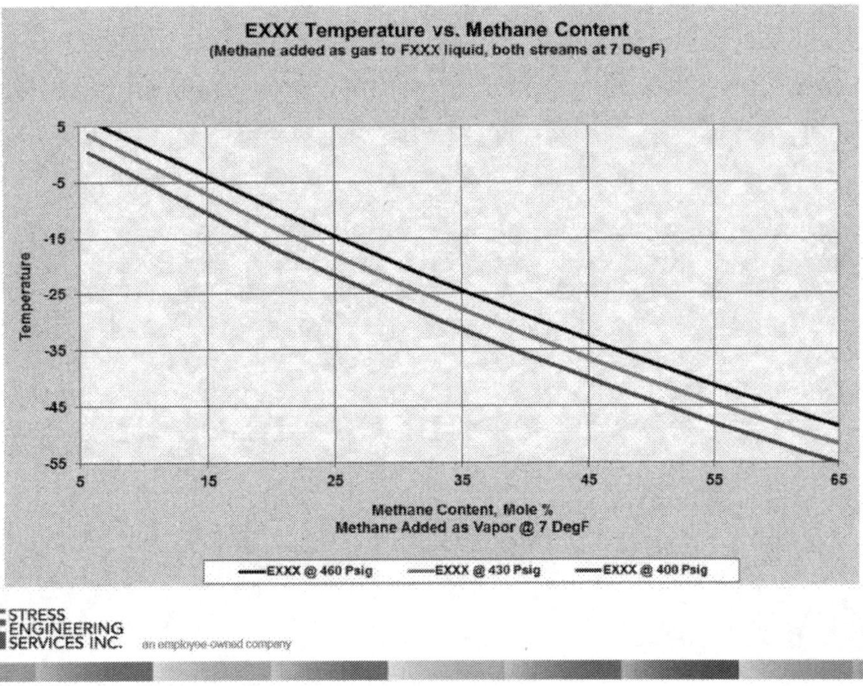

Example:
Tower Overfill – OH Hydraulic Condition

- Question – how often does company overfill a tower with the risk of uncontrolled release (BP Texas City) event?
  - It could never happen here!
  - Takes a long time to fill the tower....we would detect it and correct it way before it becoming a hazard...
- Executed a "black-box" review of tower levels for all unit startups over a 2 year period.
  - Flagged all towers with sump levels at 100%.
  - Executed a tower inventory integral (totalizing) analysis of all flagged towers .

STRESS ENGINEERING SERVICES INC. an employee-owned company

27

Example
Tower Overfill – OH Hydraulic Condition

- Out of the 100s of tower startups how many were identified as going full and hydraulic on the OH system?

STRESS ENGINEERING SERVICES INC. an employee-owned company

28

Example
Tower Overfill – OH Hydraulic Condition

- Out of the 100s of tower startups how many were identified as going full and hydraulic on the OH system?

6

- These "at-risk" conditions were not even recognized during the normal startup monitoring.
- What is the unmitigated initiating event frequency for going hydraulic on the OH of a tower?
  - ~10E0 to 10E-1.
- Is additional mitigation required?

STRESS ENGINEERING SERVICES INC. an employee-owned company

29

Auto-Refrigeration / Brittle Fracture

- API-579 / ASME FFS-1 Minimum Allowable Temp. (MAT) analysis.
- AR/BF Focus PHA Identified risk of 2-step potential.
- Frequency of enabling loss of system pressure with auto-refrigeration was significantly higher than anticipated.
  - Required mitigation protection.

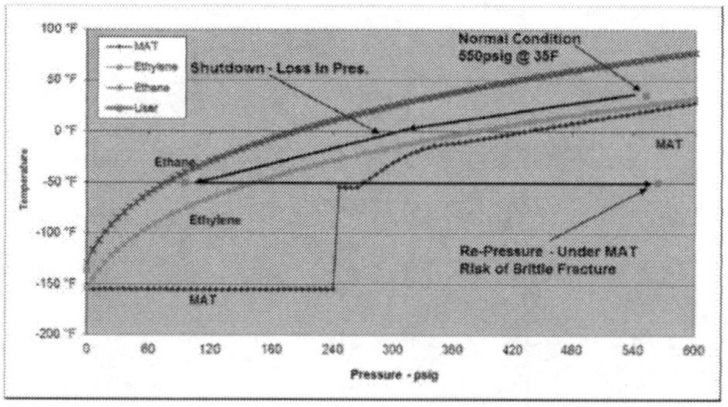

30

Chilling Train Vessels

Cross Heat Exchange with Methane/Hydrogen Streams

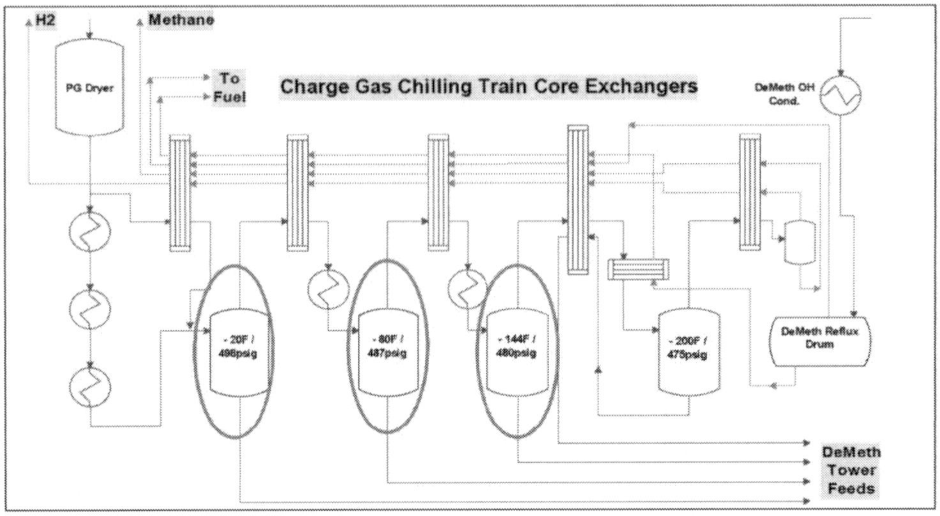

STRESS ENGINEERING SERVICES INC. an employee-owned company

Chilling Train Vessels

Continued Chilling Streams Through Core Exchangers with Condensing and Liquid Accumulation in Chilling Train Drums

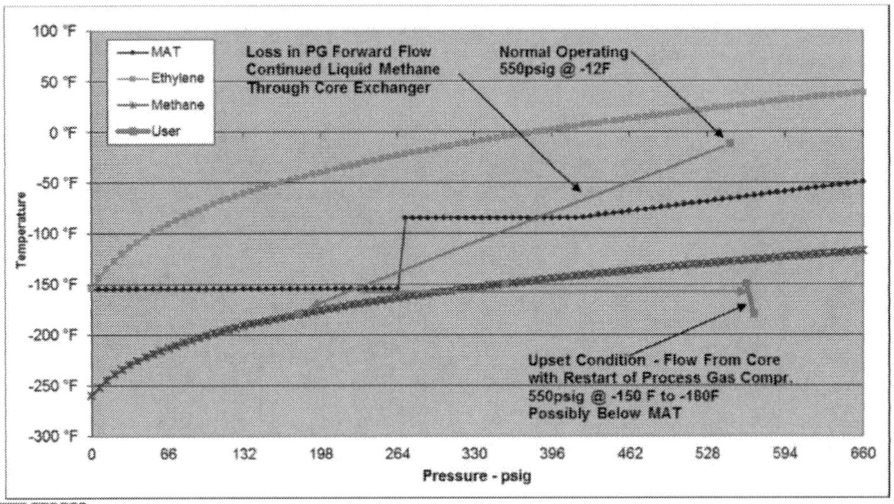

STRESS ENGINEERING SERVICES INC. an employee-owned company

Example – Brittle Fracture MAT

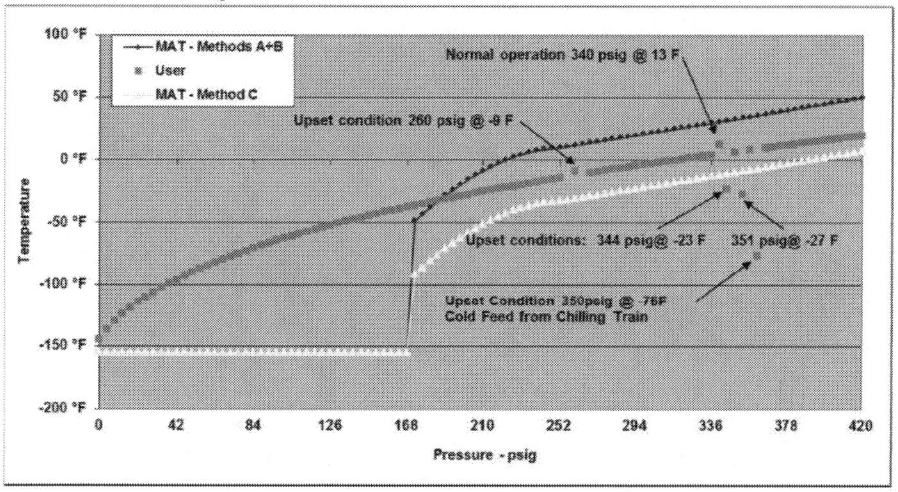

- API-579 / ASME FFS-1 Minimum Allowable Temp. (MAT) analysis.
- "Black-Box" Identified conditions UNDER minimum allowable.
- Requires mitigation via protective interlocks or vessel replacement.

STRESS ENGINEERING SERVICES INC. an employee-owned company

33

Ethylene Plant (Cold Section)

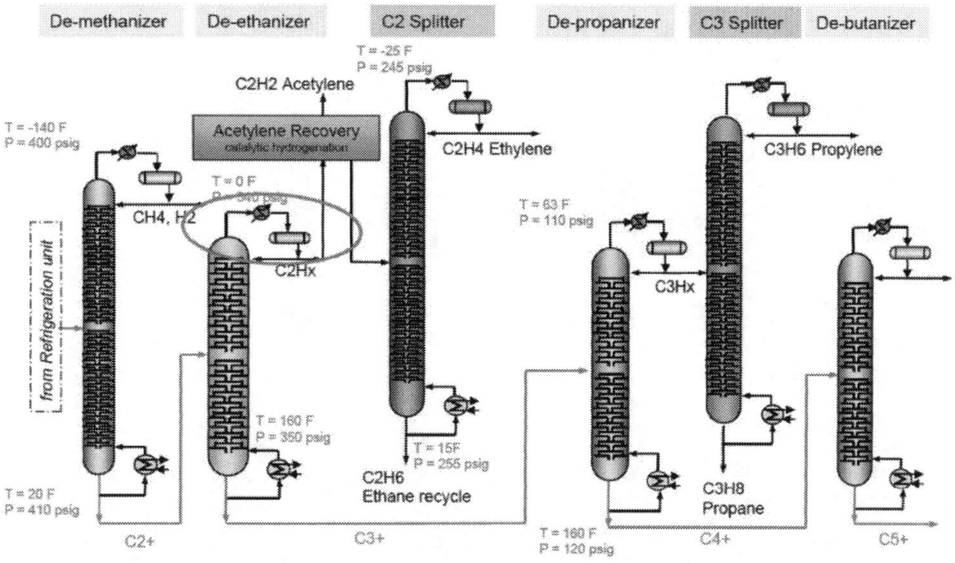

STRESS ENGINEERING SERVICES INC. an employee-owned company

Slide Courtesy of Sulzer Chemtech

34

Distillation Towers and Peripherals

De-Ethanizer Reflux Drum MAT

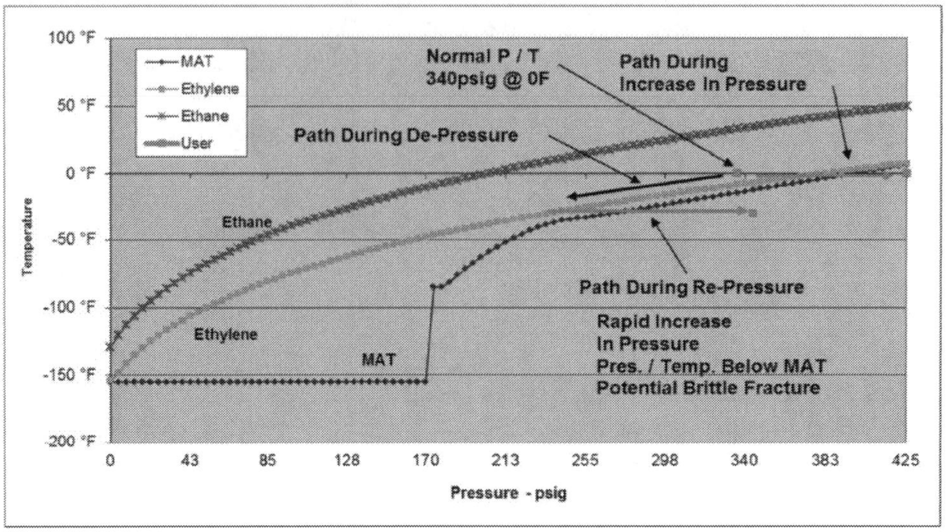

STRESS ENGINEERING SERVICES INC. an employee-owned company

35

Ethylene Plant (Cold Section)

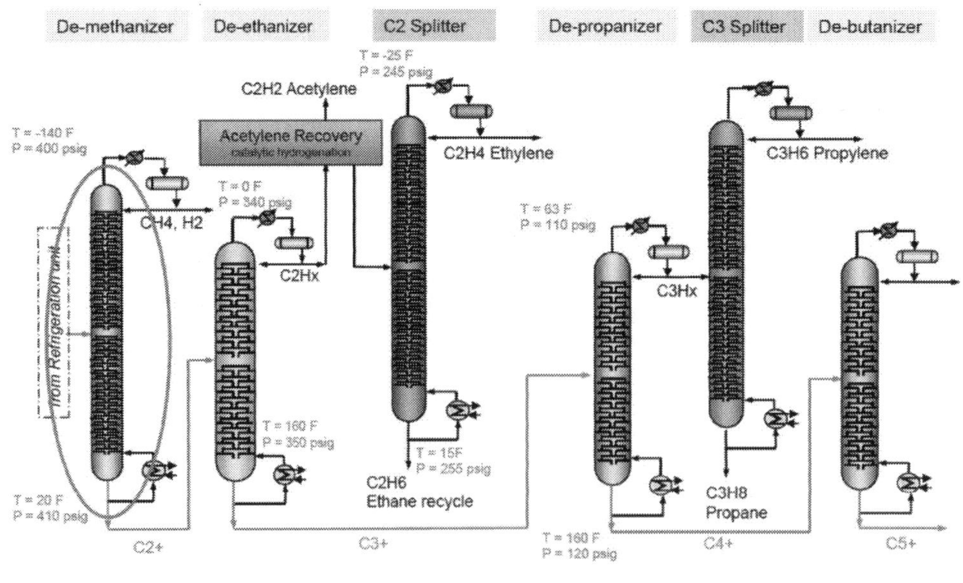

STRESS ENGINEERING SERVICES INC. an employee-owned company

Slide Courtesy of Sulzer Chemtech

36

Bimetallic De-Methanizer

Carbon Steel Bottom Section

Loss of Reboil with Slump of Tower

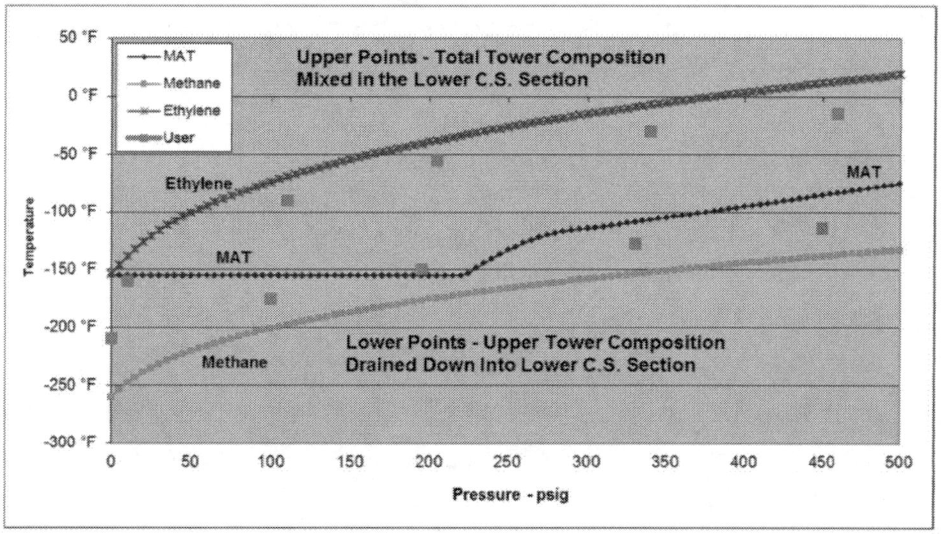

STRESS ENGINEERING SERVICES INC. an employee-owned company

37

Example – Upset Investigation

- Refrigeration system interchanger that provides unit feed vaporization.
- Major trip of the unit with associated systems upsets and major flaring.
  - Root cause investigation determined cause of unit trip and implemented corrective actions.
- Subsequent PHA "Black-Box" review two years later identified feed system over-pressure excursion.
  - Loss of feed vaporization with continued feeding of ethane liquid into the hot feed system.
  - This over-pressure scenario was not identified as part of the unit PHA and this governing relief sizing basis was not provided.

STRESS ENGINEERING SERVICES INC. an employee-owned company

38

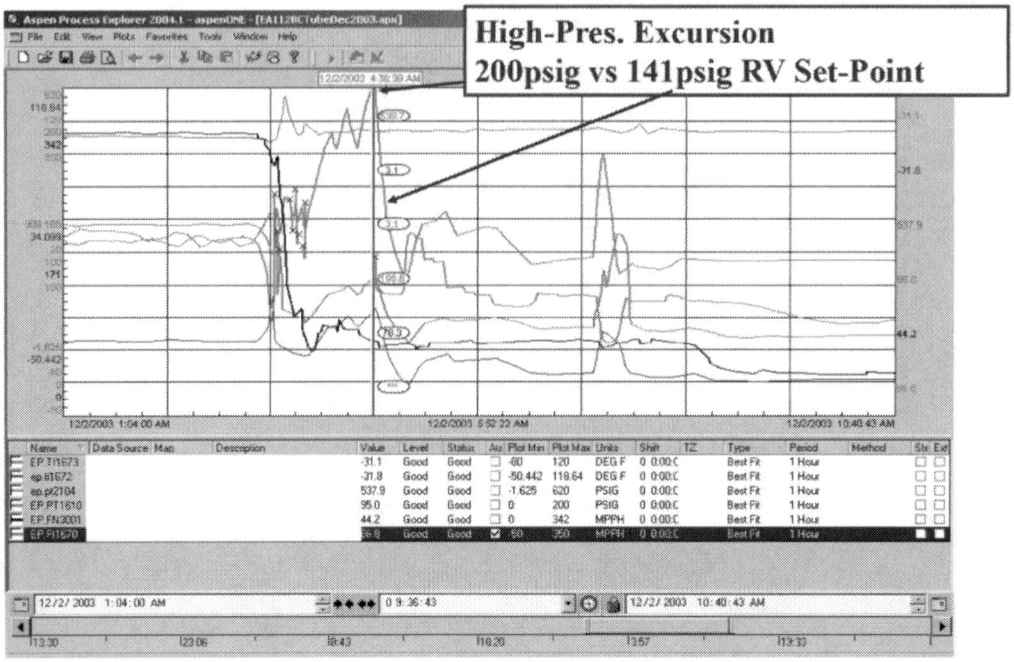

"Black-Box" Root Cause Analysis

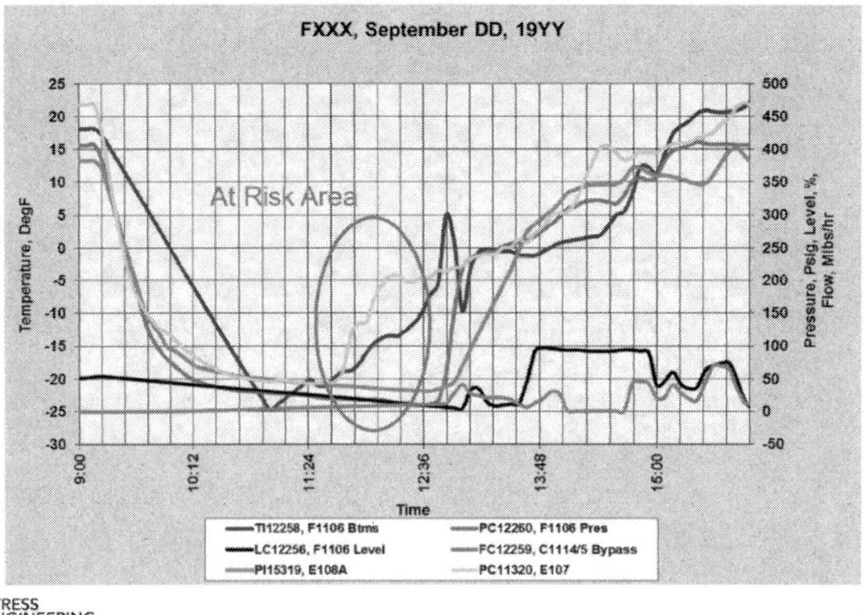

Non-Normal Operation

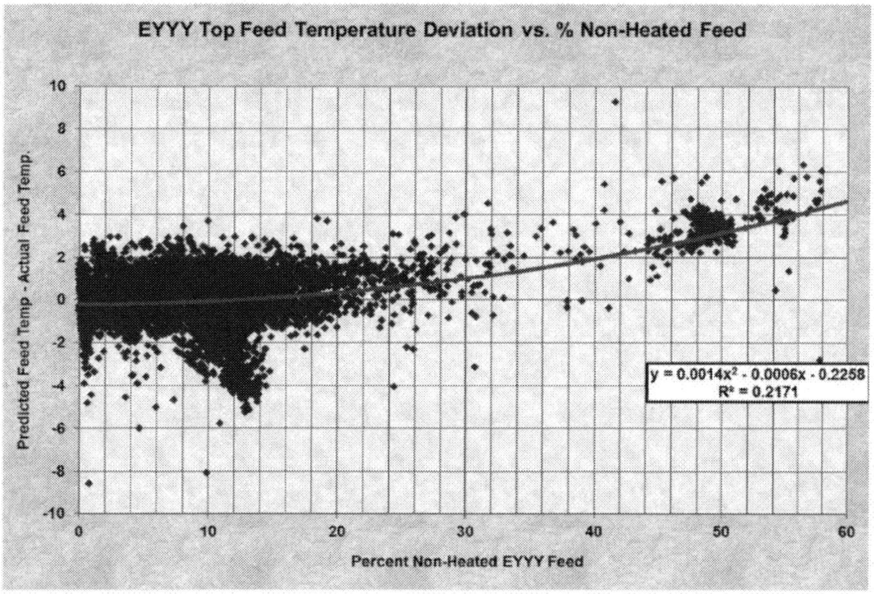

STRESS ENGINEERING SERVICES INC. an employee-owned company

41

Conclusion

- Fitness-For-Service (FFS) and Safety Critical Variable (SCV) limits can often be challenged by the process system.
- Use of "Black-Box" reviews can help identify potential process safety scenarios.
- "Black-Box" reviews can be invaluable to:
  - Determining initiating event frequencies
  - Validation of the LOPA analysis
- "Black-Box" reviews can be the determining information in accessing challenging FFS limits.
  - Can the excursion be mitigated?
  - Is equipment replacement justified?

STRESS ENGINEERING SERVICES INC. an employee-owned company

42

Conclusion (cont'd)

- We as an industry can better utilize our process data historian results to reduce the frequency and consequence of near-hit conditions via "Black-Box" reviews.

- Logic and aviation results would indicate that utilization of "Black-Box" reviews should reduce our industry incident rates while providing better insight into the abnormal response conditions of our processes.

STRESS ENGINEERING SERVICES INC. an employee-owned company 43

STRESS ENGINEERING SERVICES INC. an employee-owned company 44

Thank You!

STRESS ENGINEERING SERVICES INC. an employee-owned company

45

Process modeling requirements for the safe design of blowdown systems

Changes to industry guidelines and how this impacts current practice

Dr James Marriott
Dr Apostolos Giovanoglou
PSE Oil & Gas

Presented By: Michael Szatny
PSE Oil & Gas

2014 AIChE Southwest Process Technology Conference
October 9-10, 2014

Biography

- Michael Szatny is Sr. Business Development Director for PSE Oil & Gas based in Houston.
- A Chemical Engineer with a BS from University of Texas at Austin and MS from University of Houston
- Michael has been involved with Oil & Gas process simulation for over 20 years having held positions at Aspentech and Halliburton prior to joining Process Systems Enterprise in 2012.

 2

Process modeling requirements for the safe design of blowdown systems

Changes to industry guidelines and how this impacts current practice

Dr James Marriott
Dr Apostolos Giovanoglou
PSE Oil & Gas

Presented By: Michael Szatny
PSE Oil & Gas

2014 AIChE Southwest Process Technology Conference
October 9-10, 2014

Overview

- Introduction to process depressurization
- Importance of low temperature analysis and API 521 changes
- Single vessel depressurization
- System depressurization
- Concluding remarks

4

Pressure relief and blowdown systems

- **Pressure relief and blowdown systems**
  - is a collection of controls, valves and pipes by which pressurised gas (& liquid) contained within a process, piping, or pressure vessel, can be safely relieved
  - is designed to prevent the release of flammable or hazardous gas to the atmosphere, by routing it to one or more flare tips for controlled combustion

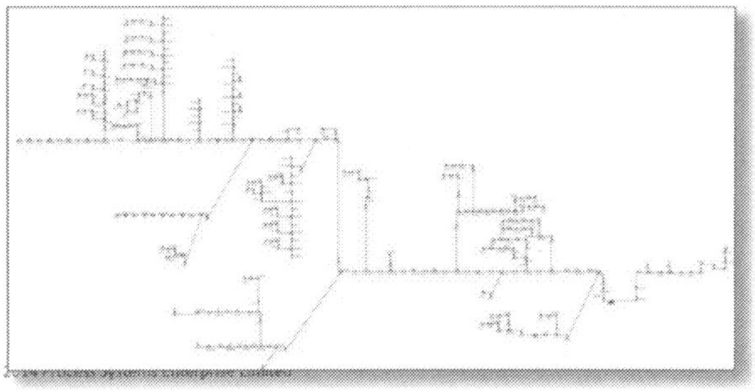

Flare systems installed on chemical plants, refineries, oil & gas processing facilities

What are they used for?

- **Over-pressure protection**
  - pressure exceeds design pressure in system as a result of an unplanned event (e.g. fire, blocked outlet..)
  - Pressure Safety Valves lift automatically or Bursting Disks blow
- **Plant depressurization (Blowdown)**
  - In an emergency situation, production is stopped and the entire plant is depressurized to remove the hydrocarbon inventory
  - The plant is segregated into a number of isolated segments
    - vented (often simultaneously) into the flare system
- **Upset & Maintenance**

6

Blowdown systems

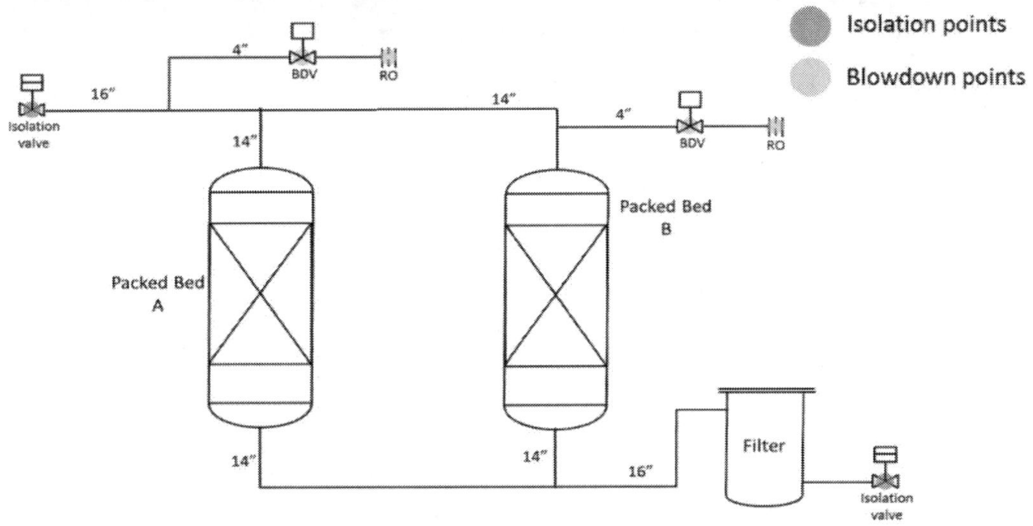

Gas sweetening system

Gas sweetening system

7

Blowdown systems - examples

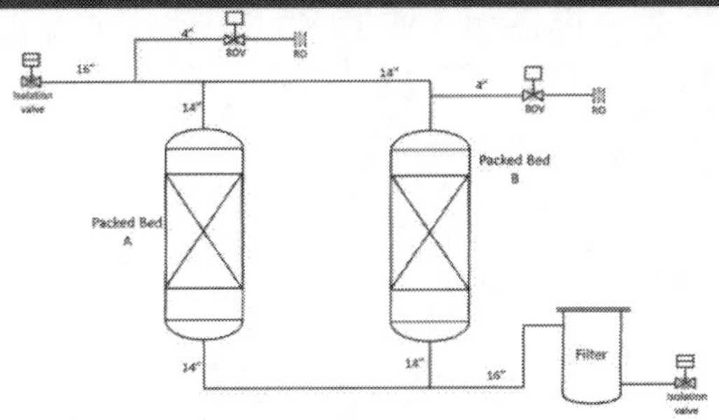

Gas sweetening system

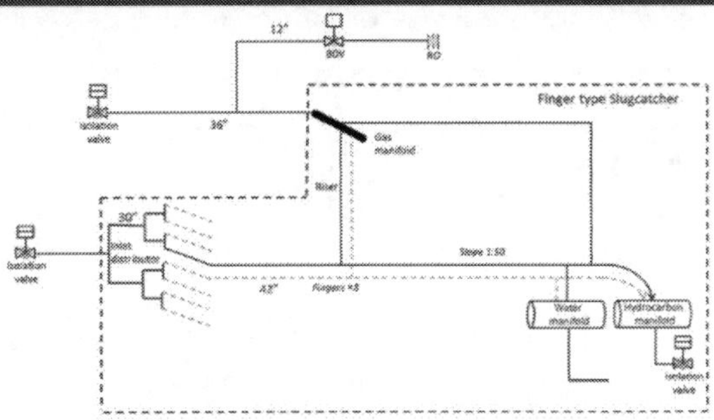

Finger type slug catcher

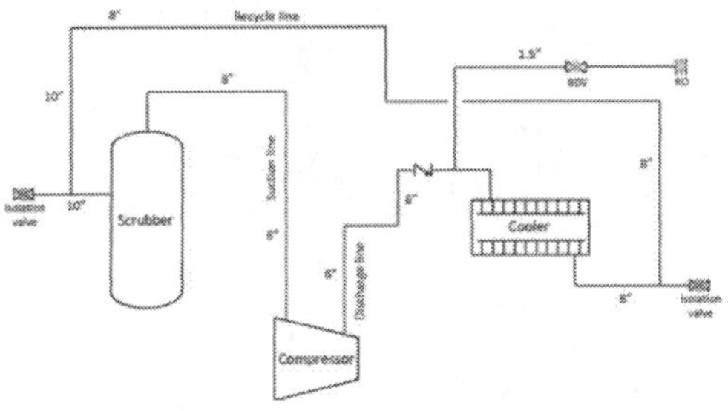

Compressor system

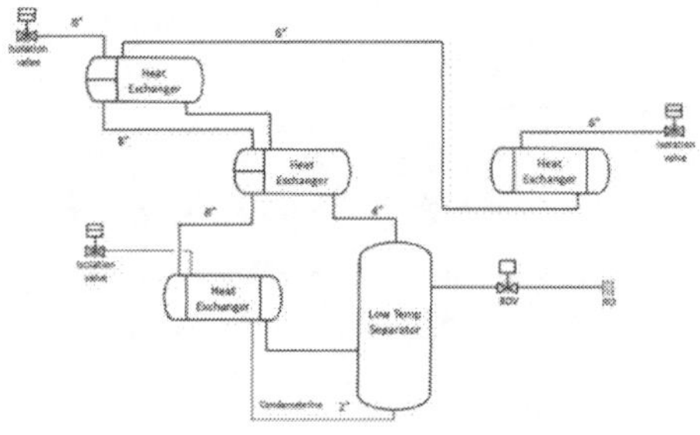

Low temperature separator

What could go wrong?

- **The blowdown operation itself may be hazardous….**

1. A large amount of material must be disposed of simultaneously

Key aspect of today's presentation

2. Fluid may become very cold due to Joule-Thomson expansion

3. New phases may form
   - water-containing solids

In most cases in O&G facilities, the blowdown operation sets the **minimum metal temperature** for the process equipment and so has critical implications on metal selection and facility cost

API changes: Low temperature assessment

…ortance of …rature / brittle … ve (in the process) … recommended, with a particular focus on cases

"A detailed analysis may be required to determine the minimum temperature occurring during depressuring."

"The liquid may accumulate in the low points (e.g. bottom of vessels, drain connections)."

"The vessel/piping wall may be at a higher temperature than the liquid, causing liquid boiling in these low points and low local temperatures."

API changes: Low temperature assessment

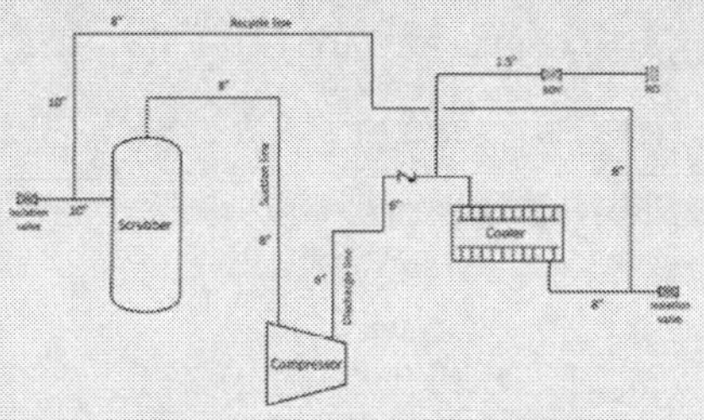

Compressor system

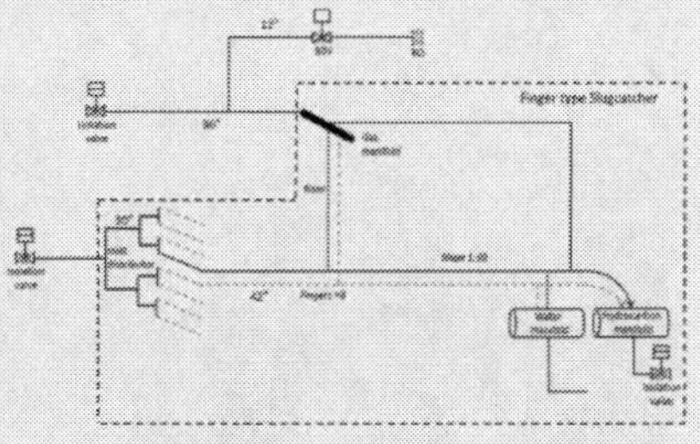

Finger type slug catcher

- **Minimum Metal Temperature (MMT) - is it important ?**
  - Benefit: Project savings of over $200 million by avoiding the need for stainless steel facility construction
  - Risk: Facility start-up delayed by 9 months by late detection of incorrect metallurgy
  - Risk: Many organisations are reassessing existing facilities as part of asset integrity reviews and process changes:
    - numerous examples of MMT violations that need to be rectified in the flare and / or process equipment

Conventional analysis methodology

Isolated blowdown segment

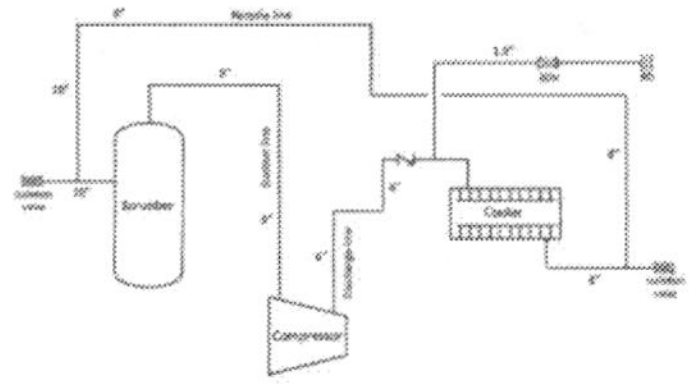

Step 1

Lumping blowdown segments into an "equivalent" vessel

Step 2

Determine MMT with an equilibrium-based simulation with tuneable parameters

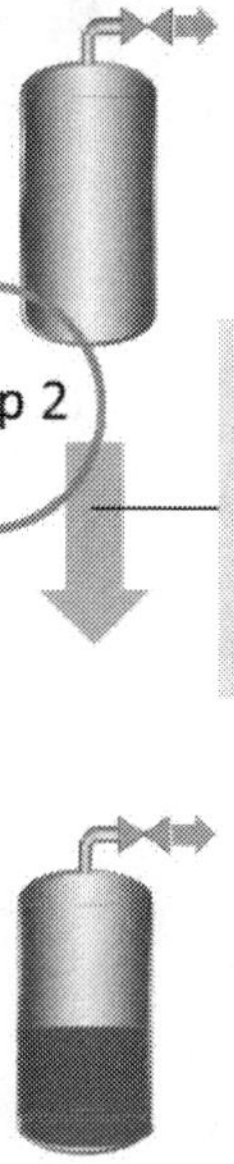

Wide spread-use of non validated models and parameters

- *Isentropic efficiency*
  - Engineering studies use values anywhere from 40% to 100%
- *Non-equilibrium*
  - Engineers recognise this and may set up a non-physical recycle stream on the vapour phase

Conventional approach – experimental validation

PSE

BLOWDOWN OF PRESSURE VESSELS
II. Experimental Validation of
Computer Model and Case Studies

INTRODUCTION

BLOWDOWN EXPERIMENTS

Vessels

- Full-scale vessel blowdown
  - Vertical: 3.2m / 1m vessel
  - Horizontal: 1.5m / 0.27m
- Mixtures:
  methane, ethane & propane
- 18 experiments: gas & liquid phase
  - well instrumented equipment

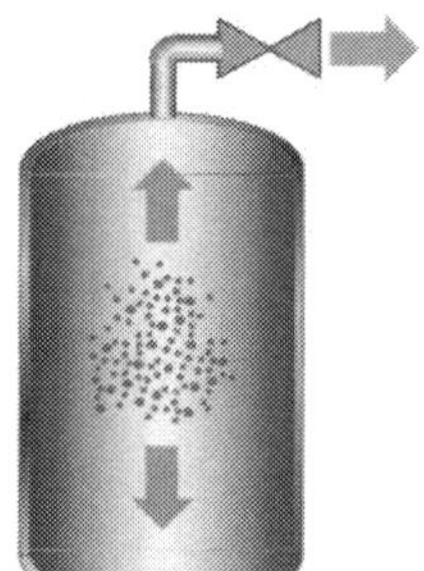

Conventional approach – experimental validation

PSE

Experiment S9: **120 bar, 300.25 K; C1: 85.5%, C2: 4.5%, C3: 10,% C4 trace**

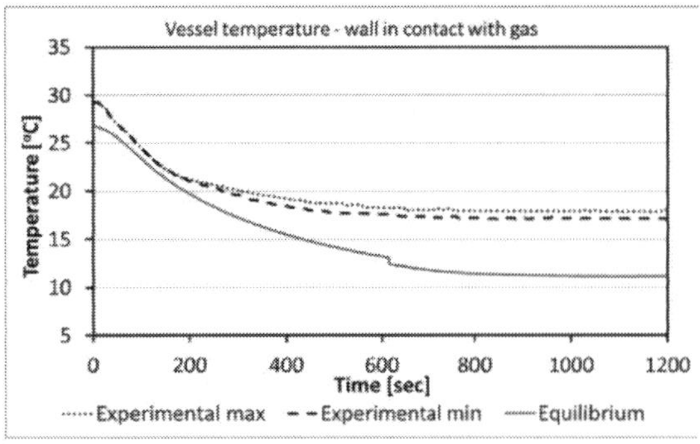

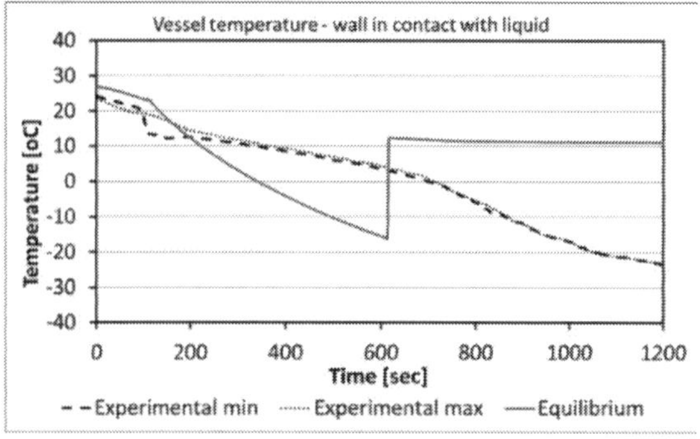

Under-predicts metal temperature in contact with gas
Inaccurate but conservative

Quantitatively and qualitatively incorrect predictions for the wall temperature in contact with the liquid – that are not conservative

Poor, potentially unsafe, predictions if/when liquid condensation occurs

M.A Haque et al., Blowdown of pressure vessels, *Trans IChemE Part B Process Safety Environmental Protection, 70 (BI) 1 and 10, 1992*

What actually happens?

- Initially gas-filled vessel
- Sudden decrease in pressure
  - rapid change in thermodynamic state of gas
  - potential nucleation of liquid droplets within the gas bulk
- Some liquid leaves as entrained droplets
  - evaporates into the bulk gas stream,
  - lowers temperature of exiting stream even further

  ➔ risk of brittle fracture of the flare system pipework

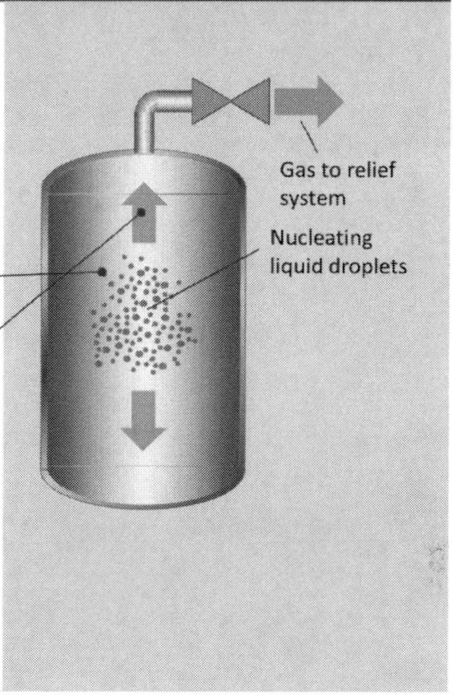

What actually happens? (continued)

- Liquid remaining in the vessel may pool on vessel floor
  - instantly starts boiling
  - through evaporation, pool cools significantly
  - potentially large temperature difference between metal in contact with liquid pool and surroundings

  ➔ threat of brittle fracture and rupture from the base of the vessel.
- Because of rapid change of conditions, <u>non-equilibrium</u> between
  - the three phases co-existing in vessel (gas, liquid droplets and bulk liquid)
  - these phases and the vessel wall

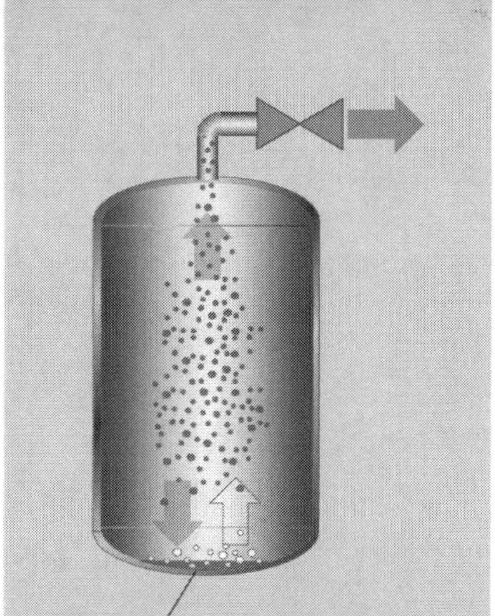

Rate-based models of vessel depressurization

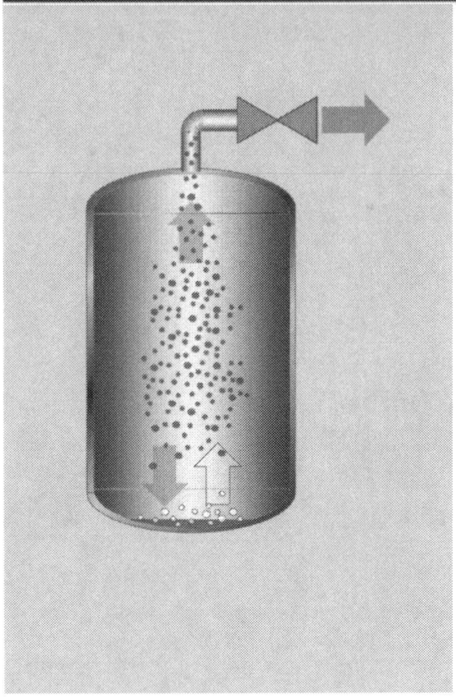

- Separate mass and energy balances for each of the phases present
- Rigorous handling of the non-equilibrium thermodynamics prevalent during blowdown, including liquid pooling
- Heat transfer between regions of wall in contact with different phases
- 3-dimensional metal wall temperature modeling

17

Rate-based approach – experimental validation

Experiment S9: 120 bar, 300.25 K; C1: 85.5%, C2: 4.5%, C3: 10,% C4 trace

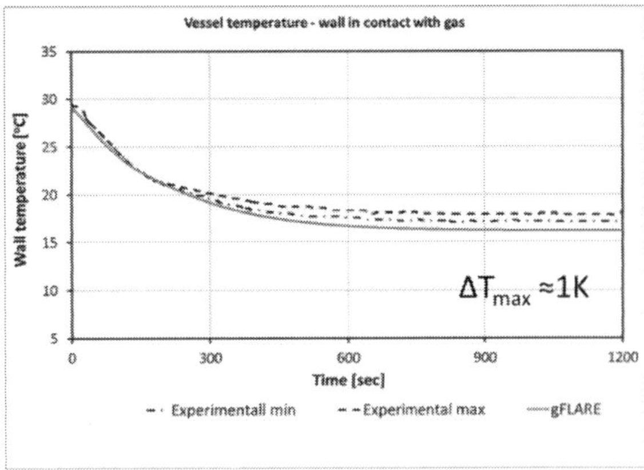

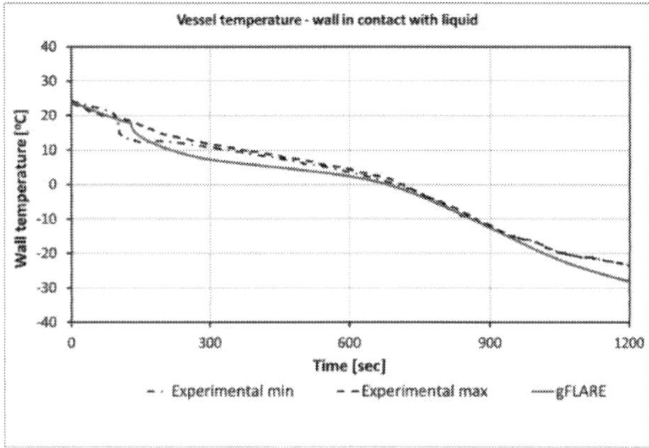

M.A Haque et al., Blowdown of pressure vessels, *Trans IChemE Part B Process Safety Environmental Protection*, **70** (BI) 1 and 10, 1992

Rate-based approach – experimental validation

~~Experiment S9: 120 bar, 300.25 K; C1: 85.5%, C2: 4.5%, C3: 10,% C4 trace~~

Experiment S12: 118 bar, 297 K, C1: 66.5%, C2: 3.5%, C3: 30%, C4 trace

Same model,
no adjustable parameters

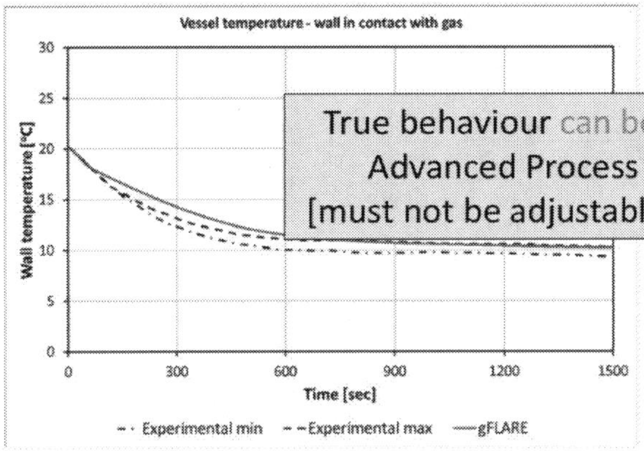

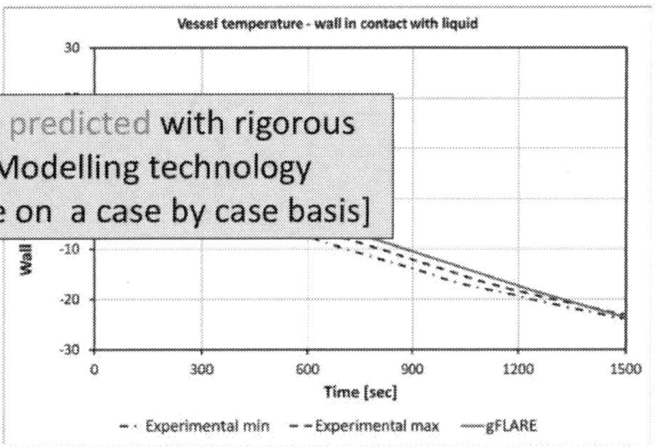

True behaviour can be predicted with rigorous Advanced Process Modelling technology [must not be adjustable on a case by case basis]

M.A Haque et al., Blowdown of pressure vessels, *Trans IChemE Part B Process Safety Environmental Protection*, **70** (BI) 1 and 10, 1992

Conventional analysis methodology

Isolated blowdown segment

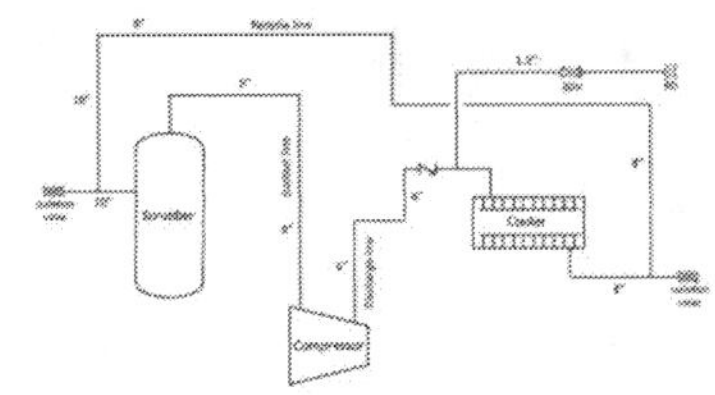

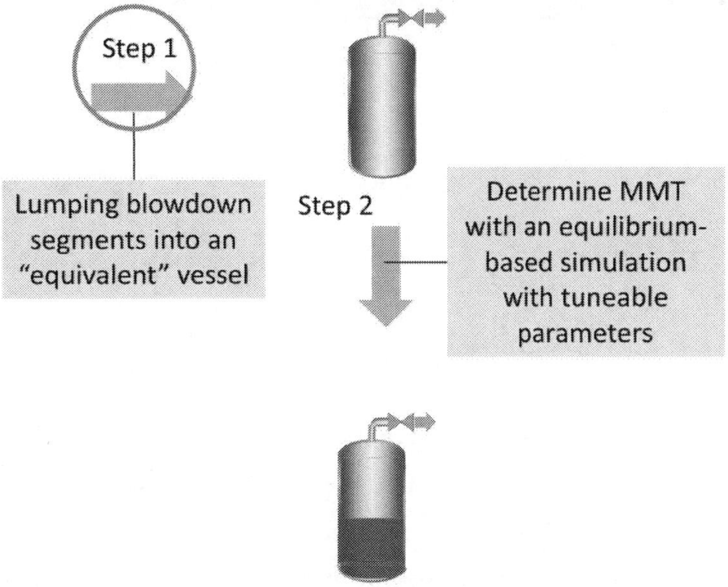

Compressor system

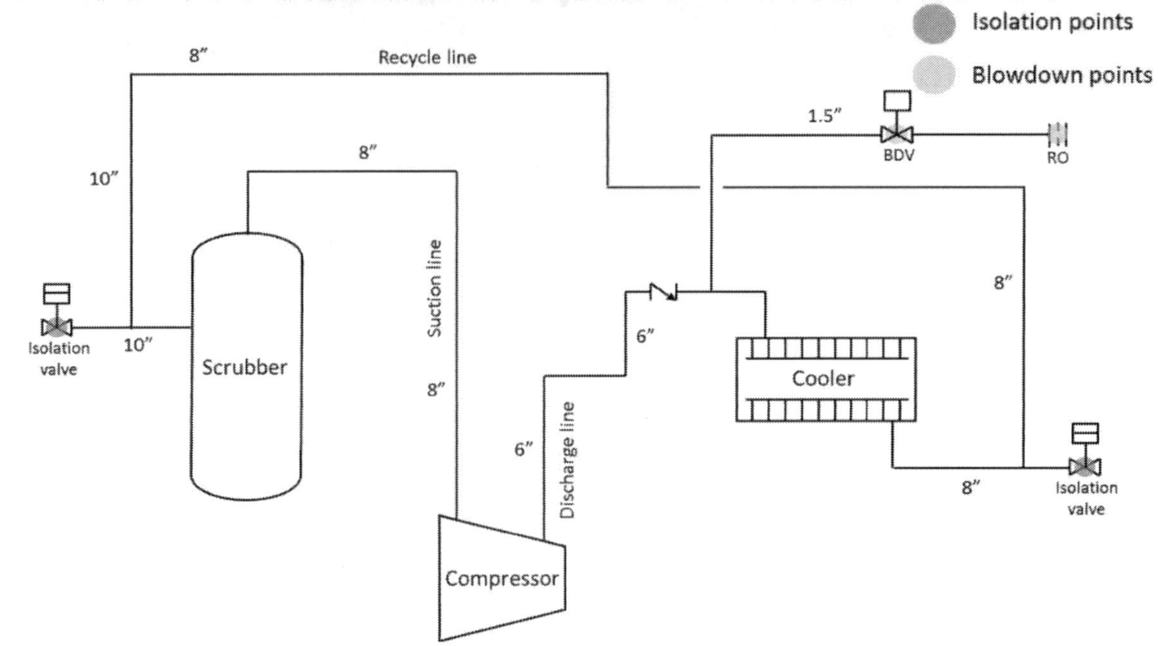

Conventional approach - system aggregation

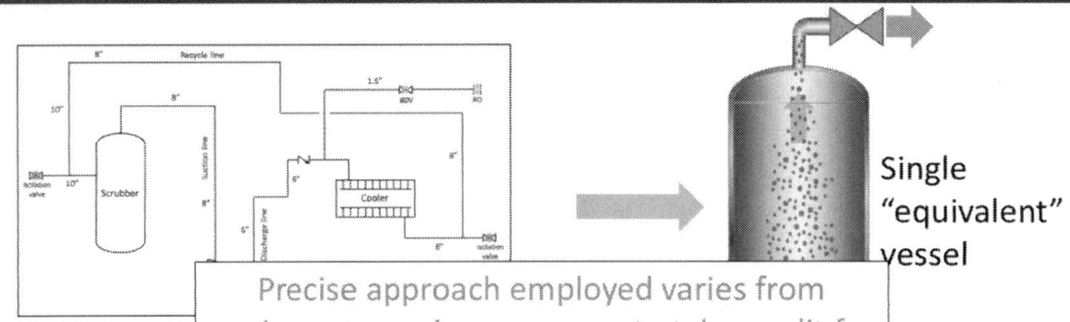

Precise approach employed varies from engineer to engineer, no way to take credit for metal mass of pipe / vessels and heat transfer with the fluid [as per API 521]

- Each system
- ...with adjust
  - volume matches total sub-system volume — unique choice
  - an appropriate aspect ratio chosen to provide... — subjective
    - either a suitable heat transfer area
    - or a representative diameter for heat transfer calculations
  - wall thickness to approximate the metal mass — subjective

Compressor System – analysis

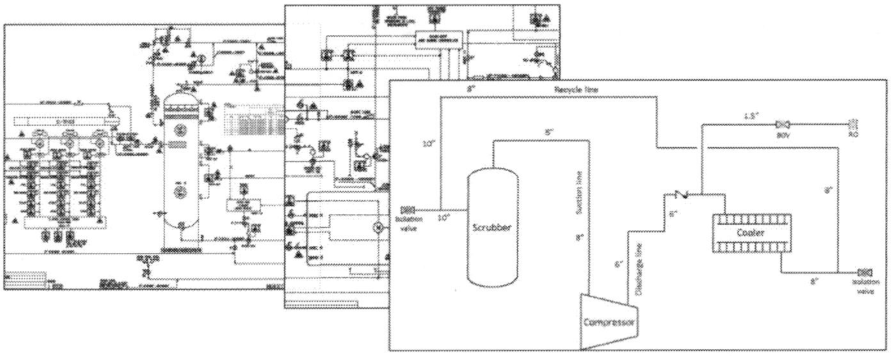

The engineer should determine

- pipe and equipment dimensions from data sheets and piping isometrics
- if and where condensate liquid can form and accumulate

Compressor System – model

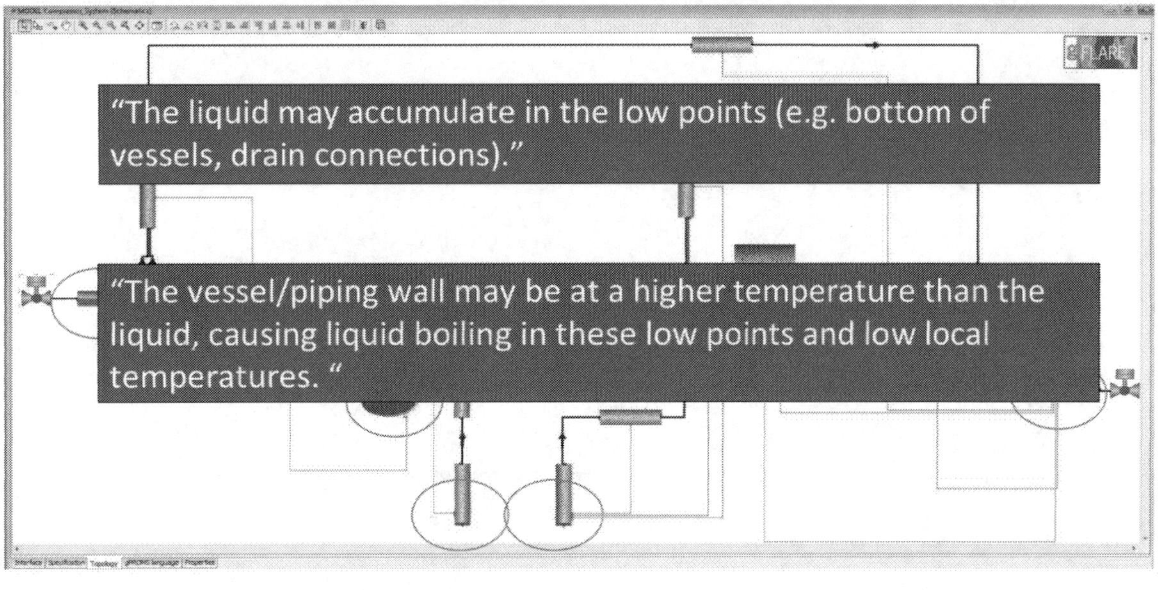

"The liquid may accumulate in the low points (e.g. bottom of vessels, drain connections)."

"The vessel/piping wall may be at a higher temperature than the liquid, causing liquid boiling in these low points and low local temperatures. "

24

Key results 1.5" blowdown line

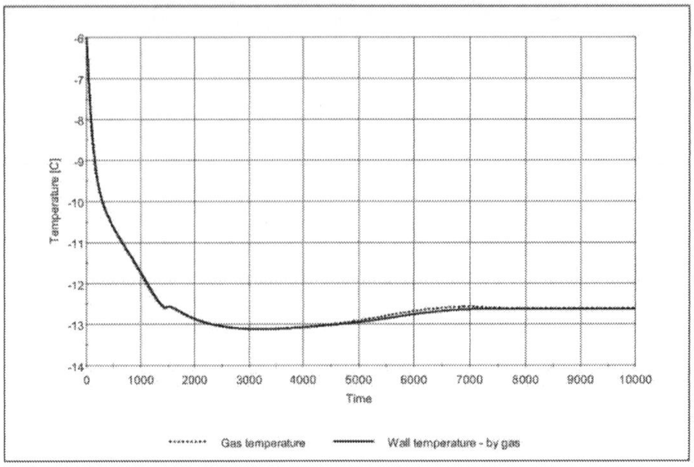

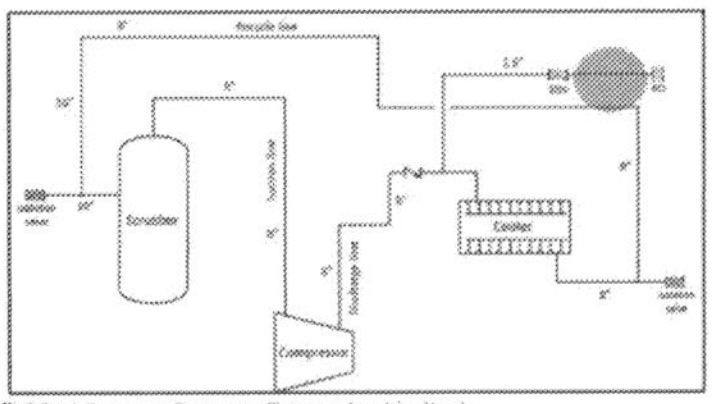

Minimum temperature = -13°C

Key results 8" line

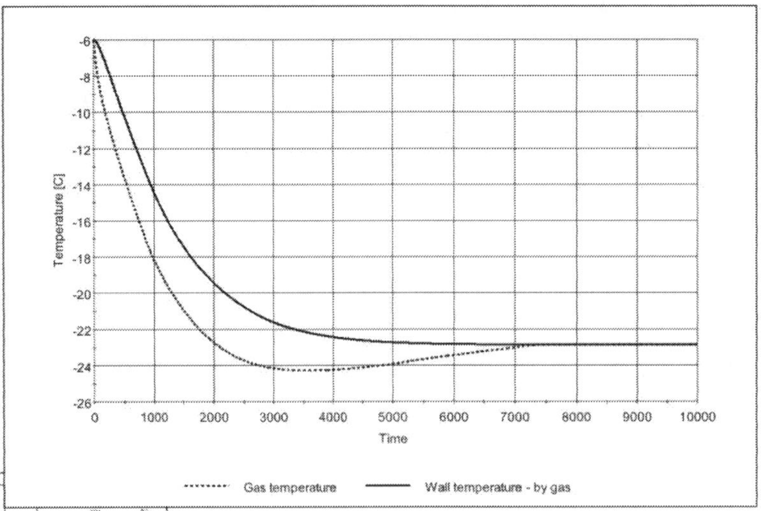

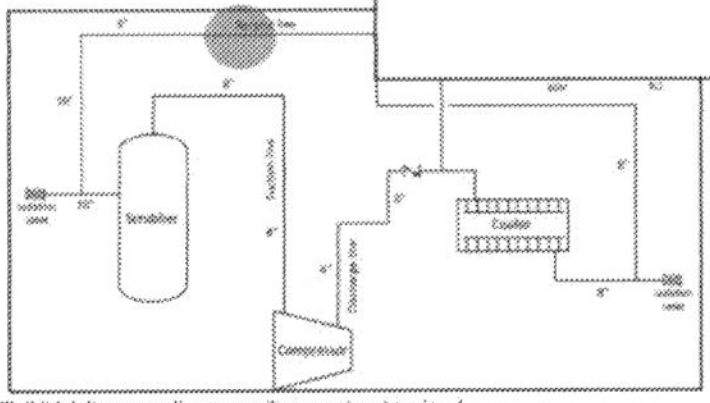

Minimum metal temperature = -23°C

Scrubber

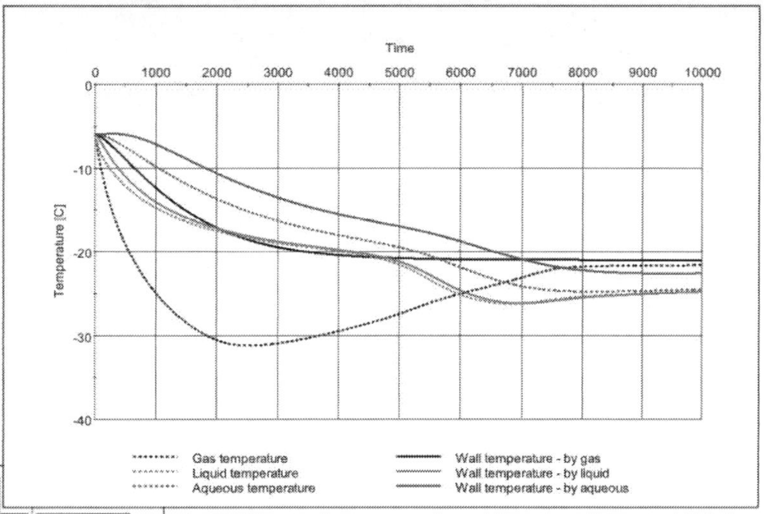

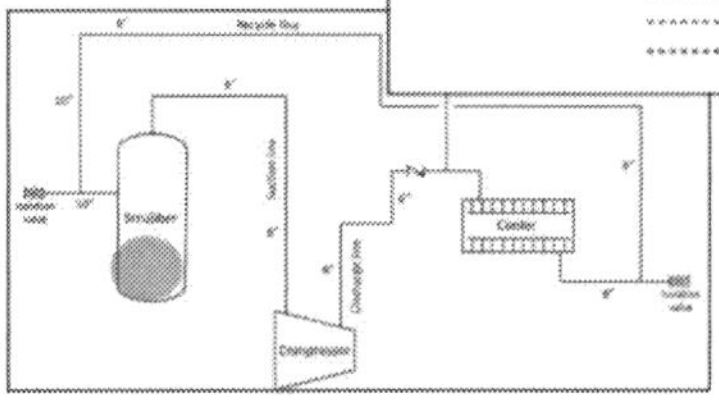

Minimum metal temperature = -26°C

Key results at a low point

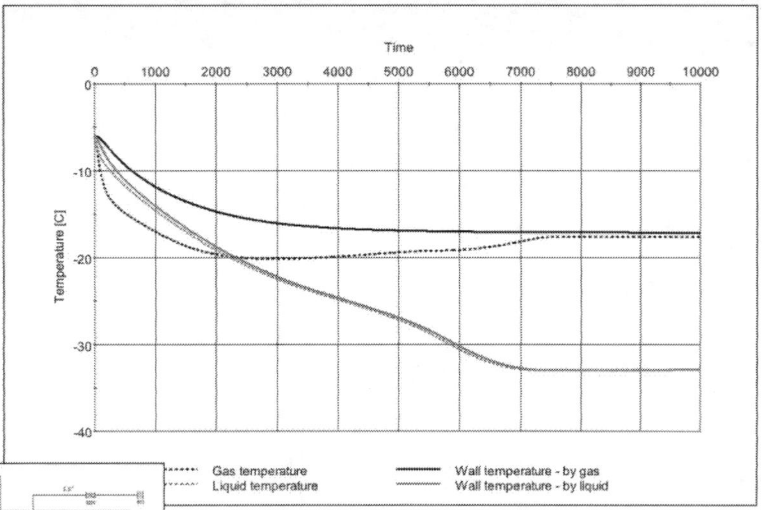

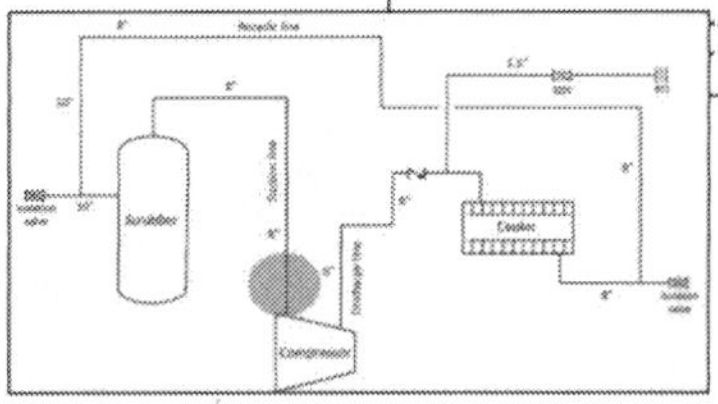

Minimum metal temperature = -33°C

Distributed approach

Different parts of the system metalwork experience significantly different temperature profiles

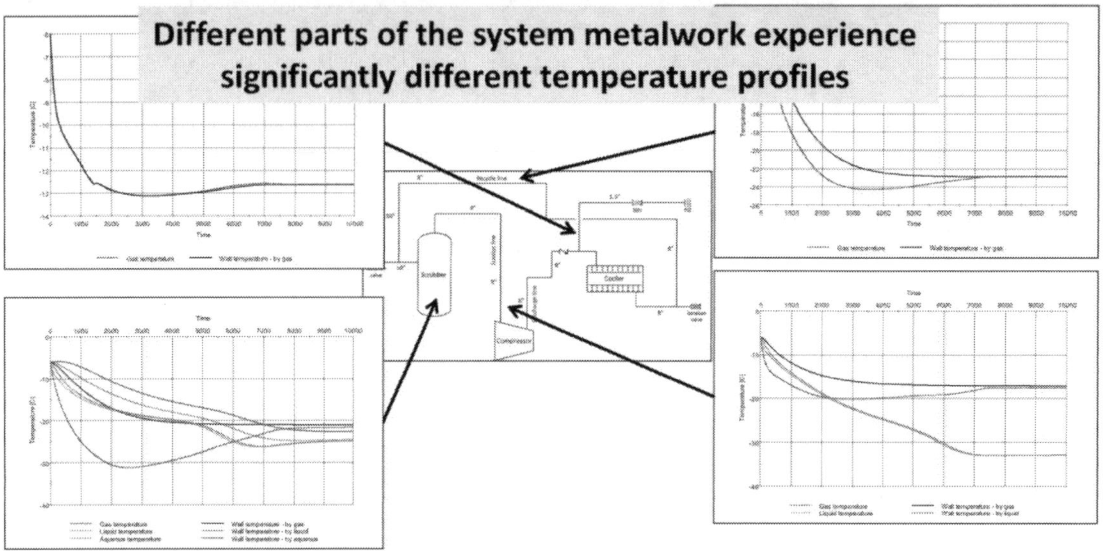

Minimum metal temperature = -33°C

Summary

Two key requirements for accurate blowdown calculations

1. Accurate prediction of physical behaviour of the system — Advanced process modeling techniques and environment

2. System level analysis and representation — Engineering knowledge and judgement
   - Systematic workflow for describing accurately specific system level information
     - geometry
     - material of construction
     - locations where liquid may accumulate

Concluding remarks

PSE

- Conventional analysis approaches for determining the minimum metal temperatures during blowdown can introduce significant errors: **differences as much as 30°C** are not uncommon
- 6th edition of the API 521 standard (Jan 2014) makes a number of **ne**… …ssessment durin…
- The a… …ogy affords consi…
  - Ex… …advocated here **saved several hundred millions of dollars** with regards to material selection for a recent FEED project
  - Example 2 – in many projects risks missed with conventional screening calculations have been identified using the rigorous modeling methodology

Determine the best possible target for safety-related capital expenditure and ensure design risks are not missed

31

WorleyParsons

resources & energy

Is It Possible to Look Out for your Company's Best Interest and Still be Ethical?

By
Kenneth E. Arnold
WorleyParsons
August, 2014

- What Do We Mean By Ethics
- Some Examples of Unethical Behavior
- Do I Have To Act In An Unethical Manner?
- Benefits of Ethical Behavior

Ethics - Religion

- In the Talmud, Rabbi Hillel said, "Do not do to others what is hateful to you. All else is commentary."
- In the Gospels, Jesus said, "Do unto others as you would have them do unto you."
- In the teaching of the Moslem prophet Mohammed, it is written, "No one of you is considered faithful until you wish upon your brother what you wish for yourself."
- Confucius said, "What you don't want for yourself, don't do to others."

Ethics - Religion

- The Swami Chinmayananda, a spiritual Hindu teacher said, "Charity is an attempt wherein I try to expand and bring into the ambit of my life, all others around me and grow to consider the other man's needs and requirements as important as my personal needs."
- The Baha'u'llah of the Baha'i Faith said, " Deal ye one with another with the utmost love and harmony, with friendliness and fellowship."

Ethics – Cultural

- The Declaration of Independence, "We hold these truths to be self-evident .... That all men are created equal ... that they are endowed by their Creator with certain unalienable Rights... That among these are Life, Liberty and the pursuit of Happiness
- The Magna Carta, "No Freeman shall be taken or imprisoned, or be disseised of his Freehold, of Liberties or Free customs...we will not deny or defer to any man either justice nor right."
- National motto of France, "**Liberté, égalité, fraternité**"
- **The preamble to the Constitution of India, "to secure all its citizens: JUSTICE, social, economic and political; LIBERTY, of thought, expression, belief, faith and worship; EQUALITY , of status and opportunity; and to promote among them all; FRATERNITY, assuring the dignity of individual the unity and integrity of nation;**

A Common Truth

- To act ethically is to take into account the wants, desires, needs and feelings of others.
- Everyone deserves the same respect that we believe should be paid to us.
  - The right to be heard
  - The right to be treated with dignity
  - The right to earn a living
  - The right to be told what is happening
- Do we really believe in this?

Examples of Unethical Behavior – Pecking Orders

- **The "Global Pecking Order"**
  - The Owner
  - Consultants
  - Vendors
  - Contractors
  - Grunts in the Field
- **The "Corporate Pecking Order"**
  - CEO
  - Vice President
  - Managers
  - Supervisor
  - Project Manager
  - Engineers

Examples of Unethical Behavior – Pecking Orders (Continued)

- **The "Engineering Pecking Order"**
  - **Engineers**
  - **Draftsmen**
  - **Procurement**
  - **Accounting**
  - **Secretaries**
  - **Other support Staff**

Examples of Unethical Behavior - Relationships

- **Relationship Building is Good … But is it unethical to :**
  - Bribe people to get work
  - Accept kickbacks to award work
  - Elaborately entertain clients
- **Giving to Charity is Good … But is it unethical to:**
  - Use purchasing power to force others to give
  - Use purchasing power to force others to sponsor costly social events

Examples of Unethical Behavior - Bidding

- **There is nothing unethical about bidding**
- **But … Preparing a bid costs money and resources**
- **Is it unethical to:**
  - Bid a job that you know may not be awarded
  - Bid someone you know you will not award the job
  - Withhold information bidders need to judge their chance of success – competition, evaluation criteria
  - Bid a job more than once without compensation
  - Take a value added idea from an unsuccessful bidder and give it to the successful bidder to implement without proper compensation

Examples of Unethical Behavior – Change Orders

- Complex engineered equipment, fabrication or construction will always have the need for a change order adjustment in price
- There is nothing unethical about protecting yourself from a vendor who is motivated to get the most out of change orders
- There is nothing unethical about negotiating strongly with your client for a change order
- Is it unethical to:
  - Use future purchasing power to blackmail a vendor to avoid paying for extra work that was requested and performed
  - Use technicalities to avoid the responsibility for paying for your mistakes or extra requirements
  - Hold a contractor to a legal requirement when circumstances have clearly changed the intent
  - Demand something when you know your client is unaware that it is not really a change in scope

Do I Have to Act In An Unethical Manner?

- Maybe
- We all consider ourselves and our actions to be ethical
- But … In the rush to do business, or because we don't have all the facts, or because we are distracted we all act unethically from time to time
- We all have bosses or clients who require us to do things we think are unethical
- When you act unethically, it is unethical to:
  - Refrain from explaining to someone that they are asking you to do something that may be construed as unethical
  - Refrain from apologizing

Benefits of Ethical Behavior

- **The management text answer: If we act ethically, in the long run we will be more profitable**
- **But … maybe not – our bosses (clients) don't always understand and reward such behavior**
- **So why be ethical**
  - Religious
  - Cultural – it is what separates us from failed societies
  - Helps create an enjoyable work climate
  - We feel good about ourselves

Conclusion

So … Is it possible to look out for your Company's best interest and still be ethical?

- **I don't know – it depends on your definition of "best interest"**
  - Return on shareholder Investment
  - Providing benefits, security and career development for staff
  - Perpetuating the enterprise
  - Supporting society
- **It is up to you**

ConocoPhillips

Upstream Day In The Life
Facilities Engineering

Joshua Bird
October 10th, 2014

About Us – Josh Bird

- 2013 UH ChemE Alumni
- Facilities Engineer
  - Eagle Ford Development (June 2013 - Current)

ConocoPhillips

UH

ConocoPhillips

Upstream Industry

- Concentrated on different types of reservoirs

Conventional:

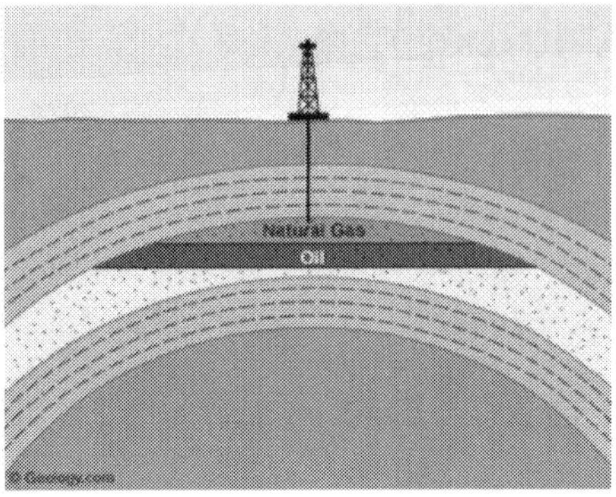

Unconventional:

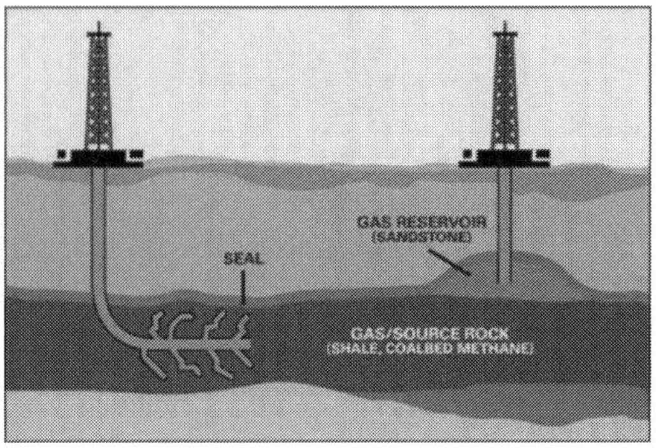

ConocoPhillips

Unconventional Reservoirs

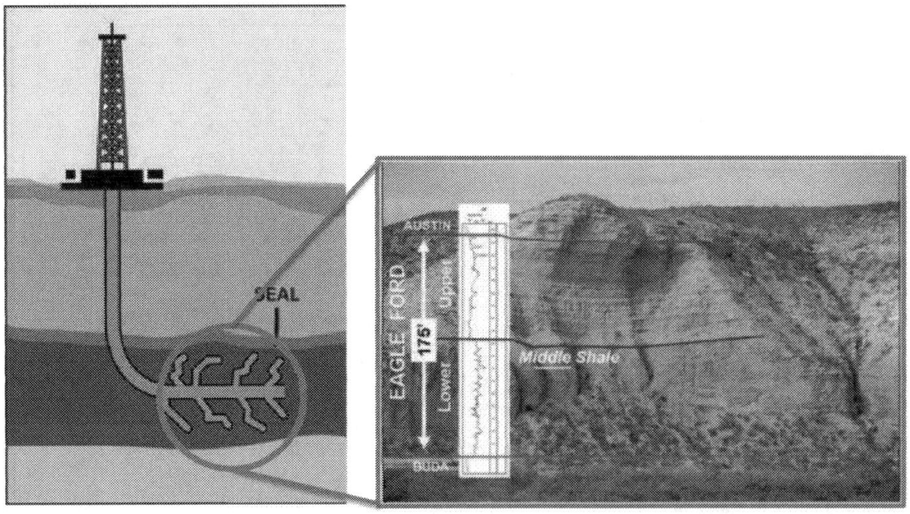

ConocoPhillips

Unconventional Revolution

- Unconventional reservoirs have revolutionized the oil & gas industry
  - Hydraulic Fracturing
  - Horizontal Drilling

- Three factors leading to booming job industry in Upstream
  - Unconventional reservoirs (Eagle Ford Shale)

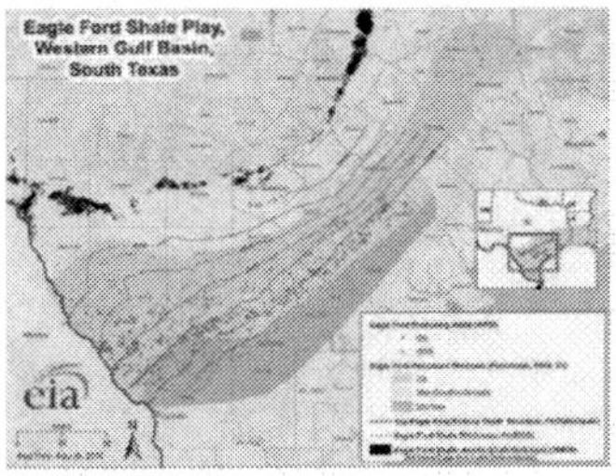

  - High oil prices and low transportation costs
  - Aging workforce

Unconventional Revolution

- Unique factors about onshore unconventional operations
  - Hundreds and even thousands of wells and small facilities
  - Production drops off quick so dominated by drilling speed
  - Full field development happens at staggering pace

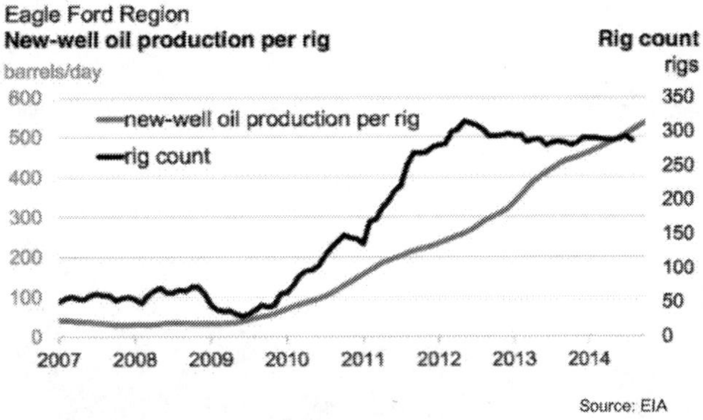

Produced Oil

- Mostly a very light condensate

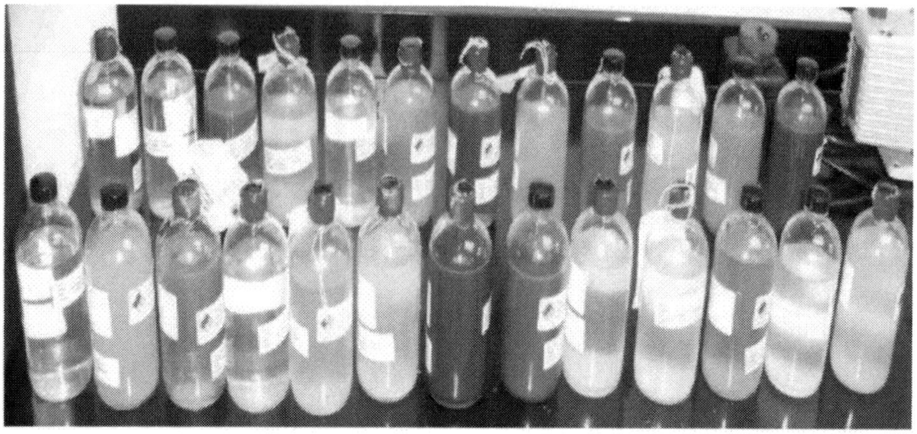

ConocoPhillips

Onshore Supply Chain

ConocoPhillips

Stabilization

ConocoPhillips

Day to Day

- Facilities Engineer in Eagle Ford Development
- Manage design and construction of facilities and pipelines from planning to production
  - Routing pipelines through the field in a way that minimizes environmental impact, pipeline length, and costs.

EAGLE FORD

ConocoPhillips

Day to Day

- Following drillings rigs to ensure that production is put online as fast as possible

ConocoPhillips

Day to Day

- Reviewing drawings to make sure all piping is correct and that everything is the correct material specification

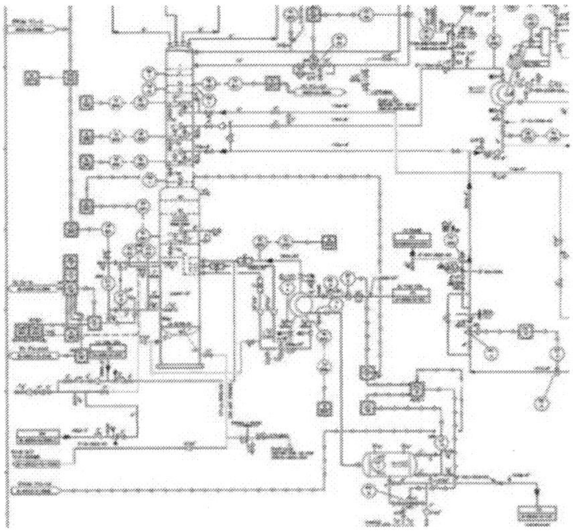

ConocoPhillips

Day to Day

- Interfacing with construction and solving any problems and managing any changes that need to be made

ConocoPhillips

Day to Day

- Perform safety reviews on newly constructed pipelines and Facilities to ensure that everything is to spec and is safe for start up

ConocoPhillips

Day to Day

- Work on other engineering development projects to improve designs and create innovative solutions to complex problems

ConocoPhillips

Why This Matters To You

- Eagle Ford shale region hired more than 260 new workers a day in 2012 (Houston Chronicle)
- Jobs in the oil and gas industry currently pay about 4 times more than the average job in Houston (Houston Chronicle)
- Professional services, including engineering, have the highest job growth rate in Houston (Houston Chronicle)
  - 4.5% per year or 8,900 jobs between December 13' and June 14'

ConocoPhillips

Career Options

- Chemical Engineers are very versatile
- Engineering Positions
  - Drilling Engineering
  - Reservoir Engineering
  - Completions Engineering
  - Production Engineering
  - Facilities Engineering
  - Field Engineering (contractor companies mostly)
- Chemical Engineers typically focus in Reservoir, Production, Facilities, and Field Engineering

ConocoPhillips

Questions

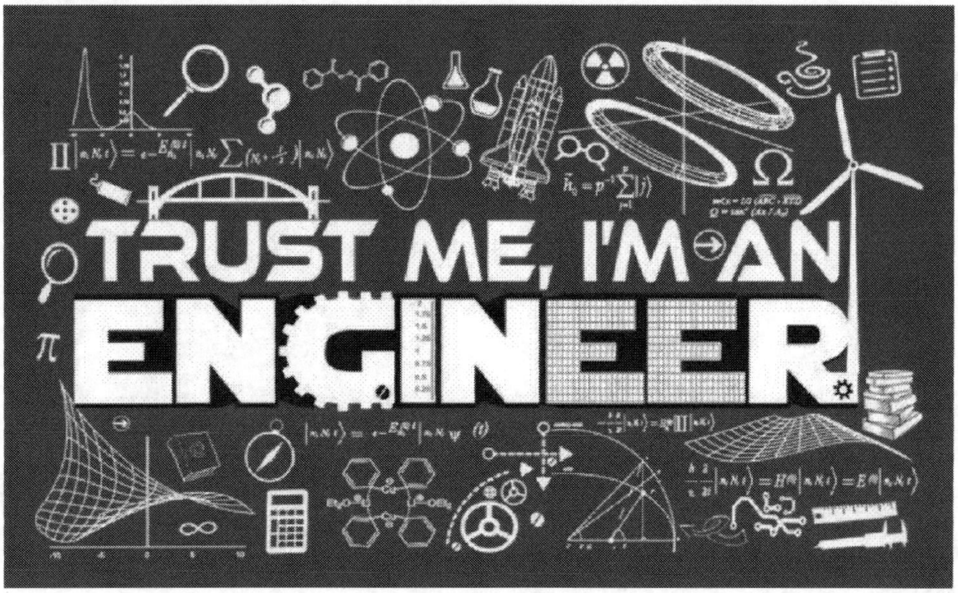

ConocoPhillips

Janie Brown - Personal

Texas A&M University, B.S. Chemical Engineering, class of 2006

Chemistry Minor, Systems Safety Specialty Certificate

8 years with ExxonMobil

Hobbies: Intramural sports, coaching basketball, volunteering, teaching Sunday School, live Texas country music, two-stepping

Why I chose Downstream

- A refinery is a Chemical Engineer's sandbox
- Fast paced and varied work leads to limitless learning
- Opportunity for advancement
- "Hands on" experience is hard to come by later in your career
- I actually use my textbooks!

Why I chose ExxonMobil

- Engineering family (mom, dad, brother, sister-in-law, sister)
- Immediate responsibility, no 'warm up' job
- Stable company with excellent benefits
- Focus on the long term – career and investment
- Community oriented and supports volunteering

1

Career

- Refinery contact engineer (3 ½ years)
  - Troubleshooting, optimization, and problem solving
  - Daily interface with plant Process personnel
  - Infinite opportunity to ask "Why" questions
- Refinery Earnings analyst (1 year)
  - How do refineries make money? Now I know!
- Regional Distillation Specialist (3 years)
  - Broadened into regional (US, Canada, and South America) support
  - Debottleneck – how to make the steel on the ground work harder
  - Turnaround, startup, and shutdown support ← the MOST FUN
- Refinery Senior Process Engineer (current)
  - "Engineering Judgment".. my decision is final!
  - Mentorship and responsibility for 2 underling engineers

2

ExxonMobil
Taking on the world's toughest energy challenges.

So…what would you say, you do here?

Short answer: identify problems & design solutions

- Refinery Morning Meeting
  - What's wrong today?
- Field Visit
  - Self-initiated to look at problematic heat exchanger with IR gun
- Plant Test Review
  - Calculate tower limitations from first principles (textbooks!)
- Morning Meeting Resolution
  - Impact of Heat Exchanger in / out of service
    - Rundown cooler (VPS resid)
    - Ovhd condenser
  - Drum is intermittently filling up with liquid
    - Faulty instrumentation – yours or someone elses
    - Tower flooding and other water problems
  - Hydraulic restriction – broken valve / plugging / salting
  - Operating conditions for completely new product

3

ExxonMobil
Taking on the world's toughest energy challenges.

Shutdown, startup, and Turnaround

- Plant runs 24 hrs a day, 365 days a year

….except, some days it doesn't, and you roll up your sleeves

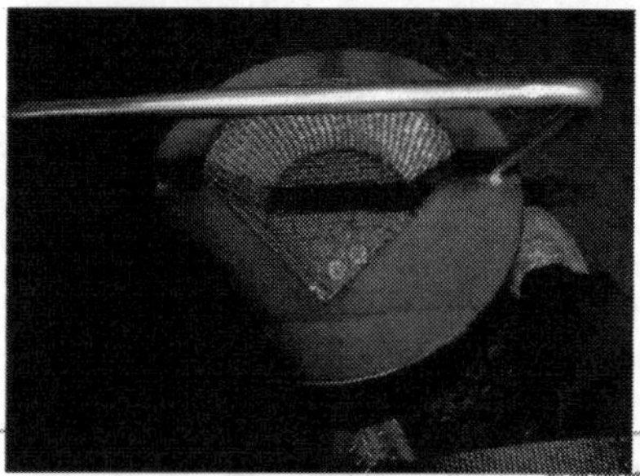

4

ExxonMobil
Taking on the world's toughest energy challenges.

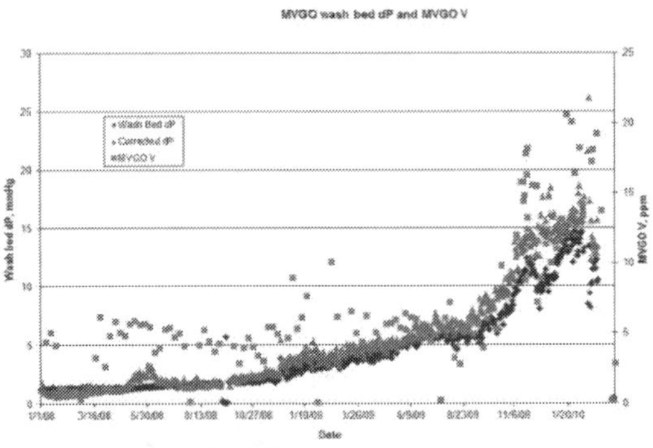
MVGO wash bed dP and MVGO V
Wash Bed dP
Corrected dP
MVGO V
Wash bed dP, mmHg
MVGO V, ppm
Date

5

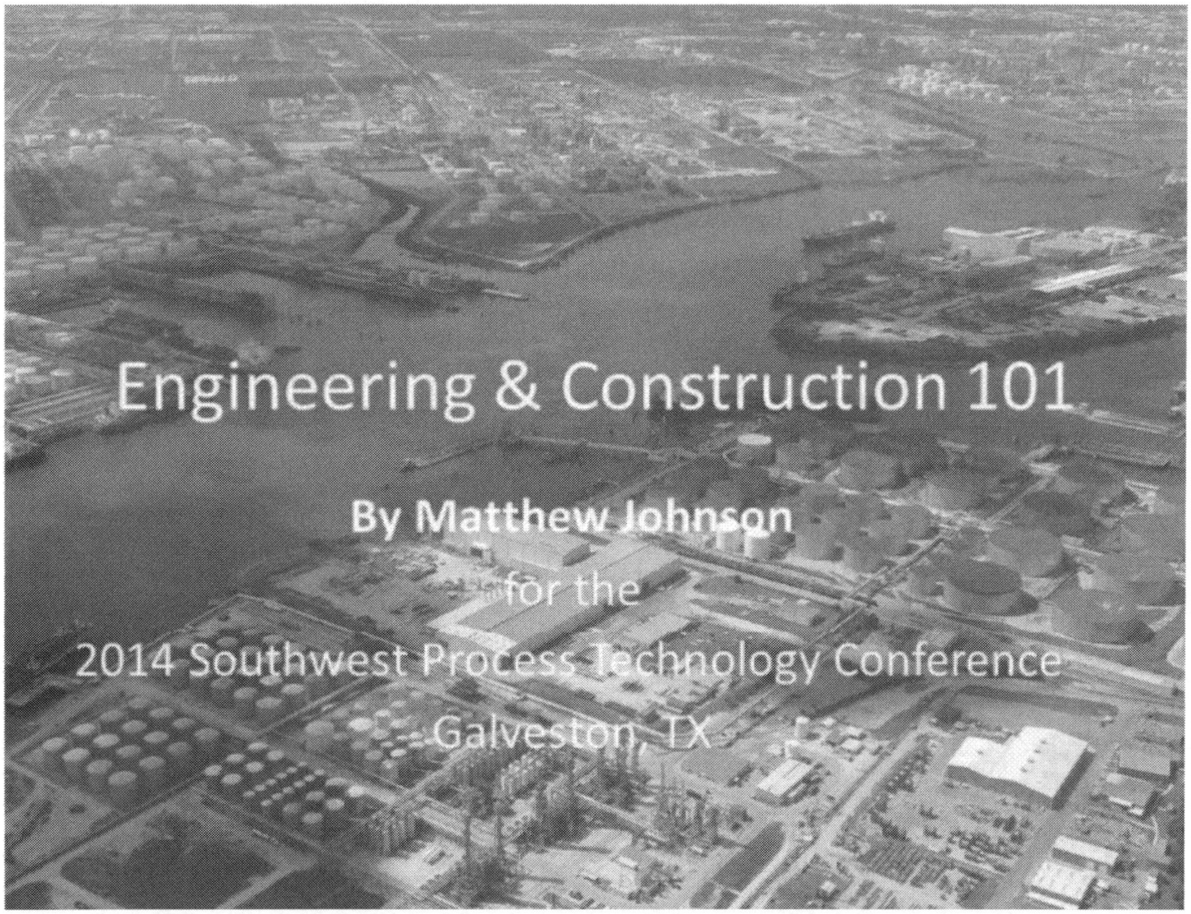
Engineering & Construction 101
By Matthew Johnson
for the
2014 Southwest Process Technology Conference
Galveston, TX

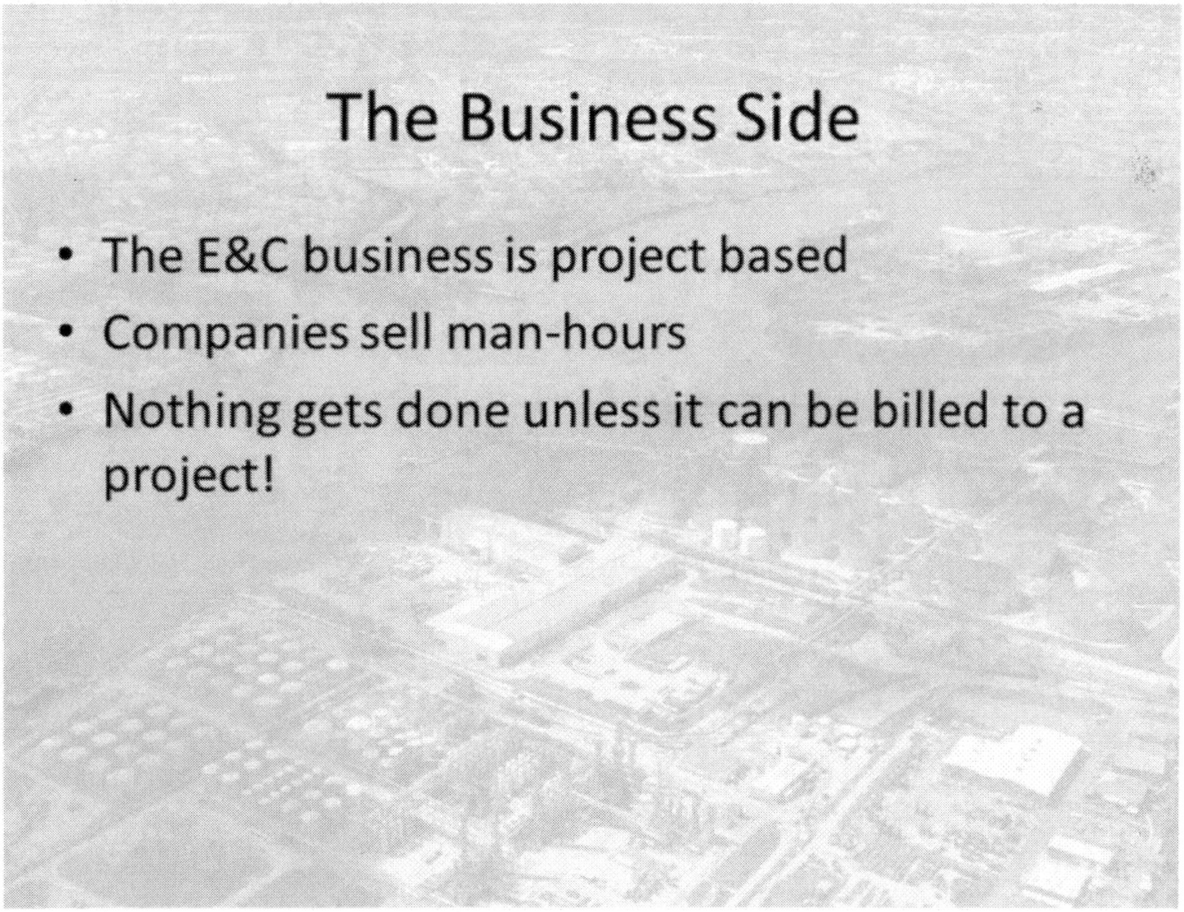
The Business Side
• The E&C business is project based
• Companies sell man-hours
• Nothing gets done unless it can be billed to a project!

Lump Sum vs. Reimbursable

- Reimbursable:
  - Profit = Man-hour Rate minus Pay Rate
- Lump Sum
  - Profit = Total Project Amount minus Cumulative Pay

Project Phases

1. Proposal / Studies
2. Basic Engineering Design
3. Front End Engineering Design (FEED)
4. Detailed Engineering
5. Procurement / Construction
6. Start-up

Typical Project

- Starts with a Request for Proposal (RFP)
  - The operating company is attempting to meet a business need/goal.
  - Often comes with "process guarantees"
- Establish a TIC estimate (+/- 30-50%)
- Operating company decides to proceed and selects the E&C contractor to perform the work.

Typical Project

- Kickoff meeting with the client establishes the **basis** for the project and any changes from the proposal.
  - Basic Engineering Design Data
  - P&ID Lead Sheets, CAD standards, etc.
  - Clarification of project **scope***

*Any revisions to the project scope must be captured in the contract.

Basic Engineering Design

- Process Engineers *drive* the projects at the beginning!
  - Simulations, Heat and Material Balances,
  - Process Flow Diagrams (PFDs)
  - Equipment Specifications (+ long lead items)
  - Piping & Instrumentation Diagrams (P&IDs)
- Other engineering disciplines get involved more as the projects proceed.

Front End Engineering Design

- More detail applied to the P&IDs & Equipment Specifications
- *May* allow Basic Engineering Design and Detailed Engineering Design to proceed more quickly.
- Technically optional, in which case more work is front and back-end loaded.

Detailed Engineering Design

- Uses information generated in earlier phases of the project to finalize all the details.
- Vendors for equipment specified
- Isometrics produced
- Still more detail applied to P&IDs
- Other disciplines *heavily* involved (Vessels, Heat Transfer, Rotating Equipment, Instruments & Controls, Piping, Civil)

Construction

- Process is put together using Detailed Engineering drawings & specifications from all disciplines.
- Issues during construction may be fed back to Process & other disciplines for review.

Start-up

- Once the client is confident in the construction it is time for “oil in”.
- Licensors may send highly experienced operations experts to assist with unit start-up.
- A performance test may be run to ensure the unit is able to meet process guarantees.

Project Closeout

- Very important to document lessons learned so that any issues seen on the project can be avoided in the future.
- Organizations should keep some form of lessons learned readily available to engineers when performing the next design.

Questions?
Photo credit: By Author

Assuring Environmental Compliance While Maintaining Operational Flexibility

2014 AIChE Southwest Process Technology Conference

October 10, 2014

Lisa Swanson, P.E.
Air Engineer

MURPHY
OIL CORPORATION

Introduction

- **Air Permits**
  - Required by state and federal regulations to minimize exposure to the public and the environment
    - New sites or modifications to existing sites
  - Construction Permits and Operating Permits
    - Different levels of construction permits
      - Depends on size of facility, types of operations, location
    - Operating permits required for certain facilities

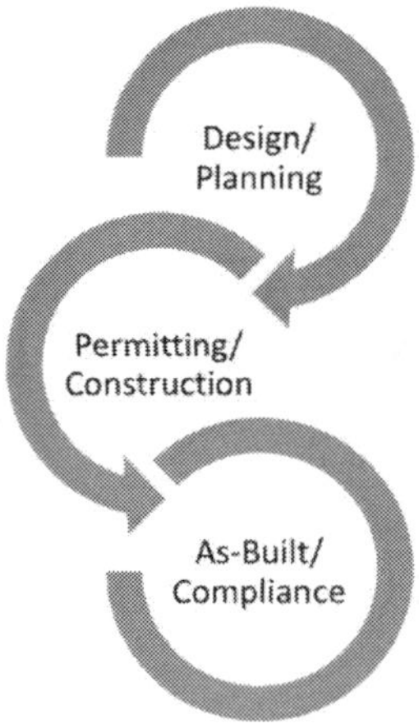

Air Permit Information

- **Always Required**
  - Process Descriptions
  - Maps
  - Emissions Calculations
  - Demonstration of Compliance with State/Federal Rules
- **Required for More Involved Permit Types**
  - Air Dispersion Modeling Analyses
  - Best Available Control Technology or Lowest Achievable Emission Rate Analysis
  - Public Notice

Types of Facility Authorizations

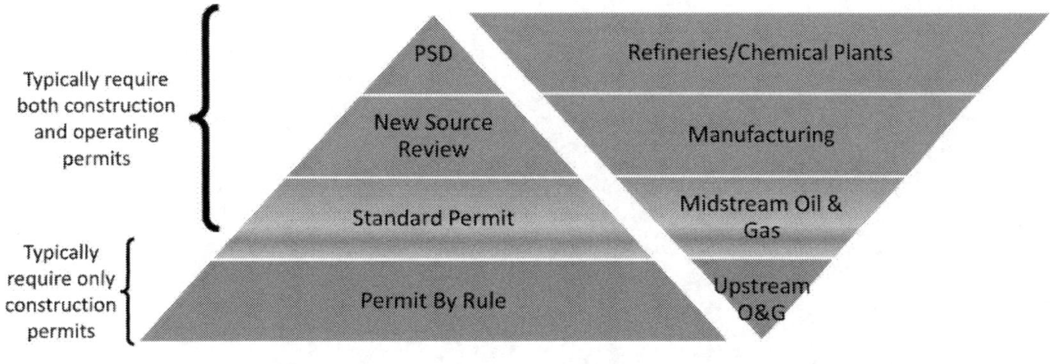

- All potential emission sources must be authorized (including de minimis sources)
- Permits listed above are specific to authorization types in Texas
  - Other states have similar programs and permit types

Types of Construction Permits

Permit By Rules (PBRs)

- **Advantages**
  - Quick authorization period
    - Allows operational flexibility in a rapidly changing environment
  - Minimal permit application requirements
  - Reduced regulatory scrutiny
  - No public notice
  - Lowest permit fees
- **Disadvantages**
  - Only certain operations and emission units allowed to be authorized
  - Lowest emission thresholds

Note: There are approximately 120 different PBRs available for various emission units and operation types.

Types of Construction Permits

Standard Permit

Advantages
- Quick authorization period
  - Allows operational flexibility
- Moderate permit application requirements
- No public notice
- Low permit fees
- Emission thresholds higher than PBR

Disadvantages
- Only certain facility types allowed
- Emission thresholds less than Case-by-Case NSR

Note: There are approximately 20 different Standard Permits available for various facility types.

Types of Construction Permits

Case-by-Case New Source Review (NSR)

- **Advantages**
  - Any type of equipment or operation can be authorized as long as state and federal standards are met
  - Significantly higher emission thresholds
- **Disadvantages**
  - Substantial permit application requirements, may include dispersion modeling
  - Long authorization period (6-9 months for permit issuance)
  - Facility modifications require permitting and approval before implementation
  - Public notice required
  - Higher permit fees (0.3% capital cost)

Types of Construction Permits

Prevention of Significant Deterioration (PSD)

- **Advantages**
  - Any type of equipment or operation can be authorized as long as state and federal standards are met
  - Essentially no emission thresholds
- **Disadvantages**
  - Extensive permit application requirements, including dispersion modeling for multiple pollutants and addition of other facilities in the vicinity
  - Equipment must meet Best Available Control Technology (BACT) standards
  - Longest authorization period (9-12 months for permit issuance)
  - Facility modifications require permitting and approval before implementation
  - Public notice required
  - Highest permit fees (1.0% capital cost)
  - EPA Review and comment

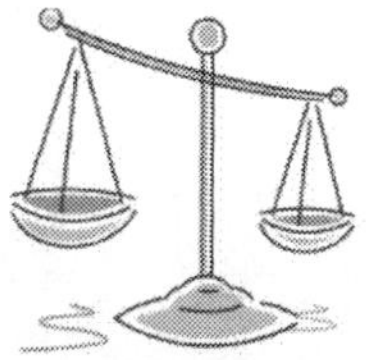

Operating Permits

- **Also known as "Title V Permits"**
- **Federal permitting program**
- **Applies if:**
  - Emissions of any criteria pollutant (NOx, SO_2, VOC, PM, CO) exceed 100 tpy; or
  - Emissions of a single hazardous air pollutant (HAP) exceed 10 tpy or a combination of HAPs exceeds 25 tpy
- **Separate permit and permitting program in some states**
- **Compliance activities and regulatory review are increased**
  - Semi-annual compliance reports required to be submitted and certified

MURPHY OIL CORPORATION www.murphyoilcorp.com NYSE: MUR 9

Air Permitting Strategies

- **Teamwork and communication promote efficiency**
  - Air permitting and emission calculations should be considered in the early design of a facility
    - Emission controls
    - Project schedule/critical path
    - Locations
    - Types of activities to include in permit

MURPHY OIL CORPORATION www.murphyoilcorp.com NYSE: MUR 10

Emission Controls

- **Required by regulations**
  - Certain types of sources
  - Specific areas (i.e. non-attainment)
  - Best Available Control Technology (BACT)
  - Lowest Achievable Emission Rate (LAER)

BOILERS

This information is maintained by the Combustion/Coatings Section and is subject to change. Last update 04/2009

Year	Source Type	Pollutant	Minimum Acceptable Control	Control Efficiency or Details
2006	Boiler >40 MMBtu/hr	NOx	0.01 lb/MMBtu when firing 75% - 100% natural gas 0.015 lb/MMBtu when firing plant fuel gas	plant fuel gas may contain up to 75% natural gas plant fuel gas specifics: <50% H2; >920 Btu/dscf fuel oil firing less than 760 hr/yr
		CO	50 ppmvd at 3% O2	
		PM	Less than 5% opacity	
		NH3	10 ppmvd at 3% O2	

Emission Controls

- **Air dispersion modeling**
  - If initial air dispersion modeling analyses indicate issues with demonstrating compliance for one or more criteria pollutants or hazardous air pollutants (HAPs), controls may need to be evaluated

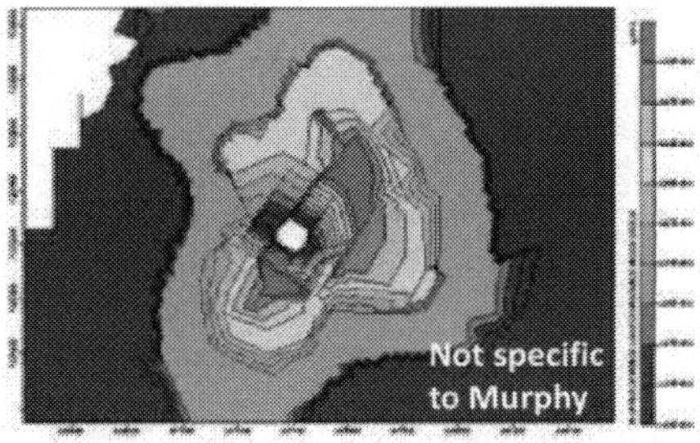

- **Voluntary, but may be a preferred path to reduce regulatory/permit requirements**

Project Schedule/Critical Path

- **Certain permits must be approved prior to construction of a new facility or modification of an existing facility**
  - "Construction" has different definitions in each state, but can be as stringent as "anything other than site clearance or site preparation"
  - Permit application preparation may take 2 to 3 months depending on the complexity of the site and permit requirements
  - Permit issuance times are typically extended for larger permits (6 to 12 months from submittal)
  - Timing for the air permit needs to be considered in overall project schedule

Locations

- **In certain cases, more than one location option may be identified for a new facility**
- **There may be advantages or disadvantages in the air permitting process by choosing one option over another**
  - Sites within ¼ mile of each other are typically required to be combined into one air permit
    - May require a different type of permit since emissions are higher
    - If the location options are far apart, one site may be preferable based on attainment designations, county requirements, different state agencies, etc.

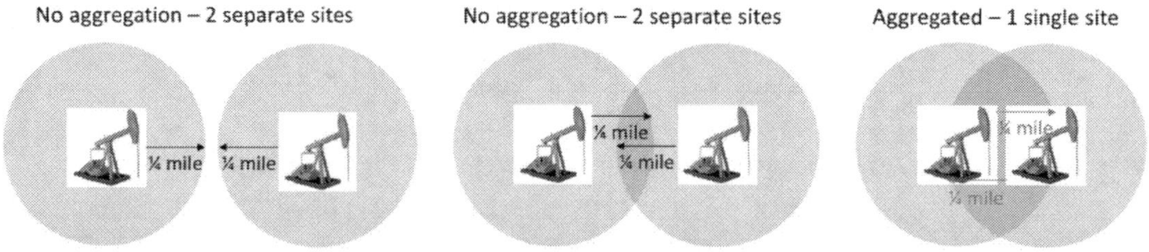

Types of Activities

- **Activities to Include in Permit Application**
  - Planned emissions
    - Startups, maintenance events, shutdowns
      - Pipeline pigging operations, compressor blowdowns, etc.
- **Activities Not Included in Permits**
  - Unplanned emissions
    - Malfunctions, equipment failures, etc.
  - However, these emissions must be reviewed to determine whether the event must be reported.
- **Some gray areas**
  - Repeat equipment issues
  - Downstream issues from other companies (i.e. pipelines)

Emission Calculations

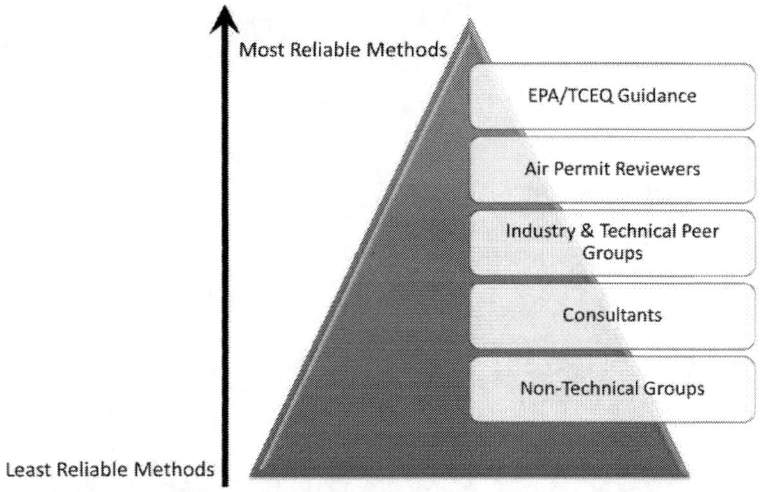

Recommendations

- **Design & Planning Stage**
  - Strategic approach to air permitting process
  - Notify environmental group of new projects or modifications as early as possible
    - Can assist with evaluating multiple scenarios to find the option that is:
      - Most cost-effective
      - Quickest
      - Lowest compliance risk
- **After Operations Begin & Permit Issuance**
  - Notify environmental group of any upset events, maintenance activities, shutdowns, etc.
    - Ensures that appropriate reports and/or notifications are in place
  - Notify of any inspections that occur at the facility

Contact Information

Lisa Swanson, P.E.
(281) 675-9048
lisa_swanson@murphyoilcorp.com

MURPHY
OIL CORPORATION

AIChE
Southwest Process Technology Conference

Effective Interviewing

Greg Yeo
ExxonMobil Chemical Company

October 10, 2014

Agenda

- Introduction
- Before the Interview
  - Preparing as if it's an important class
- At the Interview
  - Being authentic, capable and motivated
- After the Interview
  - Following up & building a network

1

Greg Yeo

- BS Chemical Engineering, Worcester Polytechnic Institute
- 3 months at company "A"
- MS Chemical Engineering, Worcester Polytechnic Institute
- 35 years at ExxonMobil Chemical
- Active in AIChE National Programming & Ethylene Producers' Committee
- Hundreds of student interviews

2

Opportunities for Chemical Engineers in the Process Industries

- Operating Companies
  - Upstream
  - Downstream
  - Chemicals
  - Research
- EPC Companies
- Technology Vendors
- Service Companies

3

Interviewer / Candidate Interaction

- What's the interviewer looking for?
  - Demonstrated Leadership
  - Demonstrated Technical Capability
  - People Aptitude
  - Experience
  - Enthusiasm, Interest
  - Flexibility
  - Communication Skills
- What does the interviewer have to go on?
  - **Academic record**
  - **Resume**
  - *Internship (maybe?)*
  - *Campus interview (maybe?)*
  - *Phone interview (maybe?)*
  - **Face to face interview**

"Does this person make my team better? Is this an engineer that I want on my next project?"

4

Prepare As You Would For An Important Class

- Research the company in advance
  - Understand the company's business
  - What they are looking for?
- Prepare a resume that people notice
  - Most resumes look alike
  - Emphasize experience and skills
  - *Make it a platform to tell your story*
- Analyze Yourself and ***prepare***
  - Know your strengths and weaknesses
  - Do you know what you want? Are you flexible?
  - Would ***you*** hire you for this position?
- Prepare examples, stories, questions
  - This is a future colleague. Be interesting, memorable, able to tell an engaging story.
  - ***I remember engaging stories from motivated, competent candidates!***

5

At the Interview

- Dress for Success
  - If you don't know, overdress. And be comfortable with it.
- Show up early
- Bring resume, folder, paper & pen
- Silence your phone in advance
- Use positive body language
- Be enthusiastic, engaging, and authentic.
  - Make eye contact. Show interest. And mean it.
- Create a two way conversation
  - You are interviewing the company as well
  - Ask questions. Take the lead when it's offered.
  - ***Be able to tell an interesting and <u>relevant</u> story.***

6

At the Interview

- Direct the conversation about your qualifications, previous accomplishments, and interest in the job
  - Highlight how previous experiences prepared you for this position
  - *If you understand the position, you can tell the right story about yourself*

- Don't be afraid to highlight areas that company may not ask about
  - Language skills
  - Interest in travel
  - Long term career goals
  - Your innate desire for team success (i.e. to "do what it takes" to accomplish the team goal)

7

Ask Questions – ***Make it a Conversation!!***

- What are the primary sources of revenue?
- Ask/talk about the key products or services
- How many of your senior leaders are engineers?
- What is a typical career path for a ChE like me?
- Do you see my experience and skills as a good fit for this position?
- What opportunities are there for training, travel, continuing education?
- How do you help employees with work-life balance?
- Is the company growing? What drives growth? Are there global operations?

8

Let Your Responses Differentiate You

Don't say :

- One word answers
- "I'm good at Microsoft Word and Excel…"

Rather, answer like a Chemical Engineer:

- "I utilized MS Excel to uncover mass balance error, and discovered a leaking vent to the flare…"

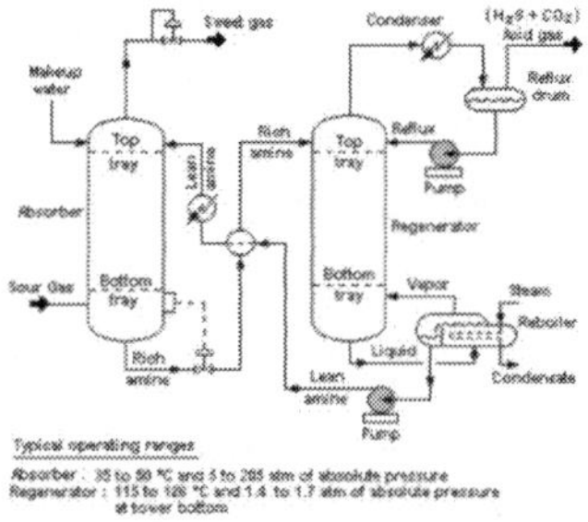

9

At the Interview

- Types of questions you may be asked:
  - What interests you about this job?
  - Leadership examples?
  - Experience working in a team? Lessons learned?
  - What things really interest you? Research, process engineering, process design, customer support?
  - What do you know about this company?
  - What are your Strengths? Weaknesses?
  - Where do you see yourself in 5 years?
  - Tell me about a situation where did not go as planned. What did you do?
  - Have you ever had to deal with difficult co-workers?
- Plus (maybe) technical questions!
  - Design project details
  - Co-op experience
  - Fluid Dynamics / Heat Transfer / Mass and Energy Balances
  - Process modeling

10

After the Interview

- Before you leave, ask "what's the next step?"
  - When will you hear from company?
- Thank everyone that you talk to
- Send a follow-up note or email
- Analyze your performance
  - What should I do differently next time?
- Follow up if you don't receive a reply

11

Start Building a Career-long Network

- Find helpful resources and build relationships
  - Within your company
  - Within your industry
  - Reputation and ethics matter
- Build bridges – don't burn them
  - Your first job may not be your last
  - Respect is a two way street
  - Get involved – Pay It Forward
- Interview widely to get to know companies, cultures, and interviewing experience
- ***Be careful with social media*** – LinkedIn, Facebook, Twitter
- Find ways to stay current in your field

12

Questions?

Or conversation later today

13

Ethylene - Ethane Splitter Optimization Control Program

Sameer Thorat, B.S. Rao

Optimized operation of Ethylene - Ethane Splitter (E-E Splitter) column is critical in ensuring ethylene product purity and simultaneously minimizing ethylene slippage in the column bottoms stream, while still utilizing the maximum separation and hydraulic processing capacity of the column. Often, the prevalent advance process control schemes do not take into account real time variation in column hydraulics due to changing plant operating conditions like feed composition, feed flow rate, etc. and the corresponding scope available in the column's feed processing capacity. Process optimization without accounting for such critical factors leads to column operation at below its maximum processing capacity and/or column experiences disturbances with operational variations. This paper focusses on a control program designed to optimize Ethylene - Ethane Splitter operation, while taking into account all possible variations in plant operating parameters.

E-E Splitter Operation & Challenges

Considering the intricacies of an ethylene plant, it is an ideal plant for application of Advance Process Control (APC). However, complexity in each unit of an ethylene plant makes application of APC more challenging. The process starts with a series of very high temperature cracking furnaces operating at low pressures, where feed is cracked to obtain the desired products along with formation of byproducts. Separation of these hydrocarbon products requires high pressure and very low temperatures, which necessitates requirement of process and refrigeration compressors. Also, as the range of byproducts formed is large, their separation requires a series of distillation columns and reactors.

An E-E Splitter is one of the downstream distillation columns of the separation train where the final product of the desired purity is obtained. However, its operation and control are equally critical as all upstream process changes impact column operation. Also, the strict product purity requirement makes its operation and control more complicated and challenging. The main function of the E-E Splitter column is to efficiently separate ethylene and ethane from the feed stream while meeting the ethylene product purity requirements and maximizing ethylene recovery.

E-E splitter column feed mainly consists of ethylene and ethane, while it may contain traces of lighters and some heavier components like propylene, depending on the upstream equipment configuration & operational efficiency. Pure ethylene is separated out as top product or side draw, while the ethane stream from the column bottom is recycled back to furnaces.

A typical control scheme has two manipulated variables, namely product draw rate or reflux and reboiler heat duty. These two control ethane carryover in ethylene product and at the same time minimize ethylene loss in the bottom recycle ethane stream. Other key instrumentation includes feed flow meter, series of temperature indicators, top and bottom stream analyzer, etc. Please refer the diagram below for more details.

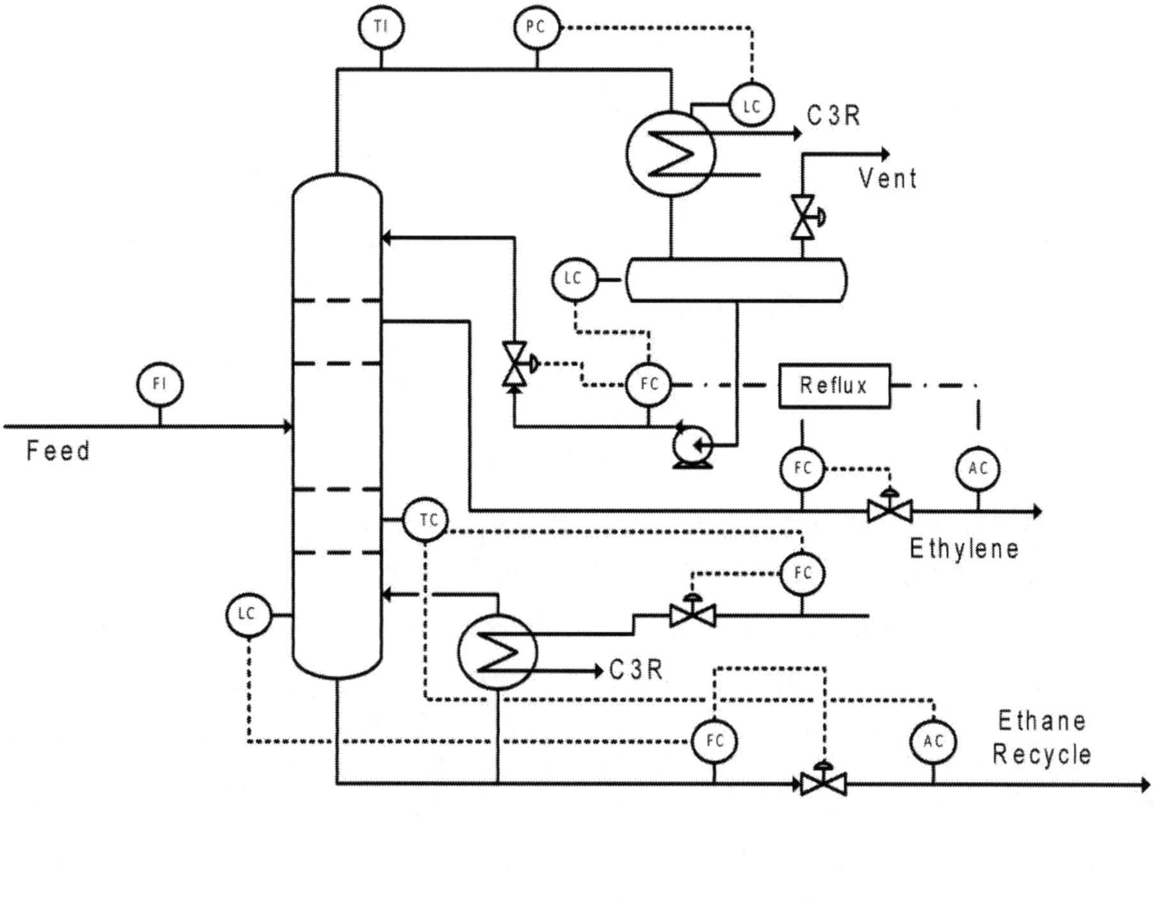

E-E Splitter Column RTO Control

Real Time Variation in Operation

In ethylene plants, usually variations in almost any unit will have an impact on E-E Splitter column operation. Plant feed flow rate is governed by several factors like downstream process requirements, the plant's inherent equipment and operational constraints, etc. This results in plant feed rate variation from time to time and thereby changes in column feed stream.

Furnace section performance is the most important factor affecting the entire plant operation. Normally, depending on severity of cracking and some other factors, furnace coils get fouled up gradually due to coke formation. This affects downstream operation in mainly two ways: first it results in an increase in furnace operating pressure, which causes decrease in product yields. Secondly, this results in gradual coke build up, which makes it necessary to take the furnaces offline regularly after a certain period of operation for decoking, thereby affecting entire plant operation during this furnace changeover process. Additionally, furnace severity control itself is a complicated process which is dependent on analyzers and several instruments. Detuning of controllers, drift in analyzers and improper instrument indications result in variation in furnace performance regularly, which impacts the furnace downstream operation, including the E-E Splitter operation.

Similarly operational changes in cold section columns, especially De-Methanizer & De-Ethanizer can result in E-E Splitter feed flow & composition variation. Another important process unit of an ethylene plant is refrigeration. Fluctuations in refrigeration load with operational changes and changing atmospheric conditions leads to variation in refrigeration stage pressures & flow rates. As E-E Splitter reboiler and condenser duties are directly linked with the refrigeration systems, any variations in refrigeration pressures are likely to impact the E-E Splitter operation. This impact is more pronounced if the E-E Splitter is directly integrated with the Ethylene Refrigeration Compressor System. Apart from these changes, variation in other units like tail end acetylene reactor operation, have an effect on Splitter operation. In some cases, ethylene storage boil off gas is directly recycled back and hence can affect splitter operation considerably.

Based on the discussion above, the challenge for process control is to optimize the column performance in the following ways:

- Absorb regular disturbances from up-stream units like furnaces, cold section, reactors, etc.
- Ensure ethylene product quality (ethane specification) all the time
- Ensure steady column operation and flow rates
- Maximize ethylene recovery
- Minimize reflux/energy requirements
- Maximize column capacity utilization

Optimization Capability of Control System

Normally, in most of the cases, to avoid any column disturbance with plant variation, the column is operated conservatively, well below its hydraulic capacity and with higher reflux rates and ethylene slippage. Hence for every system, it is critical to know if the controlling system is able to fully optimize the column performance.

Some questions that arise are:

1. Does the ethylene slippage control target remain the same for all column feed flow rates?
2. Does the column experience flooding at higher feed rates?
3. Does the ethane content in the product deviate from its set target?
4. Does the ethylene recovery vary for the same feed conditions?
5. Does the reflux/reboiling flow vary for the same feed conditions?
6. Is there frequent variation in ethylene product draw rate?
7. Is there frequent variation in recycle ethane stream flow rate?

If the answer to any of the above questions is "Yes" then it means that scope for column performance improvement exists.

Need for Real Time Optimization

What is an ideal target for ethylene slippage in the column bottom stream? For design conditions, it could be 1.0%. However, an operator is likely to fix it at 2.5% to ensure the more important specification on ethylene product purity at all times. This higher target acts as a cushion to avoid crossover of the hard constraint of product purity during process changes. Even with a normal APC, the target set point will not be varied by a control program; it will be adjusted by an operator. Hence most of the time operation will not be even close to optimum and a scope for improving ethylene recovery is bound to exist.

Even with this operating margin, any changes in the upstream plant operation affect the column operation and are likely to result in off-spec production. The frequency of these upsets will depend on the intensity of plant upstream changes, controller process efficiency, closeness to column operational limits, operator intervention, etc. Apart from the operator intervention, sometimes all these actions are not proactive.

Hence a Real Time Optimization (RTO) for E-E Splitter Column specific is absolutely essential to enhance the column operation to optimum, within given constraints. An APC with RTO will proactively change the manipulated variables and sometimes even the soft constraints so that column hard constraints are never exceeded. These proactive actions provide enough confidence to the operator to operate the soft constraints closer to optimum limits; like with robust process control an operator can set the ethylene slippage target closer to 1%, which is ideal. Further, with more sophistication, even the soft targets like ethylene slippage can be made a variable and APC can change the target set points considering the column hydraulic limits. This reduction in human interference will further optimize the column recovery and will help in maximizing column capacity utilization with minimal operational disturbances.

Optimizer will aim to maintain product purity, maximize ethylene recovery and ensure operation within hydraulic capacity limits. It will take into account process variables like E-E Splitter feed flow and composition for estimating reflux or ethylene product draw rate (to maintain product specification), column hydraulic loads, reboiling requirement (to minimize ethylene slippage) etc. for given conditions. It needs to be noted that any optimization carried out without considering column hydraulics will result in underutilization of the column's processing capacity and/or will cause operational disturbances.

Development of Multivariable Predictive Control (MVPC) for Optimizer

An advance process control system with optimizer needs to set reboiler heat duty and top product draw rate such that product specifications are met consistently. However, this is a difficult control proposition due to mainly the following reasons:

- Interaction between overhead and bottom stream specifications
- Non-linear dynamic behavior, which changes with plant operating conditions
- Slow dynamics of column
- Subsequent to changes in plant operation, frequent variation in feed flow and composition
- Criticality to ensure ethylene product specification

To handle all this criticality of process control, usage of multivariable predictive control (MVPC) methods for distillation control is recommended. MVPC applies linear dynamic models for predictive control and can deal simultaneously with column constraints & analyzer output. Accordingly, to account for real time plant variation, a Multivariable Predictive Control (MVPC) model was developed.

Key to efficient working of these models is precise process understanding and availability of valid operational data. Many-a-times, the availability of actual plant operating data is a constraint as it is difficult to find the required data for a given condition. A complete ethylene plant simulation model is vital for generating the required data that is not available directly from plant operation. Plant simulation model runs for required conditions can help in developing independent correlations. However, again validity of simulation model with actual plant operation is critical for the development of empirical equations used as basis of MVPC.

Plant operating data for substantial period was analyzed to determine the impact of process variables on column operation and column constraints. The impact of changes in the manipulated variable on overall column operation was also studied in detail. Entire plant simulation model was developed and validated with actual plant data. Simulation runs were taken for possible plant variation in

- Plant Feed Rate
- Furnace Operation
- Recycle Stream Flow
- Other Unit & Column Operation

E-E Splitter column hydraulic loads were estimated for the desired simulation cases. Later, operating data, simulation results and column hydraulic results were utilized to develop empirical equations. MVPC was programmed to generate the set points of manipulated variables for column control, considering process variables.

Functioning of Optimizer Control Program

Actual plant data, simulation model and column tray hydraulic model results have been considered to generate the empirical correlations that are used in the optimizer program. The optimizer program utilizes the plant operating data like flow, pressure, temperature, compositions, etc. and on the basis of the empirical equations, estimates the E-E Splitter feed condition, reflux & reboiling requirements and corresponding column hydraulic loads & flooding numbers. Based on these details, the column operating parameters are adjusted/optimized within the operational boundary limits by the control program, based on the prevailing needs. At higher feed rates, flooding is avoided by relaxing ethylene slippage, whereas at lower feed rates, ethylene recovery is maximized based on condenser, reboiler and column hydraulic limitations.

The optimizer program cross checks the operational data for defined constraints like column pressure drop, tray temperatures, condenser and reboiler limitation and only then changes the operational set points.

E-E Splitter Optimizer Benefits

E-E Splitter Optimizer control program provides a real time optimization and enhances the column operation to maximum possible limits. One of the major benefits is prevention of flooding and operation of column at maximum rates possible. Basically, it is observed that in some cases the E-E Splitter hydraulic limitation restricts the plant feed rate at higher loads. In such cases, the optimizer prevents flooding at higher loads by relaxing the ethylene slippage limit. This helps in reducing the reboiling load, thereby providing scope for increasing the column feed rates and maximizing the plant capacity utilization. Though relaxing the ethylene slippage results in some loss of ethylene product, the corresponding increase in plant feed rate is significant and economically a profitable option. Also, with proactive analysis and hydraulic calculations, it helps to operate this column at the maximum possible feed rate without causing operational disturbances.

The secondary constraint for ethylene column is ethylene slippage. With real time optimization, the optimizer helps to maximize the ethylene recovery for a given condition all the time. The ethylene slippage can be easily controlled at 1% -1.5% lower than the normal controllers. This not only helps in reducing the ethylene product loss and improving the ethylene recovery, but also helps in reducing the ethylene recycle to furnaces, which indirectly helps in increasing the fresh feed being processed.

The optimizer program accounts for real time plant operational variation and thereby proactively changes column operating parameters. This helps to minimize the column operational disturbances considerably and also to improve operational reliability/consistency. It helps to ensure product quality all the time due to prevention of operational disturbances.

INGENERO
Excellence Through Insight

Ethylene-Ethane Splitter Optimization Control Program

Sameer Thorat, B.S. Rao

Presented by

Dr. Pratap Nair

at

SPTC 2014 – Chemical Process Technology Session

E-E Splitter In The Ethylene Process

E-E Splitter separates ethylene and ethane streams while ensuring product purity requirements

- Feed mainly consists of ethylene and ethane
- Ethylene is removed as top/side product and ethane from bottom
  - Ethane recycled back to furnaces
- Typically tall tower – sometimes split as two towers
  - Lag times in response to changes longer than other towers in the plant

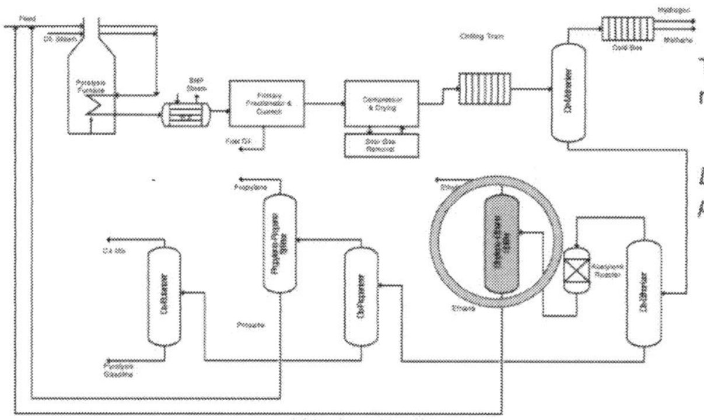

Ethylene Process Flow Diagram

Typical condensing and reboiling medium provided by:
- Ethane refrigeration (~65 psig)
- Propane refrigeration (~235 psig)

Depending on column operating pressure

INGENERO

E-E Splitter Operation Challenges

Typical challenges faced in the day to day control of the E-E Splitter

- ✓ Regular disturbances from up-stream units like furnaces, cold section, reactors etc. cause the feed to the E-E Splitter to change
- ✓ Ensure ethylene product quality (ethane spec.) all the time
- ✓ Ensure steady column operation and flow rates
- ✓ Maximize ethylene recovery
- ✓ Minimize reflux/energy requirements
- ✓ Maximize column capacity utilization

INGENERO

Causes for E-E Splitter Feed Variations

Ethylene plants rarely operate in a perfect steady state and several upstream operating parameter variations affect the E-E Splitter operation

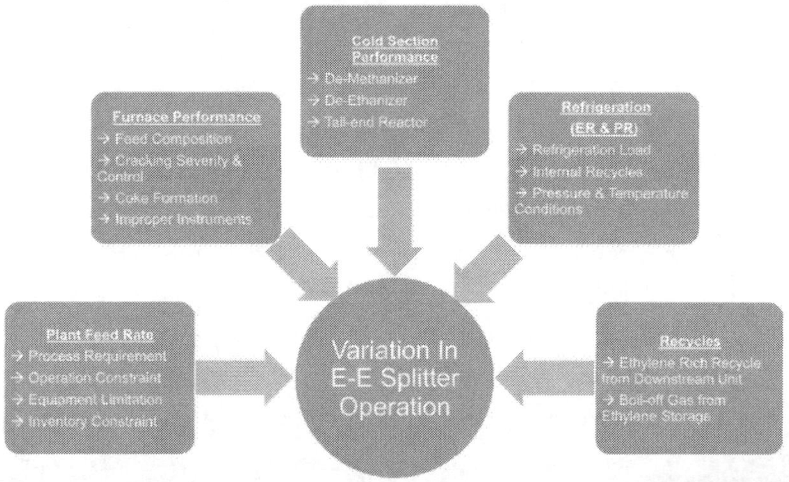

INGENERO

Optimization Potential of Control System

To avoid disturbance in the column with plant variations, the column is operated well below its hydraulic capacity, and with higher reflux rates & ethylene slippage

Quick check of existing potential

- Does the ethylene slippage control target remain same for all column feed flow rates?
- Does the column experience flooding at higher feed rates?
- Does the ethane content in the ethylene product deviate from set target?
- Does the ethylene recovery vary for same feed conditions?
- Does the reflux/reboiling flow vary for same feed conditions?
- Is there frequent variation in ethylene product draw rate?
- Is there frequent variation in recycle ethane stream flow rate?

If answers to any of the above questions is "Yes" then it means that scope for column performance improvement exists

INGENERO

Need For Real Time Optimization (RTO)

An Optimizer program for E-E Splitter Column helps enhance operation within given constraints

- The optimizer objectives
  - Maintain product purity
  - Maximize ethylene recovery
  - Ensure operation within hydraulic capacity limits
  - Optimal energy utilization
- The optimizer needs to take into account process variables
  - E-E Splitter Feed Flow & Composition
- For given conditions optimizer needs to estimate
  - Reflux or Ethylene Product Draw Rate (To Maintain Product Spec.)
  - Column Hydraulic Loads
  - Reboiling Requirement (To Minimize Ethylene Slippage)
- Any optimization without considering column hydraulics
  - Results in under utilization of column process capacity and/or
  - Causes operational disturbances

INGENERO

➔ Development of MVPC for Optimizer

Development steps for a Multivariable Predictive Control (MVPC) model that accounts for real time variations

- Plant operating data analyzed
- Plant Simulation Model developed and validated
- Simulation runs taken for possible plant variations
  - Plant Feed Rate
  - Furnace Operation
  - Recycle Stream Flow
  - Other Unit & Column Operation
- Column Hydraulic loads estimated for all simulation cases
- Operating data, Simulation & Column hydraulic results utilized to develop empirical equations
- MVPC programmed to generate the desired set points for column control considering the variables

INGENERO

Confidential & Proprietary 7

➔ The E-E Splitter RTO

The Real time optimizer utilizes plant data and a model that factors in details of the E-E splitter and critical variables from the rest of the plant and is initialized using a complete and rigorous model

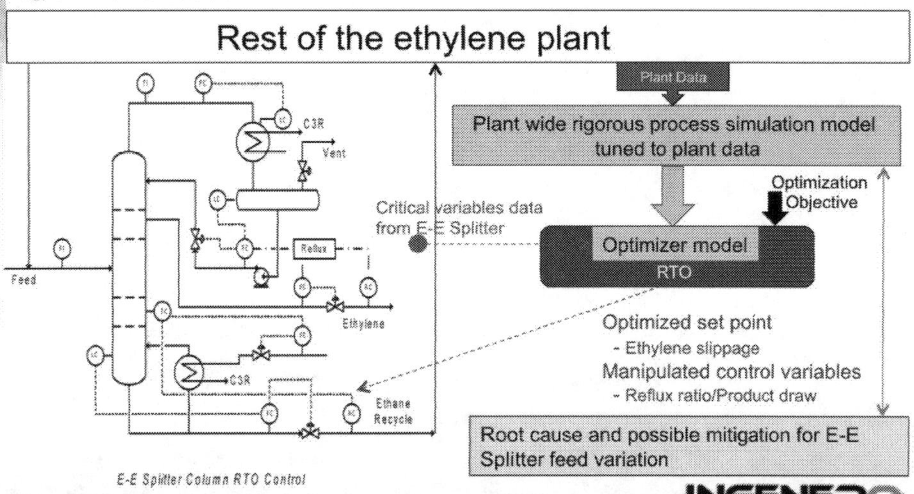

INGENERO

Confidential & Proprietary 8

→ Optimizer Functioning

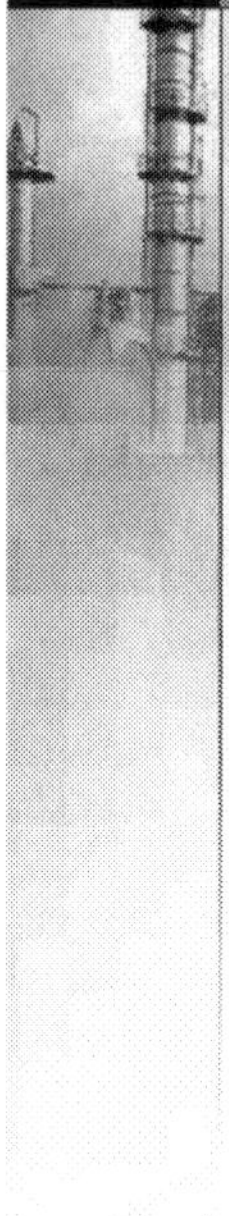

E-E Splitter Optimizer program utilizes the dynamic optimizer model for real time optimization

- Sequence of action points of optimizer program
  - Utilizes plant operating data like flow, pressure, temperature, etc.
  - Based on optimizer model equations the optimizer
    - → Estimates reflux flow requirement
    - → For prevalent reboiling rate, calculates column hydraulic loads
    - → Based on scope available in column capacity, estimates the reboiling and reflux targets
  - Optimizer program cross checks the operational data for defined constraints like column DP, tray temperatures, max. reflux flow, reboiling limitation and only then changes the operational set points

INGENERO

→ E-E Splitter Optimizer Benefits

Provides a real time optimization and enhances the column operation to max. possible limits

- Prevents flooding at higher loads by relaxing ethylene slippage
  - Helps in increasing the column feed rates, thereby maximizing the plant capacity utilization
  - Column operates at max. feed rate possible without operational disturbances
- Maximizes the ethylene recovery for given condition
- Ensures that column runs at its max. hydraulic capacity
- Accounts for real time plant operational variation and thereby proactively changes column operating parameters
  - Minimizes the column operational disturbances
  - Improves operational reliability/consistency
  - Ensures product quality all the time

INGENERO

E-E Splitter Optimizer - Case Study

Optimizer functions to utilize maximum column capacity by optimizing ethylene recovery as per plant operating conditions and demand

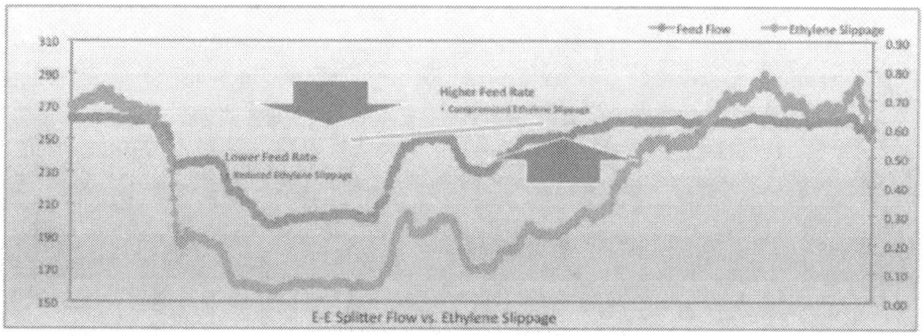

Maximized ethylene recovery at lower feed rates
Maximized processing capacity at higher feed rates

INGENERO

E-E Splitter Optimizer - Case Study

Implementation of optimizer program has delivered stellar results by increasing processing capacity and reducing ethylene slippage

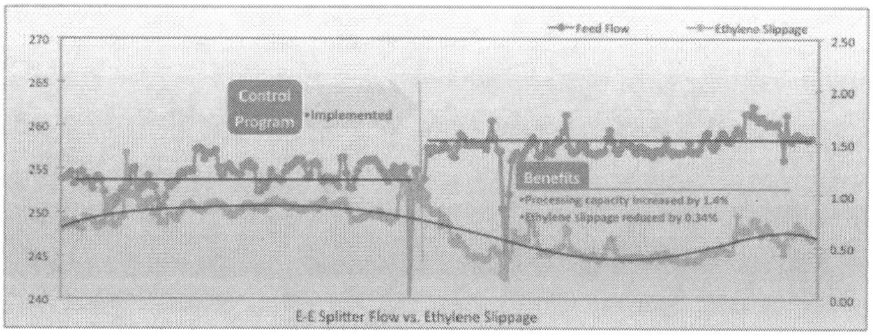

Has resulted in annual benefits of $ 14 MM due to feed increase and $ 2.2 MM due to recovery improvement

INGENERO

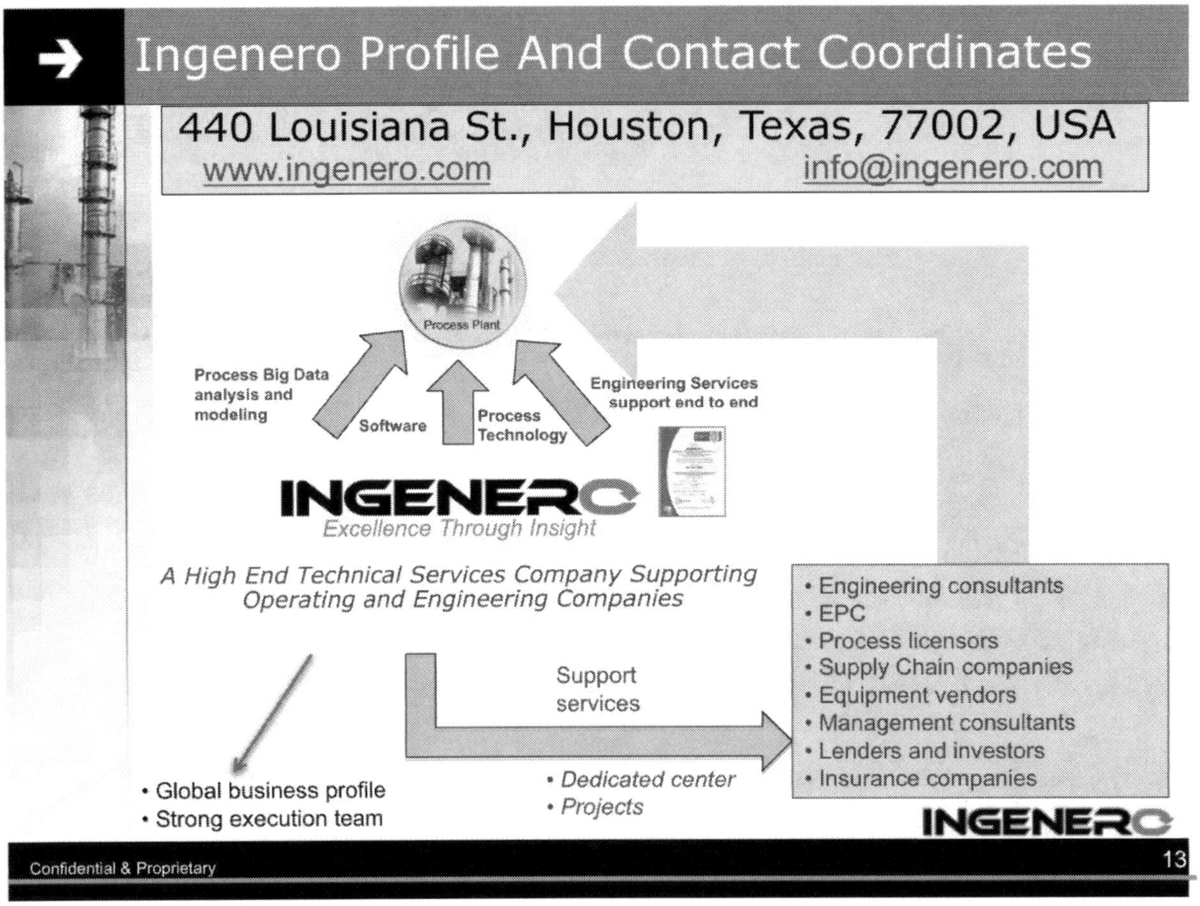
Ingenero Profile And Contact Coordinates
440 Louisiana St., Houston, Texas, 77002, USA
www.ingenero.com
info@ingenero.com
Process Plant
Process Big Data analysis and modeling
Software
Process Technology
Engineering Services support end to end
INGENERO
Excellence Through Insight
A High End Technical Services Company Supporting Operating and Engineering Companies
Support services
• Engineering consultants
• EPC
• Process licensors
• Supply Chain companies
• Equipment vendors
• Management consultants
• Lenders and investors
• Insurance companies
• Global business profile
• Strong execution team
• Dedicated center
• Projects
INGENERO
Confidential & Proprietary
13

AIChE Southwest Process Technology Conference

October 9-10, 2014

Cracker Feedstock Changes: From Naphtha to Ethane Cracking

Authors

Dr. Thomas Emmert, Manager Process Engineering Petrochemicals North America, The Linde Group,

Email: thomas.emmert@linde-le.com

Mobile: +1-281-777-4228

Address:

Linde Engineering North America

3700 West Sam Houston Parkway South, Suite #425

Houston, TX 77042

USA

Biography

2014-present	Manager Process Engineering Petrochemicals North America
2011-2014	(Lead) Process Engineer Petrochemicals
2008-2010	Linde Engineering Trainee program with various stations
2007	Dr. (equivalent to Ph.D.) for acoustics at Ecole Centrale de Lyon and Electricite de France (EDF), France
2004	Diploma in Mechanical Engineering (equivalent to M.Sc.) at Technical University Munich, Germany

Dr. Gunther Kracker, General Manager Process Engineering Petrochemicals, The Linde Group,

Email: gunther.kracker@linde-le.com

Mobile: +49-162-2004-014

Adresse:

Linde AG, Linde Engineering Division

Dr.-Carl-von-Linde-Straße 6-14

D-82049 Pullach/Munich

Germany

Biography:

2012-present	General Manager Process Engineering Petrochemicals
2011-2012	Lead Process Engineer and Process Technology Coordinator
2004-2010	(Lead) Process Engineer Petrochemicals
2004	Dr. techn. Process Engineering (equivalent to Ph.D.) at University of Technology Graz, Austria
2001	Diploma Process Engineering (equivalent to M.Sc.) at University of Technology Graz, Austria

Paul van Helmond, Director Petrochemicals North America, The Linde Group,

Email: paul.van.helmond@linde-le.com

Mobile: +1-610-203-8198

Address:

Linde Engineering North America

3700 West Sam Houston Parkway South, Suite #425

Houston, TX 77042

USA

Biography:

2013-present	Director Petrochemicals North America
2005-2013	Product Manager Ethylene Furnaces & Manager of Process Engineering with Project Management assignments at Linde Engineering North America
1990-2004	Thermal Engineer Furnaces & Project Management at Lummus Heat Transfer
2003	MBA Finance - University of St. Thomas (Houston)
1990	Ir. Degree in Physics - Eindhoven University of Technology, The Netherlands

Abstract

Driven by the US shale gas boom a lot of Ethane at fuel gas price is available on the market. Therefore cracking Ethane to Ethylene became more attractive than cracking liquid feedstocks in the last years. Therefore owners of liquid crackers envision to minimize liquid feedstocks or even to fully replace by Ethane to improve the plant economics.

The present paper briefly describes which process (engineering) challenges have to be met when changing from liquid to gaseous feedstocks in an ethylene plant. Concepts and associated modifications are explained in more detail for the hot section and C_2H_4/C_2H_6 separation unit being two key process units in an ethylene plant.

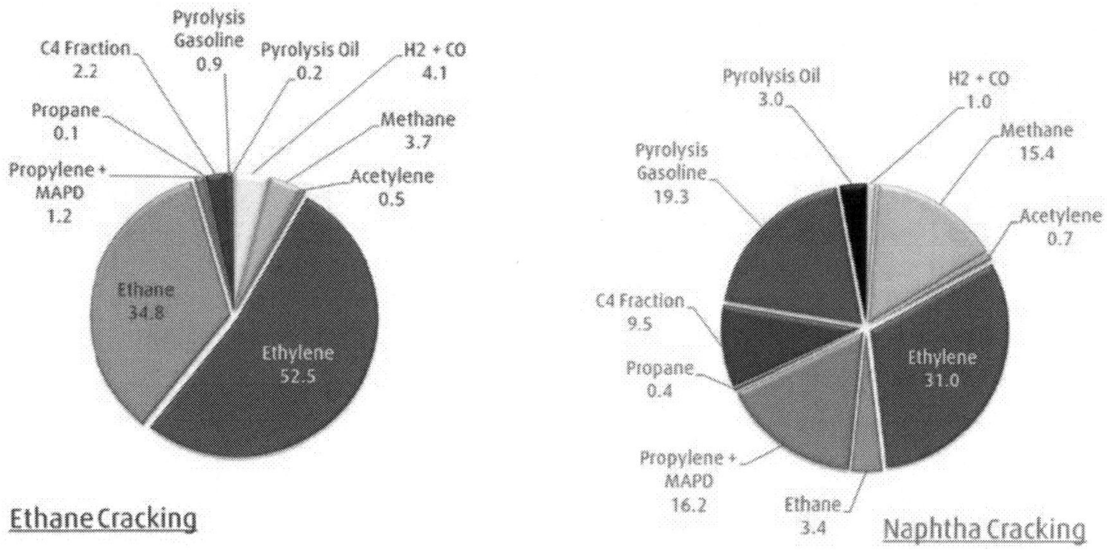

Figure 1: Once through yields/all concentration in wt%.

General

An ethylene plant consists of 2 main sections: a furnace section and a product recovery section. The product recovery section is comprised of several complex process steps: cracked gas quenching and heat recovery (hot section), cracked gas compression, acid gas removal, precooling and drying, deethanizer, cold chilling train, demethanizer, C2 hydrogenation, C_2H_4/C_2H_6 separation (C2 splitter) and C3 plus processing.

In Figure 1 the furnace yields are given for Ethane cracking and Naphtha cracking. Based on the table, the major impacts on different PU can be derived and are explained in the following.

Hot Section

A liquid cracker typically features a quench oil column where the cracked gas is quenched via quench oil (s. Figure 2). The heat removed from the cracked gas is typically used to produce dilution steam (process steam). The cracked gas is cooled to ambient temperature in a subsequent quench water column downstream. It is in this quench water column that the majority of the dilution steam is condensed. Unlike a Naphtha cracker, an Ethane cracker typically features a quench water column only, since Ethane cracking produces insufficient pyrolysis oil to operate a quench oil column. Furthermore, most heat is recovered in the furnaces and is not used to produce process steam. The necessary measures to operate a Naphtha cracker hot section under Ethane cracking conditions are described below.

Cracked Gas Compression:

By changing from Naphtha to Ethane cracking, the molecular weight will decrease, typically from 29 to 18. This impacts the head and power requirements of the cracked gas compressor. In most cases, the cracked gas compressor has to be modified (e.g. installation of new wheels).

Cold Train and Demethanizer:

Cracking Ethane reduces the amount of methane and increases the amount of hydrogen in the cracked gas. In a liquid cracker, methane acts like a refrigerant. In order to ensure proper operation of the cold section, process concept changes are expected. Several concepts, such as the addition of expander/booster units or integration of an additional open LP methane refrigerant cycle are possible.

C2H4/C2H6 Separation

Compared to Naphtha cracking, cracking Ethane will drastically increase the Ethane recycle at the same ethylene production rate. If there is no potential to increase column capacity by retraying the C2 Splitter with high capacity trays, it is recommended to reduce the load on the C2 Splitter system. This can be accomplished by using a pre-separation column, described below.

Modification options

In the following section, the Hot Section and C2 Splitter modifications necessary when changing from liquid to Ethane cracking are presented in more detail.

1. Hot Section

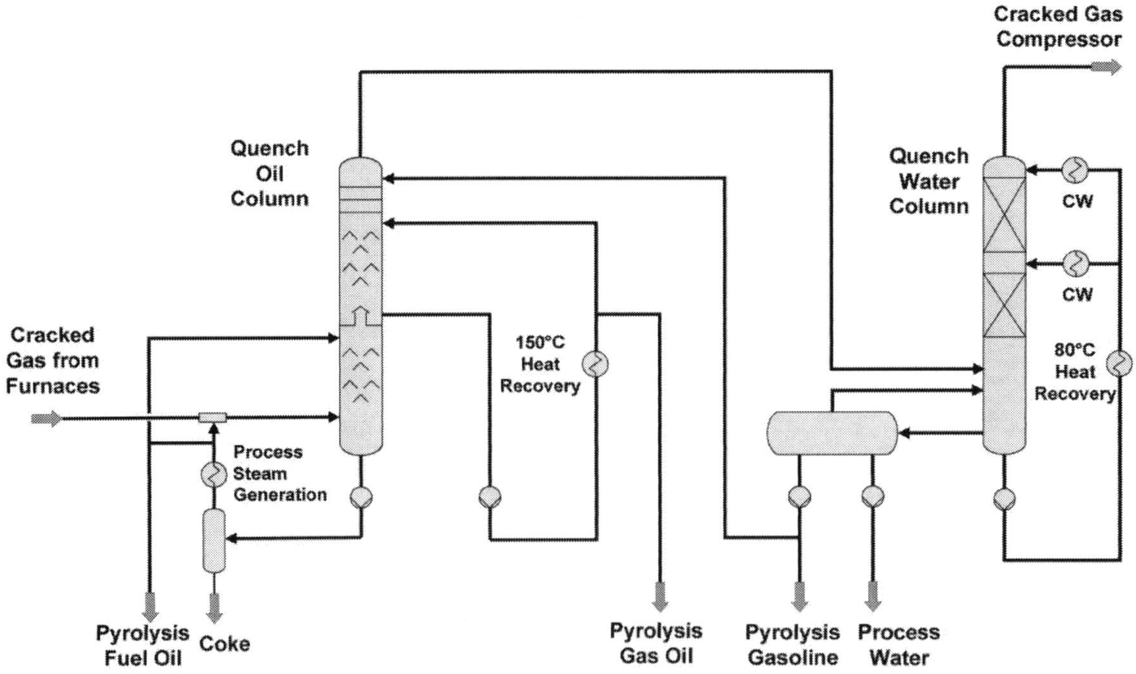

Figure 2: Process Flow Diagram of Naphtha Cracker Hot Section.

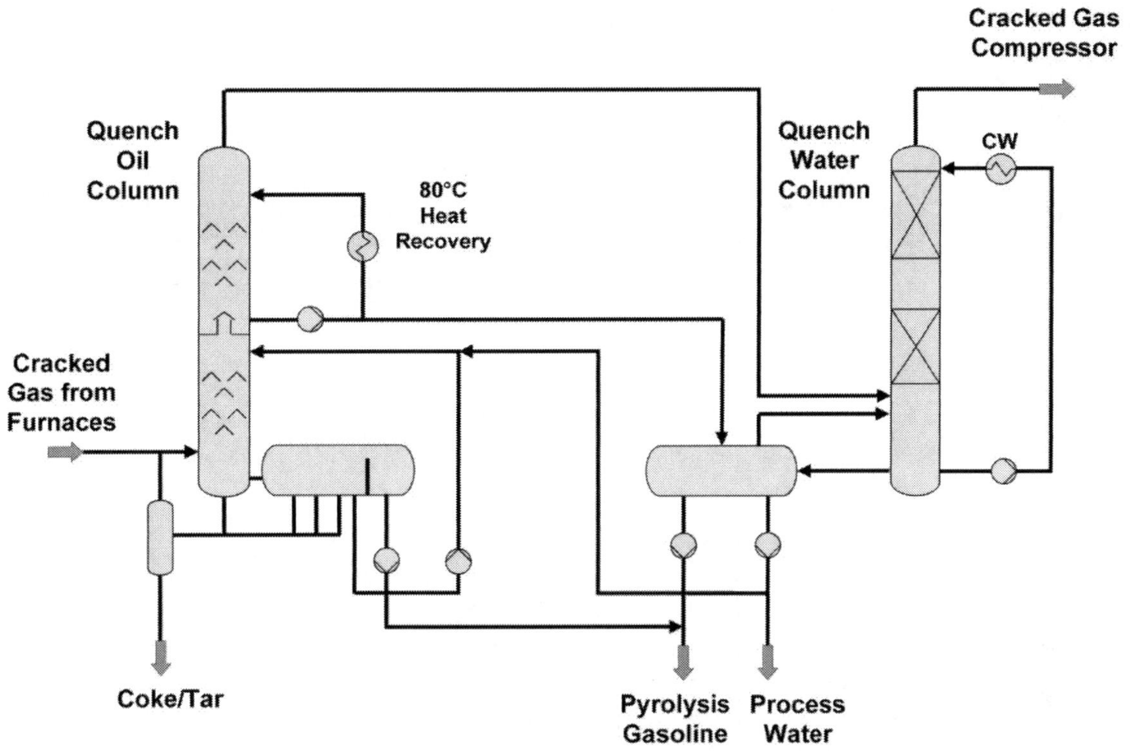

Figure 3: Process Flow Diagram of a Naphtha Cracker Hot Section, modified to crack Ethane.

Due to the lack of pyrolysis oil and heavy gasoline components in the cracked gas, the quench oil column cannot be operated. Therefore, additional flux oil (e.g.diesel fraction) has to be continuously added to the process. A sufficiently high amount of flux oil must be introduced to dissolve tar components and remove them from the system, to minimize the retention time of heavy hydrocarbons and to avoid the accumulation of fouling precursors like styrene and indene. However, the introduction of valuable flux oil can drastically reduce the plant economics. Therefore, if there is no future desire to maintain liquid cracking capability, it is recommended to transform the quench oil column into a quench water column.

Figure 2 shows the typical scheme of a Naphtha cracker hot section with a quench oil column and water quench column. Figure 3 shows the process flow diagram after modification to a Quench Water System. Note that the oil quench facilities have been removed and water quenching is now done in 3 steps. In the lower part of the quench oil column, the superheated cracked gas is quenched with water. The cracked gas leaving the lower section is saturated with water. In the upper column section, the cracked gas is quenched to an intermediate temperature (saturation and ambient temperature). The subsequent cooling to just above ambient temperature is then achieved by quench water cooled against plant cooling water. This proven three cycle concept avoids cycling heavy hydrocarbons in the cold zone of the quench tower at temperatures below the pour point. This concept is typically combined with continuous gasoline injection. Gasoline for continuous washing can be either imported or produced by plant internal recycles thanks to optimized gasoline management. This results in minimized fouling in hot section equipment and high quality process water/steam to ensure high furnace run length and high plant availability.

2. C2 Splitter System

Due to the higher amount of Ethane the C2 route (C2-Hydrogenation, C2 Splitter) is expected to be overloaded when changing Naphtha to Ethane feed, while maintaining the same C2H4 production. To keep the load on the C2 route similar to Naphtha cracking operation, a Pre-C2-Splitter Column can be implemented

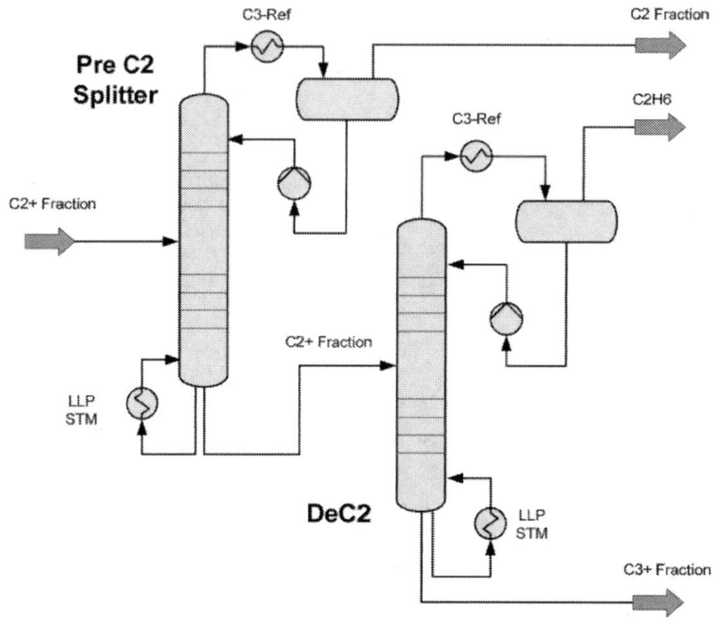

Figure 1: Process Flow Diagram Pre C2-Splitter and Deethanizer (DeC2)

Instead of being routed to the Deethanizer column, the C2+ Fraction is routed first to the Pre-C2-Splitter. The Pre C2-Splitter produces an overhead product which has the same flowrate and Ethylene to Ethane ratio as that for Naphtha plant operation. Thus, the C2 Hydrogenation system, the C2 Splitter and their associated process equipment can be operated without modifications.

The bottom product of the Pre-C2-Splitter (C2+ fraction) is acetylene free with traces of ethylene. The bottom product is routed to the existing Deethanizer column, which produces the C3+ fraction and an Ethane recycle stream to the furnaces (in addition to the recycle from the C2-Splitter column). The internal traffic of the Deethanizer column is expected to be lower than for current operation, necessitating tray modifications. Thanks to the lower load, the modified Deethanizer column can be operated at lower pressure. The condenser and reboiler are operated similar to the Naphtha cracking operation and require only minor modifications.

Conclusion

The change from Naphtha to Ethane feedstock cracking requires modifications to key process units and results in CAPEX even if no additional ethylene production is envisioned. Nevertheless, the similarity of ethane to the value of fuel gas in the US market provides an economic incentive, in most cases, for the implementation of these measures. Feedstock flexibility projects are currently ongoing, and the trend towards gaseous feeds is expected to continue not only in the US, but all over the world.

Cracker Feedstock Changes.
From Naphtha to Ethane Cracking

AIChE Southwest Process Technology Conference
Galveston, October 10, 2014

Dr. Thomas Emmert
Dr. Gunther Kracker
Paul van Helmond

Linde Engineering North America Inc.

THE LINDE GROUP
Linde

Shift from Liquid to Gaseous Feedstocks at US gulf coast

THE LINDE GROUP
Linde

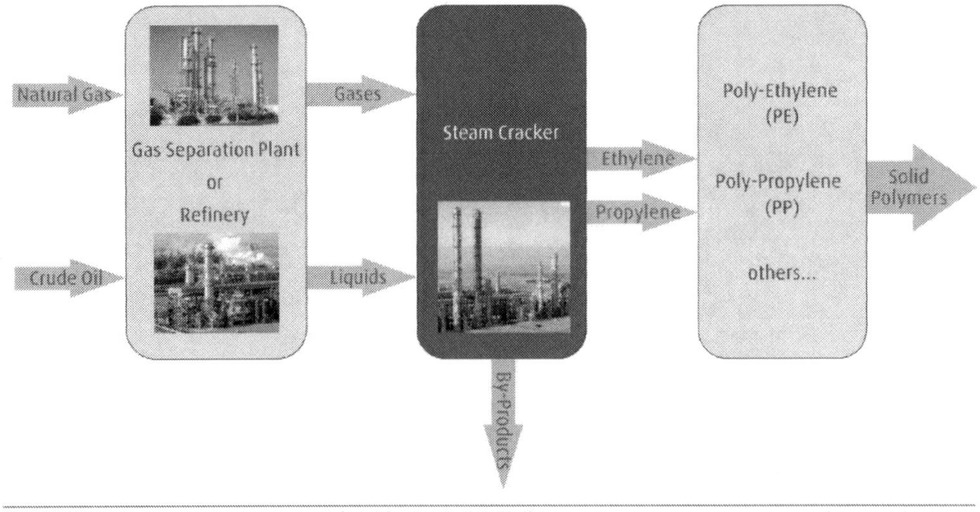

Naphtha to Ethane Cracking ?

Economics

- US shale gas boom
- Low ethane prices
- High ethane availability

Flexibility ?

- Maximise ethane cracking
- Dual Feed Cracker

Capacity Increase ?

- Increase plant capacity e.g. adding furnaces

Block Flow Diagram of Typical Naphtha Cracker

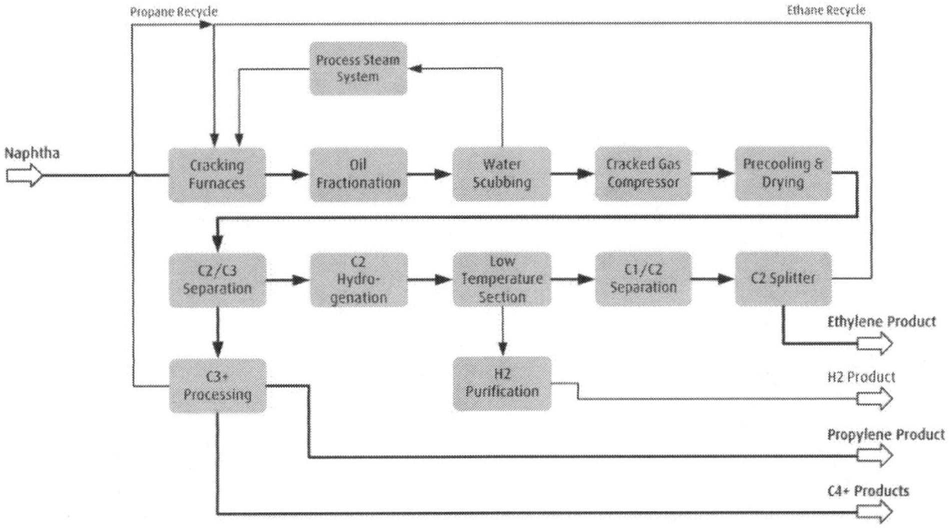

Typical Composition of Total Cracked Gas from Furnaces

Once through yields / All concentrations in wt %

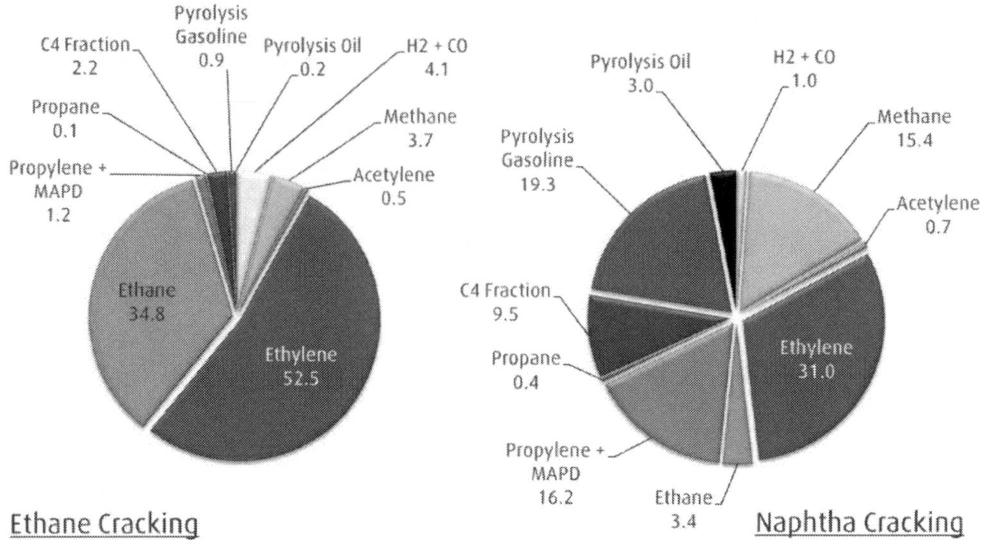

Linde Engineering North America Inc © Linde AG – Refer to Protection Notice ISO 16016 AIChE Conference 2014 / Thomas Emmert

Impact of Ethane Cracking on Naphtha Cracker

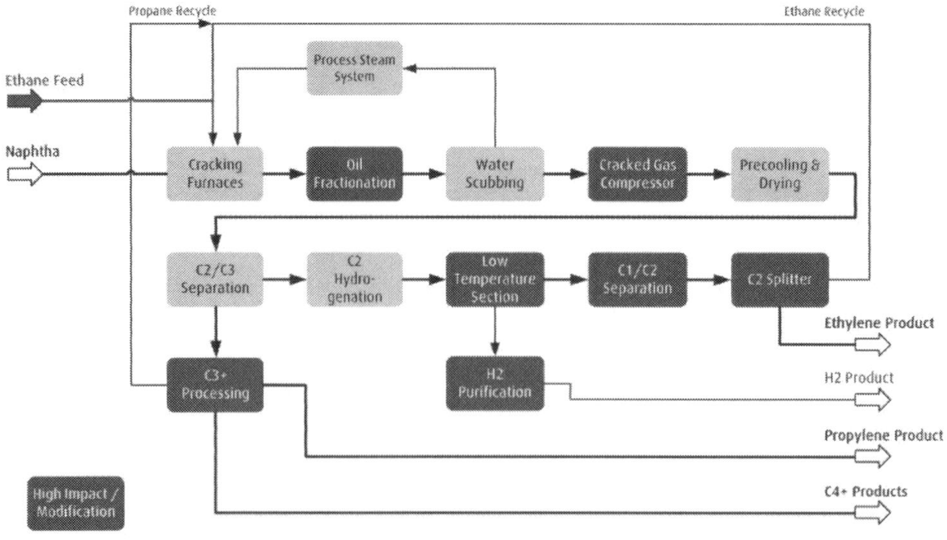

Linde Engineering North America Inc. © Linde AG – Refer to Protection Notice ISO 16016 AIChE Conference 2014 / Thomas Emmert

Impact on Cracked Gas Compressor

THE LINDE GROUP
Linde

Considerations

- Lower Molecular Weight=> Higher Polytropic Head at given pressure ratio
- Higher Polytropic Head => Higher Shaft Power

CGC unloading

- first stage suction pressure increase
- re-rotoring
- modify or replace driver
- parallel compressor (stage)

Impact on Hot Section

THE LINDE GROUP
Linde

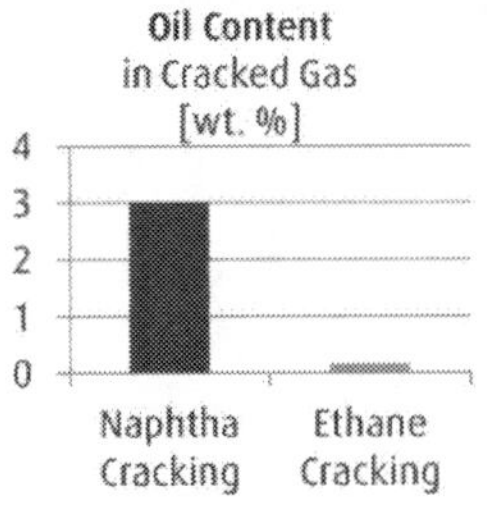

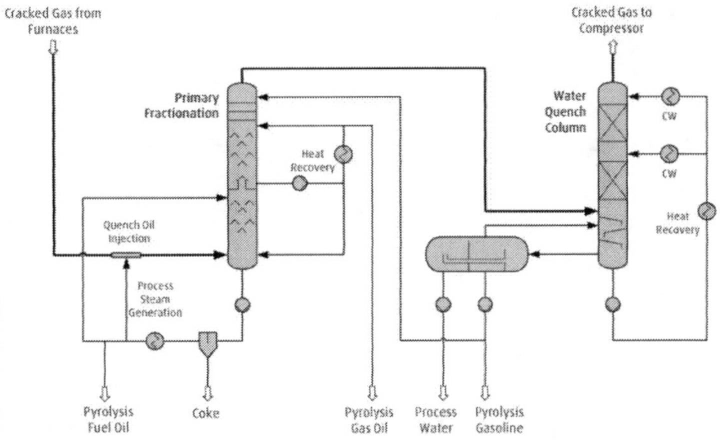

Import of Flux Oil

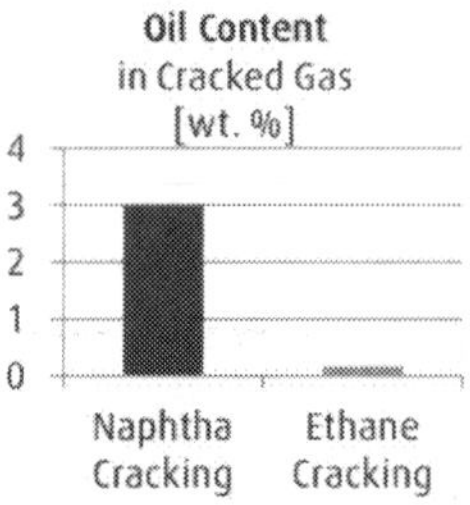

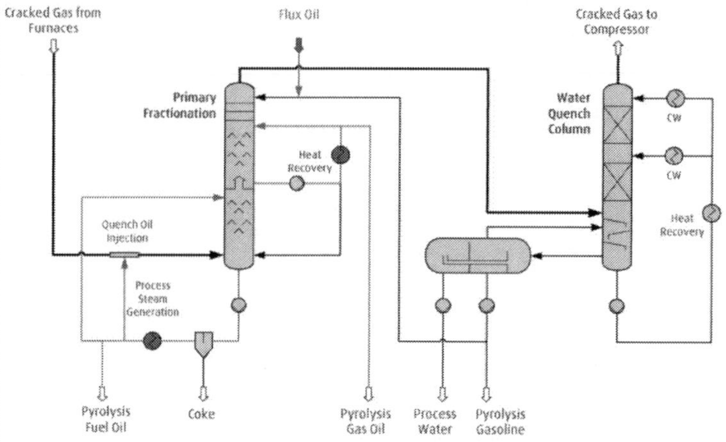

Conversion to 3-cycle Quench Water System

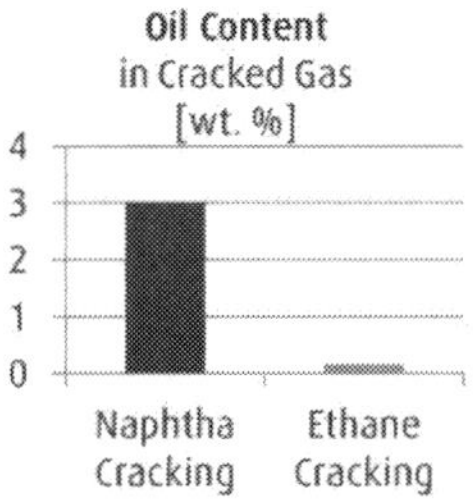

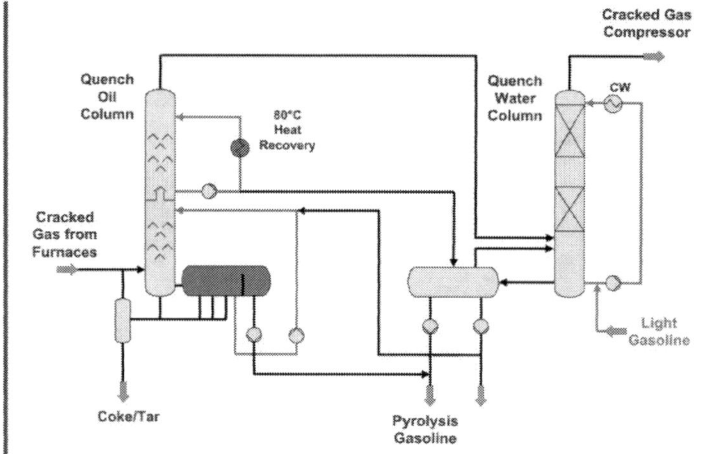

Comparison 2-cycle and 3-cycle Quench Water System

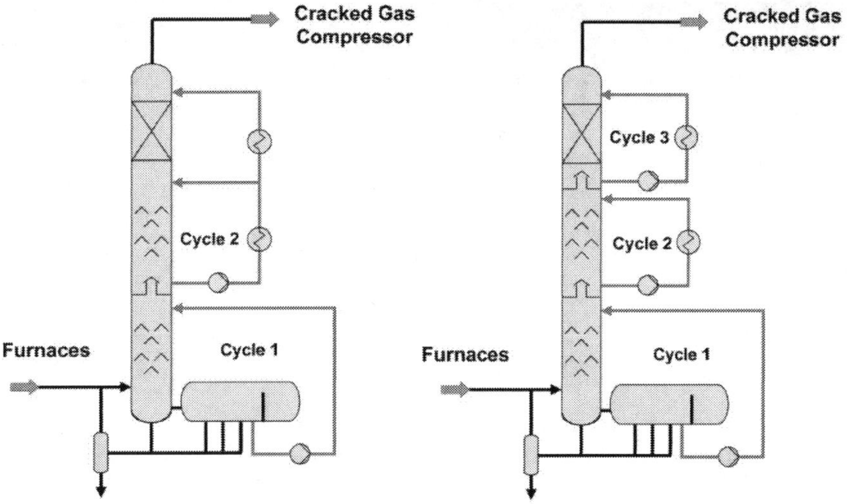

Benefits of 3 cycle Quench Water System

Fouling in Quench Water Tower and Cooler

Solidification of heavy hydrocarbons e.g tar, naphthalene when process temperature at or below pour point temperature

Minimization of Fouling

Three cycle concept avoids cycling of heavy HC into the cold zone of the quench tower where process temperatures could be below their pour point

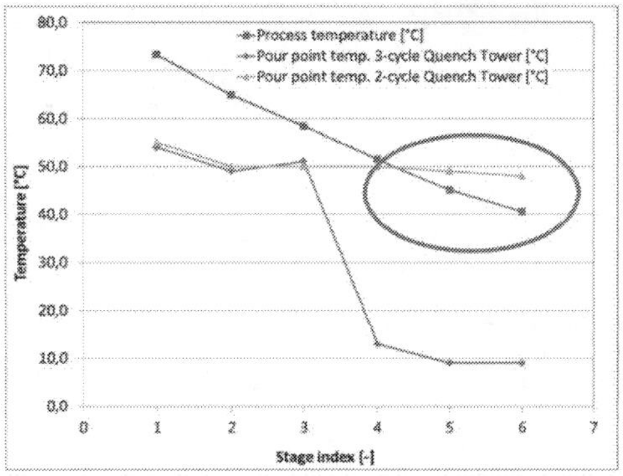

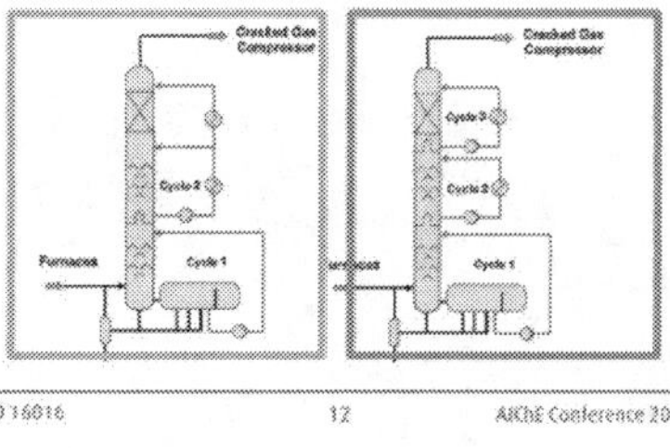

Benefits of continuous gasoline injection

Continuous Gasoline Injection

- Reduction of pour point in Packing Section
- Minimal Fouling
- High process water quality
- Maximum furnace run length

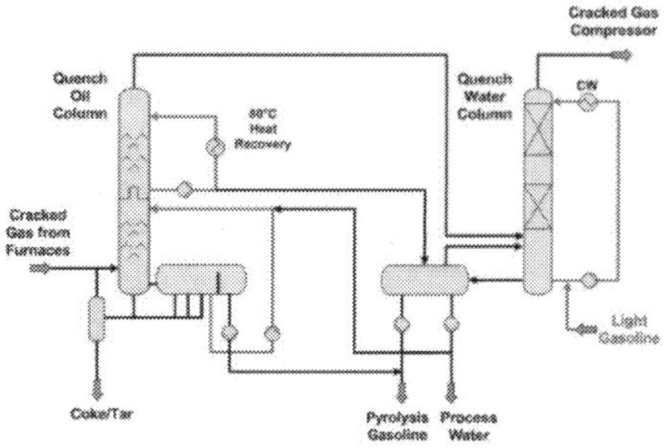

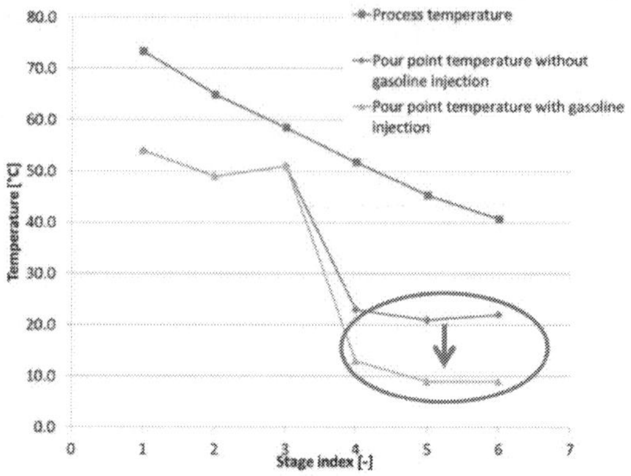

C2H4/C2H6 Separation (C2-Splitter)

	Ethane lb/hr / C2H4-production rate	Naphtha lb/hr / C2H4-production rate
Ethylene - C2H4	1.00	1.00
Ethane - C2H6	0.66	0.10
C3+ Hydrocarbons	0.09	1.56

Gas Cracking at constant ethylene yield → Hydraulic Overload of C2 Route → Plant Modification inevitable

Debottlenecking Concepts

THE LINDE GROUP
Linde

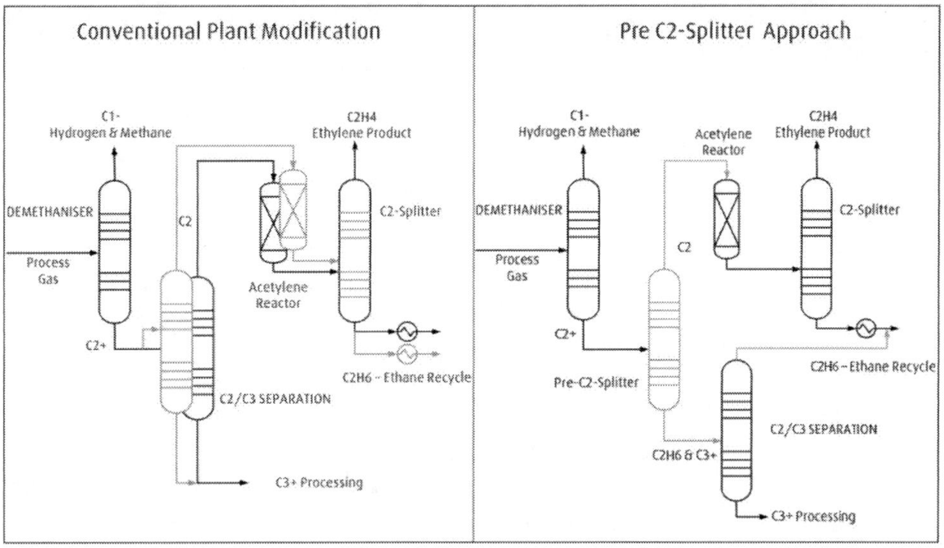

Advantages of a Pre C2-Splitter Approach

THE LINDE GROUP
Linde

- ➢ Reduction in equipment cost
- ➢ Potential for pre-fabrication and pre-installation
- ➢ Reduction of plot space
- ➢ Reduction of tie in modification
- ➢ Reduction of turn around time

Summary

THE LINDE GROUP

Linde

- Modifications to key process units necessary
- CAPEX costs for a given ethylene production rate
- Cheap ethane provides an economic incentive for implementation

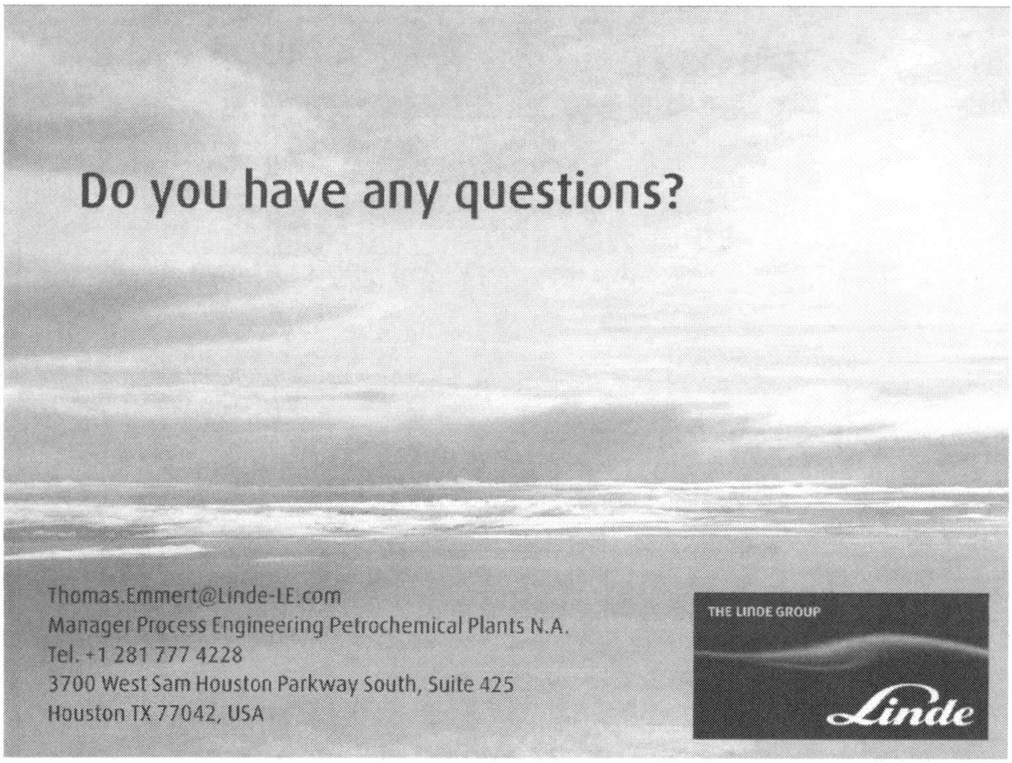

Development of Methane Run Concept in Ethane Cracker Design

Ramakrishnan Sankara-Subramaniam, Wadie Malaty and Turi Odegard

Technip Stone & Webster Process Technology

AIChE Southwest Process Technology Conference – October 10th, 2014

Introduction

- Technip implemented the concept of Methane Run for startup & shutdown of a Technip Cracker in the Middle East.
- The concept evolved on the recent USGC crackers, demonstrating continued development across the Technip offices, and effectively implementing the corporate slogan “One Technip” that was set after Technip’s acquisition of Shaw E&C (formerly Stone and Webster Process Technologies).
- Technip also has similar experience with successfully starting up a Cracked Gas Compressor on methane rich natural gas, for pre chilling in India.
- Although presented here for the reference USGC ethane cracker, the Methane Run concept applies and has been developed for the suite of ethylene technologies that Technip offers across the world.

Introduction

- **Technip evolved the Methane Run concept further from the initial application in the Middle East**
  - To reduce the HRVOC (Highly Reactive Volatile Organic Compounds) release
  - To minimize flaring of any hydrocarbons during start-up, shutdown and upsets/emergencies
  - To get all plant sections ready for initial start-up, restart after emergencies and speed-up on-spec production
  - Additional concepts for flexibility and integration have been applied to the new Gulf Coast Cracker designs
  - Concepts should be developed and incorporated early in the design during PDP/FEED Phase

Technip

Reasons and Sources of Flaring

Technip

Reasons for Flaring

- **Purge out inerts during start-up**
- **Stabilization of units during start-up**
- **Imbalance in inventory**
  - high liquid level in columns, suction drums
- **Major process upset**
- **Scheduled turnaround preparation (deinventorying)**

- **Traditional start-up is based on once-thru operation of cracking fresh feed and flaring until on-spec production is achieved.**
- **Methane Run (after development) is based on methane circulation. If off-spec Ethylene storage is available, start-up can be essentially flareless.**

Technip

Reference Ethane Cracker Design

- Front End Deethanizer and C2 Hydrogenation
- Low pressure C2 Splitter Heat Pump integrated with Ethylene Refrigeration

Flaring Locations

1. **Quench Water Tower overhead**
   CGC trip / start-up; shortage of VHP steam to start CGC
2. **Cracked Gas Compressor discharge**
   Off-spec for CO_2 / H_2S and stabilization
3. **Acetylene Reactor outlet / upstream of Demethanizer Chilling Train**
   Off-spec Acetylene Reactor effluent
4. **C2 Splitter inlet**
   Off-spec for acetylene / methane
5. **Ethane Feed Flash Drum overhead**
   CGC trip

Technip

Ethane Cracker Flaring Locations

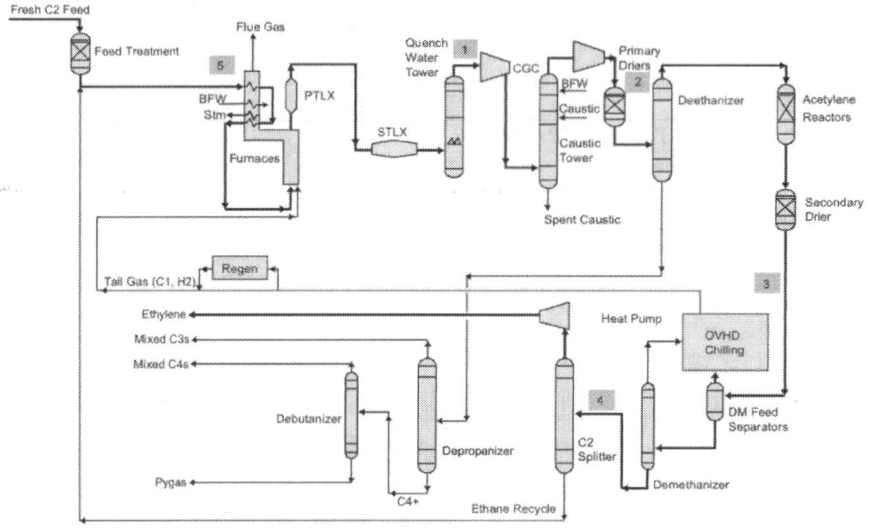

Technip

Concepts Developed to Reduce HRVOC Release

Technip

Recycle Loop Concepts Implemented

- Run CGC on methane rich natural gas
- Establish recirculation from final CGC discharge through ethane feed system to the 1st stage suction
- Clean-up of CG system inventory, free of CO_2 and H_2S
- Utilize clean compressor discharge to prechill the downstream sections
- Routing of methane rich natural gas through furnaces to generate steam as required for steam balance
- Recovery of Demethanizer bottom through feed system using ethane recycle header

9 SPTC-2014

Technip

Recycle Loop from final CGC Discharge through Ethane Feed System to the 1st Stage Suction

Loop:

1. Final CGC discharge to ethane feed system
2. Ethane feed system bypassing furnaces to QWT inlet or CGC suction
6. Natural gas inventorying

Reasons:

- Use natural gas or ethane feed for inventorying CGC while remove impurities like H_2S and CO_2
- Establish recirculation through ethane feed system with flexibility to route through furnaces to generate steam
- When on spec for H_2S and CO_2 - chill down of the systems downstream of Cracked Gas Drier

10 SPTC-2014

Technip

Recycle Loop from final CGC Discharge through Ethane Feed System to the 1st Stage Suction

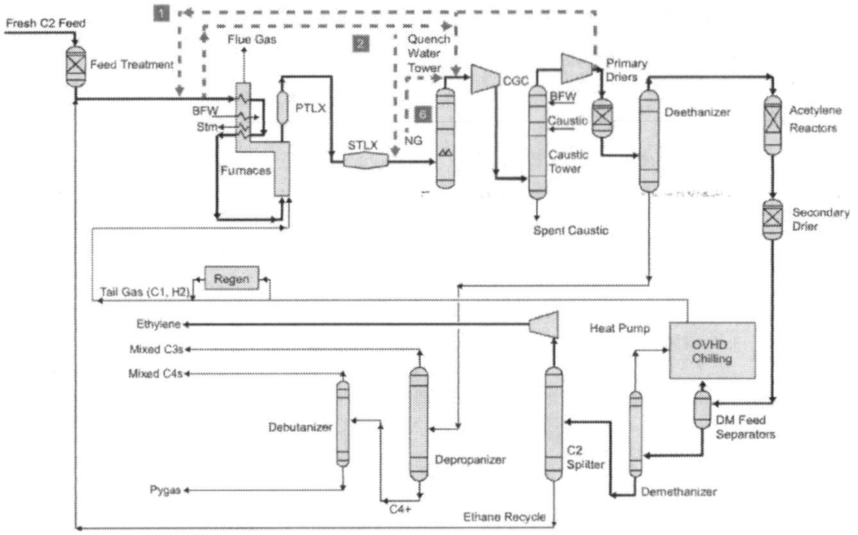

Technip

Demethanizer Bottom to Ethane Recycle Header

Loop:

3. Demethanizer bottoms to ethane recycle header
5. Two-way pipeline with existing plant / facilities

Reasons:

- Recycle from Demethanizer bottom during prechilling to the ethane recycle header to use Ethane Vaporizer
- During scheduled turnaround shutdowns, route the Demethanizer bottom inventory through CG system to:
  - Avoid cold liquid blowdown
  - Use C2 rich stream to flush the CG system free of Benzene instead of flaring
  - Complete recovery when new unit is integrated with existing Ethylene Plant
    - Expect to transfer vapor and vaporized liquid inventory
    - Keep CGC running until deinventory is complete
    - Finally purge the system including recycle lines with natural gas for Benzene freeing

Technip

Demethanizer Bottom to Ethane Recycle Header

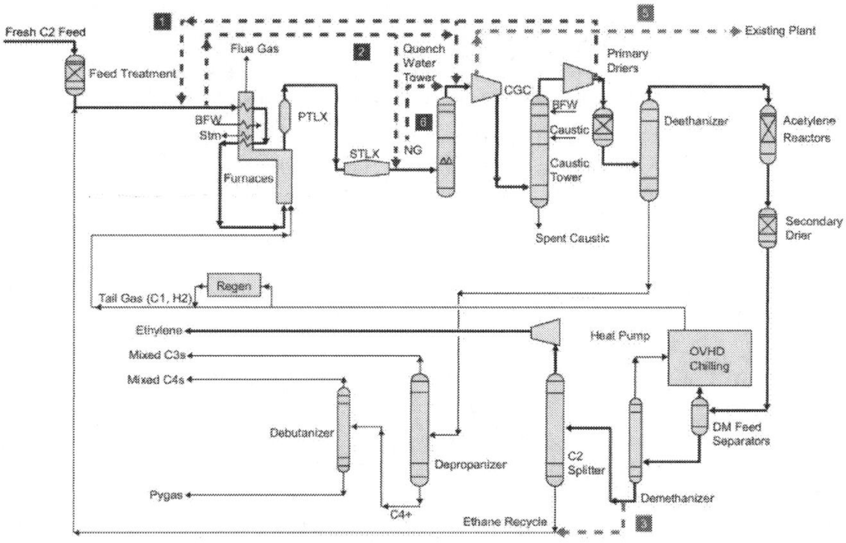

Technip

Deethanizer Bottom to CGC 3rd Stage Suction

Loop:

4. Deethanizer bottoms to CGC 3rd stage suction

Reasons:

- Recycle from Deethanizer bottom during prechilling and system loading for start-up
- During scheduled turnaround shutdowns, route the Deethanizer bottom inventory through CG system to:
  - Avoid cold liquid blow down
  - Utilize the stream to flush the CG system free of Benzene instead of flaring
  - Complete recovery when new unit is integrated with existing Ethylene Plant

Technip

Deethanizer Bottom to CGC 3rd Stage Suction

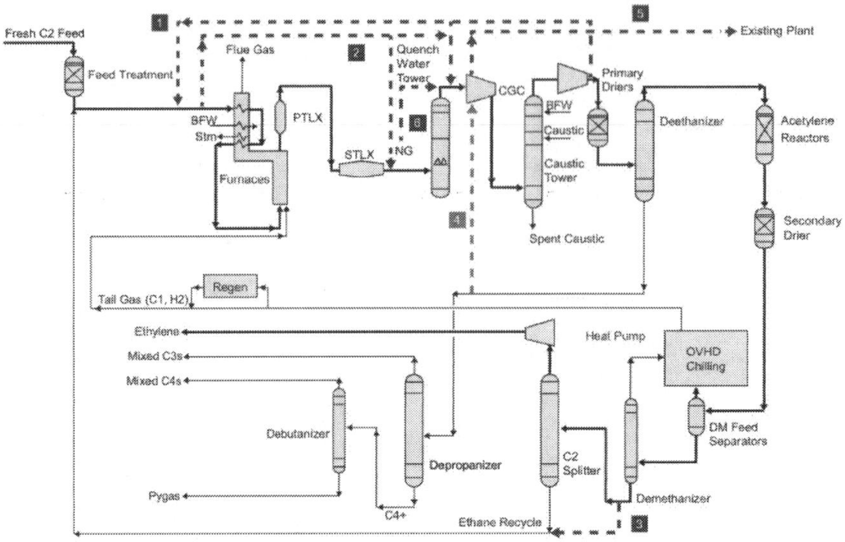

Technip

Tail Gas Recycle to CGC 1st Stage Suction

Loop:

7. Recycle tail gas to CGC 1st stage suction

Reasons:

- Allow recycling of the tail gas, while maintaining the tail gas header pressure at normal value, instead of equalizing with the CGC suction pressure
- Allow regeneration of the CG driers if required
- Allow adjustment of the C1 to C2 ratio in the CGC loop while recycling Acetylene Reactor effluent
- If nitrogen is used for drier regeneration, tail gas can be routed to fuel gas directly to avoid impact on plant operation

Technip

Tail Gas Recycle to CGC 1st Stage Suction

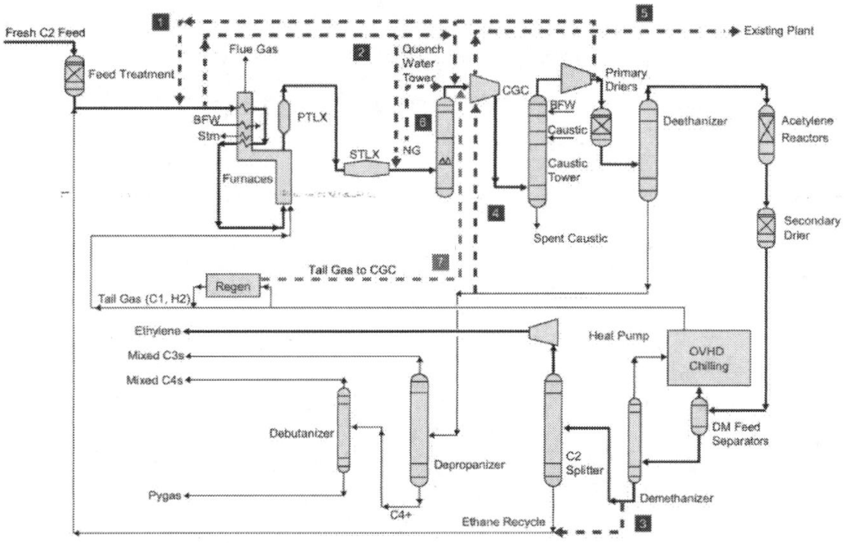

Technip

Acetylene Reactor Effluent Recycle to CGC 4th Stage Suction

Loop:

8. Recycle reactor effluent to CGC 4th stage suction

Reasons:

- Maintain minimum flow through front end Acetylene Reactor during low plant loads and avoid potential run away condition

Technip

Acetylene Reactor Effluent Recycle to CGC 4th Stage Suction

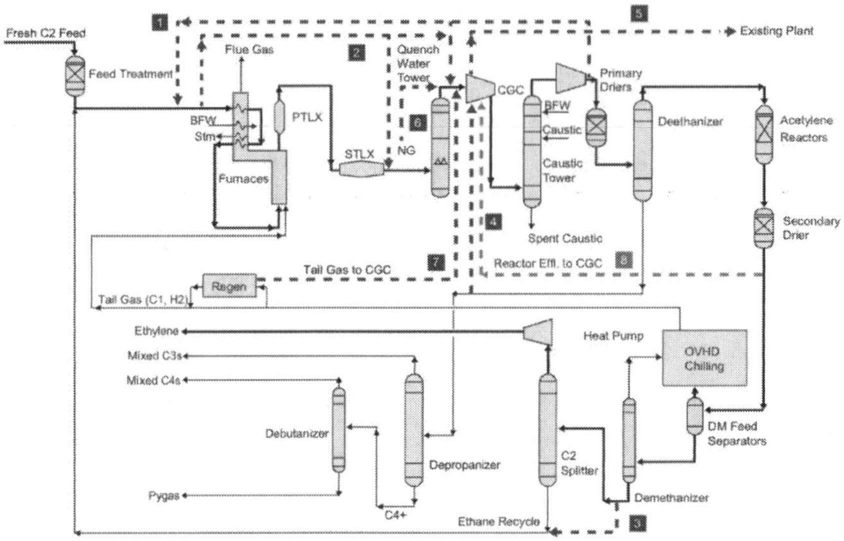

Technip

Methane Run Route Details

Advantages:

- Total circulation with gas of comparable molecular weight through most separation sections
- Close to normal operating pressure and temperatures prior to furnace feed-in
- Pressurization of the Acetylene Reactor system and establishing circulation
- Complete testing of most instrumentation and rotating equipment
- Seamless / smooth change over from chill down to processing furnace effluent
- Substantial reduction in the time required to achieve on-spec Ethylene Product

Technip

Methane Run Route Details

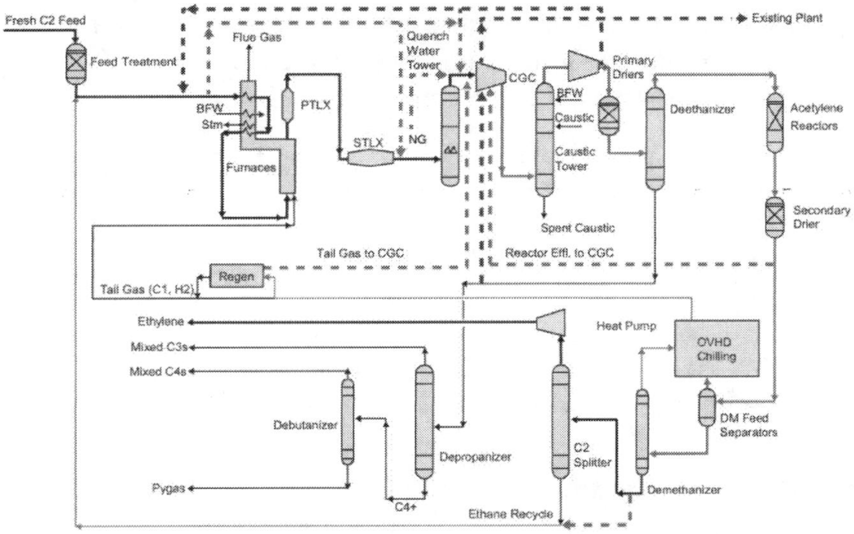

Technip

Exceptions to Minimized Flaring

- Acetylene Reactor Effluent Off-Spec Flaring
- Discharge Vent from Ethylene Compressor during Start-up

Technip

Acetylene Reactor Effluent Off-Spec Flaring

Without any off-spec storage available we need to flare either:

1. From upstream of Demethanizer Chilling Train or
2. C2 Splitter inlet

1. **Flaring upstream of Demethanizer Chilling Train will not contaminate Demethanizer system**
   - Reduce system purge time
   - But Increase system stabilization time

2. **Flaring from C2 Splitter inlet will allow**
   - Recovery of hydrogen and methane components
   - Reduce flare load and reduce external fuel gas make-up
   - Results in Acetylene contamination of Demethanizer system, while continue to operate close to normal conditions

When off-spec storage is available:

- Can store and reprocess liquids from Deethanizer Reflux Drum or off-spec liquid ethylene product
- HRVOC release can be totally eliminated

Technip

Acetylene Reactor Effluent Off-Spec Flaring

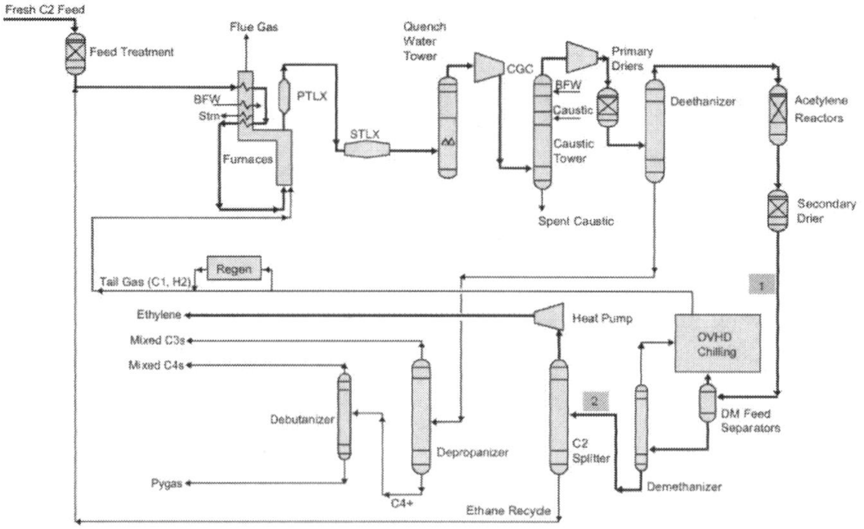

Technip

Minimization of Discharge Vent from Ethylene Compressor during Start-up

- **Potential option is:**
  - Use ethane to start C2R Compressor
    - Use ethane stream free of impurities like CO_2 and sulfur compounds from the CGC recirculation loop
  - Use ethylene for establishing "Total Reflux" operation (similar to boot strapping)

- **Advantages:**
  - Release of ethylene vapors (HRVOC rich) from C2R Compressor system to flare can be eliminated
  - Ethane can be recycled

Technip

Conclusions

- Close cooperation and open discussion between Technip and our clients can help develop and implement the Methane Run concept
- Implementation of the concept early in the design minimizes schedule and cost impact
- Current flow sheet improves start-up options and help in minimizing flaring
- Adding off-spec storage reduces emissions even further
- Testing of instrumentation and equipment avoids delays, due to improved reliability
- Effective removal of any foreign material, moisture etc., prior to fresh feed-in
- Anticipate substantial reduction in the time required to achieve on-spec Ethylene Product

Technip

Thank you!

Ramakrishnan Sankara-Subramaniam, Wadie Malaty and Turi Odegard
Technip Stone & Webster Process Technology
Date

27 SPTC-2014

Technip

Petrochemical Industry – Changing Landscape

Sanjeev Kapur
Apex PetroConsultants
Sanjeev@apexpetroconsultants.com
www.apexpetroconsultants.com

6th AIChE Southwest Process Technology Conference
October 10, 2014; Galveston, TX

APEX PetroConsultants, LLC

"The biggest risk is not taking any risk... In a world that changing really quickly, the only strategy that is guaranteed to fail is not taking risks."

\- Mark Zuckerberg

Agenda

- Industry Fundamentals
- Trends
- Summary

APEX PetroConsultants, LLC

Industry Fundamentals

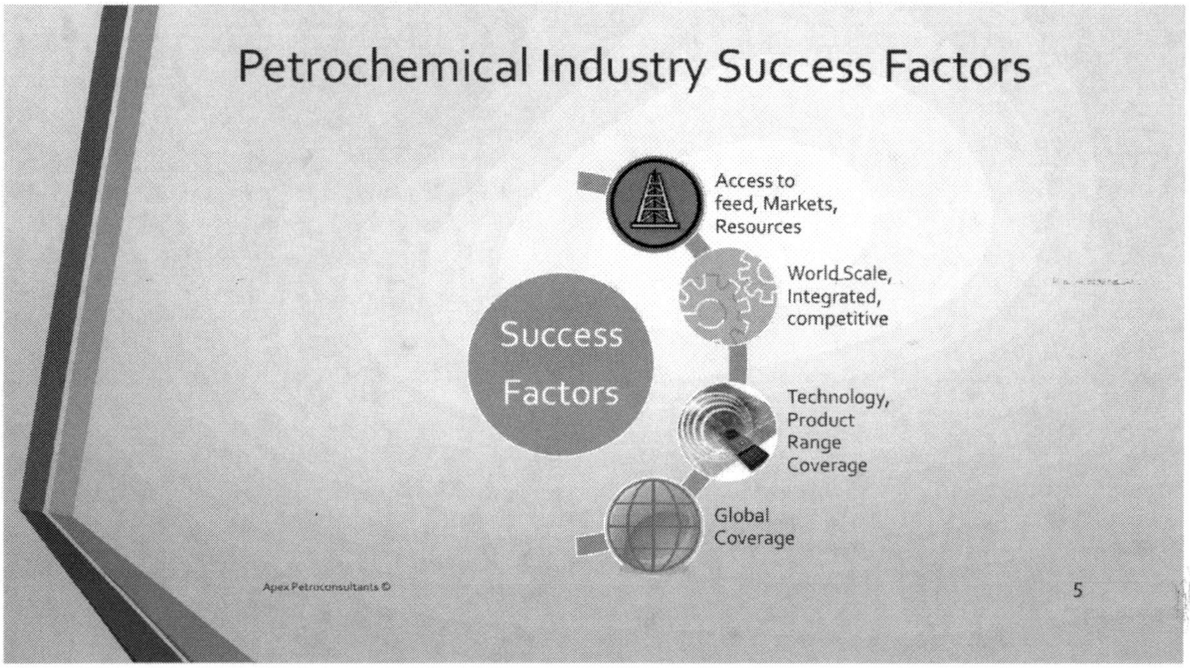
Petrochemical Industry Success Factors
Success Factors
Access to feed, Markets, Resources
World Scale, Integrated, competitive
Technology, Product Range Coverage
Global Coverage
Apex Petroconsultants ©
5

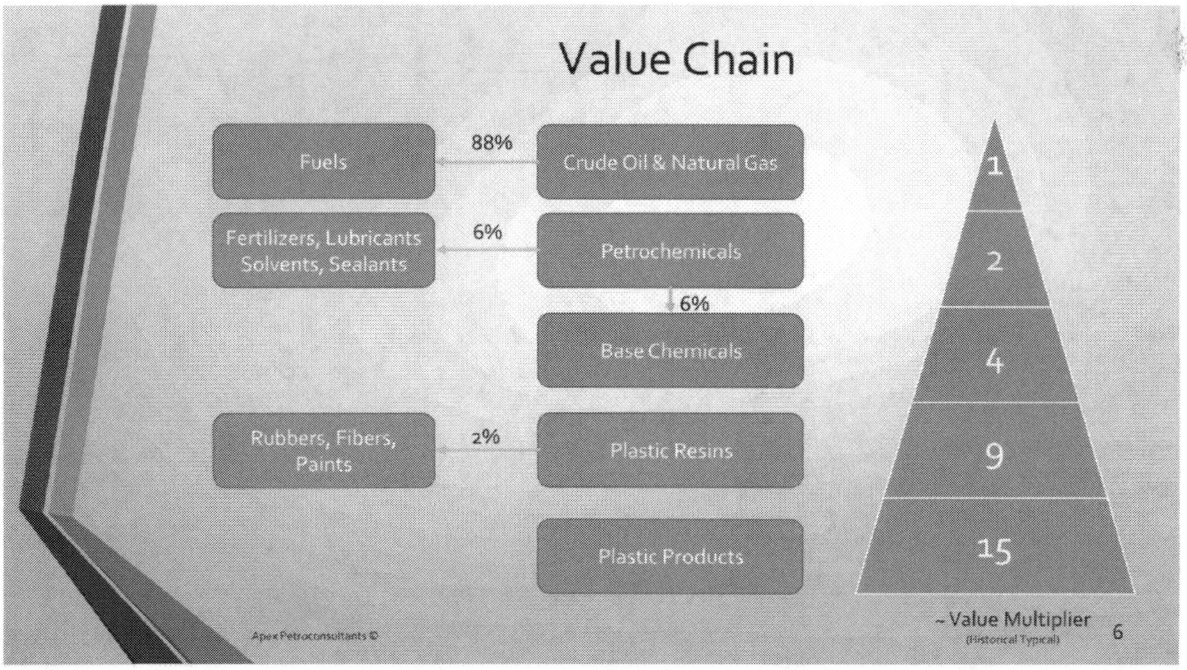
Value Chain
Fuels
88%
Crude Oil & Natural Gas
Fertilizers, Lubricants Solvents, Sealants
6%
Petrochemicals
6%
Base Chemicals
Rubbers, Fibers, Paints
2%
Plastic Resins
Plastic Products
1
2
4
9
15
~ Value Multiplier
(Historical Typical)
Apex Petroconsultants ©
6

APEX
PetroConsultants, LLC

Industry Trends

Ever Evolving Landscape – Cyclical Business

Apex Petroconsultants ©

7

Petrochemical Industry – Top Trends

- Shale driven growth
- Alternate feedstock - Coal to a large extent, potentially bio/renewables
- Middle East feed diversification and breadth of product coverage
- Demand growth – led by China, India, Brazil, Turkey, Mexico etc. driven by growing middle class and urbanization; growth of construction sectors/ automobile industry/ packaging industry
- Energy/Feedstock Prices – specifically Crude Oil Prices relative to natural gas and coal
- Technology & Innovation - On-purpose Technologies for C3s, C4s and Aromatics; material advances (lightweight, nano, functional textiles etc.)
- Environmental - Green House Gas Emission Regulations, Water management, recycling/waste treatment

Apex Petroconsultants ©

8

Shale driven growth

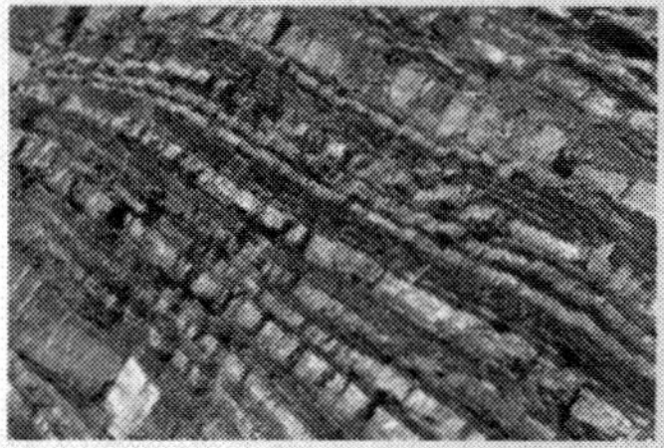

- Fracking technology unleashed the development of Shale Gas/Oil in US
- Leading to competitively priced ethane and renaissance for US petrochemical industry (even in North East where there are no petrochemical facilities - Pennsylvania, West Virginia)
- Ethane cracking leads to lack of co-products availability – propylene, butadiene, butenes, benzene
- Shale developments also increased availability of competitively priced propane and butane (NGLs) paving way for on-purpose production of propylene and potentially butadiene
- Abundant availability of NG and decoupled from oil pricing- promoting syngas based ammonia/fertilizers, methanol industries
- Shale gas reserves in rest of the world, notably China, are also likely to develop in next 10 to 20 years

Apex Petroconsultants © 9

Coal Based Petro-chemistry –
Alternate Feed Story of China

- Coal prices very competitive in China (Per ton basis - \$30 in Xinjiang, \$55 in Shanxi and rest of China in 70 – 100 \$ range), High quality bituminous coke (for gasification) in Xinjiang and Inner Mongolia
- Coal based ethylene forecasted to grow to nearly 10 million tons (nearly31% of overall); propylene from coal will grow to more than 6 million tons (nearly 15%)
- Cash cost of ethylene based on coal is lower and competitive as compared to naphthas (Europe, China, Korea)
- Potential for significant CT-MEG capacity (cash cost lower as compared to naphtha based)
- Significant amount of PVC/VCM facilities based on coal chemistry, nearly 6 million tons of PVC (64% of Chinese PVC production)

Apex Petroconsultants © 10

Middle East Feed Diversification

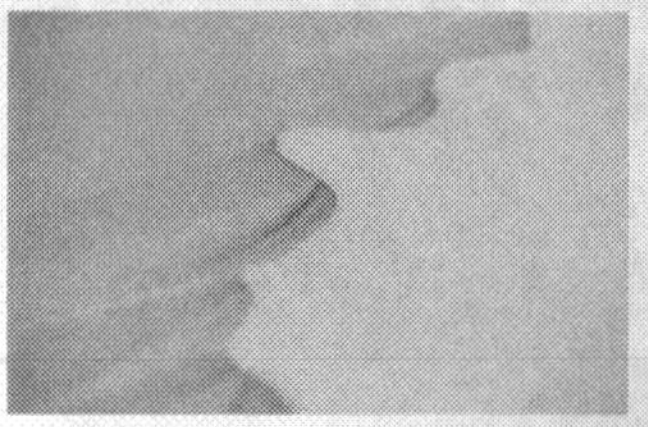

- KSA's petrochemical growth was driven by advantage ethane feedstock (based on associated gas). Additional growth is not supported by cheap ethane.
  - Growth of chemical industry will be driven by feed diversification – heavier feed mix provides synergy with broad product range strategy
  - Industry is also focusing on broader product range coverage – high value chemistry, high labor intensity
- Countries like Iraq have the potential for advantaged ethane feed – likely to develop over next ten years

Energy/Feedstock Prices

- Crude Oil – in $15 – 18/ million Btu (International – Brent)
- Natural Gas –in $4-5 /million Btu range (Henry Hub - HH)
- NGL prices (including propane) – in $10/ million Btu range (HH)
- NG delivered pricing/million Btu – US $3-5, Europe $10-12, Brazil $13-15, India $12-14, China/Korea/Japan - $15+

- Shale developments in US resulted in:
  - Growth of Propane and Butanes faster than ethane due to depressed NG prices
  - Expected surplus of Ethane for export, facilities being developed for ethane export to Europe and India

Technology Trends

Shale Gas and Coal feed driven

- Commercial Production
  - On-purpose technologies for filling the by-product gap
    - ✓ Propane Dehydrogenation for Propylene
    - ✓ MTO/MTP for Propylene (and Ethylene)
    - ✓ Butane Dehydrogenation for Butadiene
  - Coal Based MEG through Oxalate Ester route
  - Coal based EDC/PVC – environmental concerns (alternates being developed to overcome environmental issues)
- Developing Technologies
  - Methane coupling for ethylene production
  - Methanol to Aromatics
  - Bio routes for C4 chemistry

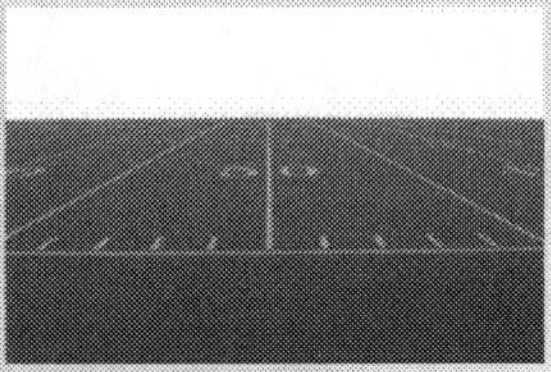

Challenges Facing Industry

Planning is Vital

- Robust business case development and careful master planning are crucial first steps
  - broad understanding of business, technology, execution and economic factors is essential
- Integration is the key
  - Hydrocarbon management, energy integration, synergies
- Successful delivery of mega capital projects
- Economics of large scale projects are based on long-term outlook – 20 to 30 years is typical
- Access to financing, complex financing structure
- Capital efficiency, project costs - given price volatility
- Regulatory uncertainty – permitting, taxation, trade barriers, changing environmental regulations
- Technical and skilled resources

Summary

Summary

"Change is the law of life. And those who look only to the past or present are certain to miss the future."

John F. Kennedy

- Petrochemical industry is changing rapidly driven by Shale developments and China's drive towards coal
- Master-planning of these businesses and facilities require in-depth understanding of contributing success factors and ever changing landscape
- Strategies for staying competitive with changing landscape through innovation and creativity

"Without change there is no innovation, creativity, or incentive for improvement. Those who initiate change will have a better opportunity to manage the change that is inevitable."

- William Pollard (English clergyman died in 1893)

Discussion Forum

Needs for the Basic Chemical Industry

Sanjeev Kapur and John Pendergast

Pendergast: For External Release 1

Introduction

- **Panel Discussion: Are there Major Advancements in the Light Hydrocarbons Industry?**
- The chemical manufacturing industry is blossoming on the Gulf Coast.
  - Investments are coming from American companies as well as foreign investment
  - Driven by the low cost of ethane.
- No industry is immune to disruption,
  - Announcements have recently been released that suggest there is potentially disruptive technology under serious study.

Beginning Topics

- What are the disruptive technologies on the horizon, if any?
  - **Siluria's Oxidative Coupling Nears Reality[1]**
  - **Direct, Nonoxidative Conversion of Methane to Ethylene, Aromatics, and Hydrogen [2]**
- Given our footprint, is there a role or a place for government sponsored research involvement?
- What are the major areas for improvement in light hydrocarbon processing?
- Is there any role for academia in the basic chemical industry, or is the lost forever in the United States?
- Will membranes ever play a significant role in processing?
- We encourage the industry experts to engage in the discussion. particularly to drive interest in encouraging research and investigation into these important topics

(1) C&EN Volume 92 Issue 27 | pp. 20-21 July 7, 2014
(2) *Science* ***344, 616 (2014);***

Industrial Control System Cyber Security

Discussion with Management

6th AIChE Southwest Process Technology Conference

October 10, 2014

The Sky is Falling!!
The Sky is Falling!!

Control System Cyber Security

- Shamoon virus takes out 30,000 computers at Saudi Aramco
- US Defense Secretary issues strong warning of cyber attacks on US critical infrastructure
- DHS issues alerts about coordinated attacks on gas pipeline operators

Management Agenda

- Management listens to the news too – expect the question
- Don't be Chicken Little – Have a plan
- If possible, be proactive

Control Systems are Vulnerable

- **Strong drivers to automation**
  - Better data
  - More integration
  - Cost savings
- **Unintended consequences**
  - Commercial technology
  - Highly connected
  - Offer remote access
  - Technical information is available
  - Hackers now targeting control systems

Management Agenda

- Don't be nostalgic – Understand the history
- Understand the economic drivers of:
  - Commercial Technology
  - Remote Access
  - Network Systems

NASA Confirms Virus On Space Station

NASA has confirmed that some laptops on the International Space Station have been infected with the W31.Gammima.AG virus, a worm that steals users names and passwords for online games. Symantec detected the virus on August 27, 2007 as a worm affecting Windows 2000, Windows 95, Windows 98, Windows NT, and Windows XP. The virus spreads by copying itself to removable media.

"This is not the first time we have had a worm or virus," said NASA spokesman Kelly Humphries. "It's not a frequent occurrence, but this isn't the first time. The bottom line it is a nuisance for us."

Due to NASA security policies, Humphries would not release details about how the virus made it to the International Space Station. Humphries said none of the previous viruses found on computers aboard the International Space Station have had any operational impact.

The laptops the astronauts use do not have an anti-virus system. To prevent future incidents, NASA said all laptops will be updated with the latest version of Norton AntiVirus. NASA says it is still investigating how the virus got on the International Space Station laptops, but may not reveal information due to secu[...] Station crew is working on preventing the virus fro[...] problems from occurring again.

Management Agenda

- Air Gap is not enough
- Isolation is a start but not a solution

Information Technology

IT is your best friend...but

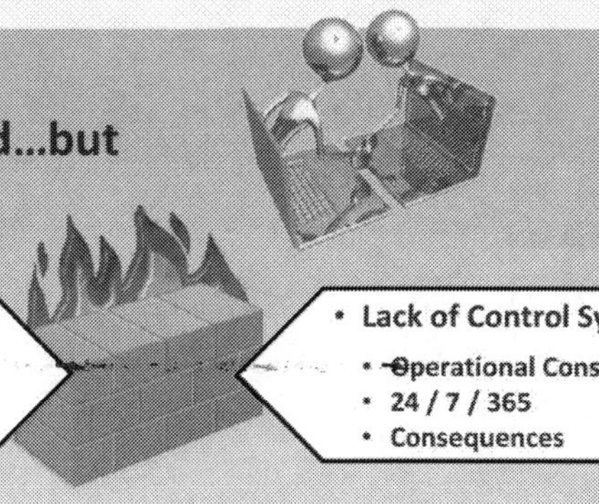

- Cyber Security Expertise
- Large Budgets
- Management Connections

- Lack of Control System Knowledge
  - Operational Constraints
  - 24 / 7 / 365
  - Consequences

- Consequences
- Environmental
  - Temperature
  - Hazardous Areas
- Architectures
- CIA – Confidentiality / Integrity / Availability
  – Availability / Integrity / Confidentiality
- Zero Down Time

Management Agenda

- Make Friends with IT
- Be prepared to discuss differences between Business Systems and Control Systems
- Know your Audience

 5

Regulations

- Department of Homeland Security
  - 6 CFR part 27: Chemical Facility Anti-Terrorism Standards (CFATS)
  - National Cyber Security Division
    - Control Systems Security Program (CSSP)
- Department of Energy
  - Federal Energy Regulatory Commission (FERC)
    - 18 CFR Part 40, Order 706 (mandates NERC CIPs 002-009)
- Nuclear Regulatory Commission
  - 10 CFR 73.54 Cyber Security R
  - RG 5.71

Management Agenda

- Proactively research regulations for your industry
  - Technical Organizations
  - US Certs
  - Department of Homeland Security

Standards

- International Society for Automation (ISA)
  - ISA 62443 Industrial Automation and Control System (IACS) Security
- International Electrotechnical Commission (IEC)
  - IEC 62443 series of standards (equivalent to ISA 99)
- National Institute for Standards and Technology (NIST)
  - SP800-82 Guide to Industrial Control Systems (ICS) Security

Actual Incident Data

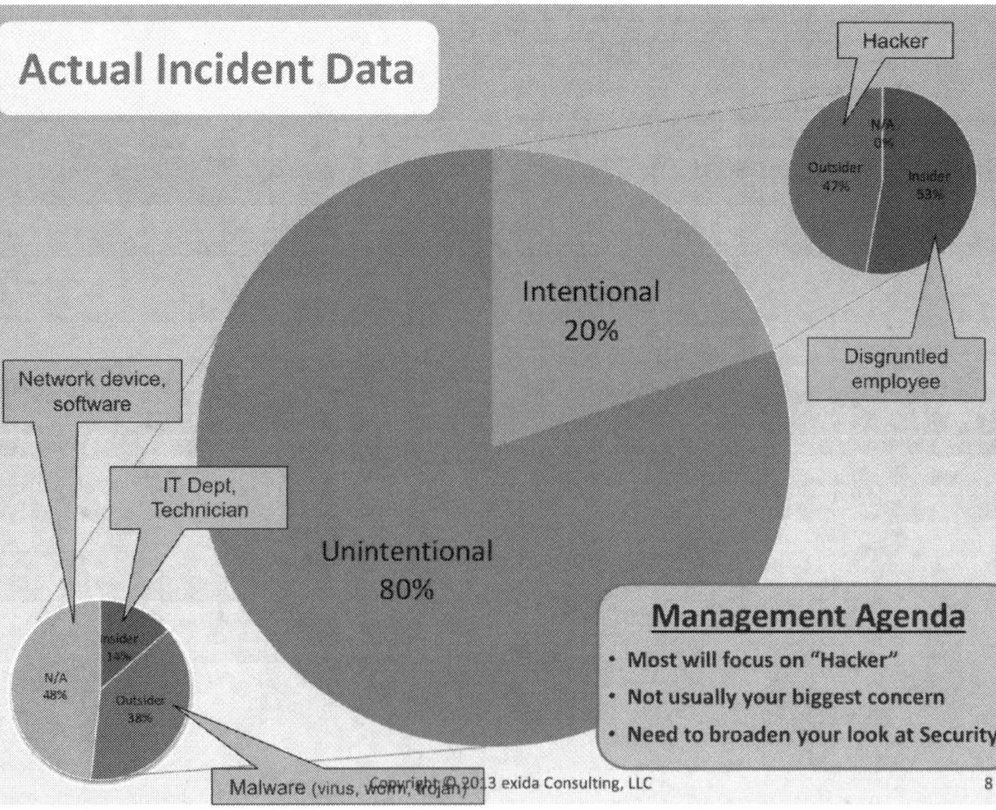

Management Agenda

- Most will focus on "Hacker"
- Not usually your biggest concern
- Need to broaden your look at Security

Assess Existing Systems

- Perform control system security assessments of existing systems
- Compare current control system design, architecture, policies and practices to standards & best practices
- Identify gaps and provide recommendations for closure
- Document! Document! Document!

Third Party Assessments
- Additional Expertise
- Different Eyes
- Management Credibility
  But...
- $$$ Additional Cost

Management Agenda
- Provides management with solid understanding of current situation, gaps and path forward
- Helps identify and prioritize investments
- First step in a security management program

 9

Cyber Risk Assessment

- Cyber Risk, Like Safety, is defined by:
  - Likelihood
  - Severity (Consequences)
- Requires Methodology for determining Risk
- Allows for
  - Vulnerability Assessment
  - Quantification of Risk
  - Prioritization of Remediation

Management Agenda
- Quantifies Exposure and Consequences
- Aids in Prioritization and Budgeting
- Documents Assessment (drawings, processes)

 10

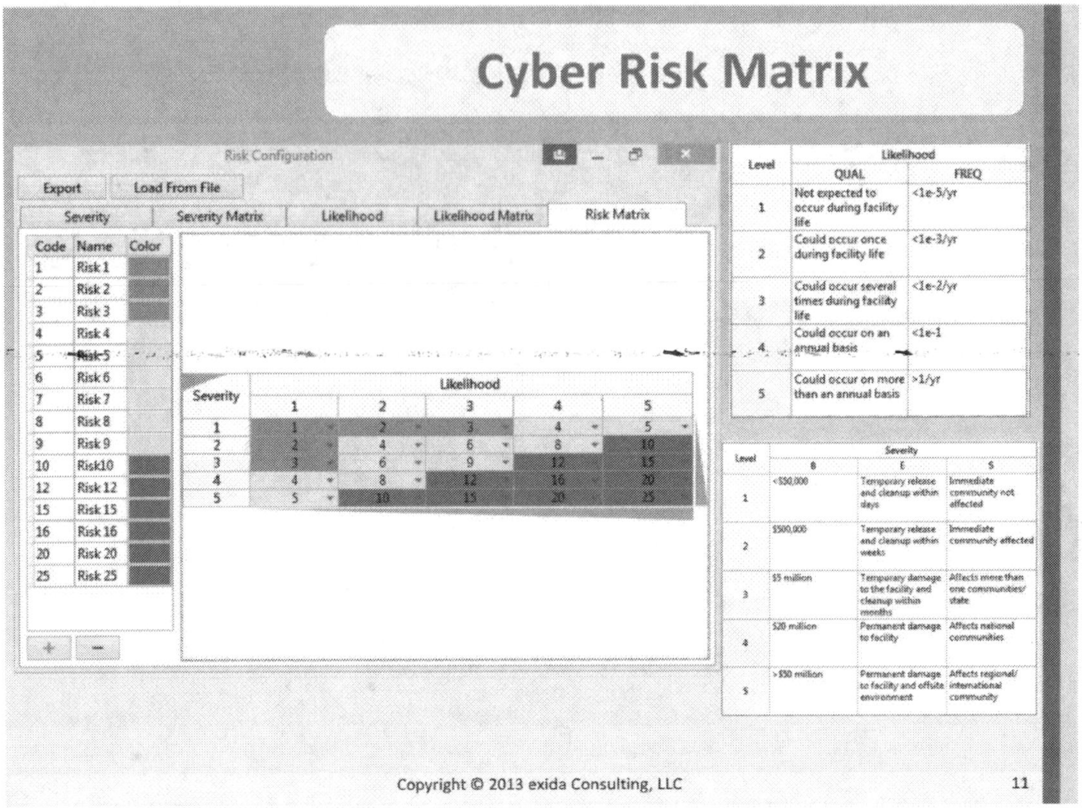

Cyber Risk Matrix
Risk Configuration
Export
Load From File
Severity
Severity Matrix
Likelihood
Likelihood Matrix
Risk Matrix
Code Name Color
1 Risk 1
2 Risk 2
3 Risk 3
4 Risk 4
5 Risk 5
6 Risk 6
7 Risk 7
8 Risk 8
9 Risk 9
10 Risk10
12 Risk 12
15 Risk 15
16 Risk 16
20 Risk 20
25 Risk 25
Severity
Likelihood
1 2 3 4 5
1 1 2 3 4 5
2 2 4 6 8 10
3 3 6 9 12 15
4 4 8 12 16 20
5 5 10 15 20 25
Level
Likelihood
QUAL
FREQ
1 Not expected to occur during facility life <1e-5/yr
2 Could occur once during facility life <1e-3/yr
3 Could occur several times during facility life <1e-2/yr
4 Could occur on an annual basis <1e-1
5 Could occur on more than an annual basis >1/yr
Level
Severity
1 <$50,000 Temporary release and cleanup within days Immediate community not affected
2 $500,000 Temporary release and cleanup within weeks Immediate community affected
3 $5 million Temporary damage to the facility and cleanup within months Affects more than one communities/ state
4 $20 million Permanent damage to facility Affects national communities
5 >$50 million Permanent damage to facility and offsite environment Affects regional/ international community
Copyright © 2013 exida Consulting, LLC
11

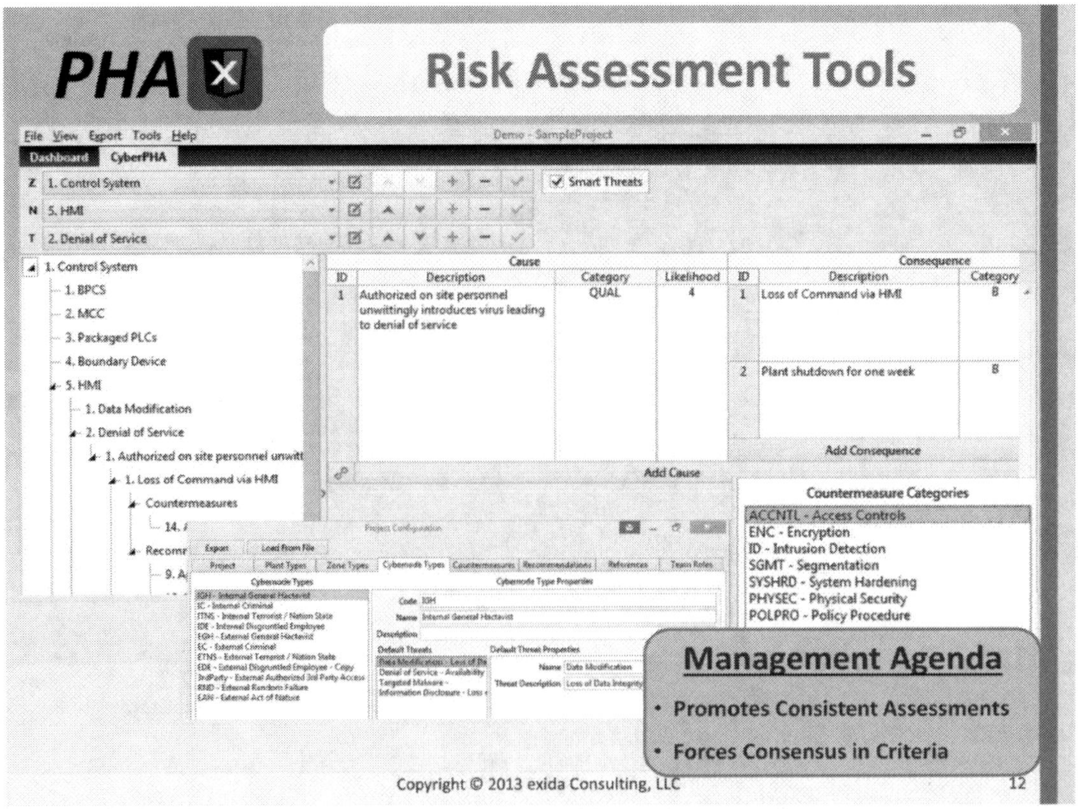

PHA X
Risk Assessment Tools
File View Export Tools Help
Demo - SampleProject
Dashboard
CyberPHA
Z 1. Control System
N 5. HMI
T 2. Denial of Service
Smart Threats
1. Control System
1. BPCS
2. MCC
3. Packaged PLCs
4. Boundary Device
5. HMI
1. Data Modification
2. Denial of Service
1. Authorized on site personnel unwitt
1. Loss of Command via HMI
Countermeasures
Cause
ID Description Category Likelihood
1 Authorized on site personnel unwittingly introduces virus leading to denial of service QUAL 4
Add Cause
Consequence
ID Description Category
1 Loss of Command via HMI B
2 Plant shutdown for one week B
Add Consequence
Countermeasure Categories
ACCNTL - Access Controls
ENC - Encryption
ID - Intrusion Detection
SGMT - Segmentation
SYSHRD - System Hardening
PHYSEC - Physical Security
POLPRO - Policy Procedure
Project Configuration
Export
Load From File
Project
Plant Types
Zone Types
Cybernode Types
Countermeasures
Recommendations
References
Team Roles
Cybernode Types
Cybernode Type Properties
IGH - Internal General Hactavist
IC - Internal Criminal
ITNS - Internal Terrorist / Nation State
IDE - Internal Disgruntled Employee
EGH - External General Hactavist
EC - External Criminal
ETNS - External Terrorist / Nation State
EDE - External Disgruntled Employee - Copy
3rdParty - External Authorized 3rd Party Access
RND - External Random Failure
EAN - External Act of Nature
Code IGH
Name Internal General Hactavist
Description
Default Threats
Default Threat Properties
Denial of Service - Availability
Targeted Malware -
Name Data Modification
Threat Description Loss of Data Integrity
Management Agenda
• Promotes Consistent Assessments
• Forces Consensus in Criteria
Copyright © 2013 exida Consulting, LLC
12

Document Policies & Procedures

Establish control system security policies & procedures

- **Scope**
- **Management Support**
- **Roles & Responsibilities**
- **Specific Policies**
  - » Access
  - » Remote access
  - » Portable media
  - » Patch mgmt
  - » Anti-virus management
  - » Change Management
  - » Backup & Restore
- **Document! Document! Document!**

Management Agenda

- **Management understands importance of policy**
- **Must have Management buy-in to be effective**

Remediation

- **Segment the Network**
  - Defense-in-Depth strategy
  - Partition the system into distinct security zones
  - Define security objectives and strategy for each zone
  - Create secure conduits for zone-to-zone communications
- **Control Access to System**
  - Review
  - Zone-by-Zone
  - Control and monitor access to control system resources
  - Logical & Physical
  - AAA ...... →
    - Administration
    - Authentication
    - Authorization
- **Harden System Components**
  - Remove or disable unused communication ports
  - Disable autorun
  - Remove unnecessary applications and services
  - Avoid file shares
  - Apply patches when and where possible
  - Consider 'whitelisting' tools
  - Use ISASecure™ certified products and systems

Management Agenda

- **Budget to get Investment Capital**
- **Know your Management's Budget**
- **Consider IT help ($$$)**

 14

Monitor and Maintain

- **Training**
  - Make personnel aware of the importance of security and company policies
  - Provide role-based training
- **Install vendor recommended anti-virus and update signatures regularly**
- **Review system logs periodically**
- **Consider Intrusion Detection (IDS) or Host Intrusion Prevention (HIPS)**
- **Pen testing (offline only)**
- **Periodic assessments**

This is a Lifecycle

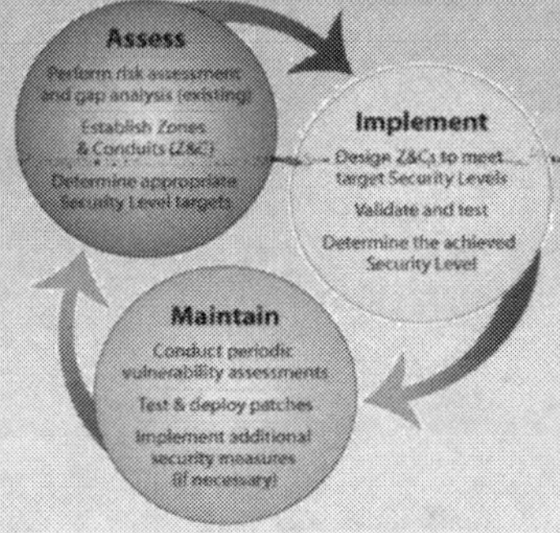

Management Agenda

- Need Continuing Management Support
- Get it in the budget – new cost of doing business
- May need additional/different resources

 15

Questions?

exida

 16

Getting Your Ducks in A Row...When Rolling out a New Product/Process

Phillips, G., Phillips & Reiter Corporate Legal Counsel
Kaiser, C., Phillips & Reiter Corporate Legal Counsel
Powell, D.,

Originally presented on: 10/10/2014 9:00:00 - 9:30:00

According to a new White House report, *Making in America: U.S. Manufacturing Entrepreneurship and Innovation*, (June 2104) new game-changing technologies are reducing the cost, increasing the speed, and making it easier for entrepreneurs and manufacturers to translate new ideas into products Made in America. These new technologies are already having an impact, with the growth rate in manufacturing at its fastest pace in over 20 years.

Texas seems to be leading the comeback. Forbes magazine recently ranked Houston as the No. 1 city for manufacturing jobs, while San Antonio comes in at No. 4. As production continues to climb with the birth of new innovation the proper use of agreements and contacts becomes vitally important to the overall health and wellbeing of the company. This talk will focus on the contracts and agreements needed to protect important research and development as well as those needed when taking the finished product to market.

I. EXAMPLES OF CONTRACTS AND TERMS

Sample contracts of some currently used in the process and manufacturing industry, for which the topic discussed in this paper could be relevant, will be briefly described. The chosen examples are not intended to present all technologies; but are merely intended to illustrate current terms now in use for manufacturing and process related operations. To include: secrecy agreement, employee contracts, sales contracts, licensing and distributor agreements and the like.

II. SHORTFALLS OF IMPROPER AGREEMENTS

News and trade publications are full of stories of lost innovation. Whether the loss occurs from inside the company, outside the company or even from foreign sources, the proper agreements can greatly benefit the company as a deterrent or they can provide a stronger position in court proceedings. These instances of adversity can also be a learning opportunity, to help us better protect the birth and creation of new technologies and the bringing to market of these new innovations.

III. CONCLUSION

History proves that after every major court battle pertaining to the innovation and marketing of new technologies there has been a flood of new agreements and safe guards put into place in hopes of preventing a reoccurrence of the same. Protecting new intellectual property is becoming increasingly important for the developers in this highly competitive International industry. U.S. patent law changes regarding disclosure will be addressed, as well as what contracts are needed, when they need to be put into place, who needs to sign them and where proceedings if needed will take place. Also to be discussed will be proper language and terms needed in sales, licensing and distributor agreements, as well as ownership issues.

Optimum Protective Systems for Critical Turbomachinery[1]

[1]*Material Extracted from TX A&M Turbo TUTORIAL # 5*
Machinery Protective Systems
October 2010

Optimum Protective Systems for Critical Turbomachinery[1]

[1]*Material Extracted from TX A&M Turbo TUTORIAL # 5*
Machinery Protective Systems
October 2010

MACHINERY PROTECTION

- Is Your Machinery Adequately Protected?
- How do You Know?
- In the event of a Mishap are You Sure your Protection System will Function as designed?

MACHINERY PROTECTION:

Example: Failed System or Insufficient Protection

MACHINERY PROTECTION:

Example: Failed System

or Insufficient Protection

MACHINERY PROTECTION:

- Latch failed/
- Result wrecked rotor

MACHINERY PROTECTION:

Example: Failed System

or Insufficient Protection

MACHINERY PROTECTION:

Example: Failed System or Insufficient Protection

MACHINERY PROTECTION:

Example: Failed System or Insufficient Protection

RECENT ROTATING EQUIPMENT FAILURES AT FIVE REFINERIES

Equipment	Refinery	Event	Description Of Incident
XXXXX Unit Charge Pump	A	Major Fire	Bearing in 2,250 hp motor overheated (due to lack of lubrication) until shaft overheated and lost strength. No bearing temperature instrument. Shaft broke causing massive imbalance that resulted in pump and motor wreck. Pump seals failed resulting in large fire. XXXXX, 2008. Pump was equipped with MOV's on suction and discharge; considering adding an MOV in the common suction line.
XXX Pump A	A	Vapor Cloud Release	Dual (Tandem) pump seals failed catastrophically on XXXXXX, 2008 due to design error. Vapor cloud release; vapors were seen entering a fired heater, but the vapor cloud did not ignite.
XXX Pump B	A	Vapor Cloud Release	Dual (Tandem) pump seals failed catastrophically on XXXXXX, 2008 due to design error. Occurred two days after above event.
XXXXX Unit Screw Compressor	A	Major Fire	Motor bearing overheated until shaft overheated (due to lack of lubrication) and lost strength. Shaft broke causing massive imbalance that resulted in compressor and motor wreck. Compressor seals failed resulting in large fire. Also, the coupling became airborne and hit a lube oil drain line on the compressor. The drain line broke releasing gas and lube oil which also ignited. XXXXX, 2008.
XXXXXXXX Pump	A	Major Fire	XXXXX, 2008. Design error in piping resulted in excessive pipe strain on pump that resulted in misalignment. Misalignment eventually caused the thrust bearing to fail. The thrust bearing overheated causing the shaft to overheat. The shaft finally broke causing the pump to wreck. When the shaft broke, the seals were destroyed resulting in a large fire.
Pump	B	Major Fire	"Double" pump seals failed catastrophically less than one minute after vibration increased. Further information not available.
Coker Wet Gas Compressor	C	Vapor Cloud Release	Following is general consensus; exact sequence of events not determined due to extent of damage: Suspect that seals had been overheated during past startup events (rubs). During normal operation, seal faces "shattered" resulting in catastrophic seal failure and large vapor cloud release. Operators had to go on compressor deck and close valves to stop release. Vapor cloud did not ignite. Excessive vibration had broken bearing cap bolts; no vibration trips. XXXXX, 2006.
XX Crude Pump	C	Major Fire	Excessive pipe strain caused repeated coupling and bearing failures. Problems also experienced with spare pump, so management made decision to continue repairing couplings and bearings until turnaround; they assumed coupling and bearings would continue to fail as in past. On XXXX, 2008, the motor and pump wrecked catastrophically; the motor was torn off its foundation. Pump seals were destroyed resulting in large fire.
FCC Auxiliary Air Blower	D	Fire	FCC Unit has two Main Air Blowers operating in parallel. The Auxiliary Blower provides about 30% of the required air for the Regenerator. Due to an instrumentation problem, the Auxiliary Air Blower severely surged. While the compressor was in surge, the steam turbine driver oversped. The compressor and steam turbine were destroyed; rotor parts were thrown through the compressor case. The subsequent lube oil fire damaged instrumentation and wiring on the Main Air Blower and its steam turbine driver.
FCC Unit Main Air Blower	E	Near Miss	A sequence of events, starting with the failure of a flow instrument associated with the FCC Unit Main Air Blower, resulted in a reverse flow between the Reactor and Regenerator. When the flow instrument failed, the compressor inlet guide vanes closed

INCIDENT NUMBER	SOURCE	YEAR	SIZE	DRIVEN OBJECT	WHY GOVERNOR FAILED TO CONTROL SPEED	WHY OVERSPEED TRIP FAILED	OTHER FACTORS/INFORMATION
109		2005	>5000 hp	Compressor	Governor system did not control speed; exact reason not stated.	Overspeed trip system did not prevent steam turbine from overspeeding; exact reason not stated. Had mechanical overspeed trip system.	Refinery has two large steam turbine driven process air compressors operating in parallel. During process upset, surge control valve closed on one compressor resulting in massive surge. During surge, the steam turbine oversped. The compressor was destroyed; parts were thrown through the case. The steam turbine was also destroyed; its parts stayed inside the case, but a bearing housing was ripped off. The ensuing lube oil fire also damaged the controls and instrumentation on the adjacent compressor.
110		2009	>5000 hp	Generator	Governor system did not control speed; exact reason still under investigation.	Overspeed trip system did not function properly; exact reason still under investigation.	Relatively small steam turbine-generator oversped to complete destruction. Plant now has to purchase the power previously supplied by the steam turbine generator.

Senior Loss Control Consultant/Underwriter

Most people in the chemical, oil, and gas industries feel that steam turbine overspeed incidents are extremely rare. The author's experience has been that these incidents are all too common. However, for reasons discussed below, it is very difficult to obtain data on steam turbine overspeed events. So, many years ago, the author decided to keep a list of these incidents.

MACHINERY PROTECTION: Failures/ Wrecks

Why is this still happening?

- Protective Systems: Lack off/ Insufficient
- Protected: Validation/ How Tested/ Frequency
- Technology: Appropriate Technology Used
- Business: Economic Condition/ Funding/ Justifiable
- Analysis: Poor Evaluation/ Available Data

MACHINERY PROTECTION: Right Level

Right Level of Protection

- The Intelligent Application of Standards, Guidelines, Procedures and Processes to keep Machinery Related Risks as close to Zero as practicable. *(Following the ALARP principle as stated in the standards..)*

MACHINERY PROTECTION: New Processes

1. Standards Bodies Driving Protection
2. Industry Standards – IEC/ API/ CCPS
3. New Hazard Identification Processes – HAZOP/ FMEA
4. New Quantification Methodology - LOPA
5. Samples of New Methodology for deciding protection
6. Examples of good and poor applications
7. Validation/ Auditing/ Testing Frequency of systems
8. Potential traps
9. Questions?

MACHINERY PROTECTION: Driving Bodies

Recognized Authorities/ Standards Bodies

- API (670-Vibration, et al)
- OSHA (1910.119 PSM, maint., etc.)
- CCPS (LOPA, QRA, interlocks)
- ANSI/ IEC (specific to interlocks, details follow)

Interlock Standards History

- **29 CFR 1910.119**, *Process Safety Management (PSM)*, February 24, 1992
- **AIChE CCPS**, *Guidelines for Safe Automation of Chemical Processes*, 1993
- **ISA S84.01-1996**, *App. of Safety Inst. Sys. for the Process Ind.*, Feb. 15, 1996
- **ANSI** approval of S84, 1997
- **IEC 61508**, *Functional Safety: Safety-Related Systems* "General" released, 1998
- **OSHA** recommends S84, Mar 23, 2000
- **IEC 61511,** *Functional Safety: Safety Inst. Systems for the Process Ind.*, 2002
- **ISA S84.00.01-2004**, Sept 2, 2004

Functional Safety Lifecycle

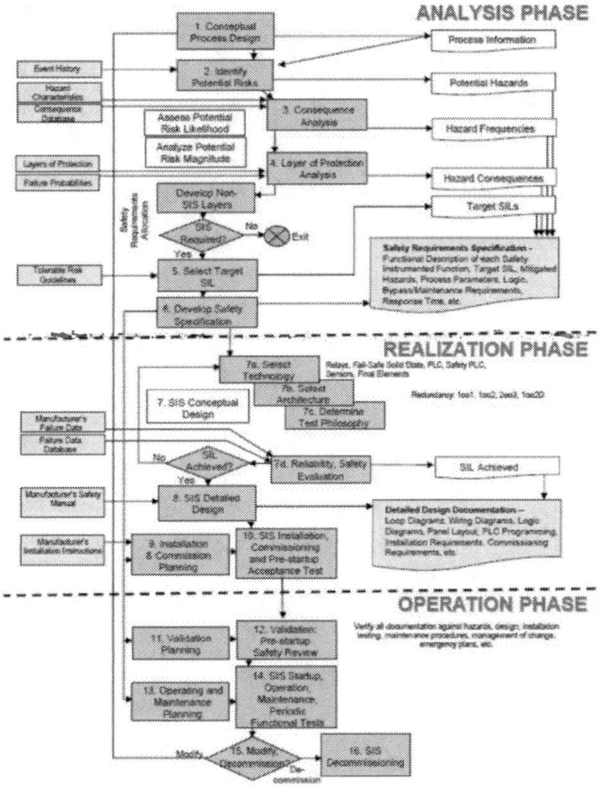

The Standards for Functional Safety (interlocks/trips)

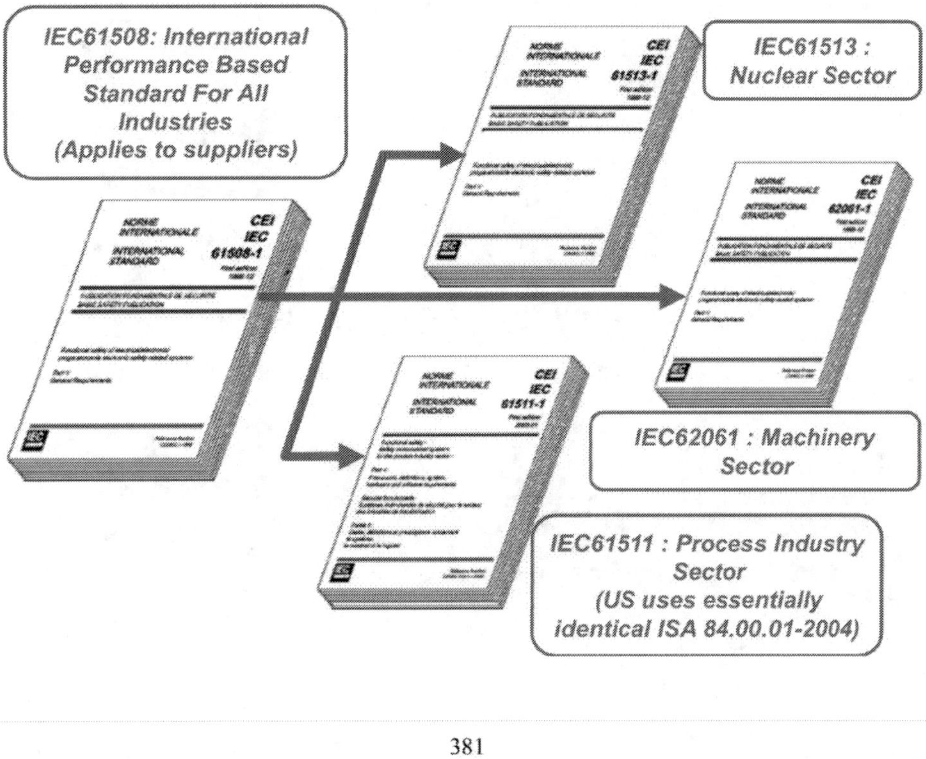

IEC 61508 Standard

- Targets Suppliers
  - Requirements for **suppliers** of process control and instrumentation for component / sub-system safety
  - **End Users** seek suppliers with products certified to this standard by reputable certifying agency

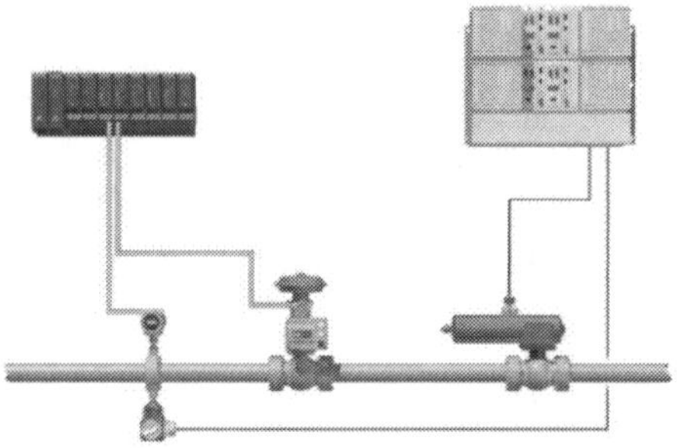

19

ISA 84 / IEC 61511 Standard

Targets **End Users, Contractors and Integrators** in process industries

Covers the entire SIS Life Cycle
- Risk Analysis
- Performance based design
- Operations and Maintenance

Performance NOT Prescription

End user applications
- Not typically certified
- Independent Functional Safety Assessments

3 Sections
- Requirements
- Guidelines
- SIL Selection

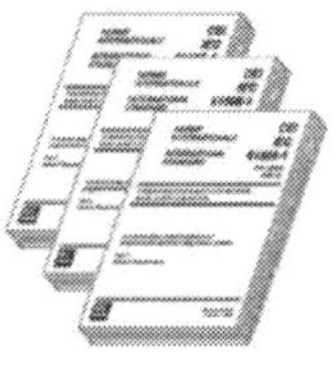

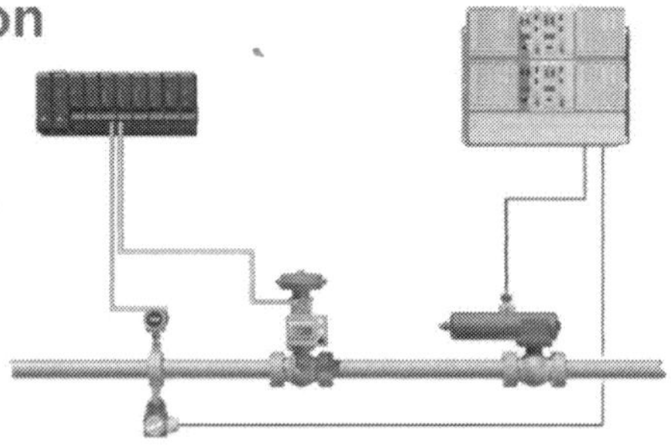

20

Safety Lifecycle - Analysis Phase

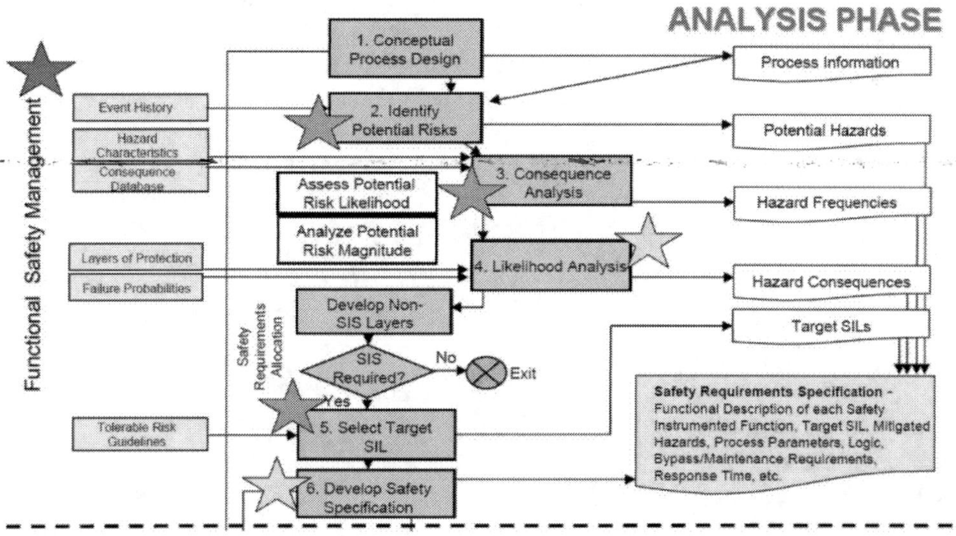

HAZOP Results

Node: Reciprocating Compressor
Parameter: Liquid Level

Deviation	Cause	Consequence	Safeguards	Recommendation	Action
Too high	Level dump fails closed	Water entrainment leading to damage & loss of containment	High level alarm	1. Consider adding high level trip if required by LOPA	J. Jones
	Maintenance valve inadvertently left closed	Water entrainment leading to damage & loss of containment	High level alarm	(repeat 1. Add high level trip)	J. Jones
Too low	Level dump fails open	Gas blow-by potentially over-pressuring produced H2O tank	Low level alarm	2. Consider addition of second thief hatch on	S. Smith

PHA Methods: Failure Modes and Effects Analysis (FMEA)

- Very systematic
- Excellent for analysis of mechanical failures
- Difficult to effectively involve a team
- Does not easily lend itself to human factors
- Scenarios involving multiple failures not readily identified
- Results in relative ranking of risks associated with the various failure modes and their effects

FMEA Results

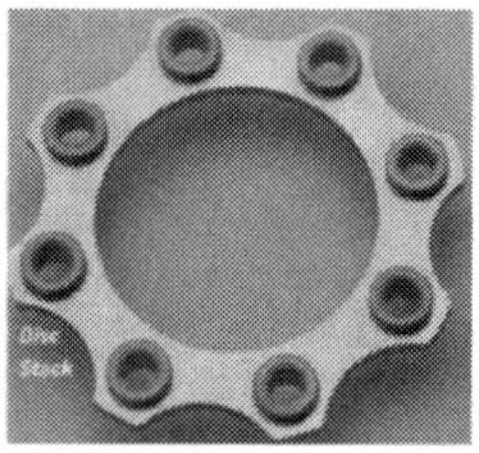

Component	Failure Mode	Possible Causes	Effects on System
Disk pack	Fatigue leading to fracture	excessive misalignment	Wreck machine with possibility to eject parts with personal injury
		Impact loading from water slug	Wreck machine with possibility to eject parts with personal injury
Gear coupling	Fatigue leading to fracture	Poor lubrication	Wreck machine with possibility to eject parts with personal injury

Safety Integrity Level (SIL) Determination

- Identify how much risk reduction is needed to attain a tolerable risk
  - **Quantitative** methods give **specific numerical targets for risk (e.g. RRF)**
  - **Qualitative** methods group numerical targets into more broad **categories of risk reduction (e.g. SIL band only)**
- A consistent method or set of methods must be used

25

SIL Determination Methods: Hazard Matrix Procedure

- Start with a matrix expression of tolerable risk

	Recordable Injury	Lost Time Injury	Permanent Injury/Death	Many Deaths
1 per 100 years	Acceptable	Moderate	Extreme	Extreme
1 per 1000 years	Acceptable	Acceptable	Moderate	Extreme
1 per 10,000 years	Acceptable	Acceptable	Moderate	Moderate
1 per 100,000 years	Acceptable	Acceptable	Acceptable	Moderate

All extreme risk will be reduced and all moderate risks will be reduced where practical.

26

Multiple Receptors per SIF

- Occasionally a set of tolerable risk levels and risk estimates gives different integrity level requirements:

  Safety IL = 2
  Environmental IL = 3
  Financial IL = 1

- Choose highest IL = 3 for specifying the system
- For the purpose of this tutorial and as in general slang, the word "SIL" will represent all three

Safety Integrity Levels (SILs)

Safety Integrity Level	Target average probability of failure on demand	Target risk reduction (RRF)
SIL 3	$>=10^{-4}$ to $<10^{-3}$	10000 to 1000
SIL 2	$>=10^{-3}$ to $<10^{-2}$	1000 to 100
SIL 1	$>=10^{-2}$ to $<10^{-1}$	100 to 10

28

Layers of Protection Analysis (LOPA)

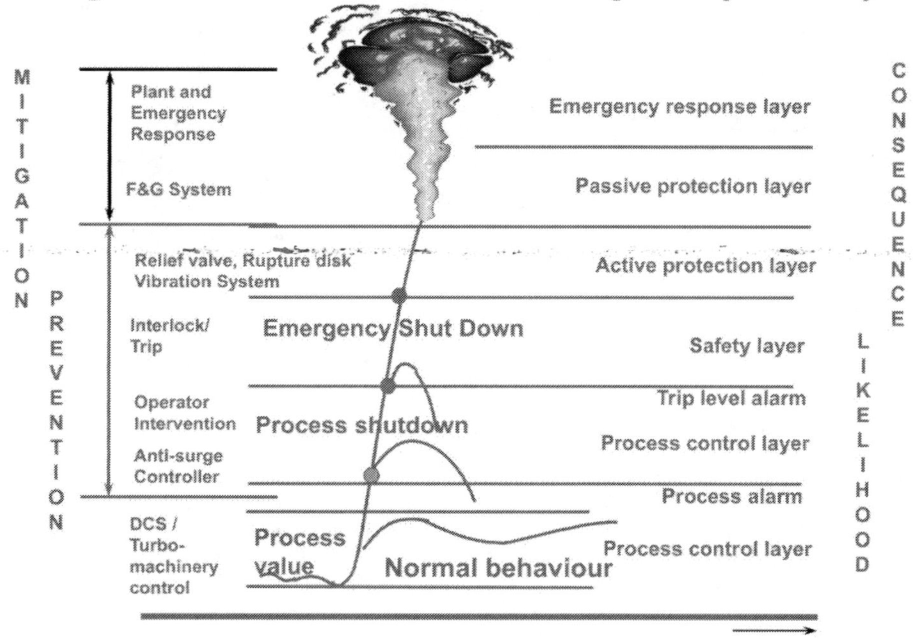

Protection Layer Attributes

Specific
- must be specifically designed to be capable of preventing the consequences of the potentially hazardous event

Auditable
- must be tested and maintained to ensure risk reduction is continually achieved

Independent
- must be completely independent from all other protection layers

Dependable
- must be capable of acting dependably to prevent the consequence from occurring (systematic and random faults)

30

SIL Determination Methods: Layer of Protection Analysis (LOPA)

Initiating Event	Protection Layer 1	Protection layer 2	Protection Layer 3		Final Outcome
			PL3 Fails	Accident Occurs	
		PL2 Fails			
	PL1 Fails				
Init Event					
			PL3 success	No Impact Stop	
		PL2 Success	No Impact Stop		
	PL1 Success	No Impact Stop			

31

MACHINERY PROTECTION: Sample – Reciprocating Compressor

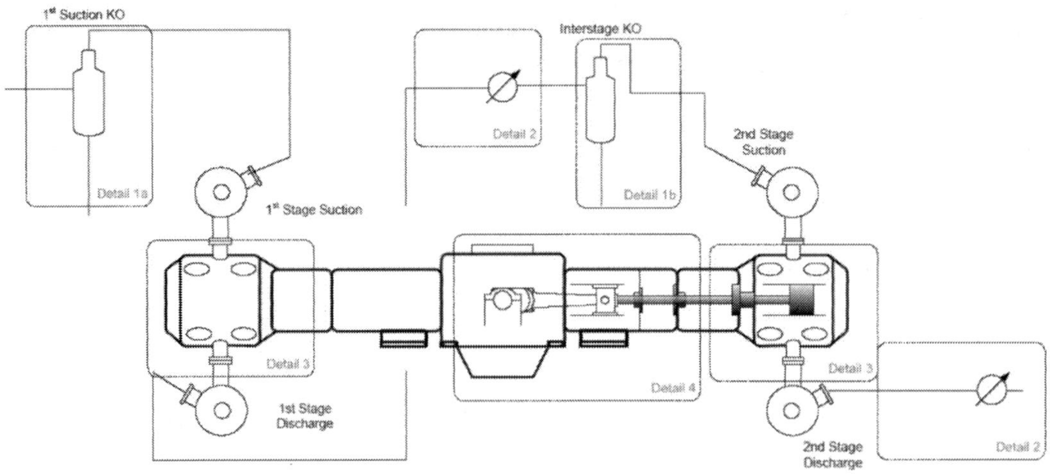

MACHINERY PROTECTION: Sample – Reciprocating Compressor

- LOPA - Assessment

Alarm and Trip Requirement - Recip			
Measurement	**Alarm**	**Safety Related Alarm**	**Trip**
Lube oil pressure low	Yes	No	Yes generally 2003
Lube oil tank level	Yes low level	No	
Case, Slide Vibration	Yes	Yes	Yes – 200N H+HH
Active KO drum H Lev	Yes	Yes	SIL1
Suction gas pressure	Yes	Yes	Unload
Outlet high temp	Yes	Yes	

MACHINERY PROTECTION: Sample – Reciprocating Compressor

LOPA Assessment

Generic LOPA Template				Unit: Machines	Date	
LOPA Scenario:	*G2 R1A*			Analyst:		
HAZOP Node No.		Page		Description	Liquid Carryover into Recip Compressor from Knockout Drum	P&ID
Deviation				Hazard	Liquid Impact to machine and lossof containment - Recip type	

TMEL			
Consequence	**Category**	**Description**	**TMEL (/yr)**
Safety	F	Permanent Disability	1.0E-03
Environmental	H	Minor localized damage to non sensitive area	1.0E-01
Commercial	H	<$50k	1.0E+00

Consequence	Detailed Description of Unmitigated Consequences
Safety	Liquid impact of machine. Some risk of a loss of containment at machine. Serious machine damage, jet fire. Immediate ignition likely.
Environmental	
Commercial	Serious machine damage

MACHINERY PROTECTION:
Sample – Reciprocating Compressor

LOPA Assessment

Initiating Causes (IC)										
			Applicable IPLs from List below							ICL * PFoDs
IC Number	Short Description	Frequency (ICL)	A	B	C	D	E	F	G	
1	Loss of level control	1E-01	1	0	0	0	0	0	0	1.0E-02
2	Liquid beyond LCV capacity	1E-01	0	0	0	0	0	0	0	1.0E-01
3			0	0	0	0	0	0	0	0.0E+00
4			0	0	0	0	0	0	0	0.0E+00
									Sum of ICLs * PFDs	1.1E-01

Independent Protection Layers (IPLs) may be common to more than one initiating event			
IPL Letter	Category	Short Description of IPL	PFoD
A	Procedural Control - includes ops response to alarm	Ops act in event of alarm	1E-01
B	Active Mechanical Safeguard	Machine loss of containment MORE LIKELY than centrifugal. Severity set to F in view of historical evidence and not all liquid events result in a loss of containment. Fire is likely to be immediate - the point implies people at risk must be very local to the machine.	1E-01
C			1E+00
D			1E+00
E			1E+00
F			1E+00
G			1E+00

MACHINERY PROTECTION:
Sample – Reciprocating Compressor

LOPA Assessment

Frequency Modifiers			
Type	Safety	Environmental	Commercial
Time at Risk	1	1	1
Occupancy Factor	0.1	1	1
Ignition Probability	1	1	1

Current SIF			
Functional Description	Sensor(s)	Controller Type	Shutdown Valve(s)

PFD of SIF			
Category	Value	LOPA Result	Resulting IL
Safety	9.09E-02	Fail	SIL1
Environmental	9.09E-01	Fail	No Special Integrity Reqmts.
Commercial	9.09E+00	Pass	Not Required

MACHINERY PROTECTION:
Sample – Reciprocating Compressor

- LOPA Assessment
- Current Condition – SIL 1
- Desired – SIL 2 (100 yrs – 1000 yrs PFD)
- Recommendation: Add 2nd Layer of Protection
  - Use Frame/ X-Head Slide Vibration as 2nd Layer

Safety Requirements Specification

Definition

IEC61511: "specification that contains all the requirements of the safety instrumented functions in a safety instrumented system"

Objective

- Specify all requirements of SIS needed for detailed engineering and process safety information purposes
- Functional Requirements
  - Description of the functions of the SIF
  - **How it should work**
- Integrity Requirements
  - The risk reduction and reliability requirements
  - **How well it should work**
  - **How quickly it should work**
- Often a contractual document prepared by one company and executed by another

38

Specification = Communication

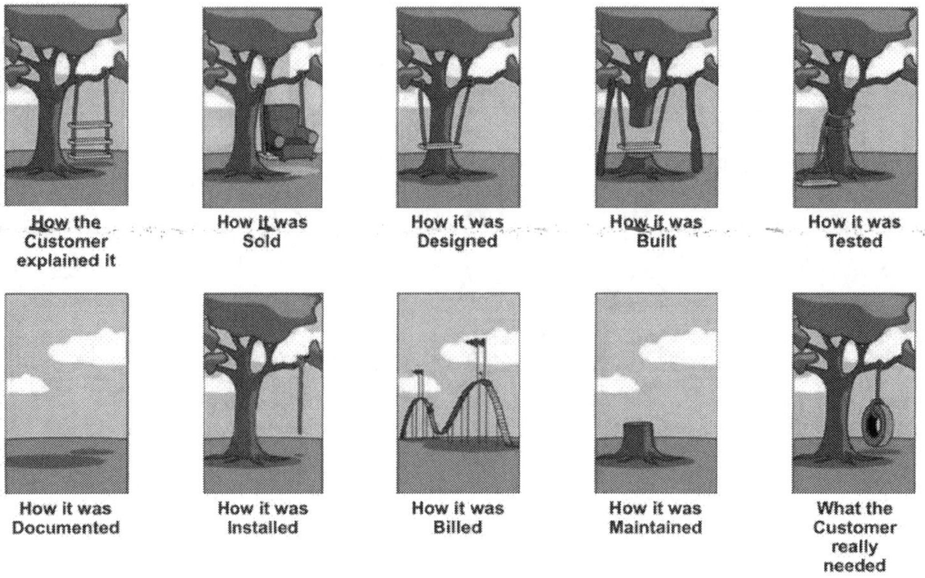

39

Safety Lifecycle - Realization Phase

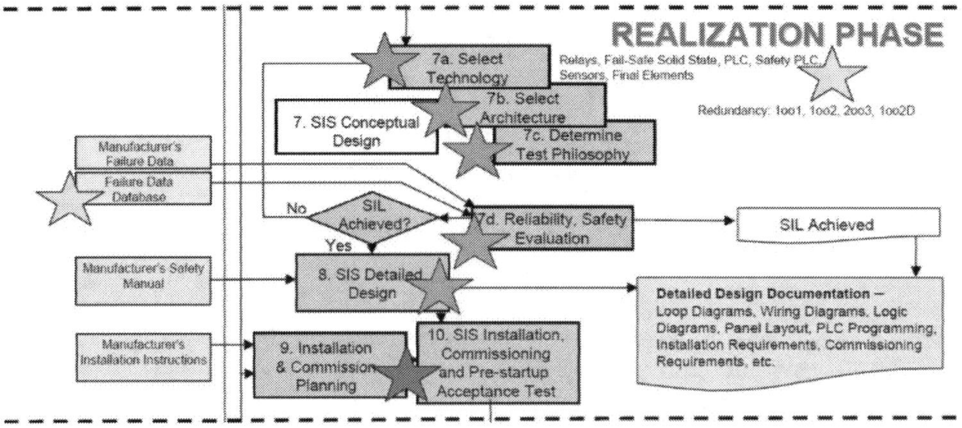

Safety Instrumented System (SIS) Definition

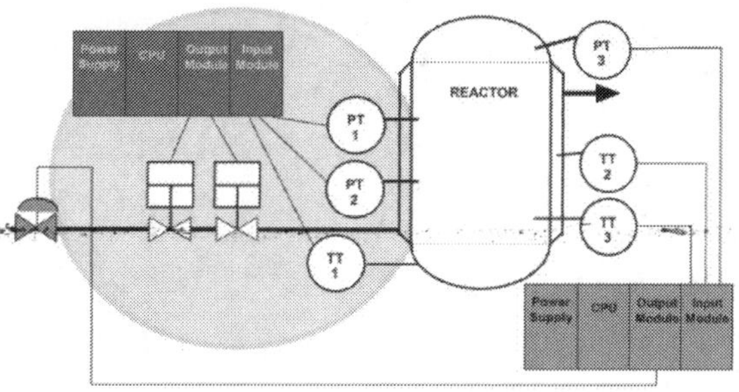

IEC 61511 defines a Safety Instrumented System (SIS) as

"instrumented system used to implement one or more safety instrumented functions. A SIS is composed of any combination of sensor(s), logic solver(s), and final element(s)." IEC 61511 Part 1 3.2.72

41

New Safety PLC Architectures

PLC Arch.	Spurious Trip Rates	Prob. of Failure (Dangerous)
1oo1	0.05 /year	0.0006
2oo3	0.00043 /year	0.00000094
1oo1D	0.069 /year	0.0005
2oo2D	0.00021 /year	0.001
1oo2D	0.00021 /year	0.0000004

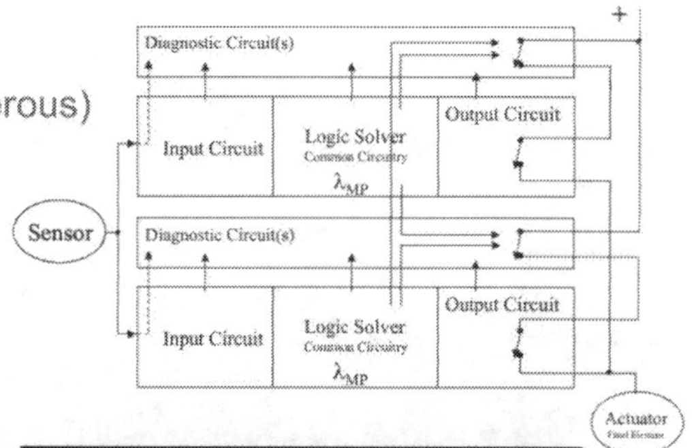

The New Safety PLCs depend highly on good diagnostics.

42

Equipment Selection

IEC 61511, Functional Safety for the Process Industries, requires that equipment used in safety instrumented systems be chosen based on either **IEC 61508 assessment** to the appropriate SIL level or **justification based on "prior use" criteria** (IEC 61511, Part 1, Section 11.5.3)

43

IEC 61508 Full Certification Enough?

- NO! A control system designer cannot simply specify 61508 certified equipment and expect a safe design!
- Equipment must match intended application
- Equipment "restrictions" must be followed
- Process connections must be included

44

Validation Objectives

- Ensure that the safety instrumented system (SIS) as installed and commissioned meets all of the safety requirement specifications (SRS)
- Validation is done using a combination of testing and inspection

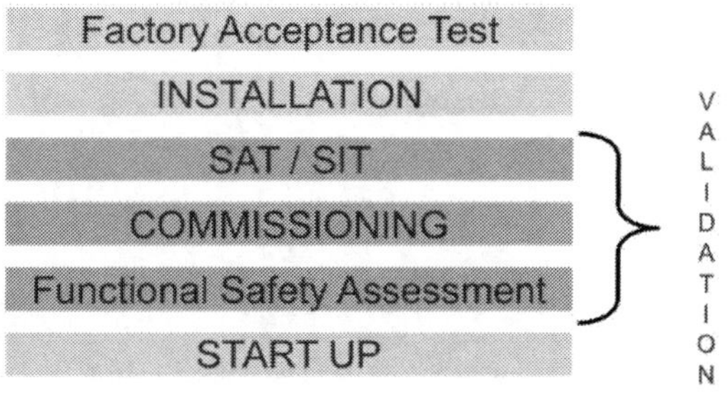

IEC 61511 Clause 15

45

H/W Design: SIF Verification Task

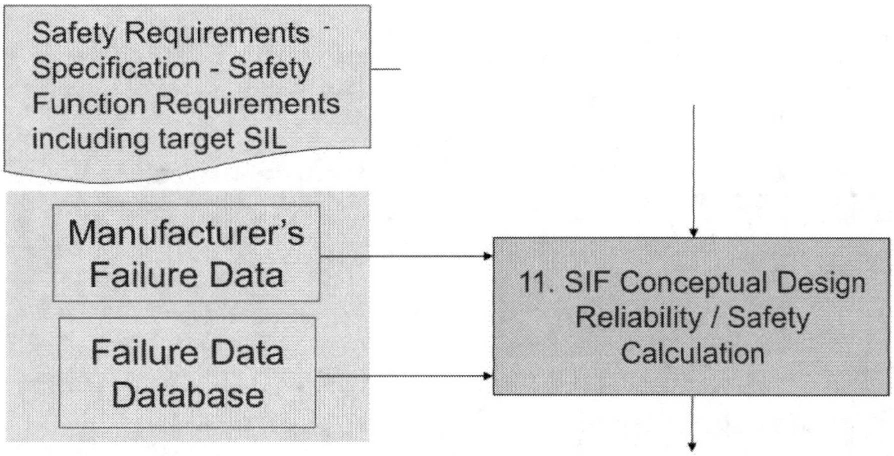

Safety Lifecycle - Operations Phase

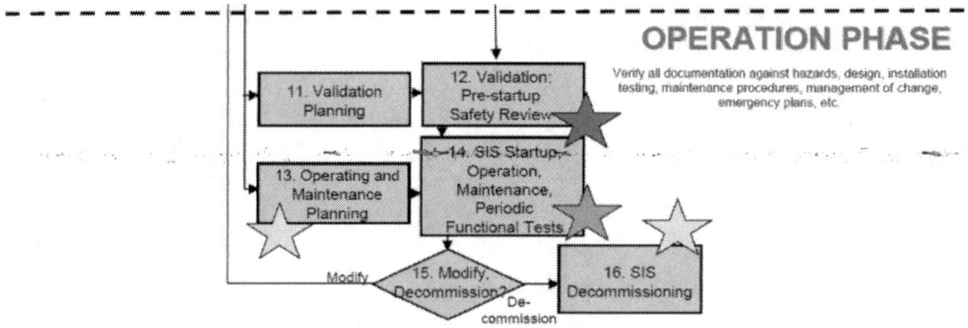

Maintenance Planning

- All tests required to verify proper operation of Safety Instrumented Function must be planned
- Proper periodic test interval that was calculated during SIF verification must be documented as part of the plan
- Online? Offline? Bypass Procedures?
- Proof test procedures must be at least as effective as planned during the SIF verification

Strategic Proof Test

The purpose of the Proof test is to detect any failures not detected by automatic on-line diagnostics.

1. We can design proof test procedures that are easier to perform, cost less and are more likely to actually get done.
2. By understanding the actual DU/AU failures in our instruments we can significantly improve our test coverage as well as lower cost.

Why Periodic Testing is Vital

Because of incomplete testing the PF never returns to its original value and the risk reduction can be significantly lower.

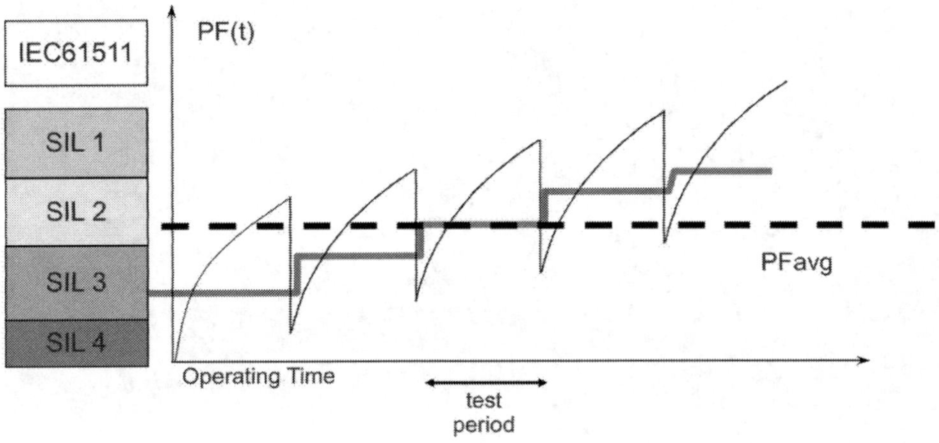

SIS Functional Safety Lifecycle

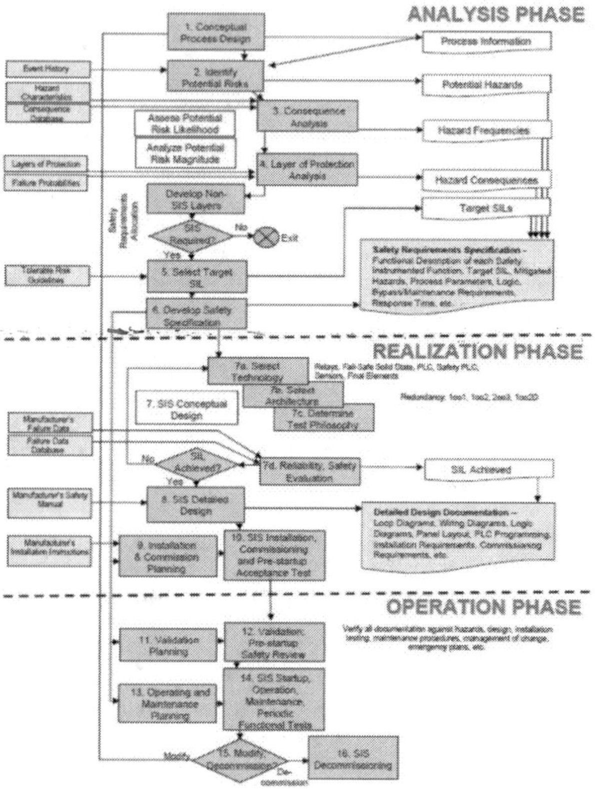

Safety Lifecycle – Example Failure

Compressor thrust bearing & shaft wrecked,
but <u>no shrapnel nor other hazards for people</u>.

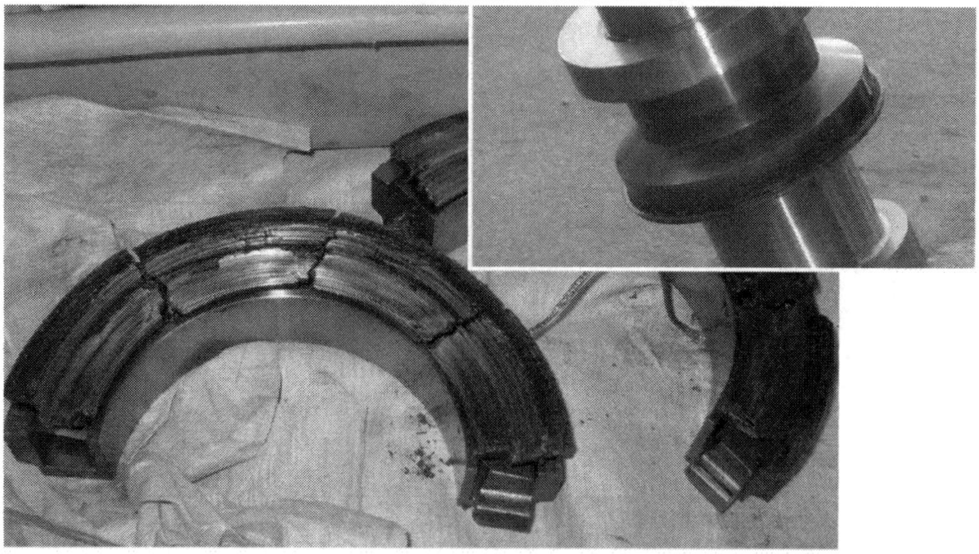

(Temperature sensors were <u>not</u> yet connected!)

Safety Lifecycle – Example Failure

This high speed compressor impeller developed a fatigue crack and broke in TWO with inadequate vibration protection. All parts were contained by the casing, **fortunately no hazards for people!**

Safety Lifecycle – Example Failure

Disk pack fatigue due to misalignment, the spool and guard were EJECTED! ***Fortunately,* no one was hurt.**

This severe wear at bolt holes took about 2 hours.

Safety Lifecycle – Example Failure

Teeth wore out, friction heating took <u>several minutes</u> before ejecting spool -- time to interlock. *No one hurt.*

Vibration interlocks <u>and</u> a low amperage warning alarm added as independent protection.

Centrifugal Compressor – Credible Failure Mode Coupling Spool Ejection

- What is the likelihood of a disk pack coupling failure that results in an ejected spool?
- What are our options to protect people, the machine and the process?
- LOPA method uses layers of protection to reduce risk, but the risk level is never zero.
- How much safety is enough? <u>Your corporate policy</u> determines the acceptable level of risk.

Centrifugal Compressor – coupling spool ejection

- Per LOPA rules, use orders of magnitude.
- Disk pack couplings can run indefinitely with proper design & care. Many do run 30+ years, so cumulative data shows 100 years can be sustained when using good practices.
- But some do fail and for multiple reasons (mostly human) which makes it unlikely to sustain 1000 years of cumulative operation.
- Conclusion: use 100 years as a reasonable disk pack failure frequency.

Exposure Risk

- Exposure for an explosion goes in all directions, so if you are in the vicinity, you risk = 1.0.
- Exposure for a turbine blade burst is in a plane. Using LOPA methodology, overall exposure risk in the plane is closer to 0.1.
- Exposure for ejection of a single element such as a shaft coupling is along the trajectory which reduces inherent risk by another factor of ten.

Centrifugal Compressor – coupling spool ejection

- Failure frequency 10^{-2} years
- Exposure (people in area?) 10^{-1}
- Exposure (plane, as a burst) 10^{-1}
- Exposure (trajectory projectile) 10^{-1}
- Unmitigated human risk $= 10^{-5}$ years

  …*not acceptable*

- Each effective layer of protection, like vibration interlock gets credit for ten fold risk reduction.
- Add independent layers to meet your corporate guidelines for acceptable level of risk. (SIL-X)

The Payoff for Testing?

- We have done field tests where a low lube oil pressure would not have shut a machine down.
- Unfortunately, we also have some examples of machines which were wrecked because a low oil pressure switch did not work.
- We also have examples where full function testing revealed errors in interlock set ups such as missing jumpers or programming errors.

Every Scheduled Shut Down is an opportunity to Test an Interlock!

- Why ever be content to use the stop switch?
- Build testing into operating procedures, test the interlock most out of date, or the most serious.
- If planned for at the design phase, full function tests of most interlocks can be straightforward.
- Many sites use digital control systems to automatically track every interlock and take credit for the tests. This reduces tests which often occur in the critical path for turn arounds.

Routine Visual Audits Can Pay Off ...provided people know what to look for

Note shards! We have avoided failures by spotting shards as an indication of a failing coupling.

Routine Visual Audits Can Pay Off ...provided people know what to look for

Is this grease from the gear coupling or oil from the bearings?

How much grease is left? How do you know?

Unique Things about Machinery vs. Other types of process Equipment

- Harmful amplitudes are often machine specific.
- X & Y vibration are **not** redundant information, often a surprise for our safety specialists.
- Machines with flexible shafts may have mode shapes with nodes – be careful about probe placement, especially retrofitting old machines.
- Vibration is a symptom, not "the problem".
- Alarm low to gain investigation time, interlock as high as possible while protecting people.

Failing Gear Coupling

Teeth were starved for grease, wore nearly through, began to shear! Did the vibration interlocks work?

Failing Gear Coupling

Vibration ramped to interlock setting in 12 hours, but didn't work! A technician got a manual shut down.

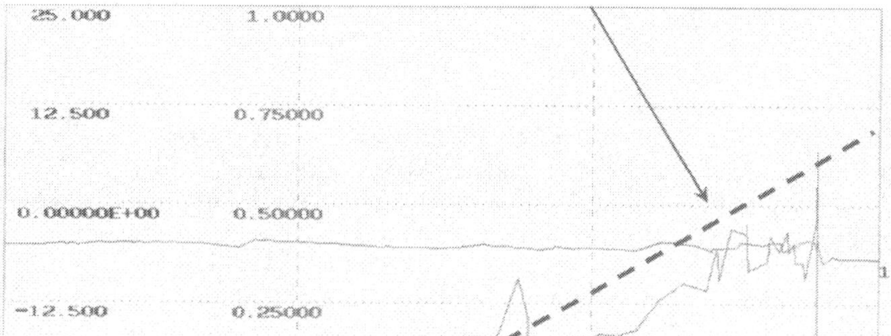

An unrealistic voting strategy required vibration *and* thrust to shut down because engineer did not understand the failure modes.

MACHINERY PROTECTION: New Processes

1. Standards Bodies Driving Protection
2. Industry Standards – IEC/ API/ CCPS
3. New Hazard Identification Processes – HAZOP/ FMEA
4. New Quantification Methodology - LOPA
5. Samples of New Methodology for deciding protection
6. Examples of good and poor applications
7. Validation/ Auditing/ Testing Frequency of systems
8. Potential traps
9. Questions?

About the Author

CFSE

- Curt Miller, an Exida partner for 10 years, opened a branch in Texas to support the Gulf, and has over 25 years of professional experience working with control & safety systems with several large corporations in the Gulf Region.
- Mr. Miller is a graduate of Texas A&M University with a B.S. in Chemical Engineering and is the ISA - Bluebonnet (Austin & San Antonio) Past President and current district program chairman
- Mr. Miller authored a book titled "Win/Win: A Manager's Guide to ISA 84 Compliance" in Jan 2008. After providing persuasive arguments to follow ISA84, it utilizes polled data from 93 participants to provide guidance in overcoming implementation gaps.

Email: cmiller@exida.com

IMPROVING THE CONTROL OF PROCESS FLOWS

Peter Ginn
author: An Introduction to Digital Minicomputers and Process Control
PeterLGinn@USAwide.net

This presentation has two main goals:

#1 To present a novel flow algorithm that gives extremely good results.

#2 To expose the cause of the deficiencies of the PID algorithm when used on a digital computer for flow control.

FOOD FOR THOUGHT

We most often accept things as they are without question:

- "Rinse and Repeat"
- On a digital computer, using the PID algorithm for flow control.

DOES THIS ALTERNATIVE FLOW ALGORITHM WORK ?

- YES
- — This alternative is not theory.
- — This alternative was put into practice
  on 12 loops in a large production plant.
- — Precision
- — Rapid response
- — Reliability

PROCESS FLOW CONTROL BY COMPUTER

- — a source of material
- — a destination for the material
- — a pipe (connecting source and destination)
- — a pressure differential (pump)
- — a control valve
- — a computer for calculations, decisions,
  and issuing changes in the valve's position
- — measurements: flow, pressure, temperature, purity, valve position
- — a set-point
- — a control algorithm

THE COMPUTER SEQUENCE & THE PROCESS SEQUENCE

- 1st computer: measures *flow, T, P, purity, valve position*
- 2nd computer: calculates
- 3rd computer: issues changes
- 4th process: control valve moves
- 5th process: the flow changes

THE PROCESS FLOW AND THE ACTUAL CONTROL VALVE POSITION ARE **RELATED**

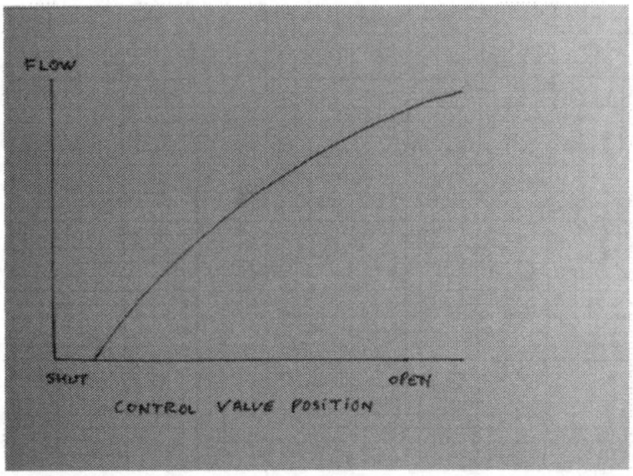

PARTICULAR FLOW   PARTICULAR POSITION

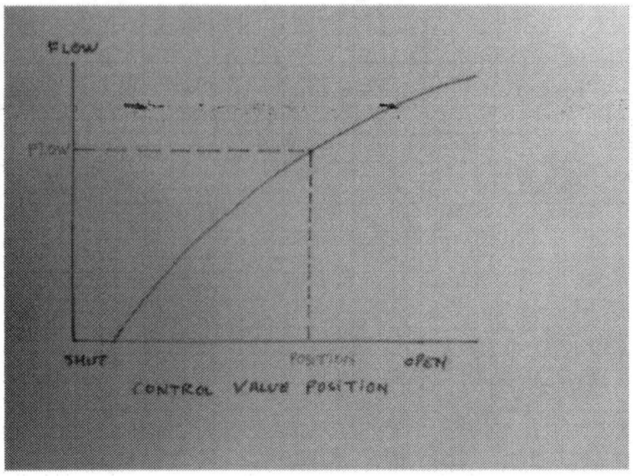

THIS IS WHAT HAPPENS WHEN CONDITIONS CHANGE

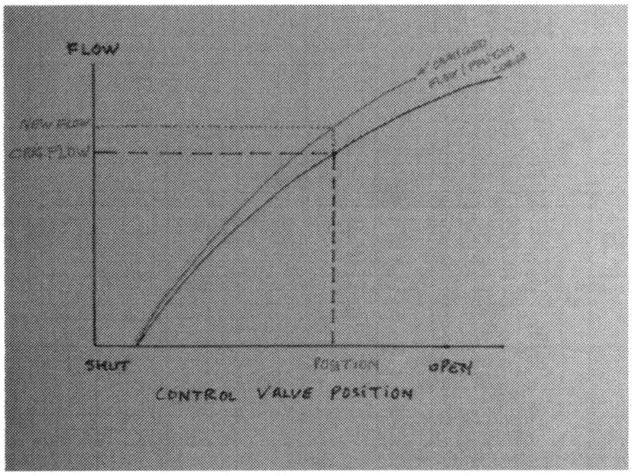

TO WHAT POSITION SHOULD THE VALVE BE MOVED ?

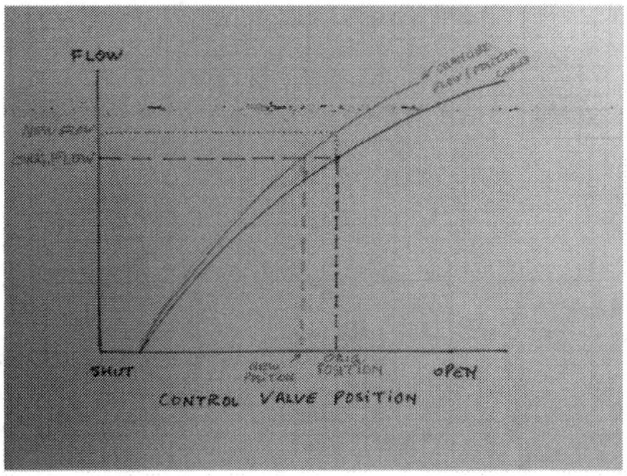

CALCULATING REQUIRED VALVE POSITION

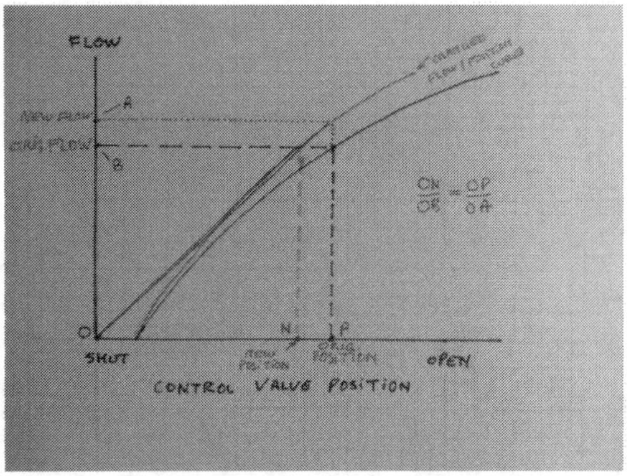

REMEMBER

For the flow system:

- — There is a family of curves of flow versus position.
  all with the same origin

- — There is one unique curve for each particular set of conditions.

- — At any **MOMENT** the flow and the valve position
  can only be on ONE of these curves.

HOW WELL DOES IT WORK ?

- Sensitivity — commissioning a GC for propylene in the vent.

- Quick acting — when a shipping packing was left in a positioner.

- Reliability — what happened one shift after 24 months of using the computer to start up reaction towers.

BACKGROUND:

CHEMISTRY

$Cl_2 + H_2O \rightarrow HCl + HOCl$

$HOCl + C_3H_6 \rightarrow CH_3.CHCl.CH_2OH$

1 mole chlorine (70.9) : 1 mole propylene (42.1)

stoichiometric molar ratio 1.684

BACKGROUND

MASS BALANCE

- "propylene" 100,000 #/day

99.5 %	C_3H_6	2,365 moles
0.5 %	C_3H_8 (propane)	12 moles

- "chlorine" 171,274 #/day

95.0 %	Cl_2	2,365 moles
5.0 %	air	124 moles

BACKGROUND

COMPUTER CONTROL SET-UP

- Executing DDC through back-up Foxboro controller.
  The computer sent up-and-down pulses to controller.
  1000 pulses full scale
- Back-up Foxboro controller integrated the pulses.
  Acted as digital to analog integrator.
  Sent 0 - 20 mA signal to valve positioner.
- Valve: shut 0 pulses, 4 mA
  open 1000 pulses, 20 mA

- Propylene C/V at 400 pulses for 100,000 #/day

PLANT'S CONTROL GOAL

MAXIMISE PROPYLENE AND CHLORINE YIELDS

- Needed to operate at a small excess of propylene.
- But never more than 10% propylene in the vent.
- For example: at a mass flow ratio of 1.713
  a typical vent composition was:

  124 moles air
  12 moles C3H8
  3 moles C3H6
- Propylene in the vent: about 0.13 % of the feed,
  about 2 % in the vent.

OPERATIONS CONTROL GOAL

OPERATE WITH A BREAKING OP

- Shoot for 0.2 < OP < 0.8

- High OP OP > 0.8 chlorine in excess.

- Low OP OP < 0.2 propylene in excess.

- Problem:
  OP could not be held between 0.2 and 0.8 for any length of time.

CONTROL LOGIC

PART 1 RATIO CONTROL

- "Propylene" mass flow was ratio controlled to "chlorine" mass flow.

- That chlorine to propylene control ratio a nominal 1.713 on the calculated mass flows.

- All control action taken on propylene control valve.

- This included changes from the feedback loop.
  This included changes in mass flow for both C3H6 and Cl2.
  (changes in: orifice plate DP, temperature, pressure, purity)

CONTROL LOGIC

PART 2 FEEDBACK CONTROL

- Feedback was from the Oxidation Potential of the liquid leaving hypochlorinator.
- Transport time of 60 seconds between the control valves and the OP cell.
- OP control logic tweaked the ratio if OP high for 6 successive cycles
  or if OP low for 6 successive cycles.
- OP had two set points: a high set point 0.8; a low set point 0.2
- OP had two counters: a high OP counter; a low OP counter.

- OP high if OP > 0.8 for the 6 cycles. Ratio cut 0.004 and high counter set to zero.
- If OP < 0.8 high counter set to zero.

- OP low if OP < 0.2 for the 6 cycles. Ratio increased 0.004 and low counter set to zero.
- If OP > 0.2 low counter set to zero.

PRECISION

Propylene in vent (part 1)

- Installing GC to measure propylene in vent.
- GC calibrated just fine in lab.
- But when installed in the field had sudden 4 % jumps in signal.

PRECISION

Propylene in vent (part 2) A change in ratio from1.713 to 1.709

Before the ratio change:

12 moles/day C_3H_8
124 moles/day air
3 moles/day C_3H_6 (2.1 % analysis in the vent)

After the ratio change:

12 moles/day C_3H_8
124 moles/day air
9 moles/day C_3H_6 (6.2% analysis in the vent)

- The change was an extra 234 #/day of propylene (10#/hr)
- **An extra 234 #/day on existing 100,000 #/day** precision to within 0.25 %

QUICK ACTING

- Had replaced a chlorine valve with a new valve and positioner.
- Ramp-up to production rates had stalled.
- Investigation showed mismatch of valve position and chlorine flow.
- Inspection found shipping packing in positioner.
- Promptly removed — a big oops.
- The tower never pressured up.
- There was no shut down.
- The ramp-up went to the full production rate without further incident

RELIABILITY

- After 24 months of being able to start up any of the four towers using the computer, there came this particular day when no operator or shift foreman on the day shift or the succeeding evening shift had ever started up a tower manually.
- The plant had to call someone out that had !

HOW WAS THIS ALGORITHM EVOLVED ?

Question: When using the PID algorithm,
what tuning factors kept the controlled flow closest to the set-point ?

- Chose standard deviation as metric.
- Worked with 64 sample populations (10 min 40 sec time periods)
- No combination of integral or proportional factors was better than when the valve was put on manual.
- That would be no surprise to Dr. W Edwards Deming.
- Whatever factors were used, the PID algorithm OVER CONTROLLED.

HOW CAN THIS BE EXPLAINED ?

- The computer sees a series of **discrete** measurements of the actual flow.
- It does not see the flow as a continuous trajectory.
- It is as if it takes a series of flash pictures.

INVESTIGATION OF THE DISCRETE MEASUREMENTS LED TO THESE OBSERVATIONS

- Simply recording a plus or a minus if a flow was larger (+) or smaller (-) than the preceding flow, a pattern emerged.
  Here is the pattern:

− +	− −	+ −	+ +	
35.9 %	12.5 %	35.9 %	15.6 %	(real data)

- The magnitudes of the unsigned differences between two successive flows was a uniform distribution of the magnitudes between the smallest and the largest. No peak. 'Small' just as likely as 'large'.

THE GOALS OF THIS PRESENTATION

- *Presentation of a novel algorithm for flow control,*
  an algorithm that gives extremely good results.

- *To bring to light two unsuspected anomalies when flow is measured at discrete intervals of time.*

- *To suggest, because of these two anomalies, that the use of the PID algorithm for flow control in a computer program, can never get the best control that it is possible to get.*

SOME CLOSING THOUGHTS

- One of the hardest things to do is to change a mind already made-up.

- As engineers, we are about change, about making improvements.

- Rinse and repeat: next time you shampoo, after that first rinse see how little shampoo you need to work up a second lather.

SCURVY LIMES THE ROYAL NAVY

1601 James Lancaster (English sea captain) commanded 4 ships sailing from England to India.

Only on his biggest ship: 3 teaspoons of lemon juice/day/sailor.
Half way: no deaths on the big ship,
110 out of 278 sailors on the 3 smaller ships had died.

1757 James Lind on the *Salisbury* conducted the **next** demonstration of limes as a cure for scurvy.

1795 The Royal Navy adopted daily lemon juice for the Home Fleet.

1800 Adopted by all the ships of the Royal Navy.

1865 Adopted by Britain's merchant marine (Merchant Navy).
But only because of an order by of the Board of Trade.

Alarm Rationalization – Optimization of Resources

Why Rationalizing in the 1st place?

- Tip#1: Develop a solid Alarm Philosophy Document
- Tip#2: Prepare well before meeting
- Tip#3: Choose the right personnel
- Tip#4: Train them
- Tip#5: Utilize "off the shelf" software
- Tip #6: Apply Rules & capture Data next PHA?

Now you can do it efficiently and effectively!

Alarm Management Drivers

- **Operational Issues from Poor Alarm Management**
- **Major cause of unplanned downtime ($$$)**
  - Polystyrene $400K
  - Refinery Cat. Cracker $500K
  - Ammonia & Urea $1M/day
  - Ethylene $1M
- **Contributor to major incidents & accidents**
  - Three Mile Island (1979)
  - Milford Haven (1994)
  - Buncefield (2005)
  - Texas City (2005)

- **Standards / Guidelines (ISA 18.2, EEMUA 191)**
- **New Regulations (Refinery NEPs, Pipelines)**

Deepwater Horizon

Alarm Rationalization

- Review existing or potential alarms against criteria in an alarm philosophy
- Define alarm attributes (limit, priority, classification…)
- Document cause, consequence, corrective action
- Record Results in a Master Alarm Database (MADB)

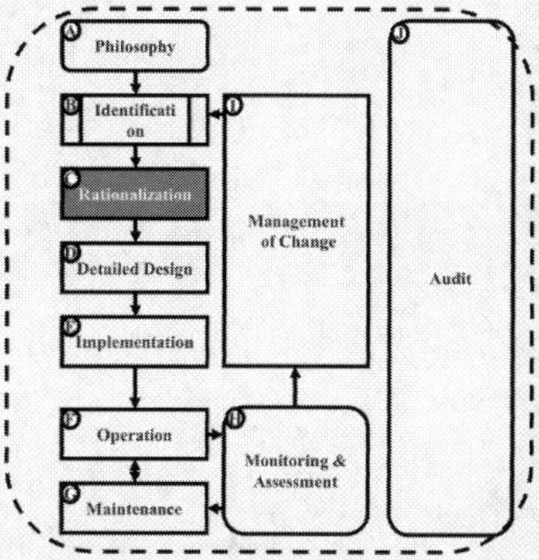

Goal - to create the minimum set of alarms needed keep the plant safe and within normal operating limits

The State of Alarm Management

	Oil & Gas	PetroChem	Power	Other	Best Practice	Standard	
Average Alarms per Day	1200	1500	2000	900	~150-300	~150-300	*Alarm Overload*
Average Alarms/ 10 Minute Interval	6	9	8	5	~ 1-2	~ 1-2	
Peak Alarms per 10 Minutes	220	180	350	180	≤ 10	≤10	*Alarm Floods*
Average Standing Alarms	50	100	65	35	<10	< 5	*Nuisance Alarms*
Distribution % (Low/Med/High)	25/40/35	25/40/35	25/40/35	25/40/35	80/15/5	80/15/5	*Incorrect Priority*
	Actual				***Recommended***		

Can Cause Operators to Miss Critical Alarms

Source: Matrikon

Key Alarm Standards / Guidelines

- EEMUA 191 Guideline
  - Launched the discipline of alarm Management (Milford Haven Explosion)
  - A collection of good engineering practices (called out by UK HSE).

- ISA-18.2 Standard
  - True auditable standard
  - Under review by IEC (IEC 62682)
  - OSHA and insurance industry adoption on the way
  - Supplemented by Technical Reports

ISA-18.2 Technical Reports (TR)

- Standards define the "What", TRs define the "How"
  - *Alarm Philosophy (TR1)*
  - *Alarm Identification & Rationalization (TR2)*
  - *Basic Alarm Design (TR3)*
  - *Enhanced and Advanced Alarm Design (TR4)*
  - *Alarm Monitoring, Assessment, and Audit (TR5)*
  - *Alarm Systems for Batch and Discrete Processes (TR6)*

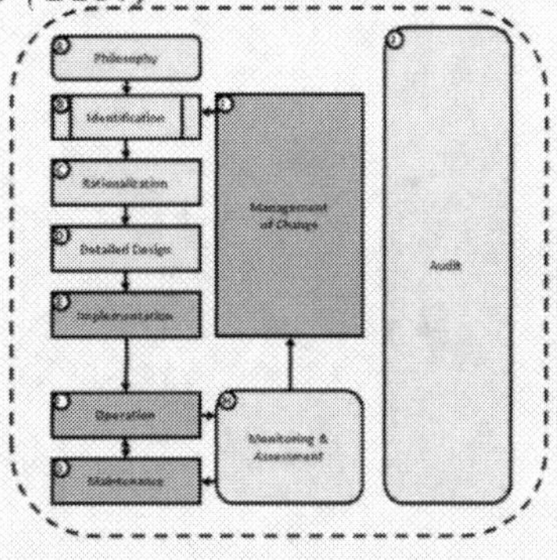

Benefits of Rationalization

- Reduces alarm load on the operator
- Reduces the chance to miss critical alarms
- Removes nuisance alarms (chattering, fleeting or stale alarms)
- Eliminates redundant alarms
- Operator response is quicker, more consistent, and more effective
- Increases system integrity (improve operator trust of alarm system)
- Alarms are prioritized for correct action
- Optimizes the risk reduction of alarms used as a safety layer of protection

TIP#1: Develop a Solid Alarm Philosophy Doc

Alarm Management Philosophy

- The guide for all alarm management activities at a site (a document)
- Establishes the standards for how you will do alarm management
- A written philosophy is necessary to maintain an alarm system over time

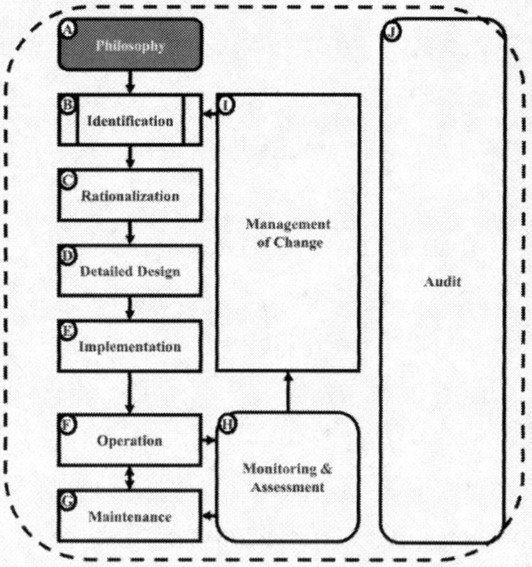

For new plants, the alarm philosophy should be fully defined and approved before commissioning.

Why Create an Alarm Philosophy Document (APD)?

Alarm Philosophy

- Critical to creating and maintaining an effective alarm system over time.
- Create common understanding of alarm management amongst personnel
- Establishes the work practices for how you will do alarm management (consistency)
- Outlines Methods for Solving your Alarm Management Issues
- Required per ISA-18.2
- Demonstrate Compliance in the event of an audit
- Provides an opportunity to create a more effective alarm management program => operational excellence

The What - Documents the Practices & Procedures you will use for…

- Identification and classification of alarms
- Justification of alarms to confirm or establish that they are required
- Prioritization of alarms
- Specification of alarm conditions and settings
- Design of alarms
- Implementation and testing of alarms prior to service
- Commissioning of alarms and alarm changes
- Training in the use, maintenance, testing & modification of alarms
- Operation and use of alarms during normal and abnormal conditions
- Maintenance and testing of alarms during service
- Management of change for alarm systems
- Performance monitoring and assessment of alarm system performance

Alarm Philosophy Content per ISA

ALARM PHILOSOPHY CONTENTS	REQUIRED	RECOMMENDED
Purpose of alarm system	Yes	
Definitions	Yes	
References		Yes
Roles and responsibilities for alarm management	Yes	
Alarm design principles	Yes	
Rationalization	Yes	
Alarm class definition	Yes	
Highly managed alarms (or site equivalent)		Yes
HMI design guidance	Yes	
Alarm setpoint determination		Yes
Prioritization method	Yes	
Alarm system performance monitoring	Yes	
Alarm system maintenance	Yes	
Testing of alarms	Yes	
Approved advanced alarm management techniques		Yes
Alarm documentation		Yes
Implementation guidance	Yes	
Management of change	Yes	
Training	Yes	
Alarm history preservation	Yes	
Related site procedures		Yes
Special Alarm Design Considerations		Yes

Required and Recommended Alarm Philosophy Content (ISA-18.2)

Key Rationalization Content

1. Roles & Responsibilities
2. Alarm Prioritization based on Impact of No Action.
3. How to treat diagnostic alarms (e.g. PV BAD)
4. Alarm Classification
5. Defining and Measuring KPIs

Reference ISA-18.2

1. Roles & Responsibilities(Typical)

7. Responsibilities in Alarm Management

The following positions within the organization are responsible for executing Alarm Management:

1. The control room operator for resolving alarms.
2. The SSV for temporarily changing the settings.
3. The process engineer, OC, and several experienced operators for identifying the alarms and determining the settings and priorities.
4. The process control engineer for the implementation, documentation and for changing the alarms.
5. The management for the procedural issues that this system entails.
6. The maintenance engineer for the knowledge of field instruments and resolving field instrument malfunctions.

RESPONSIBILITY:

The Alarm Custodian will maintain the integrity of the alarm system and ensure compliance with this standard. Area Process Engineers will be responsible for determining appropriate alarm criteria for points within their designated area(s) as well as reporting alarm system performance for their designated areas. Control Room Console Operators will be required to monitor alarm summaries as described in this document and to respond to alarms in a proper and timely manner based on their training and experience as a console operator.

Alarm System Champion (Owner)

- A critical role for successful alarm management
  - May be an operator, engineer, or supervisor depending on site culture
  - In many facilities, selecting someone from operations is the "best" choice
- Accountable for the execution of many of the alarm management activities
- Not responsible to complete the task by themselves

Example – R&R Matrix

Task	Plant Operators	Process Engineering	DCS / Control System	Alarm System Champion
Development of alarm management philosophy	Participate	Participate	Lead	Accountable
Identification of alarms	Support	Support	Lead	
Alarm rationalization (including prioritization)	R	R	R	
Implementation of alarm rationalization/configuration of alarms			Responsible	
Development of alarm system training	Consult	Consult	Responsible	Support
Completion of training on alarms and the alarm system	Complete	Complete	Complete	Support
Development and review of operating procedures / alarm response procedures	Lead	Lead	Support	
Commissioning of the Alarm System	Support		Lead	
Testing of the Alarm System	Support	Support	Lead	
Generate Alarm System Monitoring Reports			Responsible	
Review Alarm System Monitoring reports, assess vs. performance metrics and develop recommendations			Responsible	
Suggestion of alarm system changes & improvements	Participate	Participate	Participate	
Review / Authorization of changes to the alarm system	Participate	Participate	Responsible / Lead	
Implementation of alarm system changes			Responsible	
Maintain / Administer the master alarm database			Responsible	
Conduct alarm audits and develop recommendations	Support	Support	Lead	
Implementation of alarm audit recommendations	Support	Support	Lead	
Maintenance of alarms logs and alarm chronology			Responsible	

2. Example – Rules for Prioritizing Alarms

	Consequence Category 1 (Minor)	Consequence Category 2 (Major)	Consequence Category 3 (Severe)
Personnel	Minor or no injury, no lost time.	One or more severe injury(s).	Fatality or permanently disabling injury.
Environmental	Negligible	Release which results in agency notification, permit violation or fine.	Significant release with serious offsite impact.
Financial	Impact to equipment or production <$10K.	Impact to equipment or production $10K to $250K	Impact to equipment or production > $250K
Operator Urgency (Time to Respond)			
Not Urgent (> 30 mins)	**Re-engineer the alarm for urgency**		
Prompt (15 to 30 mins)	**Low**	**Low**	**Medium**
Rapid (5 to 15 mins)	**Low**	**Medium**	**High**
Immediate (< 5) mins	**Medium**	**High**	**Invalid***

*Alarm may not be appropriate, Consider automated response (interlock)

Consequence of No Action (Better)

Type of Consequence (Impact Area)	Direct Consequence of Inaction or Incorrect Action			
	NEGLIGIBLE	MINOR	MAJOR	SEVERE
Financial: Cost, Efficiency, Downtime, Lost Production, Quality, Equipment Damage	Event or lost production costing <$100K. Minor efficiency impact or quality excursion. Reporting required only at the department level	Event or lost production costing $100K to $1M Significant efficiency loss, multiple unit impact Reporting required at the site level	Event or lost production costing $1M to $10M Extensive outage (< 1 month), potential equipment damage Reporting required at the division / unit level	Event or lost production costing >$10M Long term outage (> 1 month) with loss of production Reporting required at the corporate headquarters level

3. Definition of Classes & Associated Requirements

- Classes specify requirements
  - Training frequency and documentation
  - Testing frequency and documentation
  - Alarm record retention
  - Audit frequency
  - Management of Change
- Typical Classifications
  - Safety Critical
  - LOPA Listed (Safety IPL Alarm)
  - Environmental permit alarms
  - Product Quality

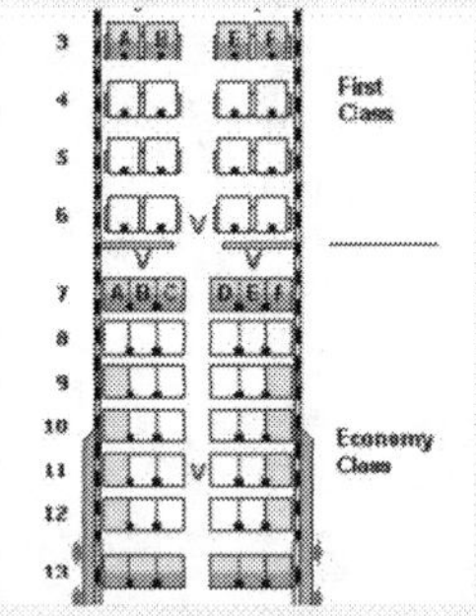

Example Classification(s)

- *Personnel protection*
- *Environmental protection / compliance*
- *HAZOP or Safeguard (e.g. alarms identified as a Safeguard in a process hazards analysis)*
- *LOPA or Independent Protection Layer (e.g., alarms credited risk reduction in a SIL selection analysis)*
- *OSHA PSM critical*
- *Equipment reliability (e.g. rotating equipment high temperature or vibration alarms)*
- *Product quality (e.g. product off spec alarms)*
- *Safety Instrumented System (e.g. SIS diagnostic alarms, SIF trip or failed to trip alarms, etc.)*
- *Control System (e.g. loss of communication alarms, module failure alarms, etc.)*
- *Fire and gas detection (e.g. smoke alarms, H2S detection alarms, etc.)*

4. Treatment of Diagnostic Alarms

Type of Diagnostic Alarm	Meaning
System diagnostic alarm	This alarm is used to indicate that a fault has occurred in the control system hardware, software or components (e.g. communication error or I/O card failure).
Instrument diagnostic alarm	This alarm is used to indicate that a fault has occurred in a field device (e.g. sensor failure, BADPV, PVBAD, BAD-IO, BAD-P, BAD-S, CNFERR, BAD MEASUREMENT). Instrument diagnostic alarms have different names and meanings between control systems.
Bad measurement alarm	A subset of instrument diagnostic alarms. They are generated by the control system to indicate that a process measurement is outside of its expected range (e.g. 3.8 mA for a 4-20 mA signal). They are also known as bad quality or Bad PV. These alarms could indicate a partial or impending failure in the sensor, a sensor out of calibration, or other degradation condition.
Fieldbus alarm	A subset of instrument diagnostic alarms that apply to smart field devices utilizing digital communication protocols. Smart field devices can provide advanced diagnostics that can be used to proactively address instrument degradation prior to failure.

Treatment of Diagnostic Alarms – Key Decisions

Operations

- Does Operator Need to Know?
- Does Operator Have an Action?
  - Troubleshoot?
  - Call Out?
  - Write Maintenance Work Order?
- What is the criticality of the lost reading?

Maintenance

- Do they have a dedicated console?
- Do they review alarms on a (daily) basis?

Rules for Prioritization of Special Situation Alarms (Example)

Alarm type	Priority
BAD PV	Equal to the highest priority alarm from the tag. Advisory in all other cases.
Time-critical DCS system alarms	Advisory at a minimum
Non time-critical DCS system alarms	Alert (Maintenance)
Fieldbus device diagnostics (other than Bad PV)	Alert (Maintenance)
Variable speed drive error messages (other than those alarming inoperable drives)	Alert (Maintenance)
Safety Shower Activation	Critical
Activation of Gas Detectors	Critical
Pre alarm for complete or partial plant trip	Critical when there is enough time to prevent trip. No alarm when there is not enough time to prevent trip.
Trip alarm	Minimum Advisory or Prioritize based on Matrix
Automatic switching of stand-by equipment (A/B pump)	Alert (Maintenance) or Log
Manually switching on stand-by equipment (A/B pump)	Prioritize based on Matrix

5. Alarm System Performance KPIs (ISA-18.2)

Alarm Performance Metrics Based upon at least 30 days of data		
Metric	**Target Value**	
Annunciated Alarms per Time:	Target Value: Very Likely to be Acceptable	Target Value: Maximum Manageable
Annunciated Alarms Per Day per Operating Position	~150 alarms per day	~300 alarms per day
Annunciated Alarms Per Hour per Operating Position	~6 (average)	~12 (average)
Annunciated Alarms Per 10 Minutes per Operating Position	~1 (average)	~2 (average)
Metric	**Target Value**	
Percentage of hours containing more than 30 alarms	~<1%	
Percentage of 10-minute periods containing more than 10 alarms	~<1%	
Maximum number of alarms in a 10 minute period	≤10	
Percentage of time the alarm system is in a flood condition	~<1%	
Percentage contribution of the top 10 most frequent alarms to the overall alarm load	~<1% to 5% maximum, with action plans to address deficiencies.	
Quantity of chattering and fleeting alarms	Zero, action plans to correct any that occur.	
Stale Alarms	Less than 5 present on any day, with action plans to address	
Annunciated Priority Distribution	3 priorities: ~80% Low, ~15% Medium, ~5% High or 4 priorities: ~80% Low, ~15% Medium, ~5% High, ~<1% "highest" Other special-purpose priorities excluded from the calculation	
Unauthorized Alarm Suppression	Zero alarms suppressed outside of controlled or approved methodologies	
Unauthorized Alarm Attribute Changes	Zero alarm attribute changes outside of approved methodologies or MOC	

Establish KPIs in a Way to Ensure that Action is taken (Example)

- Performance Metrics (at least 30 days of data)

Metric	Target	Action Limit	Review Frequency
Average Alarm Rate (alarms per day per operator)	< 300	> 600	Monthly
Percent of time the alarm system is in flood	< 1%	> 5%	Monthly
Percent of hours containing > 30 alarms	< 1%	> 5%	Monthly
Average Number of Alarms Out of Service	<1%	> 5%	Monthly
Annunciated Priority Distribution (Low Priority)	~80%	< 50%	Monthly
Annunciated Priority Distribution (Med Priority)	~15%	> 25%	Monthly
Annunciated Priority Distribution (High Priority)	~5%	>15%	Monthly

- Diagnostic Metrics

Metric	Target	Action Limit	Review Frequency
Percent contribution of top 10 most frequent alarms	< 1% to ~5%	> 20%	Monthly
Stale Alarms (number of alarms >24 hours)	< 5 on any day	> 5	Monthly
Quantity of Chattering and Fleeting Alarms	0	> 5	Monthly
Unauthorized Alarm Changes	0		Monthly

TIP#2: Prepare Well

Alarm Identification

- Define which processes / sources will be used to identify potential alarms
  - P&ID reviews
  - Operating procedures
  - HAZOPs
  - LOPAs
  - Incident investigations
  - Product Quality reviews
  - Environmental Permits
  - Technology Licensor
  - Existing Control System Configuration

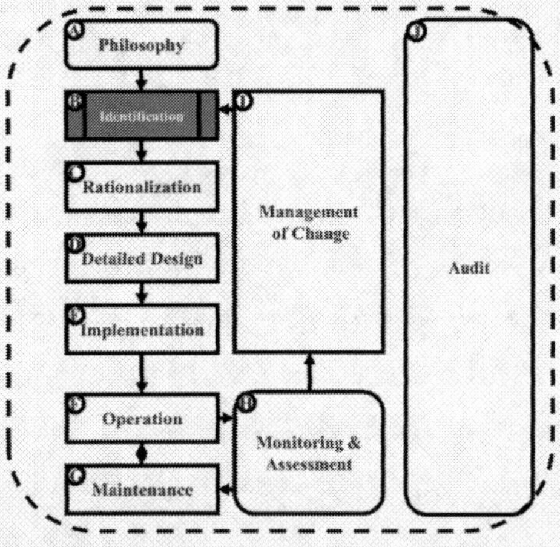

Approach May Depend on Status of System

- New facility (unit or plant)
  - No alarm history / operational experience to draw from
  - Revalidate after facility becomes operational
- Control System Upgrade (Modernization)
  - Lack of experience with new system
  - May require training on alarm capabilities of new system
- Existing System
  - Review of History can identify nuisance alarms & bad actors
  - Likely to have too many alarms configured

Master Alarm Database

- MADB: The authorized list of rationalized alarms and associated attributes
- Required Documentation from Rationalization (ISA-18.2)
  - Alarm type
  - Priority
  - Class
  - Alarm setpoint value or logical condition (e.g., off-normal)
  - Operator Action
  - Consequence of inaction or incorrect action
  - Need for advanced alarm handling techniques

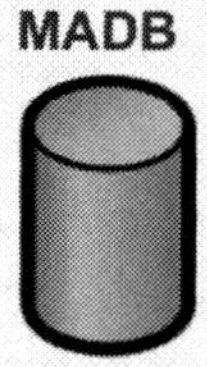

Creating a DeltaV Master Alarm Database

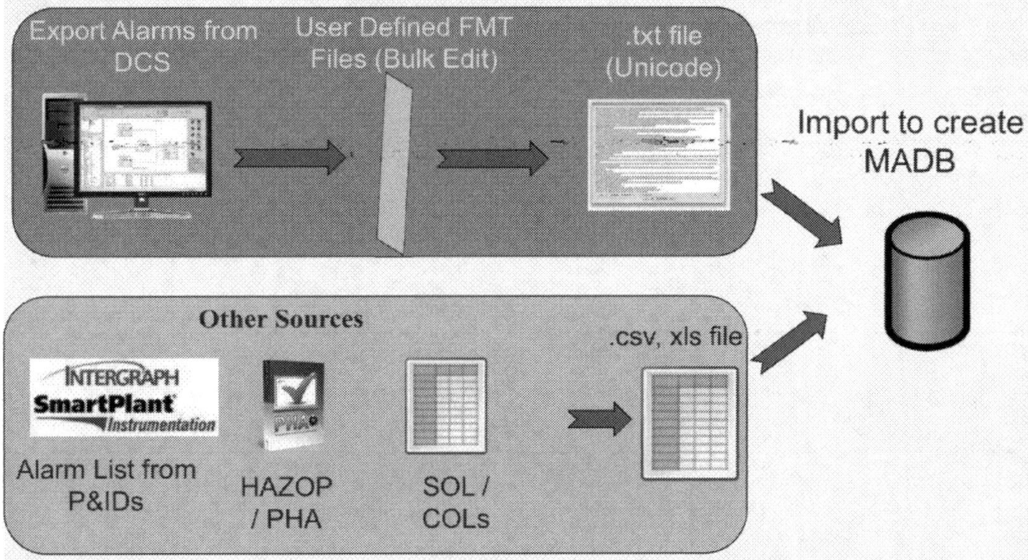

Identifying Alarms to be Rationalized

- The User 1 field can be used to help identify / segregate the alarms that are / need to be rationalized
  - Mark "Bad Actors" and Nuisance Alarms for Resolution
  - Pre-select Alarms to be Rationalized in an upcoming workshop
  - Identify Similar Alarm Conditions (that can be rationalized in bulk)

Hazard & Operability Studies are a Source for Identifying Potential Alarms

STUDY TITLE: PROCESS EXAMPLE								**SHEET:** 1 of 4	
Drawing No.:			**REV. No.:**					**DATE:** December 17, 1998	
TEAM COMPOSITION:			LB, DH, EK, NE, MG, JK					**MEETING DATE:** December 15, 1998	
PART CONSIDERED:			**Transfer line from supply tank A to reactor**						
DESIGN INTENT:			**Material:** A **Activity:** Transfer continuously at a rate greater than B						
			Source: Tank for A **Destination:** Reactor						
No.	**Guide word**	**Element**	**Deviation**	**Possible causes**	**Consequences**	**Safeguards**	**Comments**	**Actions required**	**Action allocated to**
1	NO	Material A	No Material A	Supply Tank A is empty	No flow of A into reactor Explosion	None shown	Situation not acceptable	Consider installation on tank A of a low-level alarm plus a low/low-level trip to stop pump B	MG
2	NO	Transfer A (at a rate >B)	No transfer of A takes place	Pump A stopped, line blocked	Explosion	None shown	Situation not acceptable	Measurement of flow rate for material A plus a low flow alarm and a low flow which trips pump B	JK
3	MORE	Material A	More material A: supply tank over full	Filling of tank from tanker when insufficient capacity exists	Tank will overflow into bounded area	None shown	Remark: This would have been identified during examination of the tank	Consider high-level alarm if not previously identified	EK

Ref: IEC-61882 'Hazard and operability studies (HAZOP studies) – Application guide' 2001

Exercise – Control Room Clipboard "Audit"

- Take a Clipboard to your control room
- Observe the operator and Record:
  - Number of New Alarms
  - Number of Times the Operator Acknowledges the Alarms
  - Number of Times the operator takes action (responds)
- Review Alarm Summary List
  - Compare Before / After

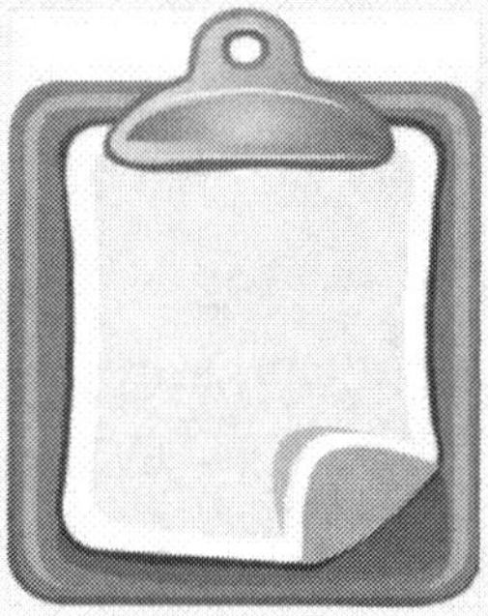

Observations

of New Alarms

# of Times the Operator Acknowledges the Alarms

# of Times the operator takes action (responds)

Identifying Nuisance Alarms

Section 7 – The Ten Modules with the Longest Alarm Activation Times

Alarm Activation Time indicates the length of time an alarm was active regardless of whether or not the alarm was acknowledged by the operator. This section reports the ten control modules with the longest overall alarm activation times as shown in the table below (time format is: DD HH:MM:SS).

Module	Description	Average Activation Time
6PALL864-15A	Oil Circulation Press	03.08:10:27
6PALL864-15B	Oil Circulation Press	03.03:19:21
6LT147-02	S-147 Lower	02.01:37:23
SIS_CCM	SIS Highest Class Fault	01.22:01:05
6XA913-05	Common Transmitter/Comms (Hard...	01.21:46:19
6PAL653-55B	G-924B Nitrogen Pressure Switc...	01.20:21:16
6XA896-37	Common Process Alarm	01.12:28:35
6XAHH655-36B	Workshop/Lab ACMV inlet HC Gas...	01.05:44:03
6PAL913-99	POV PILOT LINE PRESS LO	01.02:49:24
6FQI800-26A	Firewater from refinery system...	01.01:22:56

Monitoring & Assessment Tool identifies Nuisance (Stale) Alarms (Active > 24 hours)

Documenting which alarms are Nuisance Alarms (to be Rationalized first)

Alarm List - 33 out of 33 Alarms Shown, Search/Filter Off

Alarm Name	Tag	Tag Description	Alarm Type	Block Type	Alarm	User1
LI304-High	LI304	Separator Level	High	Analog Indicator	Enabled	Standing Alarm #2
LI304-Low	LI304	Separator Level	Low	Analog Indicator	Enabled	Standing Alarm #2
PI301-High	LI304	Separator Level	High	Analog Indicator	Enabled	Chattering Alarm # 5
PI301-Low	LI304	Separator Level	Low	Analog Indicator	Enabled	Chattering Alarm #5

Identifying Bad Actors

Top Twenty Alarm Sources

Module/Node	Description	Alarm Quantity	Average Per Hour	Sum	Percent of All
FIC-605	FF-01 Spray Reflux	11631	18.64	11631	21%
FICMP2012	MP-02 FLOW TO MP-55A	10977	17.59	22608	40%
FFC-616	RATIO - FF-01 Underflow to Xylene Ratio	1608	2.58	24216	43%
FICUP1516	UP-C-04 Flash Tank Slurry Flow to MP-50 Control	884	1.42	25100	45%
IIUP16P1	UP-C-04 Underflow P1 Pump Amps	802	1.29	25902	46%
FITST111	50 lb Steam Reducing Station Flow Transmitter	788	1.26	26690	48%
IIPX15P1	PX-C-01 PostOx Underflow Pump P1 Current	783	1.25	27473	49%
OPC_FI-N2-051	EXTERNAL: Plant Nitrogen Pipe Line Flow from PI via OPC	607	0.97	28080	50%
FIC-1252	Air Feed TO MP-02	587	0.94	28667	51%
TIFH42106	FH-35 BED #1 IN/OUT TEMP	568	0.91	29235	52%
PICFP2043	FBC Slude Return Loop Pressure Controller	438	0.7	29673	53%
TIFG41106	FG-35 BED #1 IN/OUT TEMP	434	0.7	30107	54%
FIFC1054	FC-E-22 75 lb Condensate Feed	430	0.69	30537	54%

Ref: DeltaV Analyze

Identify Disabled & Suppressed Alarms

Alarms Disabled During the Reporting Period

(Marked with exclamation point if disabled at close of reporting period.)

Module/Node	Description	Times Disabled	Avg Disable Time	Total Disable Time	
OX6RATE	Oxidizer FF-01 Production Rate Master Setpoint	1	7.09:23:18	7.09:23:19	!
FIC-752	SLURRY FLOW TO FP-50	2	4.15:53:45	9.07:47:31	!
PI-608	FF-01 Pressure	2	3.10:29:20	6.20:58:40	
FFIC-515	RATIO - FE-01 Filtrate to Xylene Ratio	2	3.05:50:53	6.11:41:48	
LICIPP421	IP-C-80 Seal Pot Level Control	3	2.22:56:02	8.20:48:06	

Top Twenty Modules with the Highest Average Suppression Time

(Marked with exclamation point if suppressed at close of reporting period.)

Module/Node	Description	Times Suprsd	Avg Suprsd Time	Total Suprsd Time	
FFC-416	RATIO - FD-01 Underflow to Xylene Ratio	2	12.07:52:43	24.15:45:28	!
FFC-616	RATIO - FF-01 Underflow to Xylene Ratio	3	7.06:15:12	21.18:45:38	!
OX6RATE	Oxidizer FF-01 Production Rate Master Setpoint	2	6.21:34:15	13.19:08:31	
FFIC-615	RATIO - FF-01 Filtrate to Xylene Ratio	2	6.21:17:53	13.18:35:47	
TI-0616	FF-01 Underflow Temperature	2	6.21:12:53	13.18:25:47	
TIC-632	Liquid Temperature in FF-30	2	6.17:17:43	13.10:35:27	
TI-611	FF-01 Temperature	2	6.17:10:23	13.10:20:47	
LIC-617	Liquid Level in FF-30	2	6.17:08:55	13.10:17:51	

Lastly, Assemble Reference Documents

- ✓ P&IDs
- ✓ Safety Hazard & Risk Assessment: HAZOP, PHA Reports
- ✓ LOPA Results
- ✓ Safe Operating Limits
- ✓ Shutdown Logic
- ✓ Interlocks / Cause-effect Diagrams
- ✓ Critical Operating Procedures
- ✓ Complex Loop Documentation
- ✓ Operating Graphics (online or hardcopy)
- ✓ Incident Reports
- ✓ Access to process historical data
- ✓ Process narrative or description
- ✓ I/O List

TIP#3: Include the Right Personnel

The Alarm Rationalization Team

- Full time
  - Production and/or Process Engineers familiar with the process, economics, and the control system.
  - Operators (2) from different shift teams with experience in use of the control system *(Ops must be "all in"!)*
  - Process Control / Instrument Engineer *(No more! No design!)*
  - Alarm Management Facilitator / Scribe
- Part time as needed
  - Safety and Environmental Engineers
  - Maintenance / Equipment Reliability
  - Instrumentation/Analyzer Specialists
  - Management (Kick-off)

"Optimum" Size is 4 – 5 people

Ref: ISA TR 18.02 (Draft)

Role of the Facilitator

- Keep the Process Moving
  - Resource intensive activity (approx 100 -200 alarms / day)
  - Maintain balance between discussion vs. story telling
- Reinforce / Interpret the rules of alarm management defined in the alarm philosophy
- Suggest better ways to handle alarms
- Capture Generic Issues / Action Items
- Ensure Consistency
- Challenge Team Decisions
- Independent (no responsibility for area being rationalized)

Ref: ISA TR 18.02 (Draft)

Roles for Core Team Members

- Check Documentation
- Track Progress
- Mark Drawing Changes
- Maintain Action Item List
- Scribe *(This is an important function, try to find a "Microsoft whiz kid")*

Ref: ISA TR 18.02 (Draft)

For Motivation, Determining How Many Alarms Have Been Rationalized in a Day

Using the Change Log

- Open Project Change Log
- Select **File > Save as CSV**
- Open in Excel

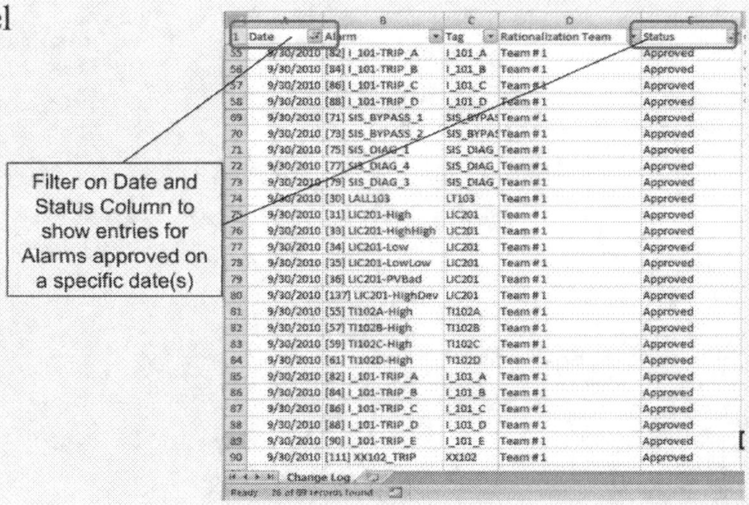

1	Date	Alarm	Tag	Rationalization Team	Status
55	9/30/2010	[82] I_101-TRIP_A	I_101_A	Team # 1	Approved
56	9/30/2010	[84] I_101-TRIP_B	I_101_B	Team # 1	Approved
57	9/30/2010	[86] I_101-TRIP_C	I_101_C	Team # 1	Approved
58	9/30/2010	[88] I_101-TRIP_D	I_101_D	Team # 1	Approved
69	9/30/2010	[71] SIS_BYPASS_1	SIS_BYPAS	Team # 1	Approved
70	9/30/2010	[73] SIS_BYPASS_2	SIS_BYPAS	Team # 1	Approved
71	9/30/2010	[75] SIS_DIAG_1	SIS_DIAG_	Team # 1	Approved
72	9/30/2010	[77] SIS_DIAG_4	SIS_DIAG_	Team # 1	Approved
73	9/30/2010	[79] SIS_DIAG_3	SIS_DIAG_	Team # 1	Approved
74	9/30/2010	[30] LALL103	LT103	Team # 1	Approved
75	9/30/2010	[31] LIC201-High	LIC201	Team # 1	Approved
76	9/30/2010	[33] LIC201-HighHigh	LIC201	Team # 1	Approved
77	9/30/2010	[34] LIC201-Low	LIC201	Team # 1	Approved
78	9/30/2010	[35] LIC201-LowLow	LIC201	Team # 1	Approved
79	9/30/2010	[36] LIC201-PVBad	LIC201	Team # 1	Approved
80	9/30/2010	[137] LIC201-HighDev	LIC201	Team # 1	Approved
81	9/30/2010	[55] TI102A-High	TI102A	Team # 1	Approved
82	9/30/2010	[57] TI102B-High	TI102B	Team # 1	Approved
83	9/30/2010	[59] TI102C-High	TI102C	Team # 1	Approved
84	9/30/2010	[61] TI102D-High	TI102D	Team # 1	Approved
85	9/30/2010	[82] I_101-TRIP_A	I_101_A	Team # 1	Approved
86	9/30/2010	[84] I_101-TRIP_B	I_101_B	Team # 1	Approved
87	9/30/2010	[86] I_101-TRIP_C	I_101_C	Team # 1	Approved
88	9/30/2010	[88] I_101-TRIP_D	I_101_D	Team # 1	Approved
89	9/30/2010	[90] I_101-TRIP_E	I_101_E	Team # 1	Approved
90	9/30/2010	[111] XX102_TRIP	XX102	Team # 1	Approved

Lastly, Select a Good Location and Creating an Effective Working Environment

- Located away from interruptions
- Room Reserved for single purpose
- Make Conducive to Group Discussions
  - Large center table (not classroom style)
  - Single computer hooked up to projector
  - Whiteboard and Flip charts
- Prevent Burnout
  - # Hours / Day
  - # Days / Weeks Continuous

Helps to keep the team focused and on Track

TIP#4: Training Required

Kickoff Meeting

- Gain Management Buy-in
  - Ensures resource availability
  - Review current alarm system performance / problems to confirm need for activity & intent
  - Set expectations for results / timetable
- Head off potentially serious issues

Train Rationalization Team

- Create common understanding of alarm management principles and alarm design practices
- Review Objectives / Goals
- Methodology / Process to be used
  - How Alarm Philosophy is Applied
  - Criteria for being an alarm
  - How alarms will be prioritized
- Roles & Responsibilities
  - Gain agreement from each member
- What's in Scope for Discussion
- How Progress will be measured

Key Design Principles

EEMUA

- Every alarm should have a <u>defined response</u>
- <u>Adequate time</u> should be allowed for the operator to carry-out a defined response
- Every alarm presented to the operator should be <u>useful</u>, <u>relevant</u> and <u>unique</u>
- Each alarm should alert, inform and guide

Ref EEMUA 191 (2007)

Operator Notifications –

Document how to handle in the philosophy

	Operator Must Act	No Operator Action Required (Informational
Abnormal	Alarm	Alert
Expected	Prompt	Message

Alert: *"An audible and/or visible means of indicating to the operator an equipment or process condition that requires <u>awareness</u>, that is <u>indicated separately from alarm</u> indications, and which <u>does not meet the criteria for an alarm</u>" (ISA-18.2)*

Operator Notifications – Cont'd

Prompt - A notification which requires an action to be taken by the operator as part of normal operation (e.g., start sequence when ready, take a sample, add material A).

Message - Provides information about the status of normal operations that does not require the operator to act. (e.g., "Moving from Online to Regeneration Phase" or heater start-up sequence moving from purge to ready for ignition).

Events - Used for automated logging of discrete changes to the system or process (e.g., plant equipment state changes, trips, alarm activation, and acknowledge). They are used primarily for review and analysis (e.g., for post incident analysis).

Equipment and process events should be elevated to an alarm if there is an associated operator action(s).

Document Cause and Confirmation

- What is the Cause?
  - List of likely events that this alarm uniquely identifies
- How can the operator verify / confirm that the alarm is real?
  - Other process measurements / conditions

Determine Consequence of No Action

- Consequence: The immediate & direct result of the abnormal situation identified by the alarm and not a possible consequence requiring other failures.
  - Check whether the consequence results in another alarm or SIS action
- Assume the alarmed condition gets worse
- Not a probabilistic risk assessment (assume no other failures)
  - SIS and Relief valves function as designed

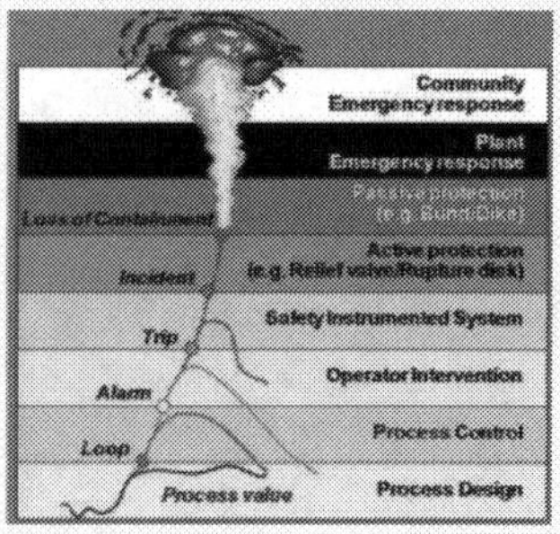

If No Consequences -> Not an alarm

When is it OK to have both a High and High-High Alarm for a Single Tag ?

- When each alarm has a different…
  - *Cause (condition)*
  - *Consequence*
  - *Operator response (in kind or degree)*

} OR

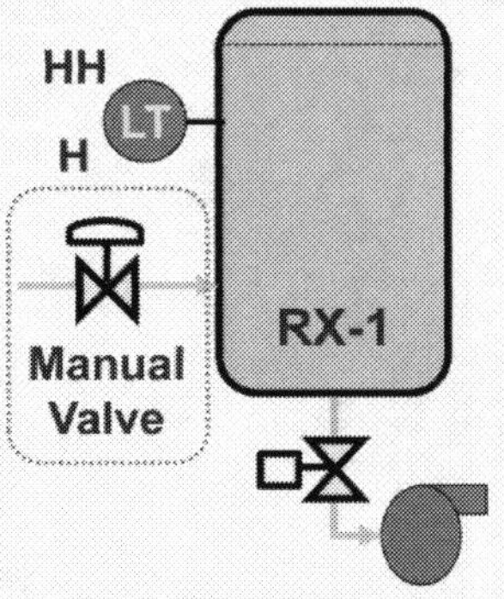

Alarm	Operator Response
High Level	Reduce Flow through Inlet Valve
High-High Level	Stop Flow through Inlet Valve (Close Valve)

Prioritize the Alarm

- Alarm Priority: The importance assigned to an alarm within the alarm system to indicate the urgency of response
- Alarms should be prioritized based on:
  - The **severity of the consequences** (e.g. personnel safety, equipment damage, environmental, economic loss)
  - The **time available** to take corrective action (aka Urgency)
- Target ("Process" Alarms)

Advisory	~80%
Warning	~15%
Critical	~5%

Document Operator Response Time

- Response Time: The time from the activation of the alarm until the last moment the operator action will prevent the consequence. (time available)

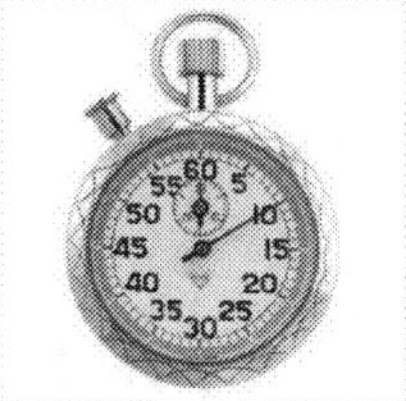

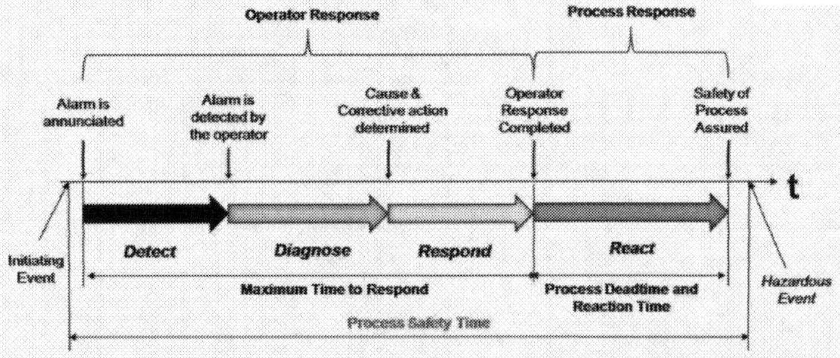

- Ensure that **time available > time needed**
  If time available is insufficient -> Automate Response (e.g. Interlock)

Prioritization – Special Cases

- Should be documented in Alarm Philosophy
- Personnel Safety (example):
  - If operator response is method to prevent harm to person then priority = High
- Diagnostic Alarms (examples)
  - Priority dictated by operator's action (immediate callout or writing of maintenance work order)
  - Priority is one less than highest priority of other alarms on the point
  - Assigned to dedicated, separate "Diagnostic Priority" or "Alert"
- Document basis for priority as part of rationalization (per Philosophy, SOP, Company Policy)

TIP#5: Use Tools

Software Tools Help Follow Lifecycle

- Identification (Import existing alarms)
- Rationalization (Justification, Alarm Objective Analysis, Limit Design)
- Basic Alarm Design (Deadbands & Delays)
- Advanced Alarm Design (Suppression, State-based Alarming, Routing)
- Operation (Alarm Response Procedure)
- MOC (Change Log, Rationalization Status)
- Audit (Compare DCS Config vs MADB)

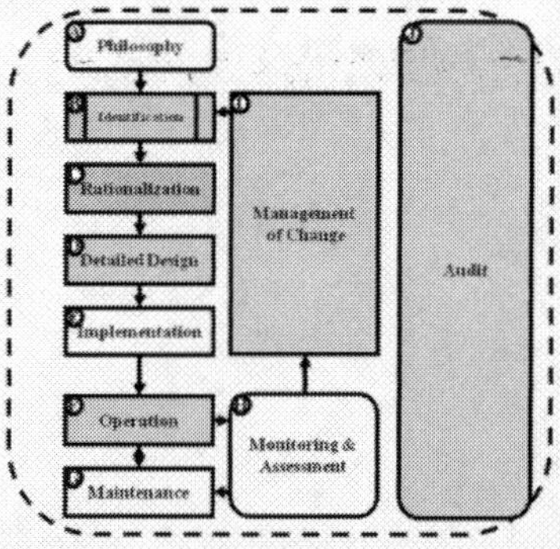

Alarm Management Lifecycle (Ref ANSI/ISA-18.2)

Creating a Master Alarm Database

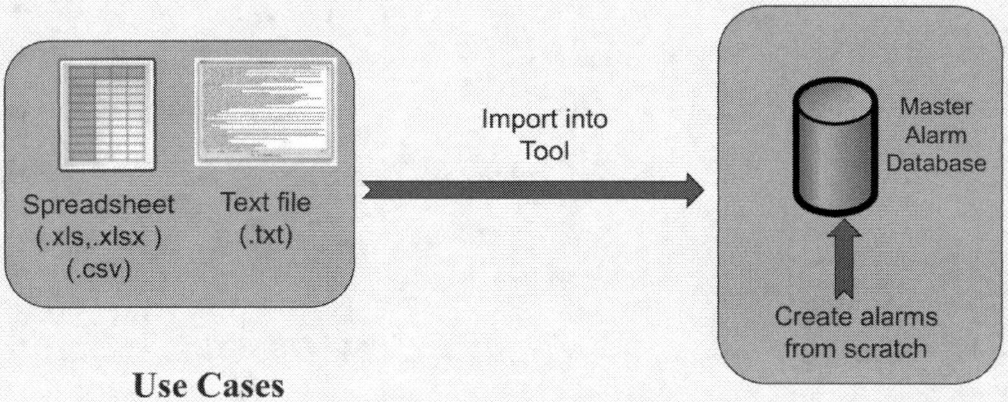

Use Cases

1. Import Alarm or I/O List (new System)
2. Create Alarms from scratch
3. Import Control System Data mapped into a generic template
4. Import A control system-specific export file

Defining the Project Defaults (Rationalization Rules)

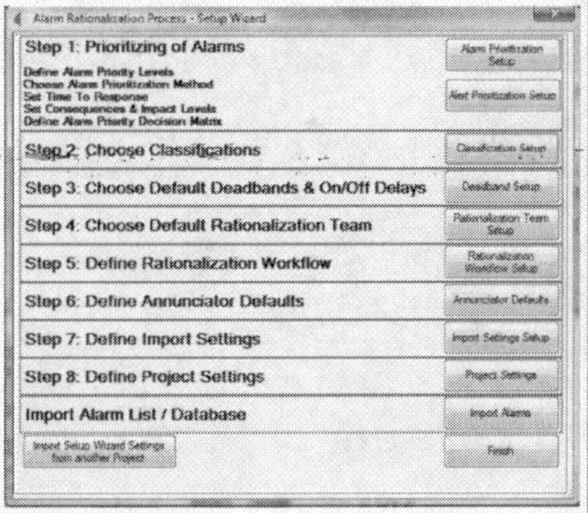

Alarm Philosophy Document

Alarm List View Functions (1)

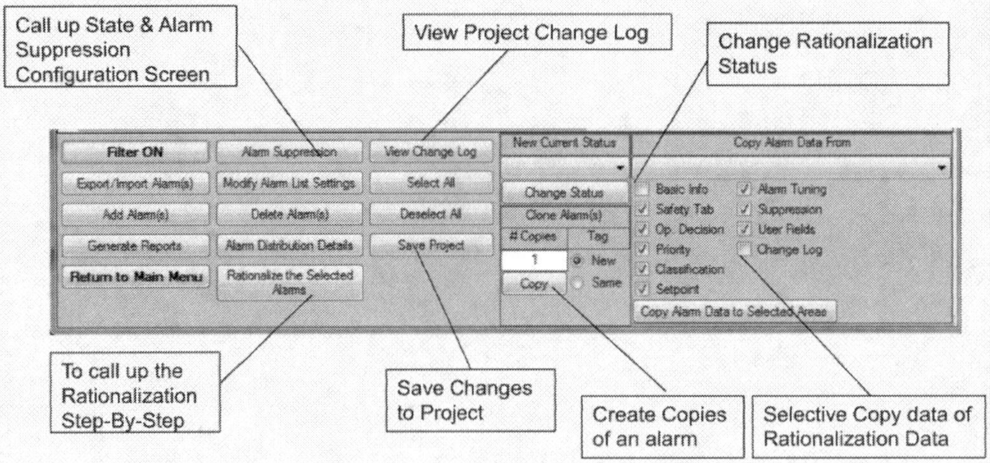

Alarm List View is the Launching Point for most Rationalization Activities

Filtering in the Alarm List View

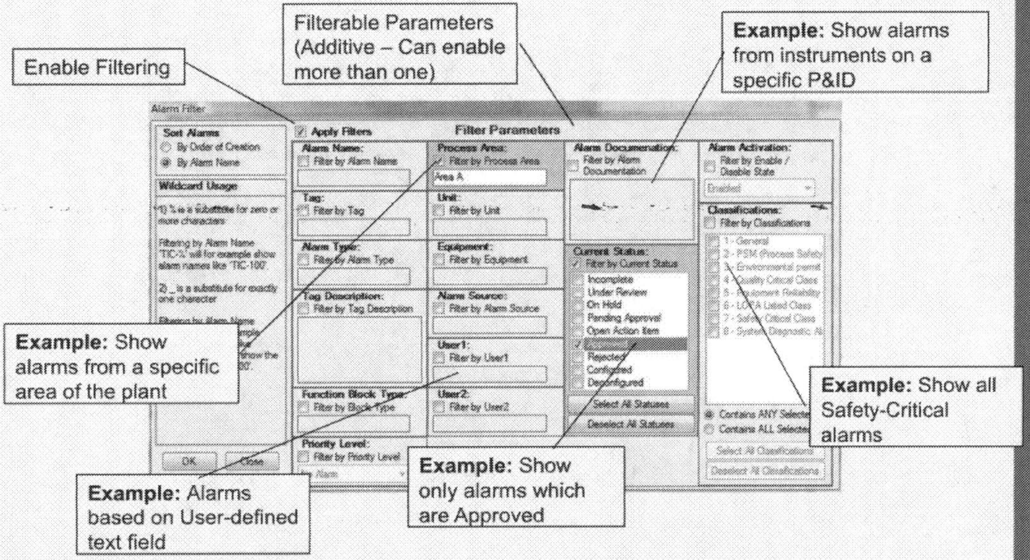

Same filter can be used to control which alarms are exported

Use of Rationalization States

- To manage change, each alarm has an individual status flag which can be set to reflect the state of the alarm.

Alarm Status in SILAlarm	Meaning	Alarm Attributes are Editable (Default)
Incomplete (Default)	Alarm has been imported from an .xls file or added from within SILAlarm but no rationalization activity has begun (no changes have been made)	Y
Under Review	Rationalization activities have begun (at least one parameter has been changed)	Y
On Hold	Rationalization activities have begun (at least one parameter has been changed) and the alarm has been placed on hold based on the user's defined process.	Y
Pending Approval	Alarm tag has been Rationalized and is waiting for approval	Y
Open Action Item	Alarm has an open Action Item that requires completion	Y
Approved	Alarm has been rationalized and approved. Now under change control (cannot be modified or overwritten)	N
Rejected	Alarm has been rationalized but was not approved.	N
Configured	Alarm Rationalization Changes have been populated into the control system (alarm has been added or modified)	N
Deconfigured	Alarm Rationalization Changes have been populated into the control system (alarm has been disabled or decommissioned)	N

The Alarm's Rationalization state effects what happens on import

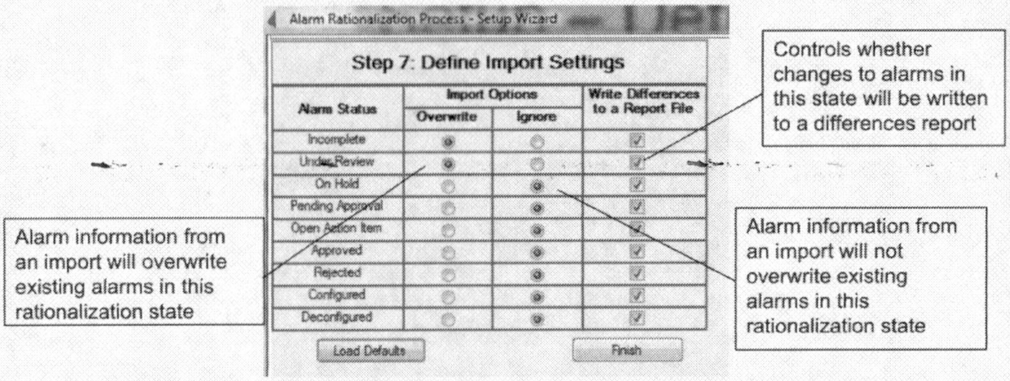

- Example: Uncontrolled Changes are made in the control system for an approved alarm
  - Do you overwrite the MADB with the new settings ? or flag for review & investigation ?

Differences Report

List of Changes to Alarms

	Alarm Name	Tag	Field Name	Original SIL Alarm Value	Import Value	Current Status	Change Status
3	FI205-Low	FI205	EU High	50000	40000	On Hold	Ignored
4	FI205-Low	FI205	Imported Priority	N/A	Warning	On Hold	Ignored
5	FI205-Low	FI205	Current Alarm Setpoint	50	45	On Hold	Ignored
6	LI304-High	LI304	EU High	20	25	Approved	Ignored
7	LI304-Low	LI304	Enabled	True	False	Under Review	Overwritten
8	LIC201-Low	LIC201	Current Alarm Setpoint	22	47	Under Review	Overwritten
9	LIC201-PVBad	LIC201	Priority Score	1.2	1.2	Approved	Ignored
10	LIC201-PVBad	LIC201	On Delay	0	5	Approved	Ignored
11	LIC201-PVBad	LIC201	Off Delay	0	10	Approved	Ignored
12	PAHH102	PT102	Imported Priority	N/A	Advisory	Under Review	Overwritten
13	PAL107_B	PX107B	Discrete Setpoint Limit	1	0	Pending Approval	Ignored
14	PAL107_C	PX107C	Discrete Setpoint Limit	1	0	Under Review	Overwritten
15	PI205-Low	PI205	Imported Priority	N/A	Warning	Under Review	Overwritten

What field was changed — Current Value in SILAlarm — Value in the Import File — Alarm Rationalization Status — Result of Import (was new value accepted ?)

- Tips & Recommendations
  - Periodically Compare Alarm Settings in the MADB vs. the control system
  - Investigate Discrepancies & Resolve
  - Export from SILAlarm to update control system configuration and re-establish approved settings (Enforcement)

Differences Report

List of Changes to Alarms					
Alarm Name	Tag	Field Name	Original SIL Alarm Value	Import Value	Current Status
AR_201-P-PVBAD_ALM	AR_201	Imported Priority	CRITICAL		Incomplete
AR_201-P-LO_LO_ALM	AR_201	Imported Priority	CRITICAL		Incomplete
AR_201-P-HI_HI_ALM	AR_201	Current Alarm Setpoint	100	0	Rejected
AR_201-P-HI_HI_ALM	AR_201	Imported Priority	CRITICAL		Rejected
AR_201-P-HI_ALM	AR_201	Current Alarm Setpoint	0	20	Approved
AR_201-P-HI_ALM	AR_201	Off Delay	0	5	Approved
AR_201-P-HI_ALM	AR_201	On Delay	0	6	Approved
AR_201-P-HI_ALM	AR_201	Imported Priority	WARNING	CRITICAL	Approved
AR_201-P-HI_ALM	AR_201	Confirmation	ffddfdffdf	Changed	Approved
AR_201-P-HI_ALM	AR_201	Corrective Actions	dfdfdfdfdf	Change	Approved
AR_201-P-HI_ALM	AR_201	Cause	gdgfdfgdfd	Change	Approved

Change List / Error List / Additions & Deletions / Priority Audit / Setpoint Audit

Excel Tab Name	Contents / Purpose
Change List	Documents what parameters are different between the SILAlarm MADB as compared to the import file, along with their associated values.
Error List	Lists any errors that were encountered during the import
Additions	Documents which alarms and tags are new (meaning they are present in the import file but not in the SILAlarm MADB)
Additions & Deletions	Documents which alarms and tags are new (meaning they are present in the import file but not in the SILAlarm MADB) as well as which alarms and tags are expected, but not present in the import file (indicates that they may have been deleted in the source)
Priority Audit	For auditing changes to the alarm priority - Documents changes in the value of the imported priority (from the control system) and identifies when it is different from the rationalized priority in SILAlarm
Setpoint Audit	For auditing changes to the alarm setpoint - Documents changes in the value of the current setpoint (from the control system) and identifies when it is different from the selected setpoint in SILAlarm

TIP#6: Initiate During PHA?

Stated Previously, Hazard & Operability Studies are a Source for Potential Alarms

STUDY TITLE: PROCESS EXAMPLE								**SHEET:** 1 of 4	
Drawing No.:			**REV. No.:**					**DATE:** December 17, 1998	
TEAM COMPOSITION:			LB, DH, EK, NE, MG, JK					**MEETING DATE:** December 15, 1998	
PART CONSIDERED:			Transfer line from supply tank A to reactor						
DESIGN INTENT:			**Material:** A **Activity:** Transfer continuously at a rate greater than B **Source:** Tank for A **Destination:** Reactor						
No.	**Guide word**	**Element**	**Deviation**	**Possible causes**	**Consequences**	**Safeguards**	**Comments**	**Actions required**	**Action allocated to**
1	NO	Material A	No Material A	Supply Tank A is empty	No flow of A into reactor Explosion	None shown	Situation not acceptable	Consider installation on tank A of a low-level alarm plus a low/low-level trip to stop pump B	MG
2	NO	Transfer A (at a rate >B)	No transfer of A takes place	Pump A stopped, line blocked	Explosion	None shown	Situation not acceptable	Measurement of flow rate for material A plus a low flow alarm and a low flow which trips pump B	JK
3	MORE	Material A	More material A: supply tank over full	Filling of tank from tanker when insufficient capacity exists	Tank will overflow into bounded area	None shown	Remark: This would have been identified during examination of the tank	Consider high-level alarm if not previously identified	EK

Ref: IEC-61882 'Hazard and operability studies (HAZOP studies) – Application guide' 2001

What if Alarm Rationalization data was captured during the PHA?

- Increased team efficiency for projects that require both?
- Currently, joint PHA LOPAs have become popular for same reason

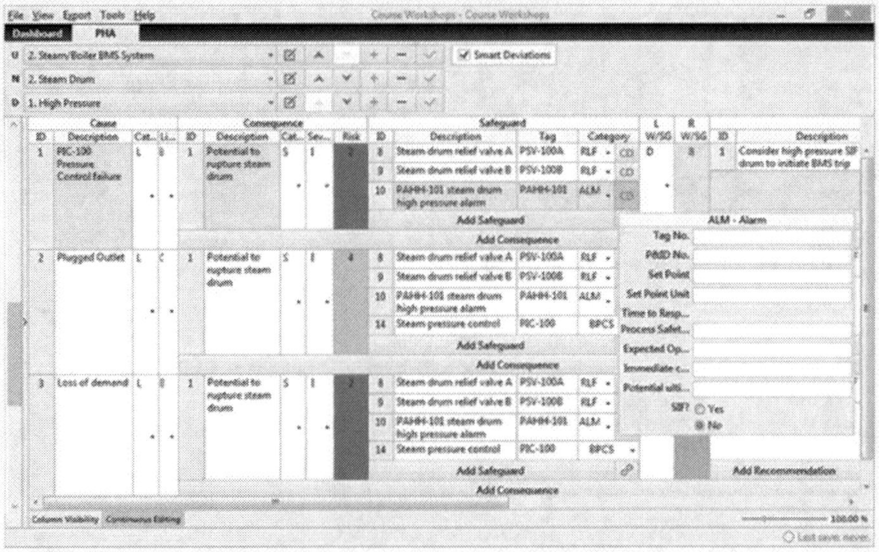

Alarm Rationalization – Optimization of Resources

Why Rationalizing in the 1st place?

- Tip#1: Develop a solid Alarm Philosophy Document
- Tip#2: Prepare well before meeting
- Tip#3: Choose the right personnel
- Tip#4: Train them
- Tip#5: Utilize "off the shelf" software
- Tip #6: Apply Rules & capture Data next PHA?

Now you can do it well!

About the Author

CFSE

- Curt Miller, an Exida partner for 10 years, opened a branch in Texas to support the Gulf, and has over 25 years of professional experience working with control & safety systems with several large corporations in the Gulf Region.
- Mr. Miller is a graduate of Texas A&M University with a B.S. in Chemical Engineering and is the ISA - Bluebonnet (Austin & San Antonio) Past President and current district program chairman.
- Mr. Miller authored a book titled "Win/Win: A Manager's Guide to ISA 84 Compliance" in Jan 2008. After providing persuasive arguments to follow ISA84, it utilizes polled data from 93 participants to provide guidance in overcoming implementation gaps.
- Email: cmiller@exida.com

10-72

Sustainable R&D for new technology in start ups

John Super
Cobroko Solutions
October 10, 2014

Introduction

- Focus is Chemicals
- No fuels product examples included
- Ten examples, the companies, identities confidential
- Measures of merit are ten, five business and five technology
- Business drivers impact technology R&D
- Each measure 1-10 score, highest total possible 100

Definitions

- Sustainable: In this context, it means renewable raw material such as corn. Not natural gas.
- Success: Technology is commercialized and profitable
- Of the ten:
  - 3 successful. GO, green
  - 4 maybe successful someday, MAYBE, gray
  - 3 not likely successful, NO GO, red

Cobroko Solutions Page 3 John Super

Business Measures of Merit and Meaning

- Economic Targets- Evaluate and chose chemicals targets that your new technology has lower cost plus than current selling prices.
- Investors have patience- Venture Capital money often wants to flip at high multiple in 3-5 years, often it takes ten years to show value.
- CEO good at raising money- Raising money and identifying partners should be 75% or more of the CEO's time in successful start ups, and this often is a natural talent.

Cobroko Solutions Page 4 John Super

Business Measures of Merit and Meaning (continued)

- Adequate Burn to Deliver- The new company needs the funds to resource the R&D with people, tools, and scale up work.
- Competitors- The comparison with your technology when competitors are working in the "space", can be a good thing.

Business Measures of Merit

	A	B	C	D	E	F	G	H	I	J
Economic Targets	8	9	8	6	5	8	3	3	5	3
Investors have patience	8	9	9	8	6	4	3	8	3	5
CEO good at raising money	9	9	9	7	8	3	8	1	6	3
Adequate Burn to deliver	9	9	6	7	8	3	8	5	4	3
Competitors	6	5	3	7	5	6	6	2	4	4
Business metrics total	40	41	35	35	32	24	28	19	22	18

Technical Measures of Merit and Meaning

- Is it novel? Start ups by definition have a novel idea or they would not have done it, novel creates buzz and motivates the employees
- Teams focused. To execute an idea, continuing innovation will not get something defined and moving.
- Patents and Applications. Granted patents and filed applications are required to hold the intellectual property and are absolutely necessary.

Cobroko Solutions Page 7 John Super

Technical Measures of Merit and Meaning (continued)

- Experienced Professionals. You need to add those with relevant experience and track records. And need to make changes.
- Good Technical Decisions. Due diligence from potential investors will tell the R&D teams if they are making the good decisions, or not.

Cobroko Solutions Page 8 John Super

Technical Measures of Merit

	A	B	C	D	E	F	G	H	I	J
Is it novel	8	8	8	3	7	8	7	8	5	7
Teams focused	9	8	9	6	6	8	5	3	5	3
Patents and Applications	10	9	9	4	5	8	8	8	5	5
Experienced professionals	9	8	7	8	8	8	8	7	5	3
Good technical decisions	9	8	7	8	5	7	3	3	4	2
Technical metrics total	45	41	40	29	31	39	31	29	24	20
Company Total	85	82	75	64	63	63	59	48	46	38

Company Business and Technical and Total Metrics

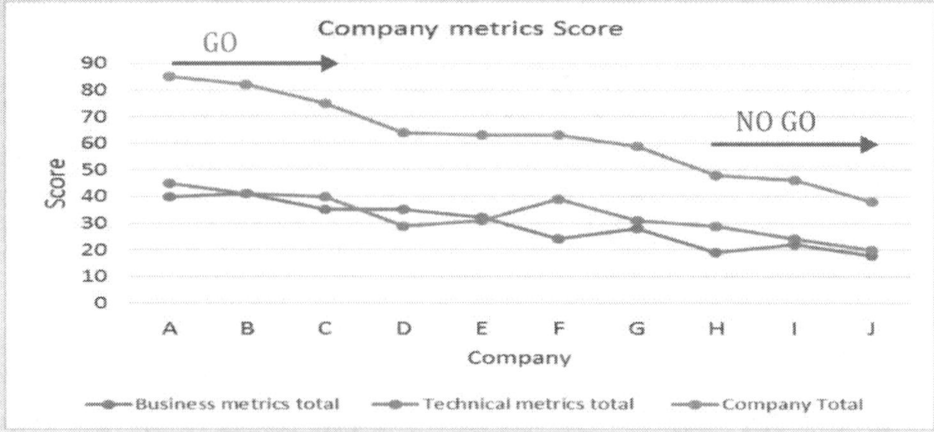

Conclusions about success

Go: A B and C
Maybe: D E F and G
No Go: H I and J

Cobroko Solutions Page 11 John Super

Remarks

- GO, high scores in both business and technical
- MAYBE, good technical and poor business with one of the four
- NO GO, one pulled down by low business score, two poor scores in both
- Four IPOs in sample
- GO Success is made by strong scores in both business and technical

Cobroko Solutions Page 12 John Super

Recent References

- Naam Wasserman, "How an entrepreneur's zeal can destroy a startup", Wall Street Journal, August 25, 2014, pages R1-R2 of section Small Business.
- Dennis McCullough, "The Path to New Technology Commercialization", AIChE, Session 94, Paper A, April 1, 2014.
- "Chemical Producers' Sustainability Strategies on the Rise", Environmentalleader.com, September 8, 2014.
- And many, many other books, papers, articles and personal experience.

Amines and Chelants

Expanded Horizons...

Market Inspired. Solutions Oriented.

Application of Statistical Methodology to Chemical Pilot Plant Processes

Xue (Ida) Chen

The Dow Chemical Company

Freeport, TX

Amines and Chelants

Problem Statement

- Efficacy study of catalysts in batch reactor and continuous process
  - Deactivation of the catalyst
  - Minimum reaction time
  - Potential treatments needed to ensure production rate
- Questions of Interests:
  - Is there an effect from the different catalysts?
  - Is there a time effect for those catalysts?
  - If there are some differences among the catalysts over the reaction time, are those differences statistically significant?

Expanded Horizons...

Amines and Chelants

Example 1

- **Goal:** Compare the efficacy of 3 types of catalysts in a batch reactor

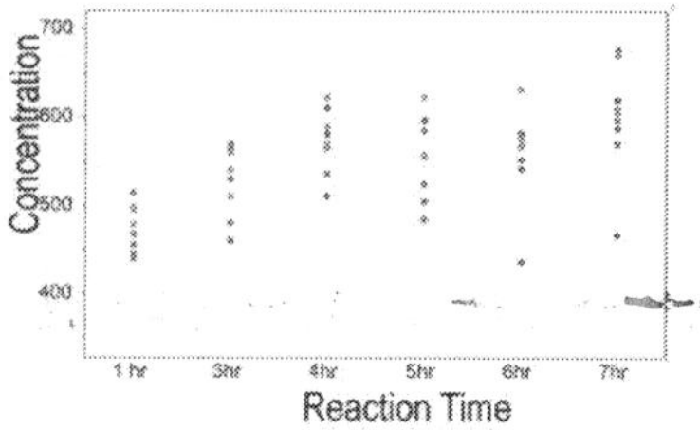

- **Experiments:**
  - Variables: 3 catalysts
  - Reaction: 3 replicates for each catalyst
  - Response: concentration of the final product in the function of reaction time from 1hr to 7hr
- **Data Analysis and interpretation**
  - Concentration is measured multiple times on the same catalyst
  - Does a serial correlation exist within the same catalyst over time?
  - Does a serial correlation violate the ANOVA assumptions?

Expanded Horizons...

Amines and Chelants

Example 1: Check Serial Correlation

Sphericity Test	
Mauchly Criterio	0.0016086
ChiSquare	26.372898
DF	14
Prob >Chisq	0.0231994

P<0.05: Concentrations within the same catalyst are highly correlated over time

- **Sphericity test**: check the serial correlation
- This correlation must be considered in the analysis
- **MANOVA** model is suitable for this case, rather than ANOVA

Expanded Horizons...

Amines and Chelants

Example 1: Bird's Eye View of the Dataset

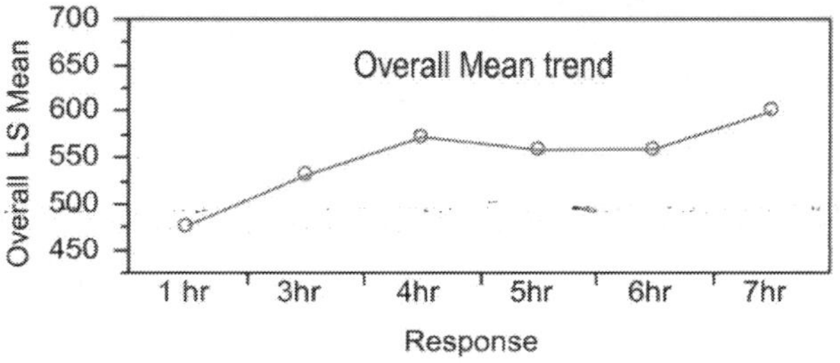

- Does average concentration change with time? (effects of time)

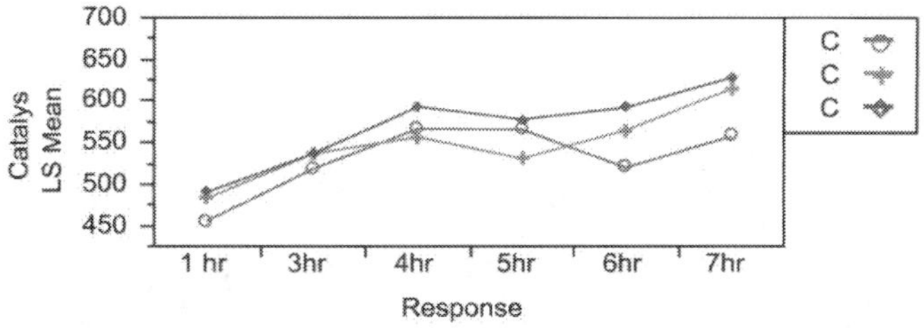

- Does catalyst type influence time effects on concentration? (Effects of catalyst types)

Expanded Horizons...

Amines and Chelants

Example 1: Data Analysis and Interpretation

All Between

Test	Value	Exact F	NumDF	DenDF	Prob>F
F Tes	0.2602224	0.7807	2	6	0.4996

Intercept

Test	Value	Exact F	NumDF	DenDF	Prob>F
F Tes	306.52872	1839.1723	1	6	<.0001 *

Catalyst

Test	Value	Exact F	NumDF	DenDF	Prob>F
F Tes	0.2602224	0.7807	2	6	0.4996

All Within Interactions

Test	Value	Approx. F	NumDF	DenDF	Prob>F
Wilks' Lambda	0.0098269	3.6351	10	4	0.1126
Pillai's Trace	1.671968	3.0582	10	6	0.0921
Hotelling-Lawley	31.381072	3.1381	10	2	0.2658
Roy's Max Root	28.987617	17.3926	5	3	0.0201 *
Univar unadj Epsilon	1	1.6896	10	30	0.1295
Univar G-G Epsilon	0.3791943	1.6896	3.7919	11.376	0.2209
Univar H-F Epsilon	0.7340953	1.6898	7.341	22.023	0.1612

Reaction Time

Test	Value	Exact F	NumDF	DenDF	Prob>F
F Test	204.28749	81.7150	5	2	0.0121 *
Univar unadj Epsilon	1	24.8410	5	30	<.0001 *
Univar G-G Epsilon	0.3791943	24.8410	1.896	11.376	<.0001 *
Univar H-F Epsilon	0.7340953	24.8410	3.6705	22.023	<.0001 *

Reaction Time*Catalyst

Test	Value	Approx. F	NumDF	DenDF	Prob>F
Wilks' Lambda	0.0098269	3.6351	10	4	0.1126
Pillai's Trace	1.671968	3.0582	10	6	0.0921
Hotelling-Lawley	31.381072	3.1381	10	2	0.2658
Roy's Max Root	28.987617	17.3926	5	3	0.0201 *
Univar unadj Epsilon	1	1.6896	10	30	0.1295
Univar G-G Epsilon	0.3791943	1.6896	3.7919	11.376	0.2209
Univar H-F Epsilon	0.7340953	1.6896	7.341	22.023	0.1612

- Effects of catalyst types: No
- Effects of time: Yes
- Effects of the interaction between catalyst and time: No

Expanded Horizons...

Amines and Chelants

Example 2

- Background:
  - Continuous reactor
  - Temperature increases to meet the final production target
  - Selectivity base line drift with time
- Goal: study efficacy of the treatments (A and C) during the process.
- Challenges for data analysis:
  - Y axis scale changes: cannot tell visually
  - Time effects (serial correlation)
  - Drifting base line with time

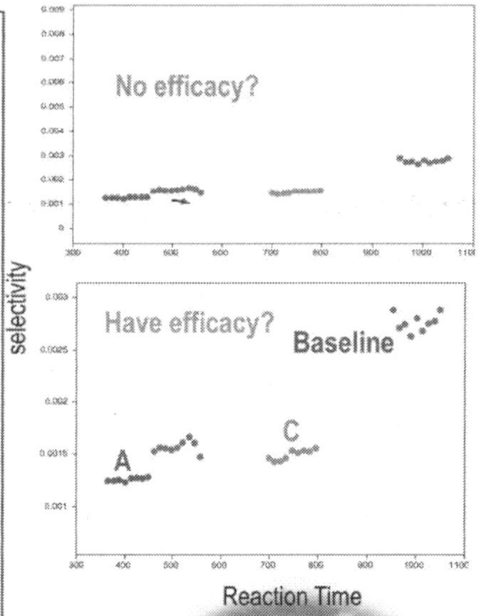

Expanded Horizons...

Amines and Chelants

Example 2: Statistical Model

- Introduce dummy variable V (indicator variable)
  - If baseline: V=0
  - If A (or C): V=1
- Set up model

$$y = \beta_1 \times t + \beta_2 \times t \times V + \beta_3 \times V + \beta_4$$

- Then:
  - Baseline: $y = \beta_1 \times t$
  - A: $y = \beta_1 \times t + \beta_2 \times t + \beta_3 = (\beta_1 + \beta_2) \times t + \beta_3$
- Comparison of A and baseline reduces to checking β_2 and β_3

Expanded Horizons...

Amines and Chelants

Example 2: Does treatment A have effects on the selectivity?

Parameter Estimates

Term	Estimate	Std Error	t Ratio	Prob>|t|
Intercept	0.0001948	6.458e-5	3.02	0.0063 *
Time	1.8048e-6	3.374e-7	5.35	<.0001 *
Dummy variable	0.0005386	0.000263	2.05	0.0529
(Time-648.962)*(Dummy variable-0.69231	1.9335e-6	1.085e-6	1.78	0.0885

(β_3 → Dummy variable estimate; β_2 → interaction estimate)

$$y = \beta_1 \times t + \beta_2 \times t \times V + \beta_3 \times V + \beta_4$$

- Both P values>0.05: treatment A does not have an effect on the selectivity

Expanded Horizons...

Amines and Chelants

Example 2: Does treatment C have effects on the selectivity?

Parameter Estimates

Term	Estimate	Std Error	t Ratio	Prob>|t|
Intercept	-2.588e-5	0.000226	-0.11	0.9096
Time	2.0211e-6	3.034e-7	6.66	<.0001 *
Dummy variable 2	0.0006584	3.435e-5	19.17	<.0001 *
(Time-753.37)*(Dummy variable 2-0.66667	1.1359e-6	8.997e-7	1.26	0.2194

(β_3 → Dummy variable 2 estimate; β_2 → interaction estimate)

$$y = \beta_1 \times t + \beta_2 \times t \times V + \beta_3 \times V + \beta_4$$

- P-value of β_3 <0.05: Treatment C has statistically significant effect on the selectivity

Expanded Horizons...

Amines and Chelants

Summary

- MANOVA model is suitable to study the dataset with serial correlation (time effects) when comparing the difference among different catalysts.
  - Identify the best catalyst
  - Minimum required reaction time to achieve the target yield
- A novel analysis method is developed to study the special case (example 2) of continuous reactor.
  - Compare the treatments with the baseline and identify effective treatment during the continuous process

Expanded Horizons...

Amines and Chelants

Thank you for your attention!

Questions or Comments?

Expanded Horizons...

Simulation Based Optimization of Stirred Tank Reactors

Aglave, R., CD adapco
Dewan, Y., CD-adapco
Newale, A., CD-adapco
Eppinger, T., CD-adapco
Originally presented on: 10/10/2014 14:30:00 - 15:00:00

The design of a stirred tank reactor is a multi□input and multi□output (MIMO) problem. Design parameters include geometric parameters (impeller & tank shape, size number, baffles.) as well as operating conditions (impeller RPM, fill level etc.). A common design objective is minimizing mixing time, maximizing mixing quality, to achieve a specific scale of agitation etc. Additionally, various constraints can be imposed on the performance of such a reactor, such as avoiding overflow, avoiding mixing `dead zones', imposing a maximum shear stress etc. Designing to these multiple objectives and constraints by modifying a large number of design parameters is a challenging task since the relationships of these multiple inputs and outputs can be highly non□linear.

Due to this reason, the design and scale□up of stirred tank reactors has traditionally been a very time consuming task. This is the primary motivation for developing a design methodology that is both autonomous, and efficient in targeting specific reactor performance. Optimization algorithms lend themselves well to being applied to solve such multi□input, multi□output problems. An optimization algorithm iteratively runs evaluations and understands the sensitivities between all inputs and combination of inputs to all outputs. The algorithm successively guesses a better design until a converged design is reached that optimizes all the required objectives while satisfying all constraints. This study details the implementation of this optimization methodology to design optimal stirred tank reactors for various design objectives and constraints. The use of an optimization algorithm for design studies requires two components: an analysis tool and an optimization algorithm. The optimization algorithm uses the analysis tool to generate the relationship between a set of inputs and outputs. The reactor is analyzed using a computational fluid dynamics (CFD) approach with STAR□CCM+. The use of STAR□ CCM+ for mixing tank performance prediction has been validated by the work of Maas et al. Results for a pareto front optimization with the goal of minimizing power and mixing intensity is presented.

The Effects of material flow uncertainties on the overall economic performance of a typical gas processing plant under various upstream conditions

By

Mesfin Getu (Ph.D)

Curtin University, Sarawak Malaysia

September 2014

Outline

- Introduction
- Model development
- Case studies
- Conclusions
- Future works

11:13 AM

Part I
Gas processing plant:
Operational aspect

11:13 AM

Introduction

- **Why natural gas?**
  - -wide availability
  - -clean burning characteristic
  - -easy transportation
  - -steady growth rate
- **Feed sources & content**
  - - associated gas
  - - non-associated gas
  - - light & heavy hydrocarbons
  - - impurities

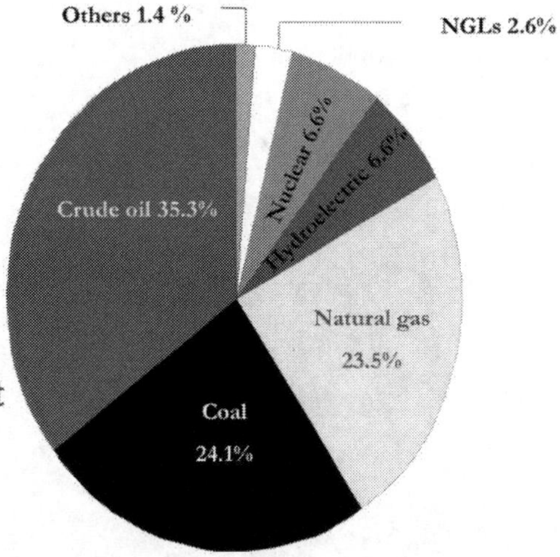

Source: Kidnay & Parish (2006)

11:13 AM

Introduction (cont')

- **Feed variations**
  - feed composition
  - feed flow rate
  - ambient temperature
  - pipe line pressure
- **Operating conditions**
  - propagation
  - problem in fine tuning
  - trial & error practice
  - workload for plant Engineer

11:13 AM

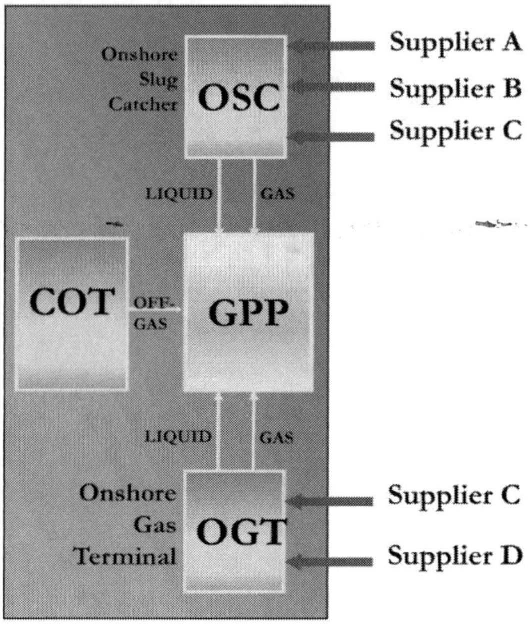

Source: PGB Kerteh Terengganu

Introduction (cont')

- **Gas processing plant (GPP)**
  - -pre-treatment unit (PTU)
  - -acid gas removal unit (AGRU)
  - -condensate treatment unit (CTU)
  - -dehydration unit (DHU)
  - -low temperature separation unit (LTSU)
  - - sales gas compression unit (SGCU)
  - - product recovery unit (PRU)
  - -Product storage unit (PSU)

11:13 AM

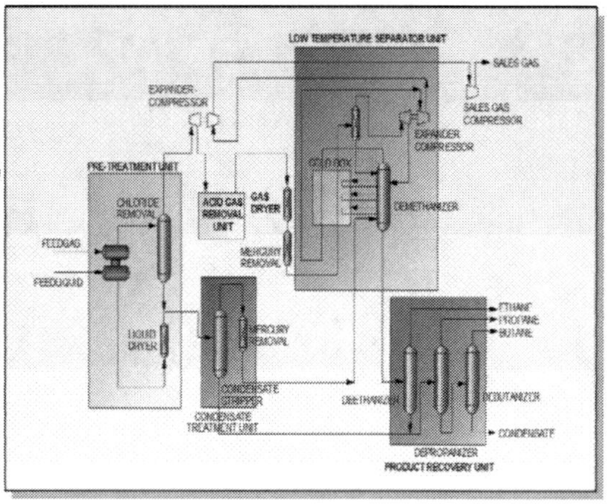

Source: PGB Kerteh Terengganu

Introduction (cont')

- **Products produced**

  -sales gas

  -ethane

  -propane/butane/

  -condensate

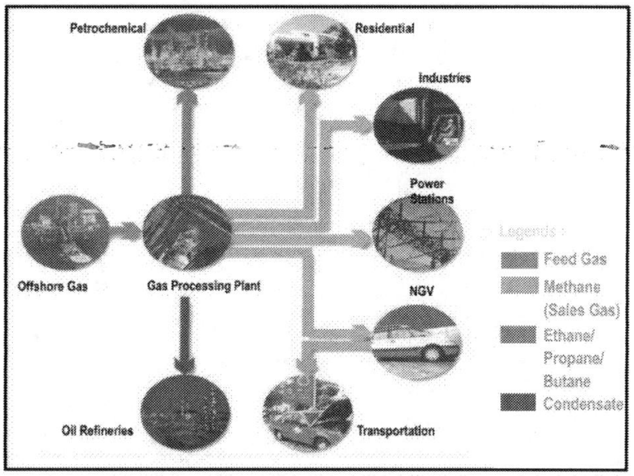

Source: PGB Kerteh Terengganu

- **Product utilization**

  - power consumption

  - petrochemical feedstock

  - refinery feedstock

  - blending agent

11:13 AM

Model Development

Optimization classification

Deterministic Optimization | **Optimization Under Uncertainty**

- **Deterministic optimization**

  - conventional process

  - constraint violation

  - non-optimal & infeasible

  - worst-case strategy

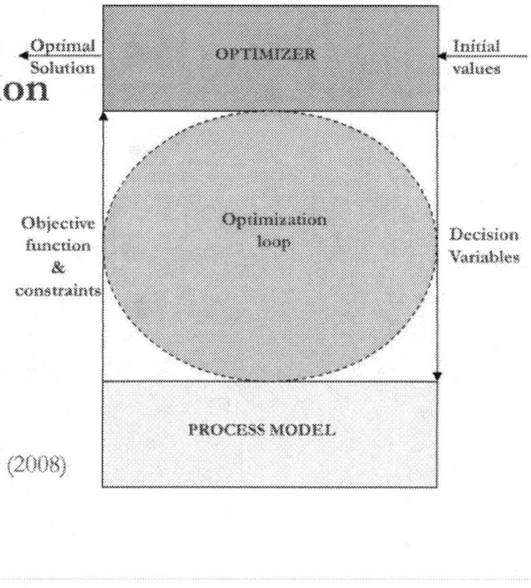

11;13 AM

Source: Diwaker (2008)

Model Development(cont')

- **Fuzzy programming**
  - -fuzzy number/set/
  - - better solution
  - - model not rigorous
  - - uncertainty modeled

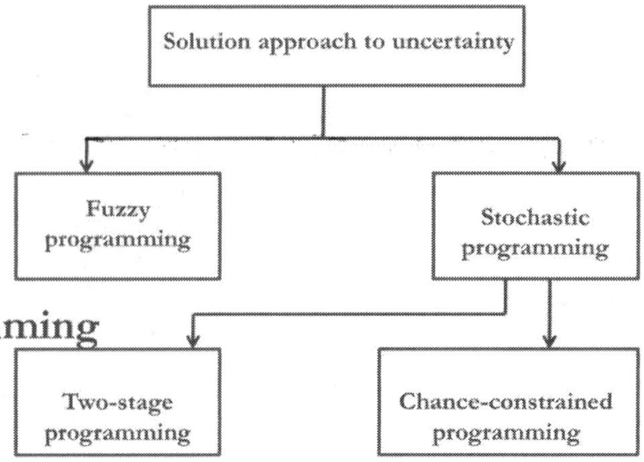

Source: Li et al. (2004b)

- **Two-stage programming**
  - - Penalty function
  - - iteration limitation
  - - computational effort
  - - no information about profit vs reliability

11:13 AM

Model Development(cont')

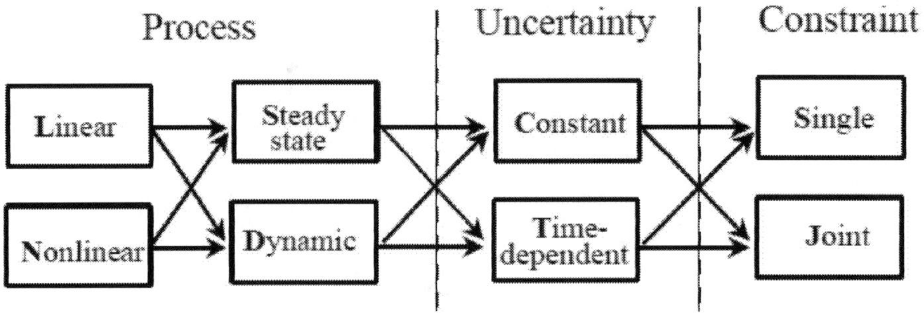

Source: Li et al .(2008)

- **Chance-constrained programming**
  - - competitive tool
  - - ensures reliability vs profit
  - - few works in process system engineering
  - - most works based on theoretical basis

11:13 AM

Model Development(cont')

- **Solution space**
  - -single chance/half space/
  - joint chance /curvature/
  - operating region outstretched
  - solution space subset
- **Feasibility test**
  - increasing confidence level
  - solution space shrinks
  - empty solution space
  - critical value (α_c)

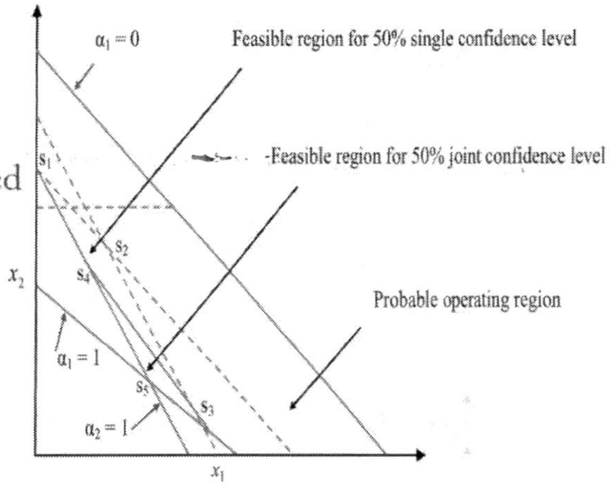

Source: Li et al. (2008)

11:13 AM

Model Development(cont')

- **Reliability vs profit**
  - -violation of constraint
  - expected profit
  - slow decreasing profit
  - difficult to determine
  - higher region of confidence
  - trade-off profit vs reliability
  - risk consideration
  - conservative decision
  - aggressive decision

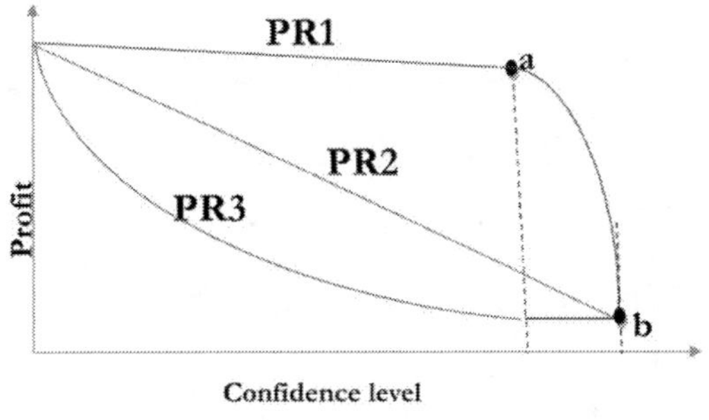

Source: Li et al. (2008)

11:13 AM

Model Development (cont')

- **Modeling material flow**

$$\sum_{j=1}^{J} a_{k,j}P_j + \sum_{l=1}^{L}\hat{a}_{k,l}\hat{P}_l = \sum_{i=1}^{I} b_{k,i}R_i + \sum_{m=1}^{M}\hat{b}_{k,m}\hat{R}_m$$

- material inflow & outflow

- begin with deterministic opt.

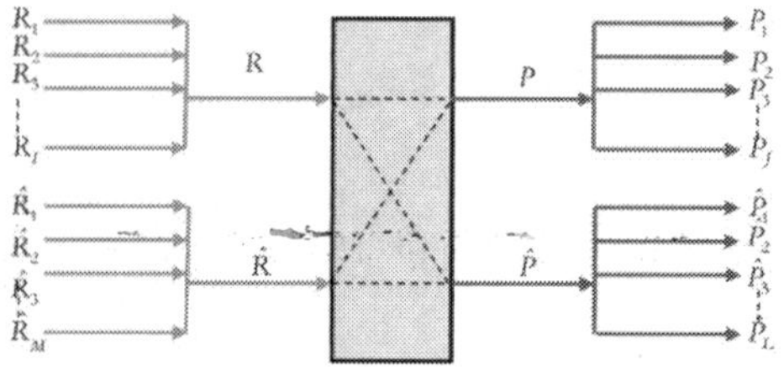

Source: Getu et al. (2013b)

- **Modeling energy flow**

$$\sum_{j'=1}^{J'} Q_{j'} + \sum_{l'=1}^{L'}\hat{Q}_{l'} = \sum_{i'=1}^{I'} U_{i'} + \sum_{m'=1}^{M'}\hat{U}_{m'}$$

- utility inflow & outflow

- apply deterministic opt.

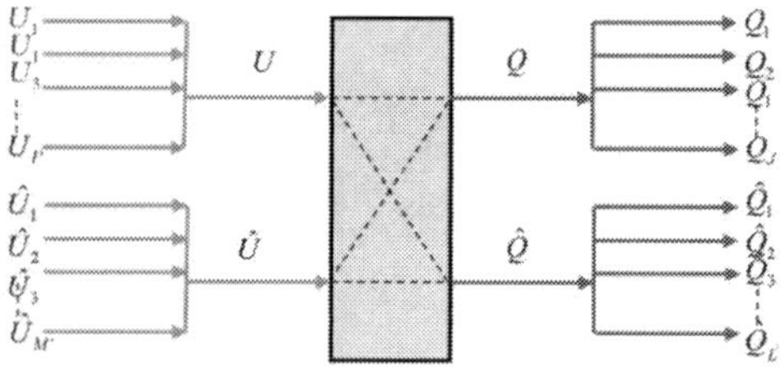

11:13 AM

Model Development(cont')

- **Uncertain feed flow**
  -raw material inflow
  -variation from suppliers

$$\text{max Profit} = \sum_{j=1}^{J}\bar{C}_j^P P_j - \sum_{i=1}^{I}\bar{C}_i^R R_i - \bar{C}^{\hat{R}}\left[\sum_{k=1}^{K} r_k^{\xi}\right]$$

- **Uncertain product flow**
  -material product outflow
  -impact from feed variation

$$\text{max Profit} = \sum_{j=1}^{J}\bar{C}_j^P P_j + \bar{C}^{\hat{P}}\left[\sum_{k=1}^{K} p_k^{\xi}\right] - \sum_{i=1}^{I}\bar{C}_i^R R_i$$

- **Uncertain energy flow**
  -energy inflow variation
  -suppliers/cogeneration plant/

$$\text{max Profit} = \sum_{j'=1}^{J'}\bar{C}_{j'}^Q Q_{j'} - \sum_{m=1}^{M}\bar{C}_m^U U_m - \left[\sum_{e=1}^{E}\bar{C}_e^u u_e^{\xi}\right]$$

- **Uncertain energy product flow**
  -energy outflow from the plant
  -looking for potential market/re-use/

$$\text{max Profit} = \sum_{j'=1}^{J'}\bar{C}_{j'}^Q Q_{j'} + \left[\sum_{e=1}^{E}\bar{C}_e^q q_e^{\xi}\right] - \sum_{m=1}^{M}\bar{C}_m^U U_m$$

11:13 AM

Model Development(cont')

- **Single chance constrained**

-higher probability

$$\Pr_k\left\{r_k^{\xi}=\sum_{j=1}^{J}a_{k,j}P_j-\sum_{i=1}^{I}b_{k,i}R_i\le\xi_k\right\}\ge\alpha_k \qquad k=1,\dots,\mathrm{K}$$

-more reliability constraint

$$\Pr_k\left\{p_k^{\zeta}=\sum_{i=1}^{I}b_{k,i}R_i-\sum_{j=1}^{J}a_{k,j}P_j\ge\zeta_k\right\}\ge\alpha_k \qquad k=1,\dots,\mathrm{K}$$

-some outputs critical

$$\Pr_e\left\{u_e^{\varepsilon}=\sum_{j=1}^{J}Q_j-\sum_{i=1}^{I}U_i\le\varepsilon_e\right\}\ge\alpha_e \qquad e=1,\dots,\mathrm{E}$$

- demand/propane & butane/

$$\Pr_e\left\{q_e^{\varsigma}=\sum_{i=1}^{I}U_i-\sum_{j=1}^{J}Q_j\ge\varsigma_e\right\}\ge\alpha_e \qquad e=1,\dots,\mathrm{E}$$

- **Joint chance constrained**

-holding constraint

$$\Pr\left\{r_k^{\xi}=\sum_{j=1}^{J}a_{k,j}P_j-\sum_{i=1}^{I}b_{k,i}R_i\le\xi_k,\ k=1,\dots,\mathrm{K}\right\}\ge\alpha$$

-safety operation

$$\Pr\left\{p_k^{\zeta}=\sum_{i=1}^{I}b_{k,i}R_i-\sum_{j=1}^{J}a_{k,j}P_j\ge\zeta_k,\ k=1,\dots,\mathrm{K}\right\}\ge\alpha$$

- independent & correlated

$$\Pr_e\left\{u_e^{\varepsilon}=\sum_{j=1}^{J}Q_j-\sum_{i=1}^{I}U_i\le\varepsilon_e,\ e=1,\dots,\mathrm{E}\right\}\ge\alpha$$

$$\Pr_e\left\{q_e^{\varsigma}=\sum_{i=1}^{I}U_i-\sum_{j=1}^{J}Q_j\ge\varsigma_e,\ e=1,\dots,\mathrm{E}\right\}\ge\alpha$$

11:13 AM

Model Development(cont')

- **Relaxed single chance constrained**

-Equivalent

$$\sum_{j=1}^{J}a_{k,j}P_j-\sum_{i=1}^{I}b_{k,i}R_i\le\Phi_k^{-1}(1-\alpha_k) \qquad k=1,\dots,\mathrm{K} \qquad \prod_{k=1}^{K}\left[1-\Phi_k\left(\sum_{j=1}^{J}a_{k,j}P_j-\sum_{i=1}^{I}b_{k,i}R_i\right)\right]\ge\alpha$$

-deterministic

$$\sum_{i=1}^{I}b_{k,i}R_i-\sum_{j=1}^{J}a_{k,j}P_j\ge\Phi_k^{-1}(\alpha_k) \qquad k=1,\dots,\mathrm{K} \qquad \prod_{k=1}^{K}\left[\Phi_k\left(\sum_{i=1}^{I}b_{k,i}R_i-\sum_{j=1}^{J}a_{k,j}P_j\right)\right]\ge\alpha$$

-solver

$$\sum_{j=1}^{J}Q_j-\sum_{i=1}^{I}U_i\le\Phi_e^{-1}(1-\alpha_e) \qquad e=1,\dots,\mathrm{E} \qquad \prod_{e=1}^{E}\left[1-\Phi_e\left(\sum_{j=1}^{J}Q_j-\sum_{i=1}^{I}U_i\right)\right]\ge\alpha$$

- confidence level

$$\sum_{i=1}^{I}U_i-\sum_{j=1}^{J}Q_j\ge\Phi_e^{-1}(\alpha_e) \qquad e=1,\dots,\mathrm{E} \qquad \prod_{e=1}^{E}\left[\Phi_e\left(\sum_{i=1}^{I}U_i-\sum_{j=1}^{J}Q_j\right)\right]\ge\alpha$$

- **Relaxed joint chance constrained**

-Independent uncertain variables

$$\prod_{k=1}^{K}\left[\int_{-\infty}^{r_1^{\xi}}d\xi_1\cdots\int_{-\infty}^{r_k^{\xi}}d\xi_k\cdots\int_{-\infty}^{\infty}d(\xi_1,\dots,\xi_k)d\xi_k\right]$$

-Correlated uncertain variables

$$\prod_{k=1}^{K}\left[\int_{-\infty}^{p_1^{\zeta}}d\zeta_1\cdots\int_{-\infty}^{p_k^{\zeta}}d\zeta_k\cdots\int_{-\infty}^{\infty}d(\zeta_1,\dots,\zeta_k)d\zeta_k\right]$$

- Not represent real-world problems

$$\prod_{e=1}^{E}\left[\int_{-\infty}^{u_1^{\varepsilon}}d\varepsilon_1\cdots\int_{-\infty}^{u_e^{\varepsilon}}d\varepsilon_e\cdots\int_{-\infty}^{\infty}d(\varepsilon_1,\dots,\varepsilon_E)d\varepsilon_E\right]$$

- Enormous/expensive/computation

$$\prod_{e=1}^{E}\left[\int_{-\infty}^{q_1^{\varsigma}}d\xi_1\cdots\int_{-\infty}^{q_e^{\varsigma}}d\varsigma_e\cdots\int_{-\infty}^{\infty}d(\varsigma_1,\dots,\varsigma_e)d\varsigma_e\right]$$

11:13 AM

Model Development (cont')

- **HYSYS process simulator**
  - -powerful engineering software
  - -accurate thermodynamic
  - -hydrocarbons & non-hydrocarbons
  - - Inherent flexibility

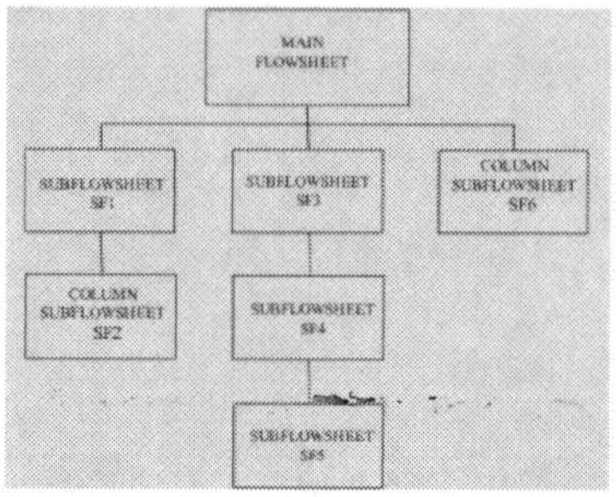

- **General algebraic modeling system (GAMS)**
  - -high level modeling system
  - - particularly for optimization
  - - high performance solvers
  - - complex & large scale model

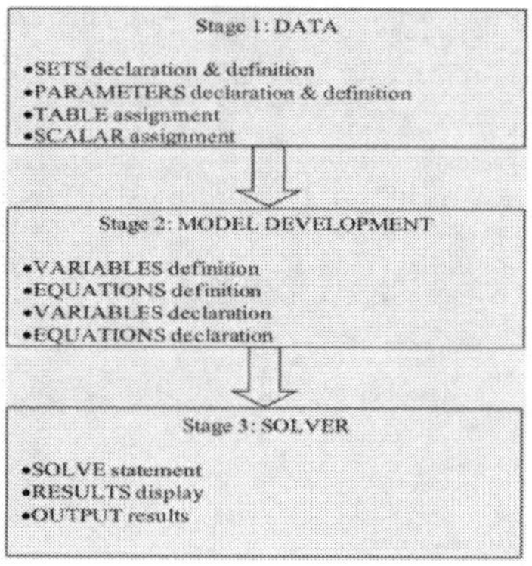

11:13 AM

Model Development(cont')

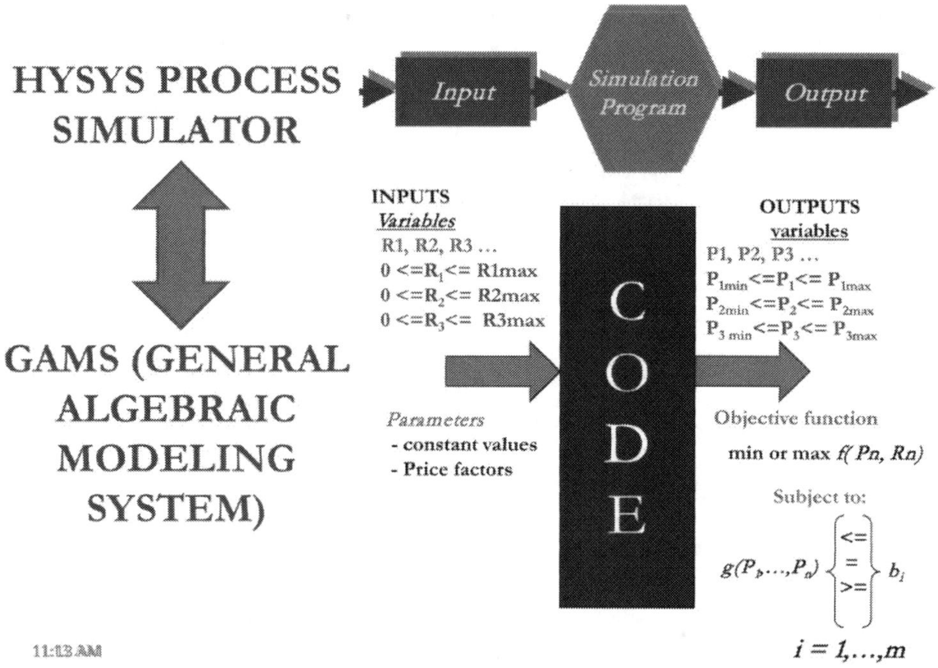

11:13 AM

Case studies

- **Case study 1: uncertain feed flow**
  - -main process involved
  - -8,787 data points
  - -hourly basis
  - - normal distribution

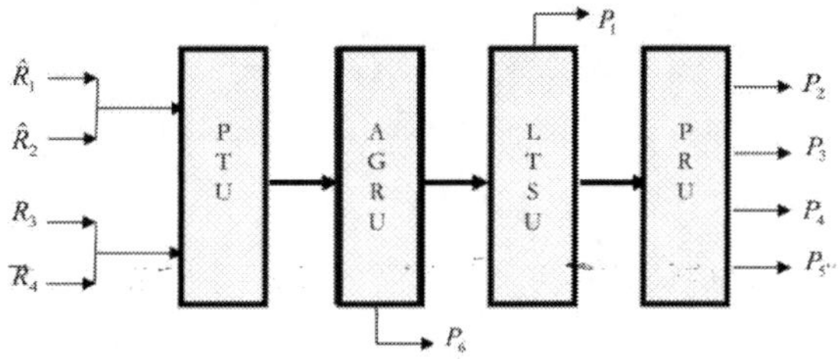

- **PRU arrangement**
  - -conventional distillation
  - - sequence as usual
  - - ethane/propane/butane/
  - - condensate-debutanizer

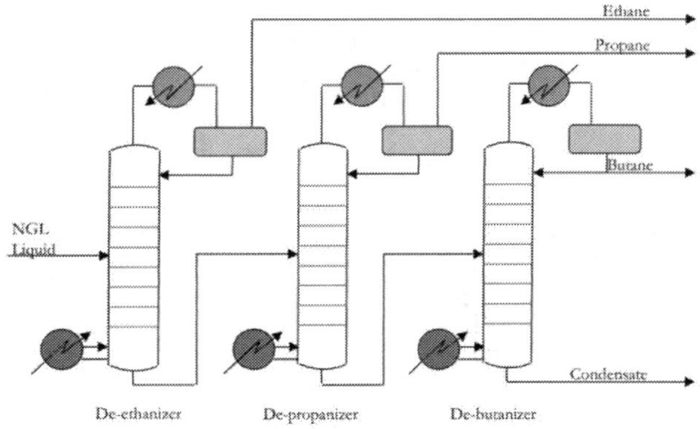

11:13 AM

Case studies (cont')

- **Reliability vs Profit**
  - -profit drops slowly initially
  - -drastic changes in profit
  - -critical confidence level
  - - trade-off of point
- **Pareto principle**
  - -reliability of process
  - - improve to certain level
  - - however expense of profit
  - - best optimal value found

Optimal profit profile

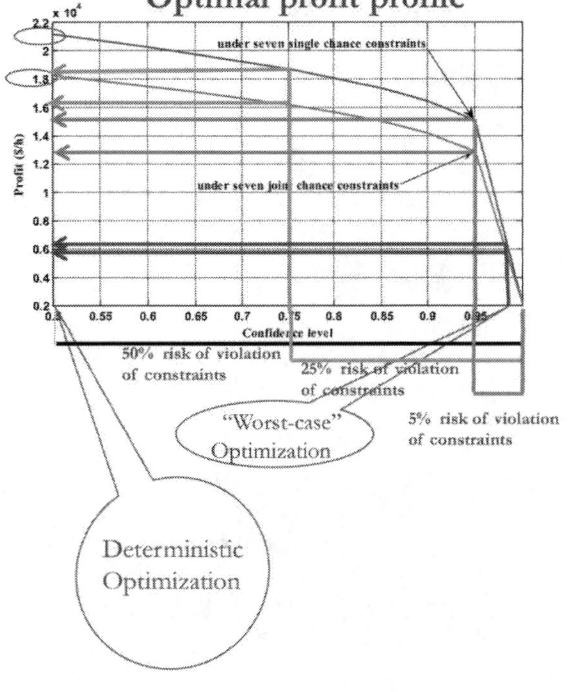

11:13 AM

Case studies (cont')

Optimal Sales gas product profile

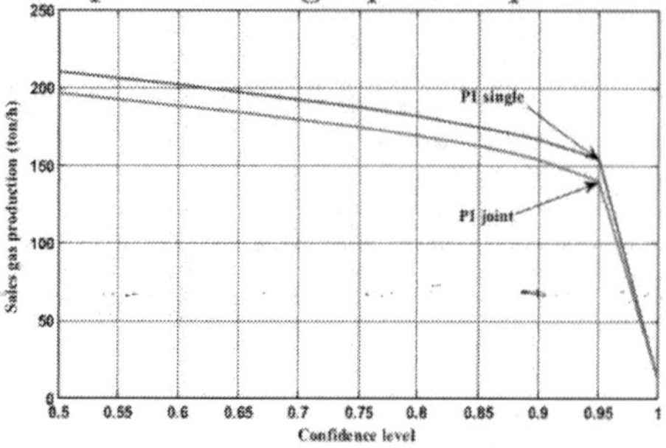

Optimal ethane product profile

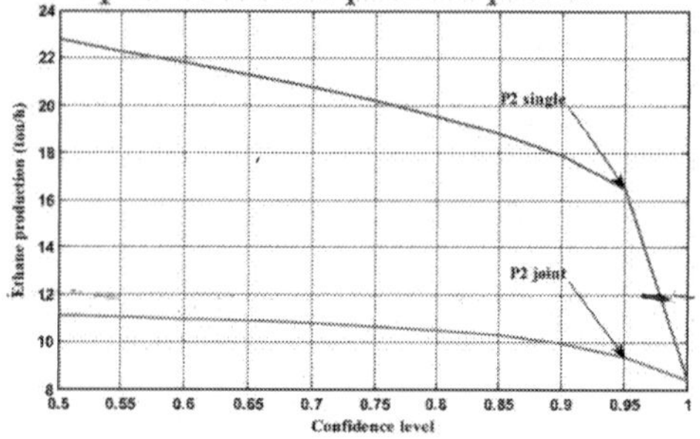

Optimal propane product profile

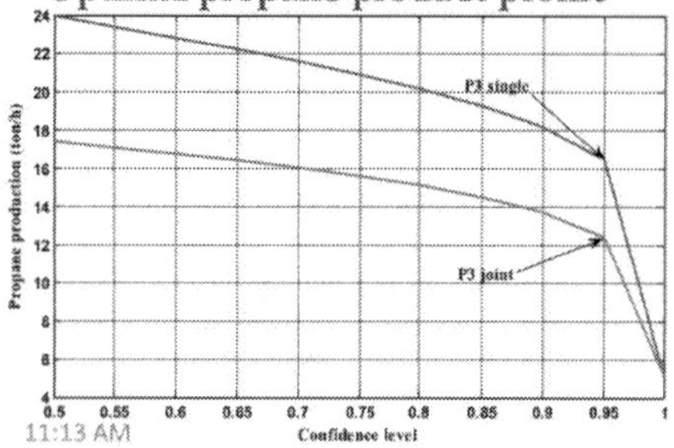

11:13 AM

Optimal butane product profile

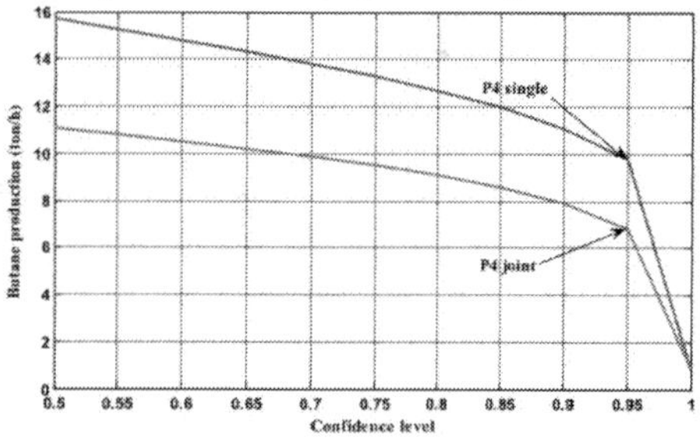

Case studies (cont')

- **Joint vs single chance**

  -sensitivity analysis

  -how optimal decision affected

  -almost 100% confident

  - some risk of violation (9%)

- **single chance vs profit**

-effect of N_2/BTU content/

- effect of CO_2/BTU, plugging, ethane/

- Environmental constraint

- C_{5+} effect-mechanical refigeration

11:13 AM

Joint vs single chance constrained

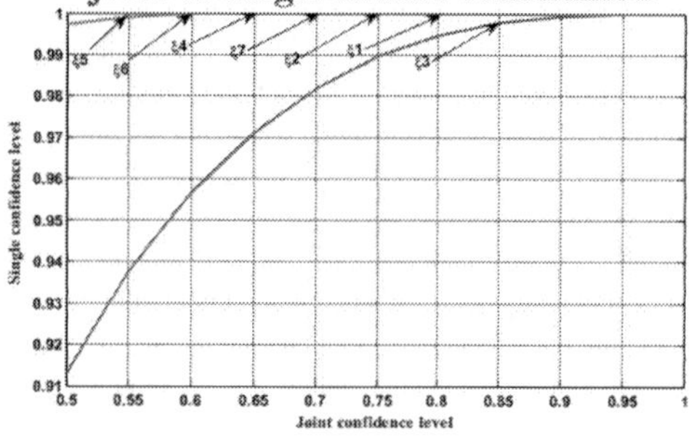

Single chance constrained vs profit

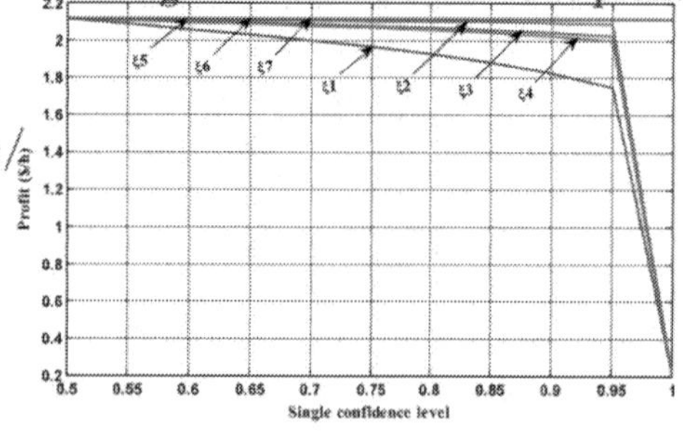

Computers and Chemical Engineering 34 (2010) 1256–1267

Contents lists available at ScienceDirect

Computers and Chemical Engineering

journal homepage: www.elsevier.com/locate/compchemeng

A chance constrained approach for a gas processing plant with uncertain feed conditions

G. Mesfin, M. Shuhaimi*

Chemical Engineering Department, Universiti Teknologi PETRONAS, Bandar Seri Iskandar, 31750, Tronoh, Perak, Malaysia

ARTICLE INFO

Article history:
Received 17 April 2009
Received in revised form 28 February 2010
Accepted 17 March 2010
Available online 25 March 2010

Keywords:
Chance constraints
Optimization
Sensitivity
Uncertainty
HYSYS
GAMS

ABSTRACT

In this work, we propose a new optimization model for gas processing plant under uncertain feed flow rate and composition. Deterministic model has initially being formulated; subsequently, linear steady state chance constrained models with time-dependent uncertainties under single and joint chance constraints have been developed. The chance constrained models have then been converted to their equivalent deterministic form so as to be solved. Case study has been taken for real plant to effectively implement the proposed model. The solution from the optimization provides prior-decision for the in-operation plant to produce the desired product at certain confidence level by satisfying the process constraints. Sensitivity analysis has been made to see the effect of each uncertain feed component flows. Final computational results ratify the need of taking optimal decisions using the chance constrained approach which comprises the profitability and reliability of the process compared to deterministic, "worst case", and two-stage programming approaches.

11:13 AM

Case studies (cont')

- **Case study 2: uncertain product flow**

-main process involved

-8,787 data points hourly basis

-normal distribution assumed

- Modification on PRU arrangement

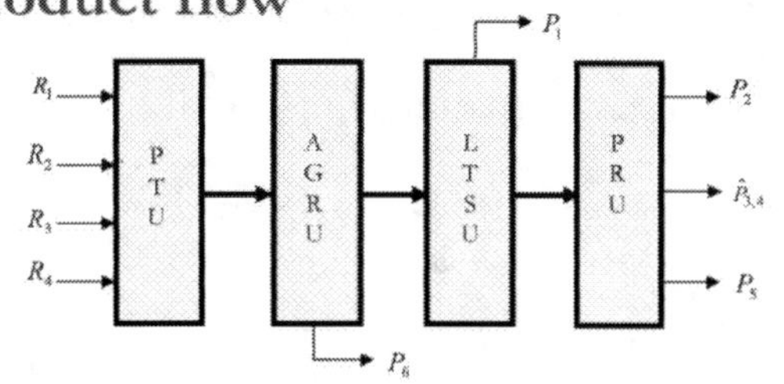

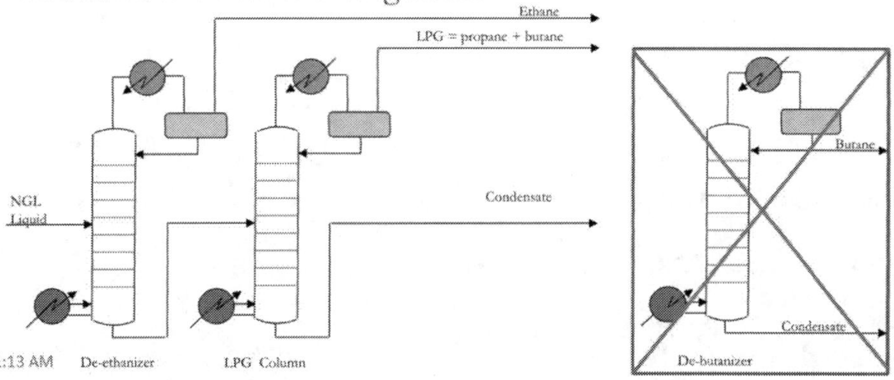

11:13 AM

Case studies (cont')

Optimal Profit Profile

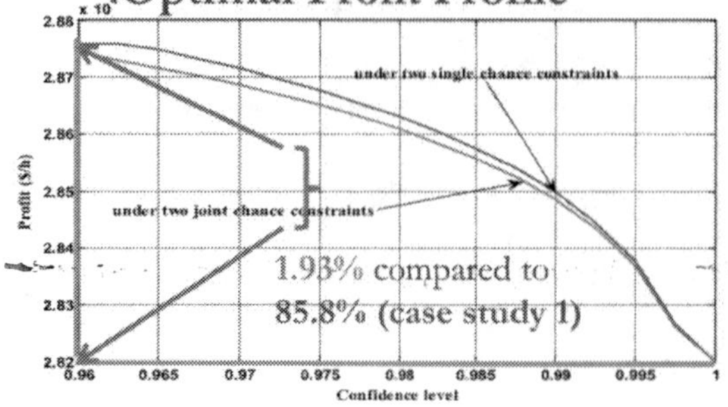

Optimal LPG product profile

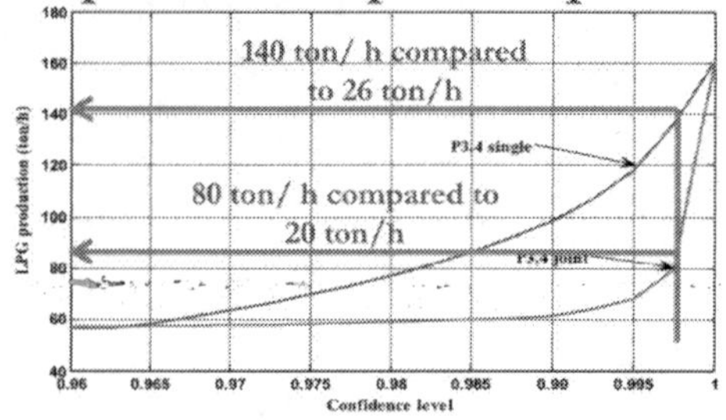

Optimal Ethane product profile

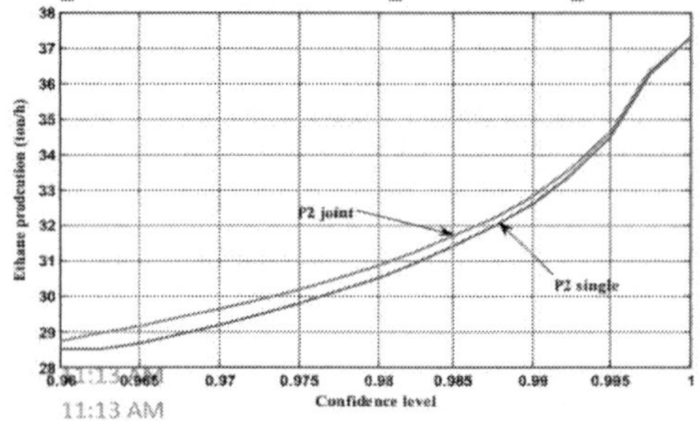

11:13 AM

Optimal Sales gas product profile

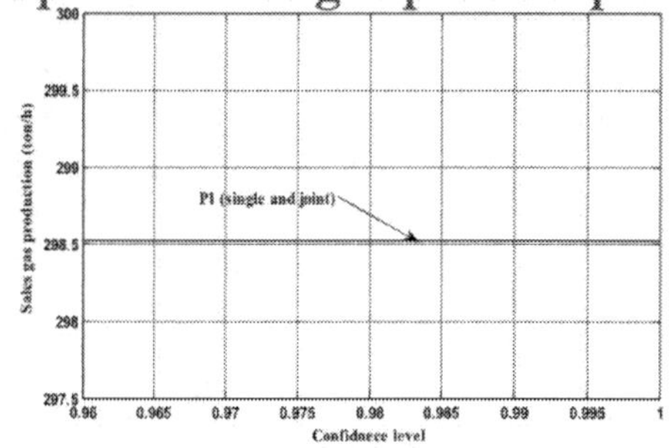

and

Case studies (cont')

- **Joint vs single chance**

  -sensitivity analysis

  -96 % to 100% conf. level

  -remaining product constant

  - risk of violation (3.5%)

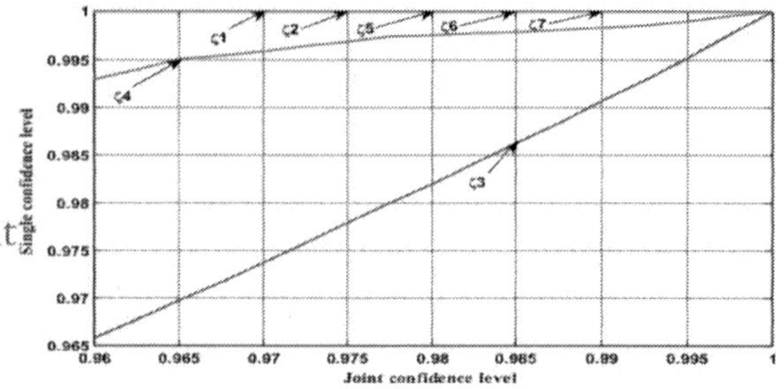

- **single chance vs profit**

-effect of C_3 and C_{4s}

- effect of other small

- LPG product market

- Decision can be made

11:13 AM

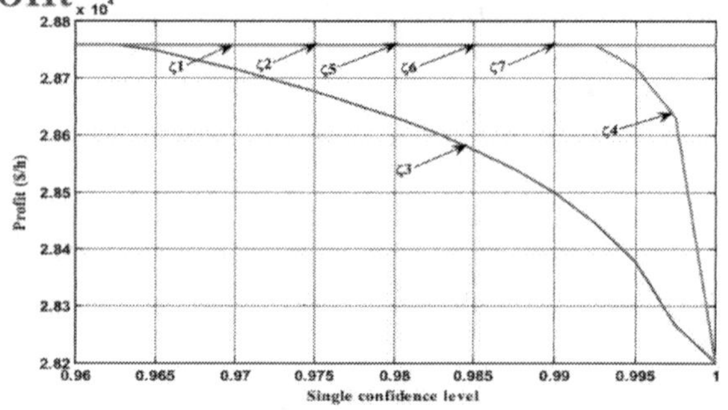

Contents lists available at ScienceDirect

Computers and Chemical Engineering

journal homepage: www.elsevier.com/locate/compchemeng

Profit optimization for chemical process plant based on a probabilistic approach by incorporating material flow uncertainties

Mesfin Getu[a,*], Shuhaimi Mahadzir[b], Moonyong Lee[c]

[a] Chemical Engineering Department, Curtin University of Technology, CDT 250, 98009 Miri, Sarawak, Malaysia
[b] Chemical Engineering Department, Universiti Teknologi Petronas, Bandar Seri Iskandar, 31750 Tronoh, Perak, Malaysia
[c] School of Chemical Engineering, Yeungnam University, 214-1 Dae-dong, Gyeongsan, Gyeongbuk 712-749, Republic of Korea

ARTICLE INFO

Article history:
Received 8 October 2012
Received in revised form 28 March 2013
Accepted 8 May 2013
Available online 18 June 2013

Keywords:
HYSYS
GAMS
Optimization
Profit
Uncertainty

ABSTRACT

This paper reports how the economic performance of a chemical process plant is affected by material flow uncertainties from the plant inlet and outlet. Two chance-constrained optimization models were proposed. The models were tested using case studies of an existing gas processing plant. Profit optimization for the case studies was made with respect to the reliability of holding the process constraints at a certain confidence level [0.5, 1]. The optimal profit change for uncertainty from the plant inlet within the confidence interval [0.96, 1] was 86%. On the other hand, the optimal profit change for uncertainty from the plant outlet was only 2% for the same confidence level interval considered. This suggests that the uncertainty from the plant inlet has a major impact on the overall economic performance of the plant. Sensitivity analysis showed how uncertain parameters from both plant sides can affect the overall profit significantly.

Computer Aided Chemical Engineering

Volume 30, 2012, Pages 832–836

22nd European Symposium on Computer Aided Process Engineering

Analyzing the effects of uncertainties on the economic performance of a chemical process plant using a probabilistic optimization technique

Mesfin Getu[a,] ♣, Shuhaimi Mahadzir[b], Moonyong Lee[c]

11:13

11:13 AM

Part II
Gas processing plant:
Design aspect

11:13 AM

Industrial single stage (ISS)

- **Lean oil absorption**
  - -early technology used
  - - energy intensive
  - - less NGL recovery
- **Turbo-expander**
  - - industrial single stage (ISS)
  - - gas piped high press
  - - refrigeration effect
  - - reuse expander energy

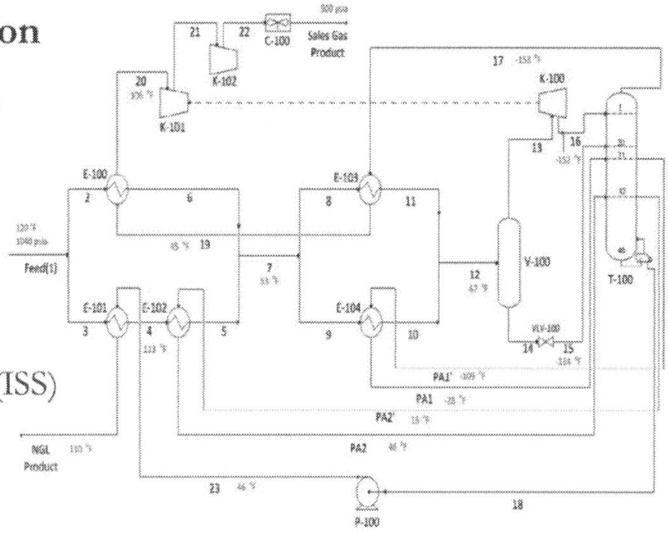

Source: US Patent 4,278,457

11:13 AM

Gas sub-cooled process (GSP)

- **GSP process scheme**

-state-of-art improvement

- ideal operation

- practical point of view

- demethanizer column

- stripping column

- lowering top temp

-CO2 icing/solidify/

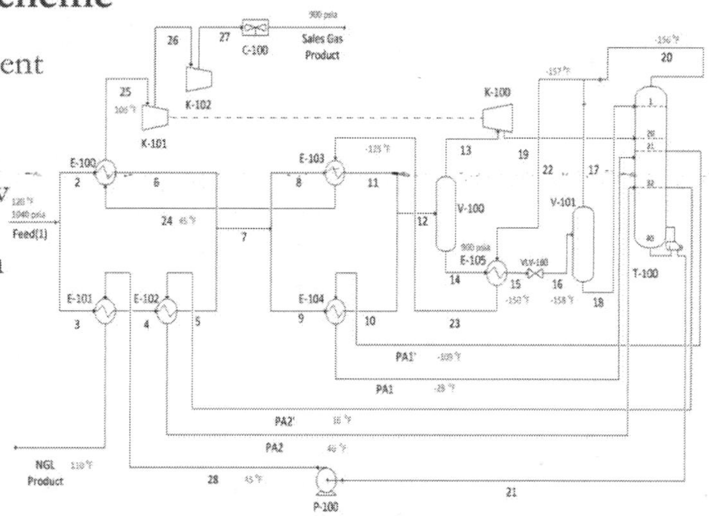

Source: US Patent 4,278,457

11:13 AM

Cold residue-recycle (CRR)

- **CRR process scheme**

-retain all the GSP design

- improving residue purity

- top reflux stream

- rectify vapors

-reduce amount of ethane

- new compressor

-high ethane recovery

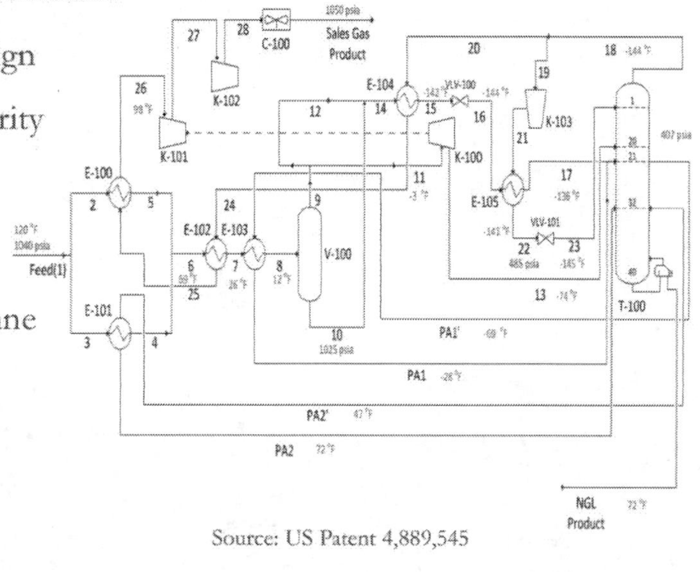

Source: US Patent 4,889,545

11:13 AM

Recycle split-vapor (RSV)

- **RSV process scheme**

-similar concept CRR

-reflux gas stream

- recompressed residue gas

- condensed & cooled

-high pressure advantage

-more condensing & cooling

-Bulky ethane recovery

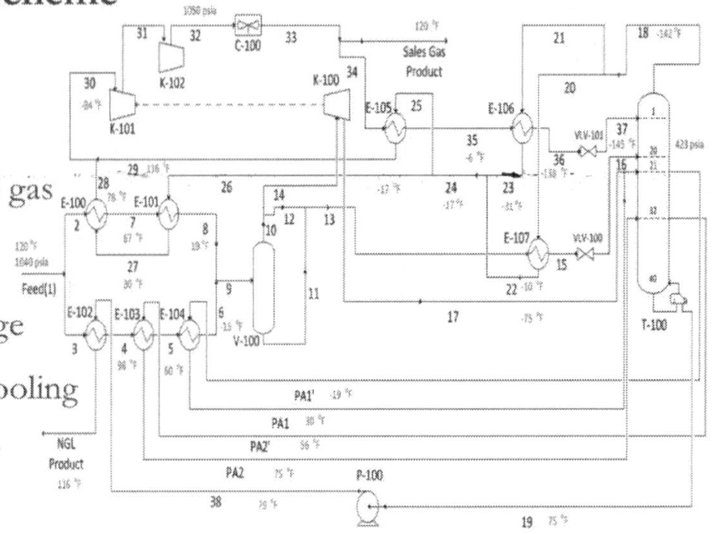

Source: US Patent 5,568,737

11:13 AM

Enhanced NGL recovery (IPSI-1)

- **IPSI-1 process scheme**

-state-of-art development

- liquid hydrocarbon stream

- Withdrawn from bottom

- reintroduced to column

-stripping gas

- reduce energy/reboiler/

-reduce external refrigeration

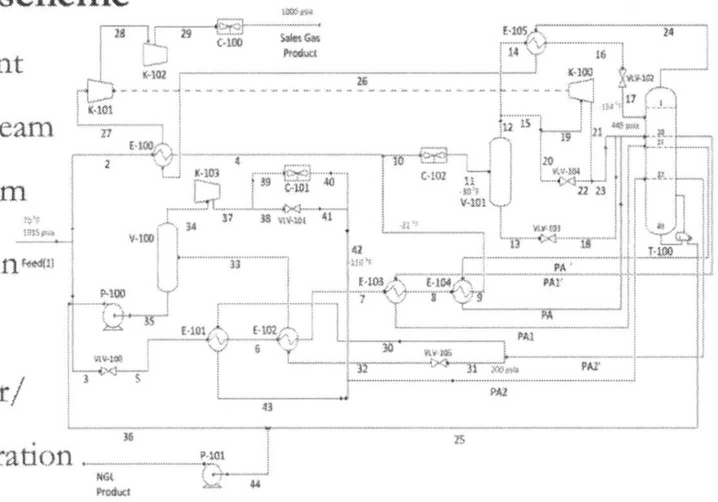

Source: US Patent 5,992,175

11:13 AM

Internal refrigeration system (IPSI-2)

- **IPSI-2 process scheme**

-similar to IPSI-1

- liquid hydrocarbon stream

- Withdrawn from bottom

- reintroduced to column

-stripping gas

- closed refrigeration cycle

-open refrigeration cycle

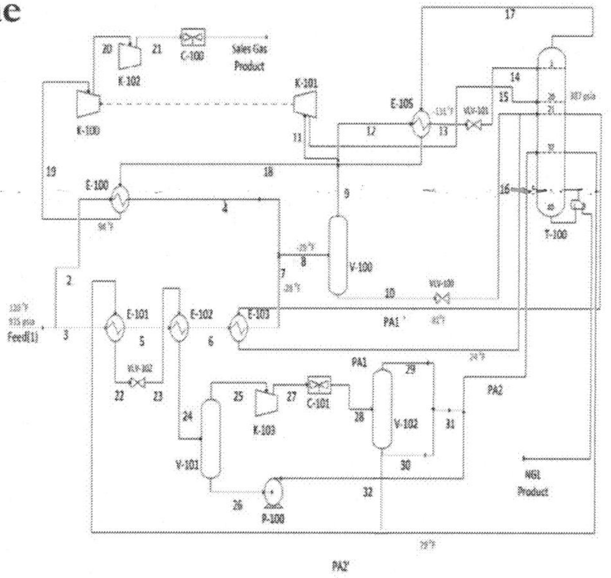

Source: US Patent 7,257,966

11:13 AM

Techno-economic analysis

- **Feed composition**

-vital factors/NGL recovery/

- leaner feeds/winter/

- rich feeds /summer/

- **Operating conditions**

- fixing certain conditions

- fair comparison

- key process parameters

- feed & prod spec

Components	Feed 1	Feed 2	Feed 3	Feed 4	Feed 5	Feed 6	Feed 7	Feed 8
C_1	0.9717	0.8457	0.8400	0.8034	0.7260	0.6507	0.6539	0.5952
C_2	0.0189	0.0820	0.0760	0.0693	0.0494	0.1050	0.0504	0.0536
C_3	0.0029	0.0340	0.0200	0.0259	0.0176	0.0497	0.0297	0.0471
iC_4	0.0013	0.0058	0.0000	0.0043	0.0038	0.0095	0.0090	0.0203
C_4	0.0012	0.0086	0.0000	0.0073	0.0067	0.0187	0.0172	0.0239
iC_5	0.0007	0.0028	0.0070	0.0018	0.0031	0.0077	0.0085	0.0180
C_5	0.0005	0.0021	0.0000	0.0015	0.0031	0.0078	0.0084	0.0161
C_6	0.0004	0.0018	0.0030	0.0016	0.0046	0.0125	0.0160	0.0260
C_7	0.0024	0.0012	0.0000	0.0017	0.0167	0.0149	0.0214	0.1998
C_8	0.0000	0.0005	0.0000	0.0018	0.0000	0.0156	0.0243	0.0000
C_9	0.0000	0.0004	0.0000	0.0019	0.0000	0.0108	0.0194	0.0000
C_{10}	0.0000	0.0005	0.0000	0.0005	0.0000	0.0352	0.0652	0.0000
N_2	0.0000	0.0055	0.0540	0.0619	0.1325	0.0349	0.0464	0.0000
CO_2	0.0000	0.0091	0.0000	0.0171	0.0365	0.0270	0.0302	0.0000
Total	1.0000	1.0000	1.0000	1.0000	1.0000	1.0000	1.0000	1.0000

Feed flow (kgmole/h)	5000
Plant inlet gas pressure (psia)	1040
Plant inlet gas temperature (°F)	120
Salesgas outlet pressure (psia)	730
Salesgas outlet temperature (°F)	120
Product specification (C_1/C_2 ratio)	0.02
Theoretical trays including reboiler	41
Compressor/Expander/pump efficiency (%)	75
Demethanizer top stage pressure (psia)	334
Demethanizer bottom stage pressure (psia)	337
Pressure drop across the heat exchangers, shell and tube side s (psi)	10
Property fluid package	Peng Robinson

11:13 AM

- **Demethanizer Rebduty**
  - -lean feed lower reboiler duty
  - rich feed higher reboiler duty
  - presence of heavier hydrocarbons
  - external refrigeration required

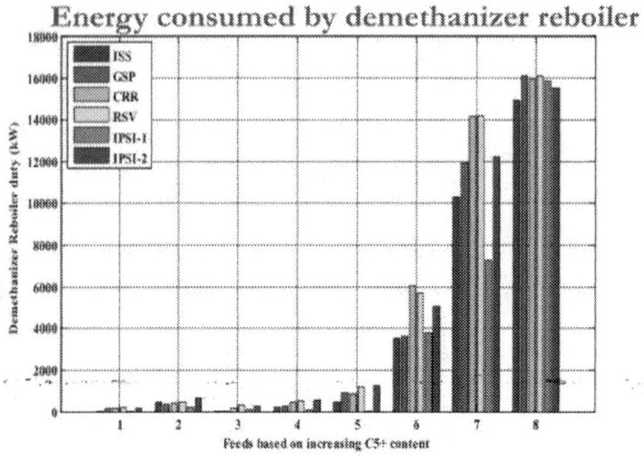

- **Recompression duty**
  - RSV higher compression
  - requires large recycle stream
  - increases horse power
  - high reflux to condense

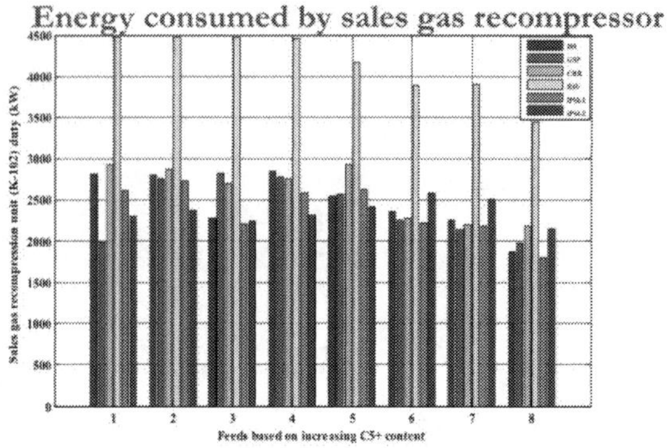

11:13 AM

Source: Getu et al. (2013a)

- **Total capital cost**
  - -equipment sizing information
  - Bare module cost applied
  - preliminary cost estimate
  - sum = direct + indirect cost

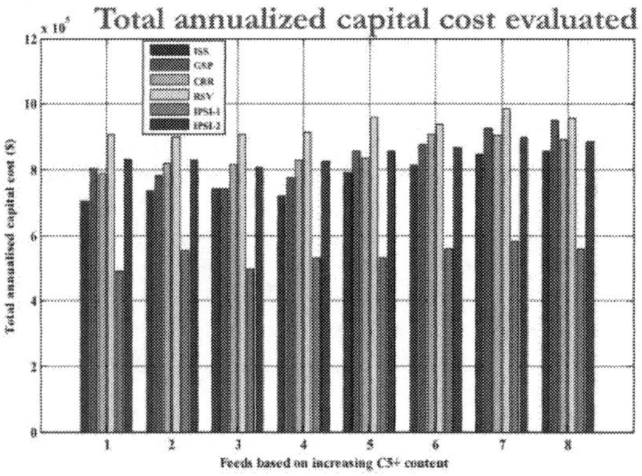

- **Total operating cost**
  - utility cost calculation complex
  - depends on inflation & energy
  - exact value difficult to determine
  - main: rebduty and sales gas duty

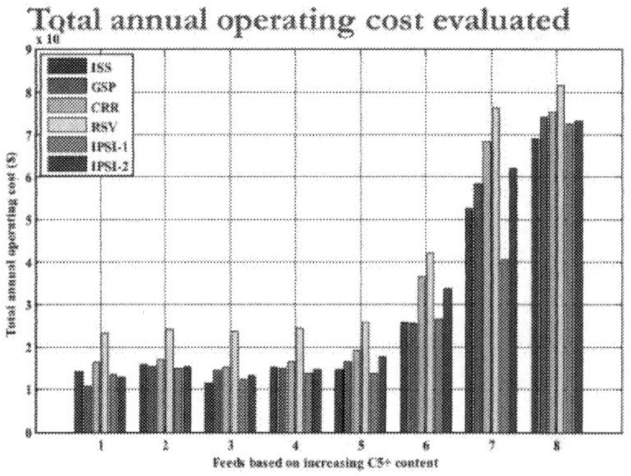

11:13 AM

- **Profitability analysis**

-TAC = annualized cost

- sum of Oper & Capt

- high TAC RSV

- lean & rich feed cases

- profitability analysis

- Shortest payback-IPSI-1

	ISS	GSP	CRR	RSV	IPSI-1	IPSI-2
Total annualized capital cost ($)						
Lean feed	737,537	790,812	815,975	917,608	520,292	828,681
Rich feed	837,850	918,241	902,605	960,539	564,947	884,197
Total annual operating cost ($)						
Lean feed	1,433,227	1,440,697	1,684,507	2,435,678	1,371,146	1,485,587
Rich feed	4,907,276	5,272,607	5,999,535	6,659,960	4,649,075	5,630,725
Sales gas product revenue ($/year)						
Lean feed	64,075,454	63,817,423	67,269,148	68,052,653	63,822,215	67,406,086
Rich feed	51,428,826	51,919,872	53,916,695	54,359,207	48,226,101	52,629,502
NGL product revenue ($/year)						
Lean feed	8,291,994	8,672,281	3,585,051	2,434,822	8,664,181	3,379,560
Rich feed	26,944,069	26,220,147	23,272,916	22,624,974	31,667,400	25,173,139
Gross profit ($/year)						
Lean feed	70,934,221	71,049,008	69,169,691	68,051,797	71,115,251	69,300,059
Rich feed	73,465,618	72,866,612	71,190,076	70,324,222	75,244,426	72,171,916
Net profit ($/year)						
Lean feed	49,653,955	49,734,305	48,418,784	47,636,258	49,780,675	48,510,041
Rich feed	51,425,932	51,006,629	49,833,054	49,226,955	52,671,099	50,520,341
Payback time (year)						
Lean feed	0.015	0.016	0.017	0.019	0.010	0.017
Rich feed	0.016	0.018	0.018	0.019	0.011	0.017

11:13 AM

Contents lists available at SciVerse ScienceDirect

Chemical Engineering Research and Design

IChemE

journal homepage: www.elsevier.com/locate/cherd

ELSEVIER

Techno-economic analysis of potential natural gas liquid (NGL) recovery processes under variations of feed compositions

Mesfin Getu[a], *Shuhaimi Mahadzir*[b], *Nguyen Van Duc Long*[c], *Moonyong Lee*[c,*]

[a] *Department of Chemical Engineering, Curtin University of Technology, CDT 250, 98009 Miri, Sarawak, Malaysia*
[b] *Department of Chemical Engineering, Universiti Teknologi PETRONAS, Bandar Seri Iskandar, 31750 Tronoh, Perak, Malaysia*
[c] *School of Chemical Engineering, Yeungnam University, 214-1 Dae-dong, Gyeongsan, Gyeongbuk 712-749, Republic of Korea*

ABSTRACT

This paper presents the different process schemes used for known NGL recovery methods with respect to their economic performance. The original turbo-expander (ISS) was considered as base case plant. The GSP, CRR and RSV process schemes focus on improvement at the top of the demethanizer column. The IPSI-1 and IPSI-2 schemes focus on the bottom of the demethanizer column. All the process schemes were initially built using Aspen HYSYS with a common set of operating criteria. Numerous simulation runs were made by taking various typical feed compositions classified as lean and rich. The economic assessment for each process scheme was later made by considering the capital cost, operating cost and profitability analysis. Results showed that the IPSI-1 process scheme gives the best economic performance with lowest TAC and payback time compared to the other process schemes. On the other hand, the RSV process gives higher TAC and payback time compared to others.

11:13 AM

Chemical Engineering and Processing 82 (2014) 54–64

Contents lists available at ScienceDirect

Chemical Engineering and Processing: Process Intensification

journal homepage: www.elsevier.com/locate/cep

Energy saving opportunities in integrated NGL/LNG schemes exploiting: Thermal-coupling common-utilities and process knowledge

Mohd Shariq Khan[a], Yus Donald Chaniago[a], Mesfin Getu[b], Moonyong Lee[a,*]

[a] School of Chemical Engineering, Yeungnam University, Dae-dong 712-749, Republic of Korea
[b] Chemical Engineering Department, Curtin University of Technology, CDT 250, 98009 Miri, Sarawak, Malaysia

ARTICLE INFO

Article history:
Received 26 September 2013
Received in revised form 12 May 2014
Accepted 2 June 2014
Available online 7 June 2014

Keywords:
Natural gas
Liquefaction
Thermal coupling
NGL
LNG
Knowledge based optimization

ABSTRACT

Novel integrated and optimization schemes for natural gas (NG) liquid recovery and liquefaction step are presented. The liquefaction of NG was carried out using the newly developed KSMR liquefaction cycle while the separation of NG liquids was performed using energy efficient thermally coupled distillation schemes. The main highlight of integration are, (i) feed splitting to provide reboiler duty in the methane scrubber, (ii) common refrigeration utilities required all over the plant; and (iii) flexibility of integrated plant for easy transition between ethane recovery or ethane rejection. These integration steps minimize the overall plant-specific power requirements and the duplication of processing equipment. After successful integration, the MR cycle was optimized for the compression energy requirement by varying the refrigerant composition and operating pressure levels with the aid of an in-house established knowledge-based optimization (KBO) methodology. The KBO approach is robust in application and gives consistent results. Compared to the base case, 9% improvement in plant compression energy requirement was obtained.

11:13 AM

Computer Aided Chemical Engineering

Volume 31, 2012, Pages 405–409

11th International Symposium on Process Systems Engineering

Studying the effect of feed composition variation on typical natural gas liquid (NGL) recovery processes

Mesfin Getu[a], Mohd Shariq Khan[b], Nguyen Van Duc Long[b], Moonyong Lee[b]

Computer Aided Chemical Engineering

Volume 31, 2012, Pages 680–684

11th International Symposium on Process Systems Engineering

A Study of Complex Distillation Arrangements for Improved Depropanizing, Debutanizing and Deisobutanizing Fractionation of NGL

Youngmi Jung, Nguyen Van Duc Long, Mesfin Getu Woldetensay, Moonyong Lee

Computer Aided Chemical Engineering

Volume 31, 2012, Pages 1196–1200

11th International Symposium on Process Systems Engineering

Modeling and Simulation of Multi-stream Heat Exchanger Using Artificial Neural Network

Mohd Shariq Khan, Yuli Amalia Husnil, Mesfin Getu, Moonyong Lee

Conclusions (Part-I)

- A first effort made to apply chance constrained programming to a gas processing plant. Uncertain variables explicitly incorporated in the formulation & their impact taken into account in the solution.

- The plant inlet uncertainty (upstream plants) has a significance impact on the overall plant performance. The profit changed by 85.8 % (Case study 1) case compared to 1.93 % profit change for the plant outlet uncertainty (Case study 2).

- Optimum trade-off point can be found for the reliability of the process and profitability of the plant. The reliability of the process can now be measured just "unit" called confidence level.

- Flexible decision can be made by shifting the production line. Decision maker can make prior-decision for the in-operating plant to produce the desired products.

- Using the systematic approach developed, process engineers can benefit by avoiding the common practice of trial and error approaches. This solution method can also be applied to other chemical process plant.

11:13 AM

Conclusions (Part-II)

- Simulation study for comparing various NGL recovery process conducted by taking various feed types classified as lean and rich.

- ISS, GSP, CRR & RSV requires more refrigeration for rich feeds.

- RSV recompression cost directly related with the amount of reflux stream flow applied to the demethanizer column.

- IPSI-1 gives best performance based on the economic analysis with lowest TAC & payback time.

- Self-refrigeration system that exist in IPSI-1 & IPSI-2 can significantly reduce the total annualized cost.

11:13 AM

References

Aspen Technology, 2009. A User Guide Manual for Aspen Physical Property V7.1.

Diwaker, U.M.(2008). Introduction to applied optimization. Springer Optimization and Applications, pp. 1-36.

Kidnay, J., Parish, R.W.(2006). Fundamentals of natural gas processing.

Li, M. (2004b). Modeling oil refinery for production planning, scheduling and economic analysis. *Hong Kong University of Science and Technology, Hong Kong*, pp. 155-200.

Li, P., Arellano-Garcia, H., & Wozny, G. (2008). Chance constrained programming approach to process optimization under uncertainty. Computers & Chemical Engineering, 32, 25–45.

Li, P., Wendt, M., &Wozny, G. (2004a). Optimal production planning for chemical processes under uncertain market conditions. Chemical Engineering & Technology, 27, 641–651.

Mesfin Getu, Shuhaimi Mahadzir, Nguyen Van Duc Long, Moonyong Lee. (2013). Techno-economic analysis of potential natural gas liquid (NGL) recovery processes under variations of feed compositions. Chemical Engineering and Research Design Journal, Chemical Engineering and Research Design, 91(7),1272-1283.

Mesfin Getu, Shuhaimi Mahadzir, Moonyong Lee.(2013). Profit optimization for chemical process plant based on a probabilistic optimization approach by incorporating material flow uncertainties. Computers & Chemical Engineering Journal (Elsevier), Accepted in press, Corrected proof.

Rosenthal, R. E. (2006). A user's guide manual. Washington, DC, USA: GAMS Development Corporation., pp. 55.

11:13 AM

Curtin University
Sarawak Malaysia

11:13 AM

Thank you

11:13 AM

Panel Discussion: Is Pilot/Demo scale validation an obsolete concept?

Chen, X., The Dow Chemical Company
Aglave, R., CD adapco
Super, J. D., Cobroko Solutions
Getu, M., Curtin University of Technology
Originally presented on: 10/10/2014 15:30:00 - 16:00:00

Building a pilot/demo plant is often expensive and time consuming. Due to advances in computing technology there are increasingly sophisticated mathematical and process models available. These models, once validated with lab scale data, could then be used for scale up to commercial plant. When can we skip the pilot/demo scale and when is it absolutely required? We encourage audience to participate and share their views and experience.

Interpreting Crude Assays

Andrew W. Sloley

CH2M HILL
21 Bellwether Way, Unit 111
Bellingham, WA 98225

Prepared for Presentation at the
American Institute of Chemical Engineers
2014 Southwest Process Technology Conference
Galveston, TX October 9 – 10, 2014

"UNPUBLISHED"

1.0 Abstract

The refining industry trades many different crude oils. Properties for crude oils vary dramatically between production fields. The geochemical history of the crude makes an enormous difference in crude oil properties. At the limit, every oil well produces crude with a different composition. While known for many years, the differences between wells in the same field were relatively minor. Modern production techniques for tight oil production may have large differences in crude oil character between adjacent wells. They may also have rapid changes in quality over time. Trading crude, predicting products, operating plants, and designing plants all require some way to characterize crude oil. The crude oil assay fills this purpose.

Different users of assays have different needs and expectations. What may be an acceptable assay for one purpose may not meet the needs of other purposes. This paper explores the different assay purposes, compares some typical crudes, and examines the meaning and significance of common assay tests.

This discussion covers what information crude assays reveal. The major emphasis is on distillation properties of the crude and how they affect the crude distillation unit products.

Assay information may be available for the entire crude or specific distillation cuts from the crude. With enough cuts, curves can be generated to allow for interpolation of properties across different possible product ranges.

The most important inspection properties of the crude include API gravity (or density) and sulfur content. The most important distillation information is the crude TBP (True boiling point) curve. This gives a breakdown of distillation versus boiling point for the crude.

Cuts on specific distillation range may have a variety of properties. The properties measured will vary depending upon the cut. Assays with narrower cuts improve estimates of product properties. Many crude oil properties blend in complex ways. A too-wide cut can lead to significant miss-prediction of product properties.

Understanding the crude assay and how it affects refinery economics and operation is critical to a profitable refinery.

CH2MHILL.

Interpreting Crude Assays

Andrew W. Sloley – CH2M HILL, Bellingham, WA

Prepared for Presentation at the AIChE 2014 6th Southwest Process Technology Conference
Galveston, TX, 9-10 October, 2014

Crude Assays

- Define crude oil composition
- Combination of
  - Bulk properties
  - Distillation properties
  - Physical properties
  - Performance properties
  - Tests on the entire crude
  - Tests on fractions of the crude
- Used by
  - Producers
  - Midstream
  - Refiners
- Used to
  - Value crude
  - Design plants
  - Define operating expectations
  - Monitor operations

Image © Can Stock Photo, Inc. / focalpoint

CH2MHILL.

Why Assays are Needed

- Many crudes available
- Every well produces a different composition
- Conventional wells in the same field have small differences
- Composition changes gradually over time
- Tight oils come from relatively impervious formations
  - Large well-to-well differences
  - Large differences over time
- Assays give a consistent method to understand what is in the crude
- Specific test methods developed over time

CH2MHILL.

Basic Classification

Basic Classification

- Density
  - Very light to very heavy
  - Labeled by specific gravity, density or API gravity
- Sulfur
  - Low-sulfur or high-sulfur
  - Sweet or sour

$$^{o}API = \frac{141.5}{SG_{60}^{60}} - 131.5$$

$$SG_{60}^{60} = \frac{141.5}{131.5 + {}^{o}API}$$

$$SG_{60}^{60} = \frac{density\ oil\ at\ 60\ ^{o}F}{density\ water\ at\ 60\ ^{o}F}$$

Image © Andrew W. Sloley

CH2MHILL.

Significance

- Density
  - Inferred hydrogen content
  - Lower density, more hydrogen
  - Higher density, less hydrogen
- Lower density crudes
  - More fuels products
- Sulfur
  - Higher sulfur, more refining
  - Higher sulfur, more hydrogen consumed
  - Higher sulfur, more corrosion

CH2MHILL.

Some Representative Crudes

Crude	°API	Specific Gravity	Classification
Sleipner	58.7	0.7440	Condensate
Gippsland	49.7	0.7810	Very Light
Bakken	42.1	0.8151	Light
WTI	39.9	0.8254	Light
Brent Blend	38.5	0.8324	Light
Arab Light	32.8	0.8610	Medium
Zuata Sweet	32.2	0.8641	Medium (Syncrude)
Tia Juana Light	31.6	0.8676	Medium
BCF 24	23.3	0.9141	Heavy
Kern River	13.3	0.9772	Very Heavy
Boscan	10.1	0.9993	Very Heavy

CH2MHILL. 7

Representative Crudes – Range of Sample Data

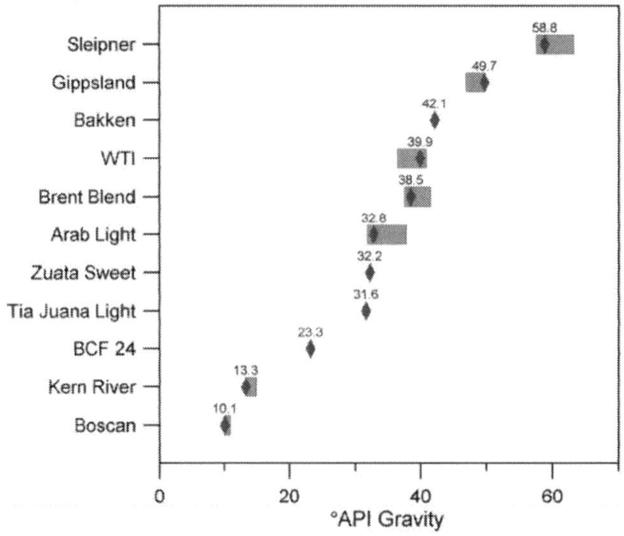

CH2MHILL. 8

More Crudes – Range of Sample Data

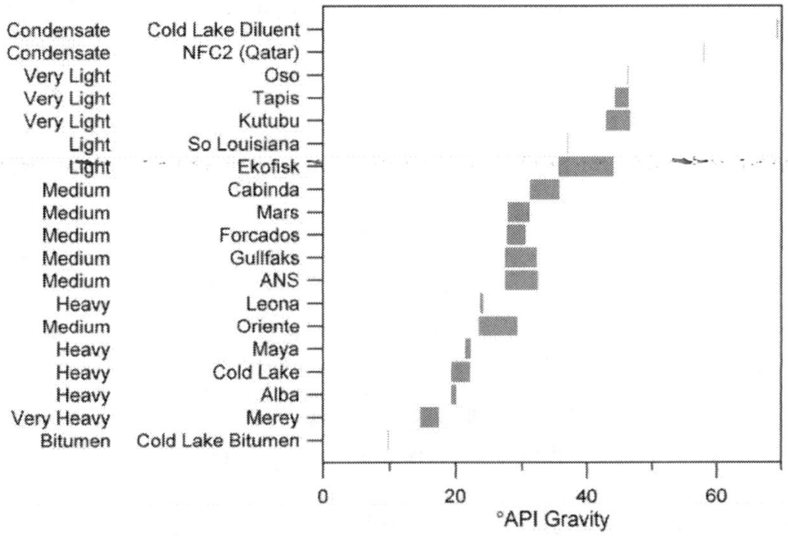

CH2MHILL.

Range of Crudes – API and Sulfur

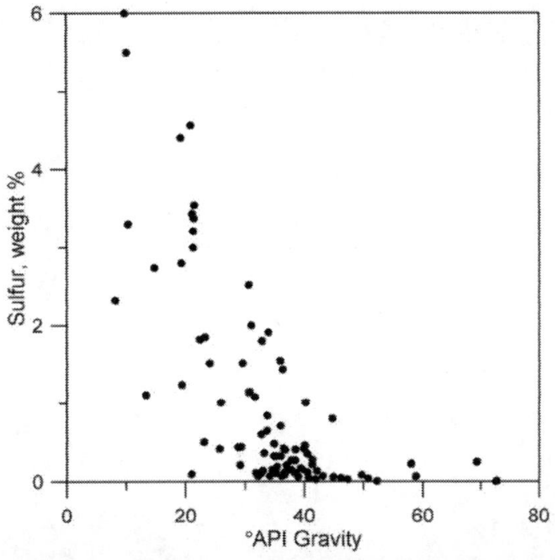

CH2MHILL.

Range of Crudes – API and Sulfur – Classified

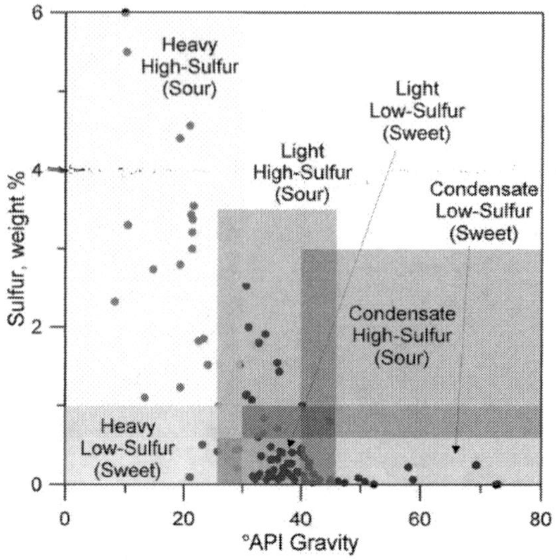

CH2MHILL.

Sulfur Ranges

Gravity, °API	Sulfur, weight percent
~20	0.1 to 4.6
~30	0.2 to 2.5
~40	0.0 to 1.1

CH2MHILL.

Uses for Basic Assays

- Often used for general discussion of crudes
- Many traders use the API and sulfur
- May include some other measurements
  - RVP or TVP
  - Total Acid Number (TAN)
  - Light ends content
  - Other bulk properties

Image © Can Stock Photo, Inc. / Jochen

Vapor Pressure – RVP & TVP

- RVP – Reid Vapor Pressure
- TVP – True Vapor Pressure
- Vapor formation in transit and storage
- RVP usually measured
- TVP needed
- Correlations between them based on 1950s work
  - Unclear basis
  - Not suited for tight oil crudes
- Tests and correlations being revised

Image © Can Stock Photo, Inc. / sframe

Chemical Classification

15

Chemical Composition

- Elemental composition
- Chemical Grouping
  - Paraffins
  - Iso-paraffins
  - Naphthenes
  - Aromatics
  - Asphaltenes
- Some older terminology described crudes as aromatic, naphthenic, asphaltic, paraffinic, or other descriptions
  - Very little value today
  - Still used
  - But only as general terms

CH2MHILL.

16

Elemental Composition

Element	Range Weight Percent
Carbon	83.0-87.0
Hydrogen	10.0-15.0
Sulfur	0.01-6.0
Nitrogen	0.00-2.0
Oxygen	0.05-1.5
Others	0.00-1.0
Bitumen, oxygen	0.2-4.5

Chemical Groups

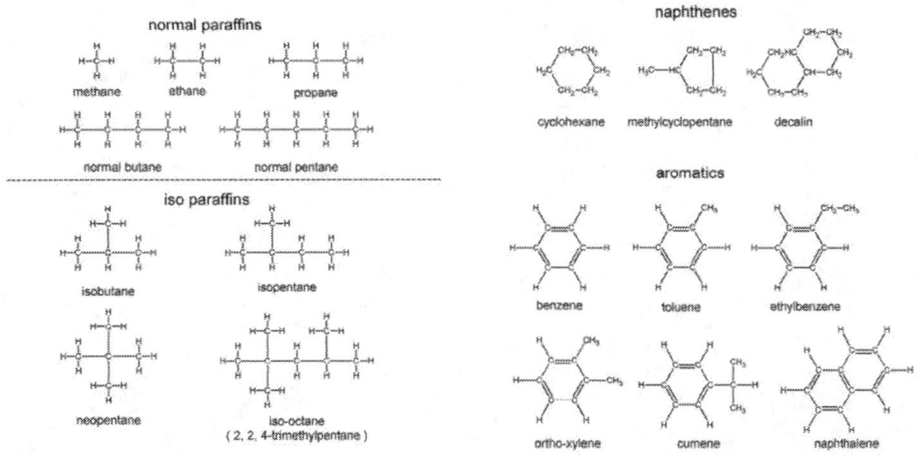

Chemical Group Distribution – An Example

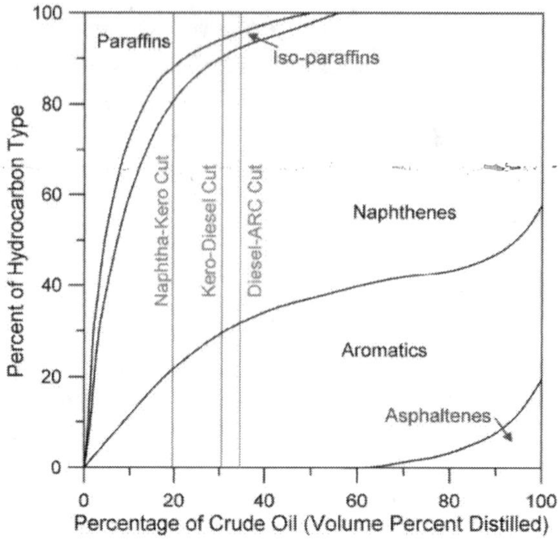

CH2MHILL.

19

Distillation – TBP

20

TBP Distillation

- True Boiling Point Distillation
- Equivalent to ASTM D2892 or equivalents
- Batch distillation
- 15 theoretical distillation stages
- 5:1 reflux ratio
- Used to cut the crude into samples for further work
- Assays combine TBP distillations with properties on cuts
- Example ⇨

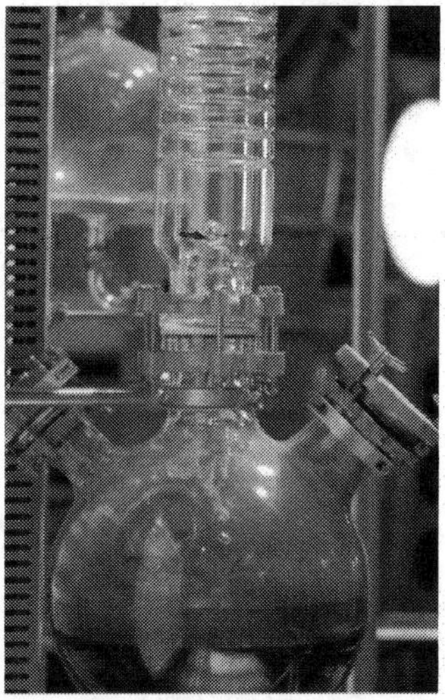

Image © Can Stock Photo, Inc. / timchen

CH2MHILL. 21

Cut Ranges

Product	From °F	To °F	From °C	To °C
LPG	IBP	Butane	IBP	Butane
Full range naphtha	90	400	32	204
Light naphtha	90	220	32	104
Heavy naphtha	180	400	82	204
Kerosene	370	520	188	271
Diesel	420	650	216	343
AGO	610	800	321	427
Vacuum gas oils	700	1170	371	632

CH2MHILL. 22

TBP Comparison – Differences at Same Gravity

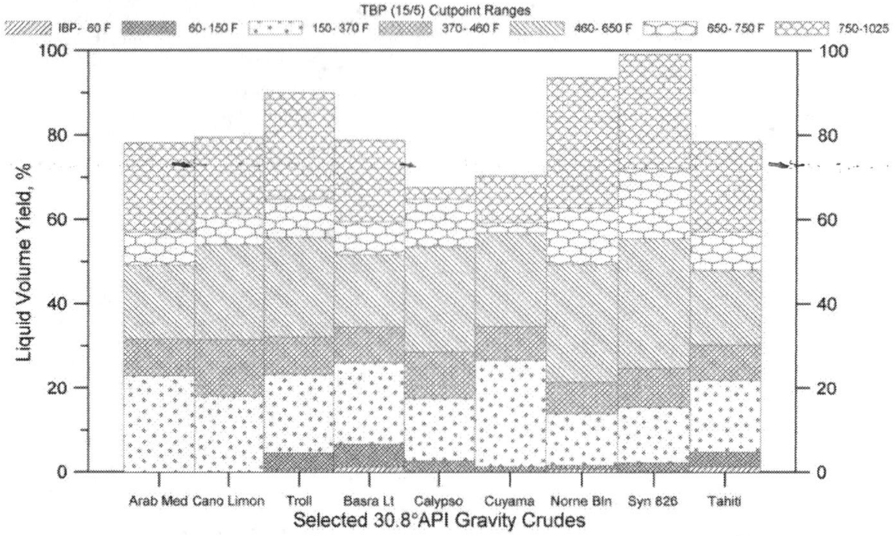

CH2MHILL.

TBP Comparison – Differences at Same Gravity

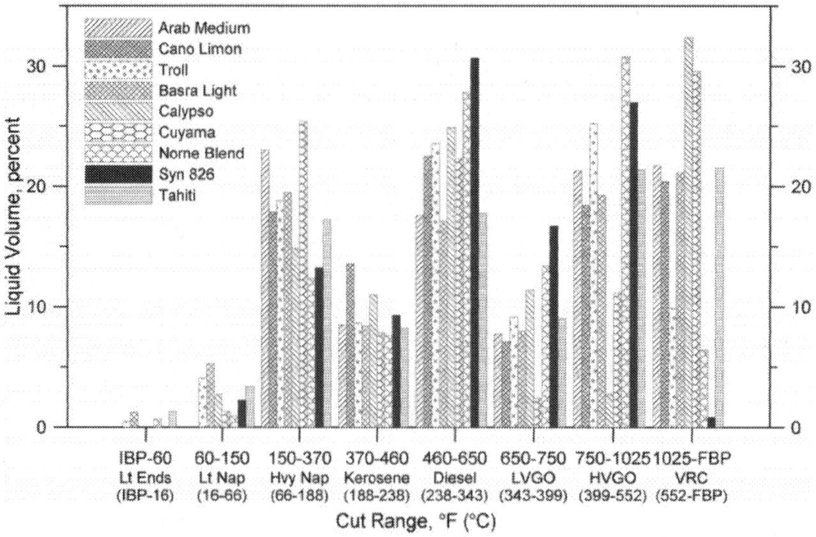

CH2MHILL.

TBP Curve – Link Together Cumulative Fractions

- Temperature versus yield
- TBP temperatures instantaneous measurements at the endpoints of fractions

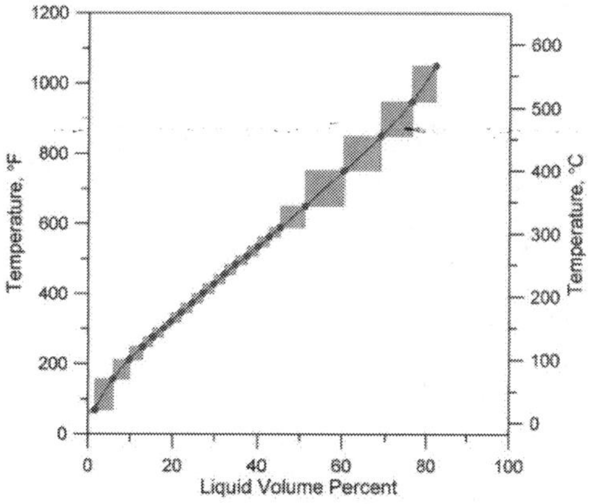

Tia Juana Light

Property Curve – Link Together Average Fractions

- Property versus yield
- Properties are measured on a sample
- Properties are the midpoints of fractions
- Fractions can overlap
- Fractions with wider temperature ranges give biased results

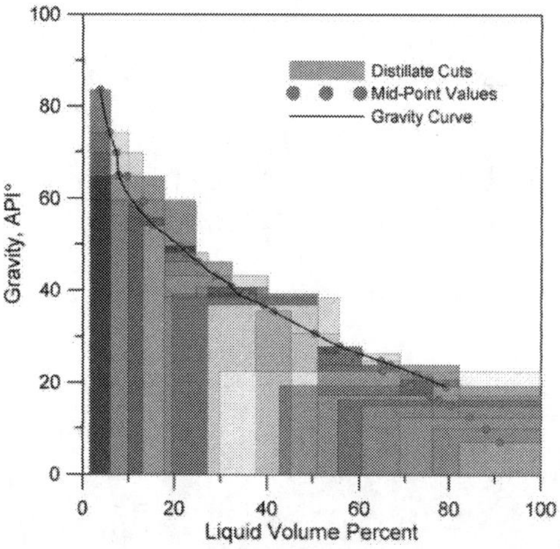

Tia Juana Light

Property Curve – Width of Cuts

- Property versus yield
- For very wide cuts
  - Blending effects bias measured value
  - Depends upon the property
- Properties very sensitive to front-end or back-end need narrow cuts
- Front-end
  - Flash
- Back-end
  - Freeze, cloud, pour

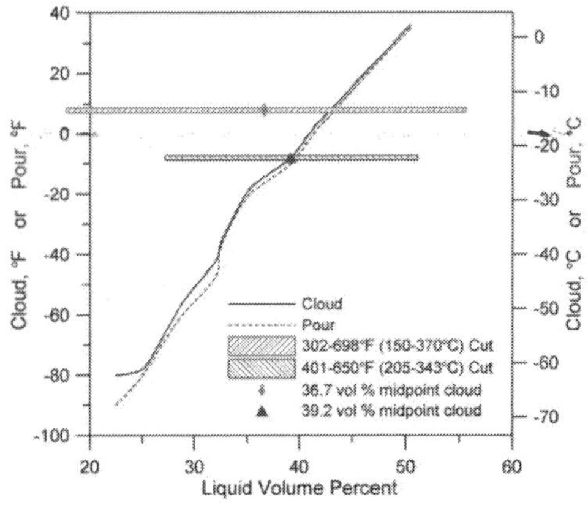

Tia Juana Light

CH2MHILL.

Predicting Yield Cutpoints

- Incremental yield and total yield can be derived
- Results stated as a cutpoint temperature
- Either liquid volume percent or liquid mass percent possible

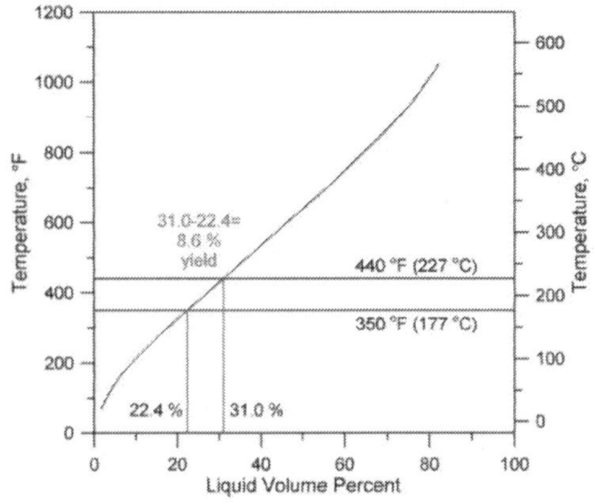

Tia Juana Light

CH2MHILL.

TBP Curves – Representative Crudes

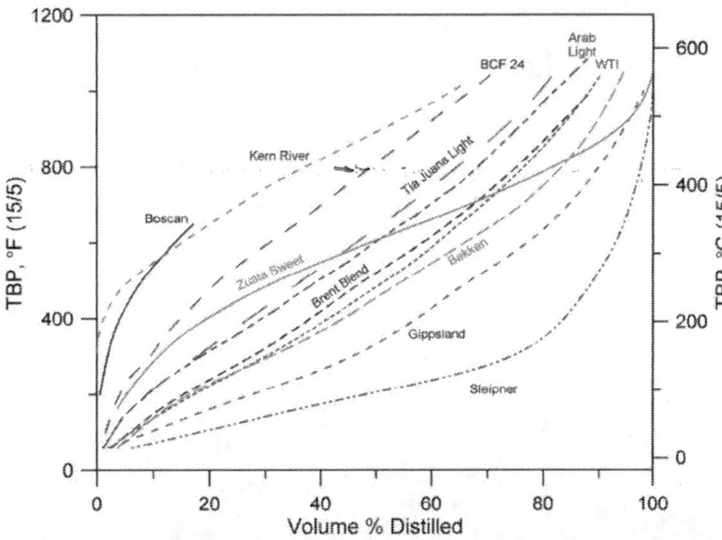

CH2MHILL.

Some Observations

- Assay curves can cross
  - Boscan has more light ends than Kern River
  - Boscan averages out heavier than Kern River because of the back end
- Synthetic crudes can have unusual TBP curves
  - Zuata Sweet has little light ends and little back ends, it has a large middle distillate yield
- Bakken
  - Very sample specific
  - This sample had more light ends and more kerosene-diesel but less naphtha than similar gravity conventional crudes
- Gippsland
  - Getting to limits of conventional crudes
  - Little bottoms
- Sleipner
  - Nearly no bottoms

CH2MHILL.

TBP – Typical Trends

Typical Crude Properties

- As cutpoint increases
  - API gravity decreases
  - Specific gravity increases
  - Sulfur increases

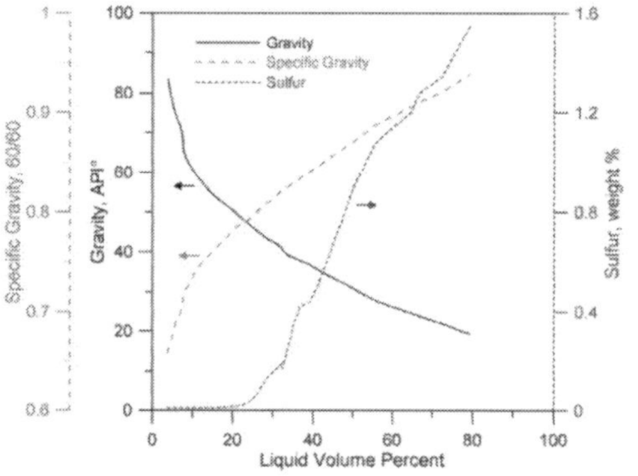

Tia Juana Light

Typical Naphtha Properties

- As cutpoint increases
  - RVP decreases
  - Sulfur increases

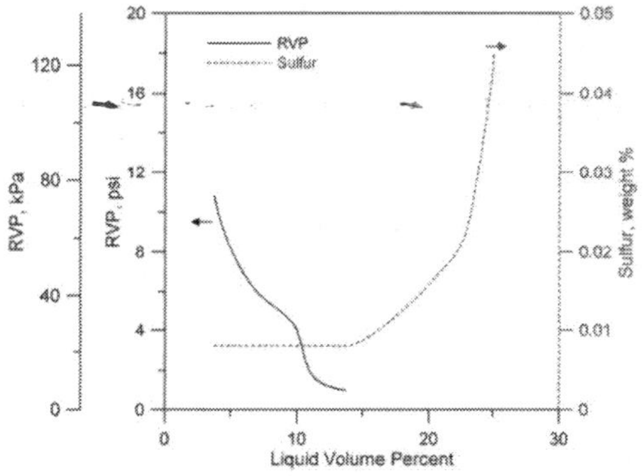

Tia Juana Light

Typical Kerosene Properties

- As cutpoint increases
  - Smoke decreases
  - Luminometer decreases
  - Freeze increases

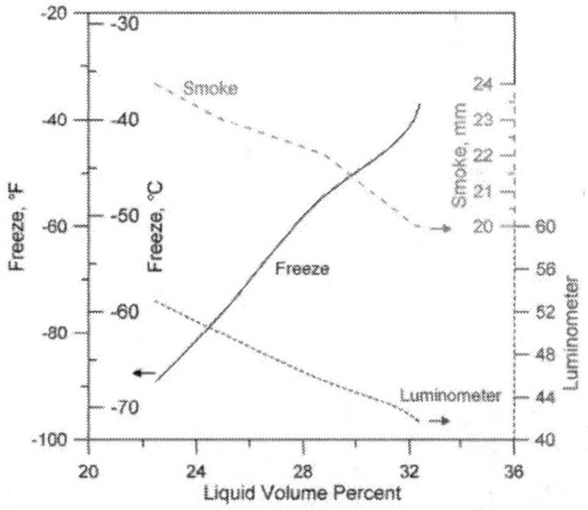

Tia Juana Light

Typical Middle Distillate (Kerosene-Diesel) Properties

- As cutpoint increases
  - Freeze increases
  - Cloud increases
- Freeze and cloud sensitive to small amounts of heavy material
- Wide temperature range cuts will bias towards high-values of freeze and cloud

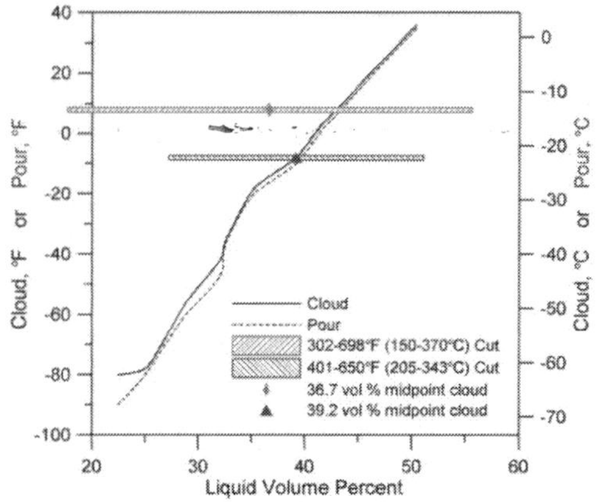

Tia Juana Light

Typical Gas Oil Properties

- As cutpoint increases
  - Nitrogen increases
  - Residual carbon increases

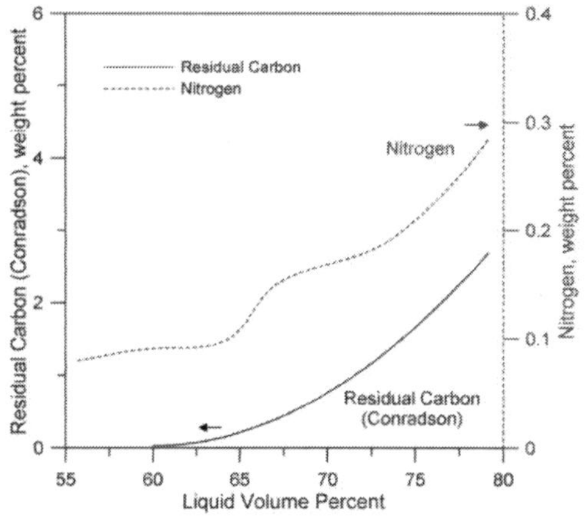

Tia Juana Light

Typical Gas Oil Properties

- As cutpoint increases
  - Vanadium increases
  - Nickel increases
- Older data often has great uncertainty at low metals values
- Small amounts of entrainment greatly increase metals in the gas oil in high-metals crudes

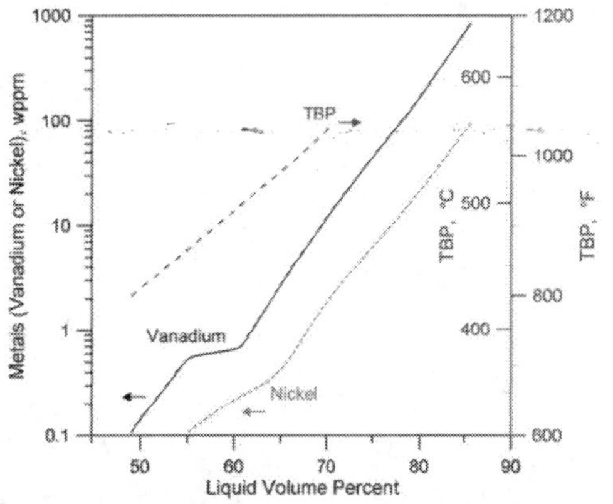

BCF 24 (Bachaquero Field 24)

Typical Asphalt Properties

- As cutpoint increases
  - Softening point increases
  - Penetration decreases

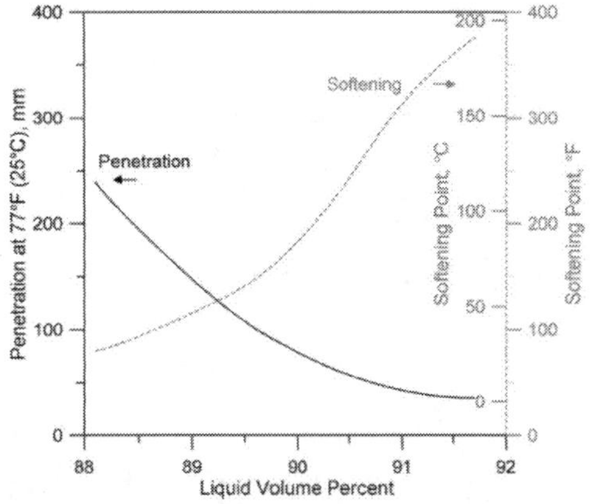

Tia Juana Light

Summary

TBP Summary

- Multiple users for assays
- Level of detail required depends upon user
- Results normally reported as a TBP distillation
- Properties measured on cuts depend upon the boiling range of the cut
- Watch for typical properties
- If the property moves in unexpected conditions, find out why

CH2MHILL.

Questions ?

41

Interpreting Crude Assays

Andrew W. Sloley – CH2M HILL, Bellingham, WA

Prepared for Presentation at the AIChE 2014 6th Southwest Process Technology Conference

Galveston, TX, 9-10 October, 2014

Appendix A – Example Assay – Bayou Choctaw Sour

SPR CRUDE OIL COMPREHENSIVE ANALYSIS

Sample ID: Bayou Choctaw, Sour — Sample No.: BC 15 (31%), BC 17 (22%), BC 19 (23%), BC 101 (24%)

Laboratory No.: none — Date collected: 10/17/2006 — Date results reported: 10/17/2006

Sediment by Extraction, mass %: 0.02 — **Crude** — Water, mass %: 0.04 — Salt, mass %: 0.0039

Relative Density, 60/60° F	0.8631	Ni, ppm	10.7	RVP, psi @ 100° F	3.68
API Gravity	32.4	V, ppm	37.6	Acid number, mg KOH/g	0.16
Sulfur, mass %	1.46	Fe, ppm	1.16	Mercaptan Sulfur, ppm	28
Nitrogen, mass %	0.157	Cu, ppm	0.02	H_2S Sulfur, ppm	17
Micro Car. Res., mass %	3.8	Org. Cl, ppm	0.4	Viscosity: 77° F	9.354 cSt
Pour Point, °F	20	UOP "K" Factor*	11.9	100° F	6.356 cSt
Wax, mass %	0.12	Asphaltenes, mass %	1.28		

Fraction	Gas	1	2	3	4	5	6	7	Residuum	Residuum
Cut Temp.	C_2 - C_4	C5 - 175° F	175° - 250° F	250° - 375° F	375° - 530° F	530° - 650° F	650° - 850° F	850° - 1050° F	650° F+	1050° F+
Vol. %	2.3	5.5	7.2	13.2	16.2	11.9	14.9	13.8	43.6	14.9
Vol. Sum %	2.3	7.8	15.1	28.3	44.5	56.4	71.3	85.1	100.0	100.0
mass %	1.6	4.2	6.1	11.9	15.5	11.9	15.6	15.2	48.8	18.0
mass Sum %	1.6	5.8	11.9	23.7	39.2	51.2	66.8	82.0	100.0	100.0
Relative Density, 60/60° F		0.6615	0.7270	0.7714	0.8250	0.8671	0.9061	0.9486	0.9679	1.048
API Gravity		82.4	63.1	51.9	40.0	31.7	24.7	17.7	14.7	3.6
Sulfur, mass %		0.0109	0.0134	0.0597	0.377	1.10	1.65	2.27	2.59	3.66
Mercaptan Sulfur, ppm		55	52	63	14					
H_2S Sulfur, ppm		6	18	29	6					
Organic Cl, ppm		1.2	1.3	2.3	2.3					
Research Octane Number*		63.3	59.5							
Motor Octane Number*		62.1	57.6							

Appendix A – Example Assay – Bayou Choctaw Sour

Research Octane Number*	63.3	59.5							
Motor Octane Number*	62.1	57.6							
Acid Number, mg KOH/g				0.08	0.09	0.10	0.08	0.15	0.08
Cetane Index*				46.2	52.3	55.7			
Aromatics, Vol. %				22.0					
Naphthalenes, Vol. %			0.00	4.17	10.90				
Wax, mass %						3.53	5.14		
UOP "K" Factor*				11.7	11.7	11.7	11.7		11.4
Hydrogen, mass %			14.2	13.8	13.3	12.8	12.1	11.6	10.3
Carbon, mass %			85.3	86.0	86.2	86.2	85.9	85.8	85.5
Nitrogen, mass %				0.0016	0.0168	0.0867	0.217	0.317	0.600
Refractive Index, 60° C						1.4857	1.5099		
Viscosity, cSt 77° F				2.393					
100° F				1.900	5.229				
130° F					3.524	13.12	107.2	234.8	
180° F						5.873	33.81	58.14	22790
210° F									4813
275° F									
Aniline Point, ° F			124.1	142.9	159.0	174.8	188.0		
Smoke point, mm			27.8	19.1	14.4				
Freezing Point, °F				-31					
Cloud Point, °F				-40	26	80	126		
Pour Point, °F				-45	19	75	112	71	
Ni, ppm							0.26	22.3	68.4
V, ppm							0.68	75.8	229.45
Fe, ppm								3.61	11.5
Cu, ppm								0.02	0.08
Micro Car. Res., mass %						<0.01	1.13	7.20	18.40
Asphaltenes, mass %								2.86	10.29

* Data are calculated

GT-LPG MAX - A Technology based on Top Dividing Wall Column to Recover LPG Product

Bhargava, M.,

Originally presented on: 10/10/2014 13:30:00 - 14:00:00

Refinery off gas is a mixture of hydrogen and hydrocarbons coming from various units within the refinery, pooled together to be used as a fuel for heating purposes. The fuel gas contains valuable components like propane, which can provide additional revenue to the refinery. The economics of LPG recovery fluctuates with LPG pricing and energy cost. GT-LPG Max^{SM} provides a cost-effective solution for recovering LPG product using a novel process concept of absorption plus distillation within the same fractionating vessel. The technology is well suited for low-pressure refinery off gas as well as associated natural gas streams. The process is capable of recovering 99% propane without the use of refrigeration. The process is based on avoiding thermodynamic inefficiencies in conventional absorption and distillation columns by incorporating the unit operations in a single top dividing wall column. The feed side of the top dividing wall column recovers the C_{3+} components from the fuel gas through absorption. The other side of top dividing wall column uses distillation to split between C_3 and C_4components to produce LPG product.

The presentation will cover the following topics:

1) Economics of LPG recovery

2) Available processes for LPG recovery

3) Concept of GT-LPG Max^{SM} as subset of advanced distillation systems

4) GT-LPG Max^{SM} case studies

5) Application areas for GT-LPG Max^{SM}.

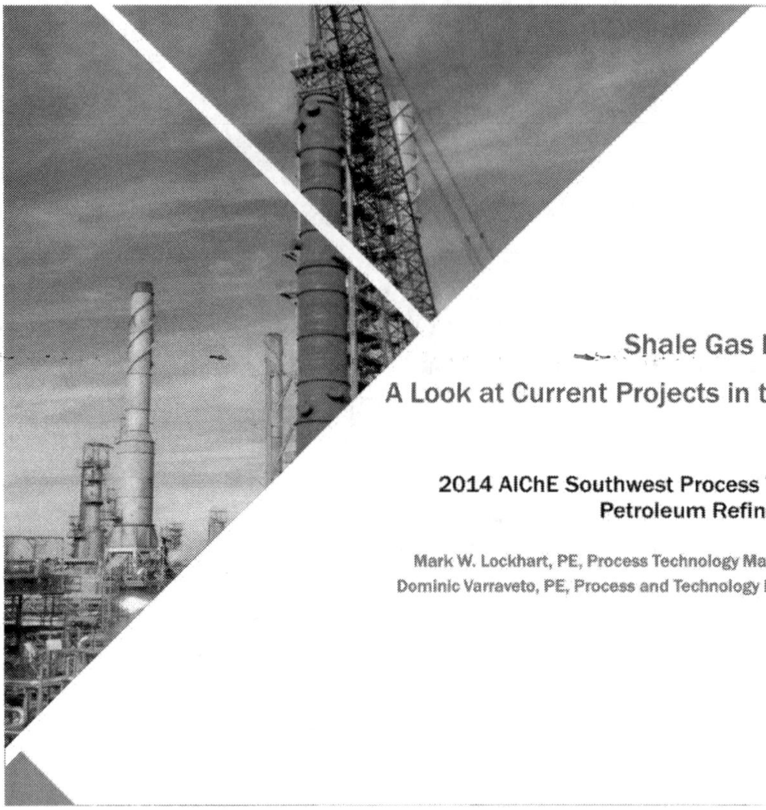

Shale Gas Boom & Tight Oil – A Look at Current Projects in the Refining Sector

Presented By:

2014 AIChE Southwest Process Technology Conference
Petroleum Refining Technology Session

Mark W. Lockhart, PE, Process Technology Manager, Refining and Chemicals
Dominic Varraveto, PE, Process and Technology Department Manager, Refining

October 10, 2014

Burns & McDonnell SINCE 1898

Introduction / Overview

- **Shale gas plays – Locations and current production**
- **Tight oil and condensate – Composition**
- **Tight oil and condensate – Production forecast**
- **A tale of increasing production and refinery processing options:**
  - ➢ Maximize tight oil with no modifications/no capital
  - ➢ Removing bottlenecks with minimal capital
  - ➢ Adding a New Condensate Splitter
  - ➢ Adding a New Crude Unit
- **Tight Oil and condensate – Design considerations**
- **Conclusions**

Shale Gas Plays Location & Current Production

- **Shale Gas Plays - lower U.S. 48 Map:**
  - Niobrara
  - Bakken
  - Monterrey
  - Haynesville-Bossier
  - Utica
  - Wolf Camp & Bone Springs
  - Eagle Ford
- **Largest liquid "Tight Oil" producers:**
  - Bakken in Montana / North Dakota
  - Eagle Ford in South central Texas
- **Production:**

– Bakken	2012E:	0.6 mmBPD	2014E:	1.21 mmBPD
– Eagle Ford	2012E:	0.4 mmBPD	2014E:	0.94 mmBPD

Shale Gas Plays | Locations

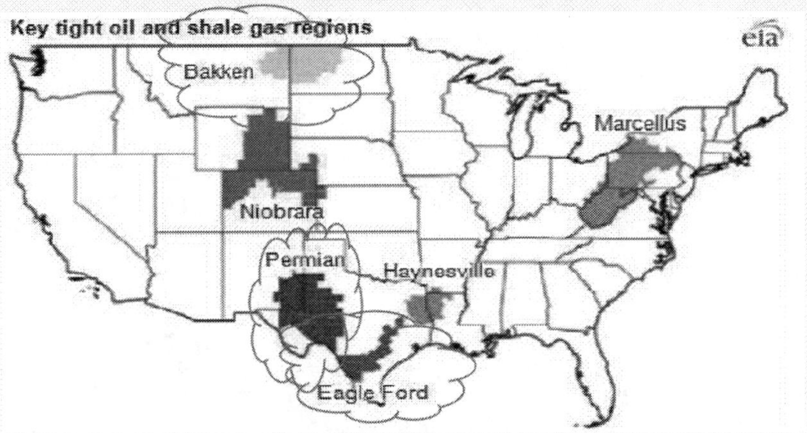

Shale Gas Plays | Locations & Production

OIL&GAS JOURNAL

In February 2014, 63% of US tight-oil production came from two basins: the Eagle Ford in South Texas with production averaging 1.21 million b/d and the Bakken shale in North Dakota and Montana with production averaging 0.94 million b/d.

Tight Oil & Condensate – Current Production (mmBPD)

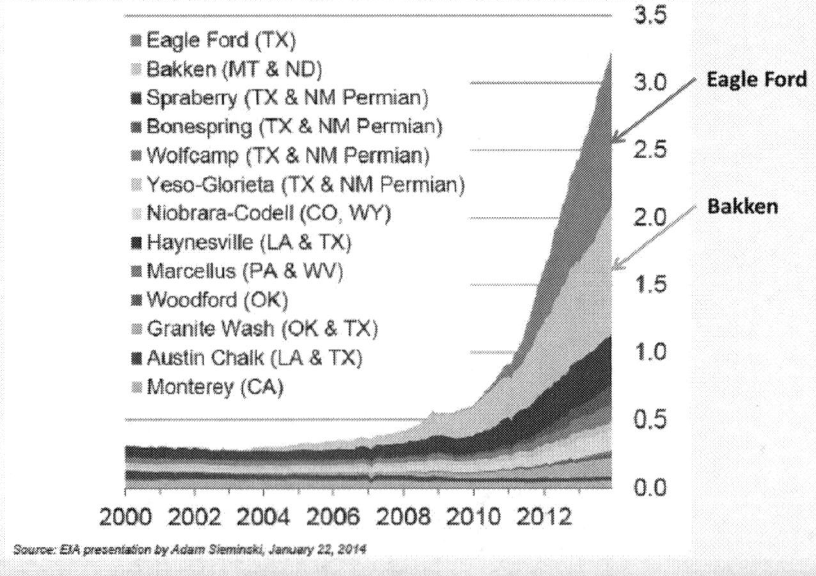

Source: EIA presentation by Adam Sieminski, January 22, 2014

Tight Oil & Condensate Composition

- **Tight oil and condensate are:**
  - Light and sweet
  - Sulfur: < 0.25 wt%
  - High naphtha content
  - Paraffinic
  - Heavy metals are low
  - High filterable solids
- **Tight Oil:** ~ 40 - 45 API gravity
- **Condensate:** ~ 55 - 65 API gravity
- **Processing Challenges from:**
  - Paraffinic nature - Wax deposition, asphaltene precipitation when blend crudes, exchanger and fired heater fouling
  - Filterable solids - exchanger and fired heater fouling
  - Refiners lack of ability to clean equipment on-line

Burns & McDonnell

Tight Oil / Condensate Properties

Physical properties of Eagle Ford shale oil samples

Parameter	Yellow	Red	Black
API	55	44.6	52.3
TAN, g KOH/g	<0.05	0.07	<0.05
Sulfur, wt%	<0.2	<0.2	<0.2
Na, ppm	1	1.6	1.6
K, ppm	0.3	0.4	0.5
Mg, ppm	3.4	2.9	3
Ca, ppm	2.6	2.8	3.8
Asphaltenes, wt %	0	0	0.1
Resin, wt%	0.5	3.2	1.6
Filterable solids, PTB	176	295	225

Source: Hydrocarbon Processing, July 2013

Burns & McDonnell

Tight Oil & Condensate Composition

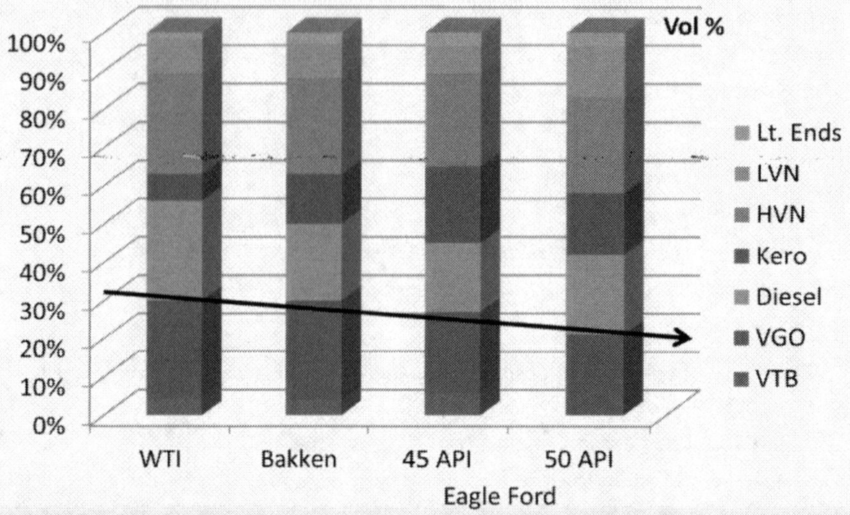

Tight Oil & Condensate Production Forecasts

- **Producer focus – initially natural gas**
- **Resulting in high production with declining prices**
- **Move to "wet" fields with higher liquid-to-gas ratios**
- **Liquids production has increased over last few years**
- **Tight Oil and condensate production forecasted to increase <u>dramatically</u> through 2020**
- **Refiners projected to see plentiful supply of "Tight oil" and "Condensate" for years to come**

Tight Oil & Condensate Production Forecasts

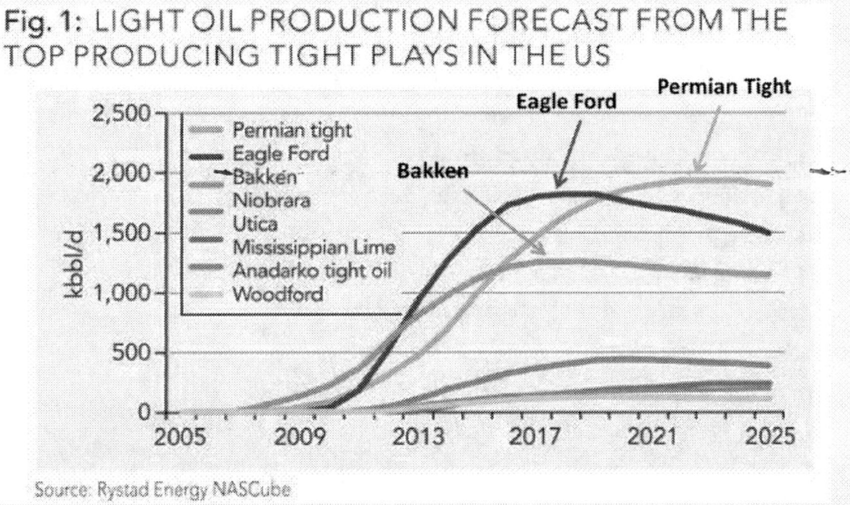

Tight Oil & Condensate Production Forecasts

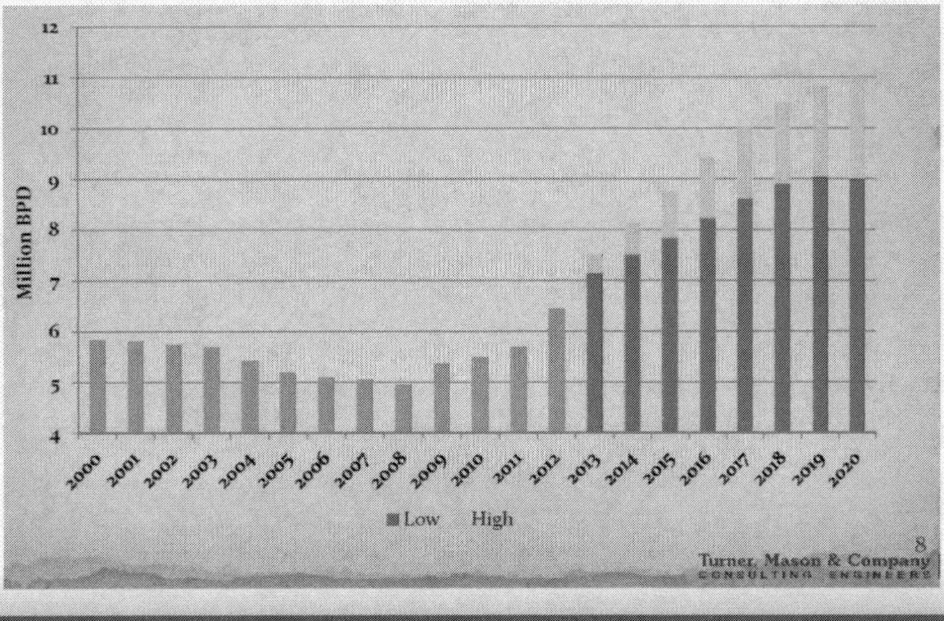

Refinery Processing Options | Trade Out Crude to Refinery Limits

- **Consideration of potential limitations**
- **Supply side and demand side – availability of capacity and routing:**
  - Tankage, pipeline, ship, rail and truck,
- **Existing Refinery Configuration Crude unit limits:**
  - Crude unit Heater, Tower, Overhead condensing and hydraulics
  - LPG processing in Saturates gas plant
  - Naphtha separation in Splitter, Hydrotreater(s), Isom and Reformer units
  - Keeping Conversion and Bottoms processing full
- **Product specifications:**
  - Gasoline RVP, Octane, MSAT 2 – Benzene, Tier 3 – Sulfur
  - Distillate flash, sulfur, pour point, cloud point

Refinery Processing Options | Trade Out Crude to Refinery Limits

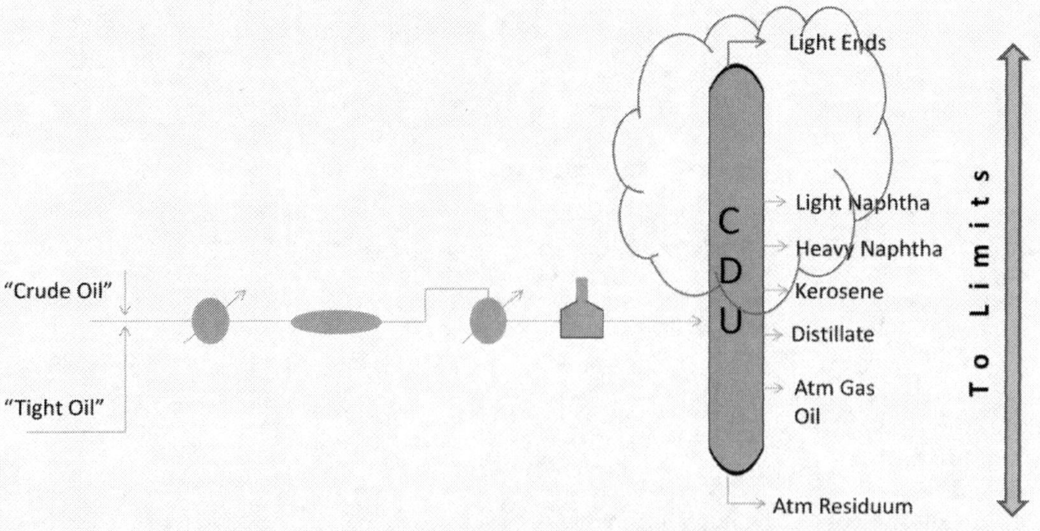

Refinery Planning – Creative Utilization of Assets

- **Possibilities for increasing capacity of existing refinery assets**
- **Crude Unit:**
  - Install new or change location of existing Preflash drum
  - Increase preheat energy recovery, re-tube existing Heater, use high capacity trays or packing in Crude Tower,
  - Increase operating pressure or upgrade overhead condenser system
  - Minimize downtime of existing equipment by completing work during scheduled turnarounds
- **Bypass excess naphtha & distillates to intermediate sales or product blending**
- **Purchase gas oil and residuum to keep conversion units full**

Refinery Planning – Creative Utilization of Assests

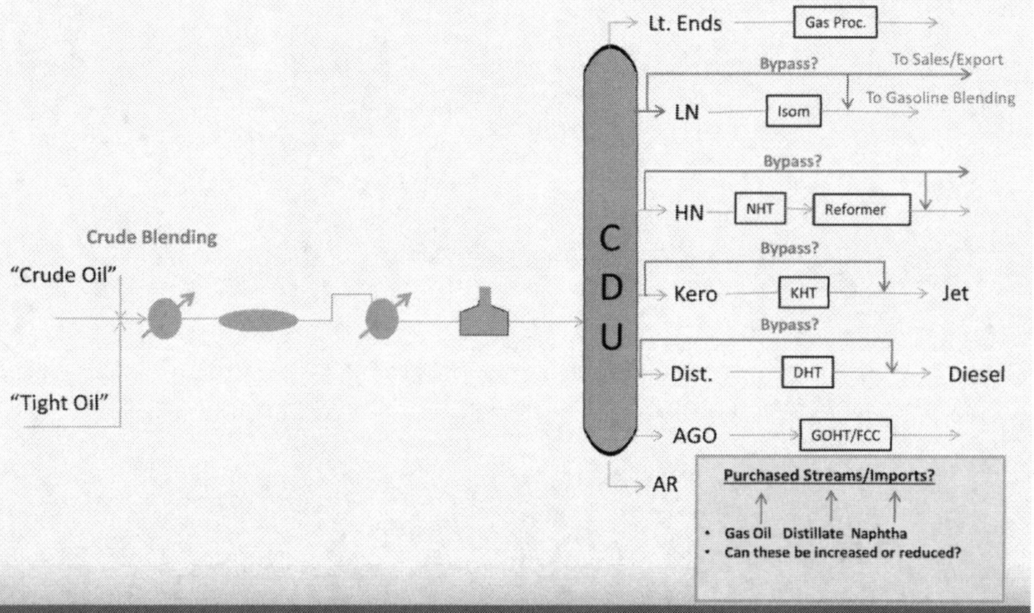

Refinery Planning – Identify and Eliminate Bottlenecks

Possibilities for increasing capacity of existing refinery assets:

- **Saturates Gas Plant:**
  - Revamp, reconfigure or replace existing Sats gas plant
- **Naphtha Processing:**
  - Increase Naphtha Splitter capacity
  - Debottleneck Naphtha Hydrotreater(s), Isomerization and Reformer units
- **Combine revamps with creative re-use of existing refining assets**
- **Implement projects coincident with turnarounds to minimize down time**
- **Fast-track schedule to reap early benefit of evolving tight oil pricing advantage**

Refinery Processing Options | Identify & Eliminate Bottlenecks

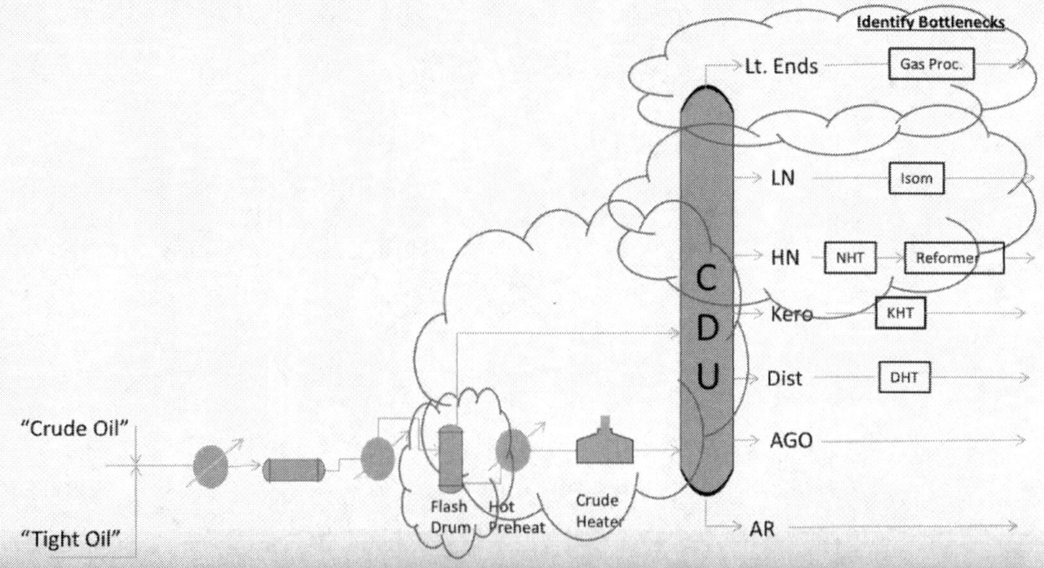

Refinery Planning – Adding a New Condensate Splitter

Considerations for various locales and product slates:

- Stand Alone - Coastal location, ship and barge loading, focused on exporting intermediates
- Co-located at Refinery - focused on exports, takes advantage of existing infrastructure (i.e. tankage, utilities, loading docks)
- Integrated into Refinery – remove LPG w/light naphtha and process in Debutanizer. Send heavy naphtha to Hydrotreater and Reformer. Process "reduced" crude with other feedstocks
- Variations - purchase adjacent facilities, joint ventures

Refinery Processing Options | Adding a New Condensate Splitter

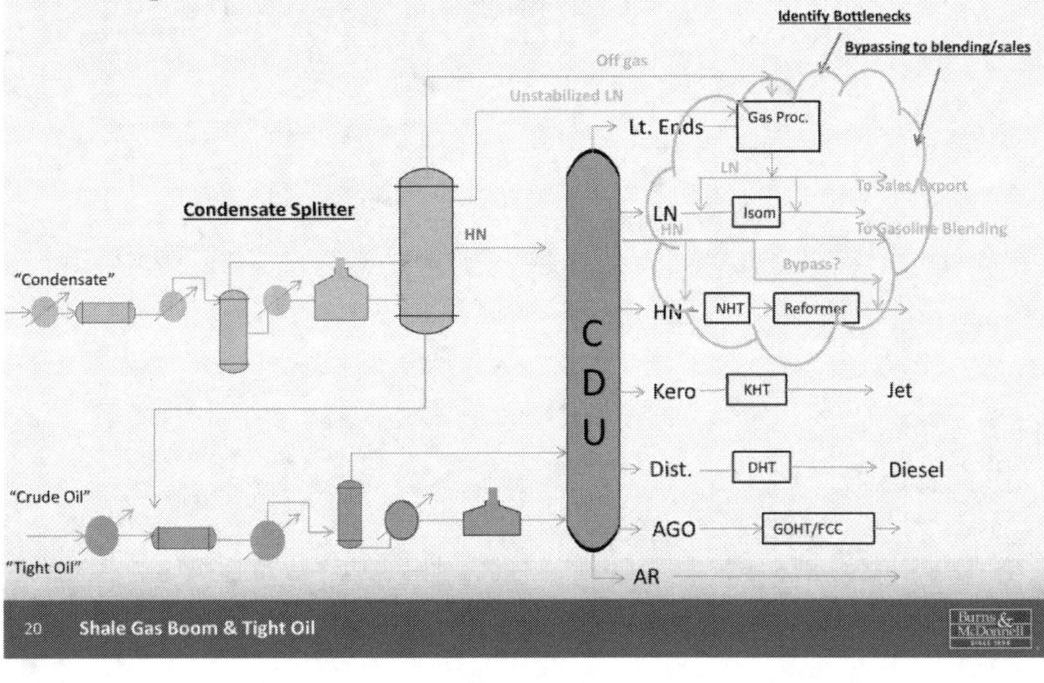

Refinery Planning –
Adding a New Crude Unit

- Special purpose small Field Topping units for local fuel consumption
- Large Crude units to replace existing, inefficient, older units
- Reduce refinery purchase of gas oil or residuum for conversion or bottoms processing or for capacity expansion
- Include flexibility to process a broad range of tight oil, condensate and cost advantaged crudes

Refinery Processing Options | Adding a New Crude Unit

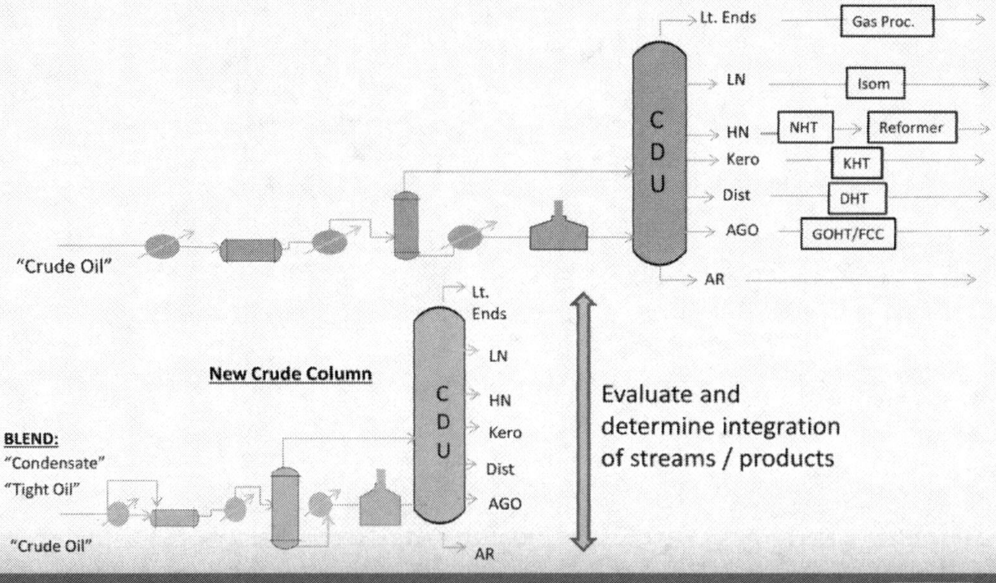

Refinery Planning – Design Considerations for a New Integrated Crude Unit

- Transportation and storage
- Exchanger fouling and design
- Desalter design and operation
- Flash drum placement and design
- Fired Heater fouling and design
- Equipment corrosion
- Tight oil effect on product quality and processing

Tight Oil and Condensate | Design Considerations

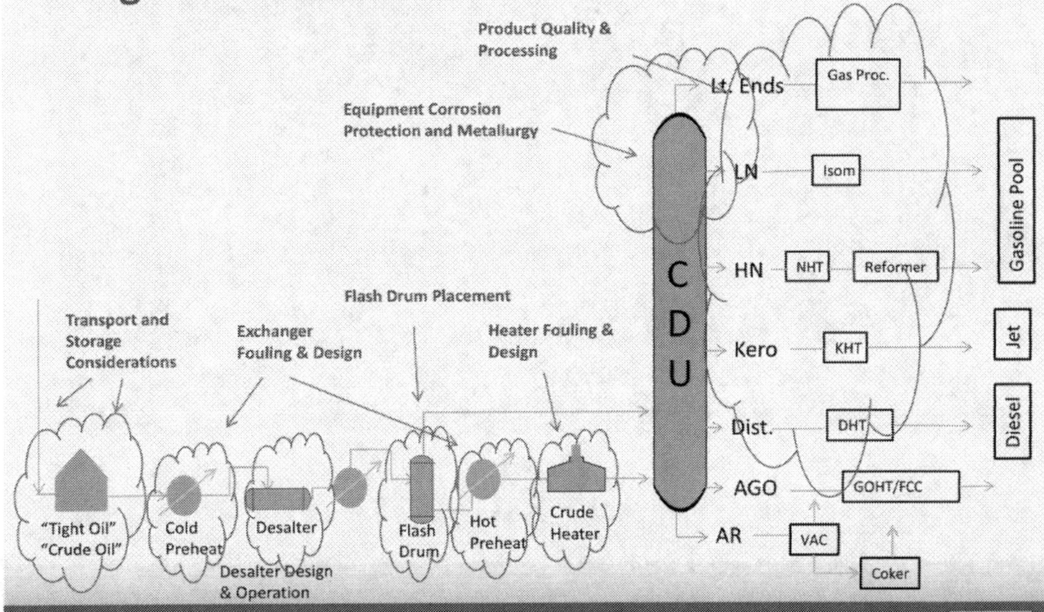

Conclusion

- Tight oil and condensate forecasted to be in high supply for years to come
- These light, low sulfur liquids are displacing traditional crudes in refineries across the U.S.
- Refiners are increasing the proportion of tight oil into their facilities within current limits
- Low cost, quickly implemented solutions are following
- Higher capital expenditures including new Condensate Splitters and new Crude units are now in different phases of study and project implementation
- Strong evaluation capability and creative solutions are necessary to determine cost-effective debottlenecking and grass-roots implementation
- Offsites and utilities are an important part of these plans

Q&A

Burns & McDonnell – 100% Employee Owned!

Mark W. Lockhart
mlockhart@burnsmcd.com
832.214.1975

Dominic Varraveto
dvarraveto@burnsmcd.com
816.822.4282

Are Unconfined Hydrogen Vapor Cloud Explosions Credible?

J. Kelly Thomas, Ph.D.,
Craig Eastwood, Ph.D. &
Martin Goodrich, P.E.
Baker Engineering and Risk Consultants
San Antonio, TX

2014 AIChE Southwest Process
Technology Conference
Galveston, TX, October 10, 2014

BAKERRISK

AIChE Southwest Process Technology Conference

Outline

- Background
- Purpose
- Accidental H_2 VCEs
- H_2 VCE Tests
- Dispersion of H_2 Releases
- Ignition of H_2 Releases
- H_2 VCE Blast Loads
- Conclusions

BAKERRISK

2

Background (1 of 3)

- Owner/operators of chemical processing & petroleum refining sites often ask whether unconfined H_2 VCEs can actually occur:
  - Normally in connection with FSS or QRA
- Generally recognized that an H_2 release within enclosure can lead to explosion
- Questions arise for external releases:
  - Doesn't H_2 just float upwards, so that it can't form a significant flammable gas cloud within the process area (i.e., congested volume)?
  - Doesn't H_2 always promptly ignite so that it just forms a jet fire?

BAKERRISK

3

Background (2 of 3)

- Chemical processing & petroleum refining site owner/operators:
  - Some do not consider unconfined H_2 VCEs as credible events
  - Some consider such events as credible, but only for relatively small release sizes
  - Some treat same as hydrocarbon release (i.e., treat as credible and consider large releases)
- Range of treatments recommended by consultancies span same range

BAKERRISK

4

Background (3 of 3)

- FM Global data sheet for VCE evaluation (DS 7-42, 10/13):
  - Excludes VCEs due to gaseous H_2 releases
  - Calls for VCE evaluation for liquid H_2 releases
  - Notes 3% of VCEs in referenced database were H_2 or synthesis gas releases
- Goal of FM Global for VCE evaluation under DS 7-42 may be different from goals of FSS and QRA

BAKERRISK

5

Purpose

- Provide an answer to questions posed by owner/operators of chemical processing & petroleum refining sites:
  - Are unconfined H_2 VCEs credible?
  - Should they be considered in a FSS, QRA or similar explosion consequence or risk assessment (e.g., API RP 752/753)?

BAKERRISK

6

Accidental H_2 VCEs (1 of 10)

- "The proof is in the pudding":
  - If unconfined H_2 VCEs happen relatively often, reasonable to conclude they are credible
  - "Relatively often" is in the eye of the beholder
- Many not reported (as with all explosions)
- Some incidents reported in summary tables in literature, but with insufficient information available to conclude given incident is an unconfined H_2 VCE

BAKERRISK

7

Accidental H_2 VCEs (2 of 10)

- Zalosh and Short (1978):
  - Reviewed > 400 H_2 accidents (1965 – 1977)
  - Slightly > ½ of incidents were explosions
  - 3/4 of incidents involved H_2 gas
  - H_2 gas explosions significant portion of reported incidents in database
- Ordin (1974):
  - Reviewed incidents from NASA operations
  - 62% of releases to environment ignited (i.e., 38% did not) & some ignitions were delayed (explosions)
  - Prompt ignition not assured

BAKERRISK

8

Accidental H_2 VCEs (3 of 10)

- Two widely reported unconfined H_2 VCEs:
  - Polysar Sarnia (MacDiarmid & North, 1989)
  - Jackass Flats (Reider, Otway & Knight, 1965)
- Other unconfined H_2 VCEs:
  - Hydrogen unit, Watson, CA, 1975 (300 kg)
    - Identified by Lenoir & Davenport (1993) & Gugan (1979)
  - Silver Eagle refinery, Woods Cross, UT, 2009
  - Muskingum River Plant, Beverly, Ohio, 2007

BAKERRISK

9

Accidental H_2 VCEs (4 of 10)

- Other unconfined H_2 VCEs (continued):
  - H_2 Incidents database
  - MCA Case History collection
  - NASA operations (Ordin)
  - Hanau, Germany
  - Sodegaura, Japan
- Not considering:
  - H_2 explosions within enclosures
  - Explosions involving mixture with hydrogen

BAKERRISK

10

Accidental H_2 VCEs (5 of 10)

- Sarnia:
  - Polysar petrochemical complex Sarnia, Canada
  - Release of H_2 from partially failed gasket on compressor in open-sided shed
  - 700 psi (48 bar), 10 to 15 sec before ignition, released 30 kg of H_2
  - Building damage at 500 ft (150 m) consistent with 1.1 psi (0.076 bar) overpressure
    - Consistent with detonation of 26 kg of H_2 (BST)
    - Large fraction of released H_2 detonated

BAKERRISK

11

Accidental H_2 VCEs (6 of 10)

- Jackass Flats:
  - Rocket motor test program
  - H_2 test run without ignition (by design)
  - Release from 3,400 psi (23.6 MPa) upward through convergent-divergent nozzle
  - Flowed 13 sec. before unintentional ignition
  - Estimated 200 lb_m (90 kg) of H_2 (10% of that released) involved in VCE
  - VCE determined to be deflagration
    - Flame acceleration due to jet turbulence

BAKERRISK

12

Accidental H_2 VCEs (7 of 10)

- Silver Eagle refinery (Woods Cross, UT, 2009):
  - H_2 at 630 psi (43 bar) released by 10-inch (25 cm) pipe failure, ignited "quickly" (open-flame furnace near release)
  - VCE caused severe damage to 2 homes & minor damage to other homes near refinery
- Muskingum River Plant (Beverly, Ohio, 2007):
  - Rupture disc failure on outdoor hydrogen storage tank vent line during tank filling operations
  - Release continued for 3-10 sec. before ignition
  - Killed driver, heavily damaged adjacent buildings

BAKERRISK

13

Accidental H_2 VCEs (8 of 10)

- H_2 Incidents database
  - 7 events clearly unconfined H_2 VCEs
  - Sarnia and another similar incident
  - Several similar to Muskingum River Plant incident
- MCA database
  - Unconfined H_2 VCE due to release from vent stack
- NASA incidents review (Ordin)
  - At least 9 unconfined H_2 VCEs
  - Releases through vent stacks & failed components
  - Some reported to be detonations, up to 20 lb_m (9.1 kg) of TNT-equivalent

BAKERRISK

14

Accidental H_2 VCEs (9 of 10)

- Hanau, Germany (1991):
  - 100 m^3 H_2 storage tank failed at 45 bar
  - Significant blast damage to process unit, broken windows off-site

BAKERRISK

15

Accidental H_2 VCEs (10 of 10)

- Sodegaura, Japan (1992):
  - H_2 released from heat exchanger
  - Significant damage to plant
- Conclusion:
  - Unconfined H_2 VCEs are not "rare events",
  - They have happened regularly in the past,
  - They have happened recently, and
  - They will continue to happen in the future.

BAKERRISK

16

H_2 VCE Tests (1 of 3)

- Numerous unconfined H_2 VCE tests have been performed
- Generally used plastic tent or similar approach to initially confine flammable mixture
- No reports of "premature ignition", although flow rates are usually low
- Tests do show can achieve deflagration-to-detonation transition with lean mixtures at moderate congestion levels

BakerRisk

17

H_2 VCE Tests (2 of 3)

- BakerRisk performed an internal research project in 2009 to determine H_2 conc. For DDT in "medium congestion" VCE rig:
  - 48 ft x 12 ft x 6 ft (14.6 m x 3.7 m x 1.8 m)
  - 3D (no confinement, no roof, no side walls)
  - Medium congestion
    - 2-inch (5 cm) vertical pipes
    - Area blockage ratio = 22%
    - Volume blockage ratio of = 4.1%
    - Pitch-to-diameter ratio =4.5

BakerRisk

18

H_2 VCE Tests (3 of 3)

22% H_2 Concentration Test
(stoichiometric = 30%)

BakerRisk

19

Dispersion of H_2 Releases (1 of 7)

- Doesn't H_2 just float upwards, so that it can't form a significant flammable gas cloud within the process area (i.e., congested volume)?
- Buoyancy does not exert significant influence until dispersing mixture has slowed sufficiently for momentum forces to weaken.
- Consider following release & conditions:
  - 2-inch (5 cm) hole size
  - 1,400 psig (97 bar) at 550 °F (288 °C)
  - Gives release rate of 8.4 kg/s

BakerRisk

20

Dispersion of H_2 Releases (2 of 7)

- Assumed:
  - Pasquil stability class D, 3 m/s wind velocity
  - Horizontal release
- Two dispersion evaluations performed:
  - BakerRisk's SafeSite$_{3G}$® code
  - FLACS CFD code
- Two dispersion evaluations performed:
  - No congestion present ("open field")
  - Release within congested module (FLACS only), at edge and center of module at selected heights

BAKERRISK

21

Dispersion of H_2 Releases (3 of 7)

- No congestion (flammable cloud side view)

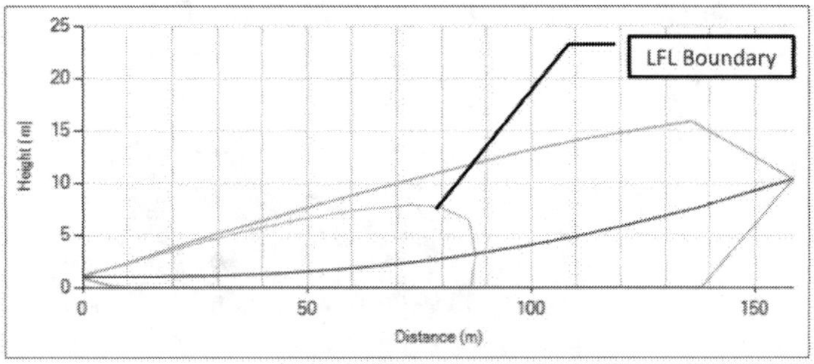

SafeSite

FLACS

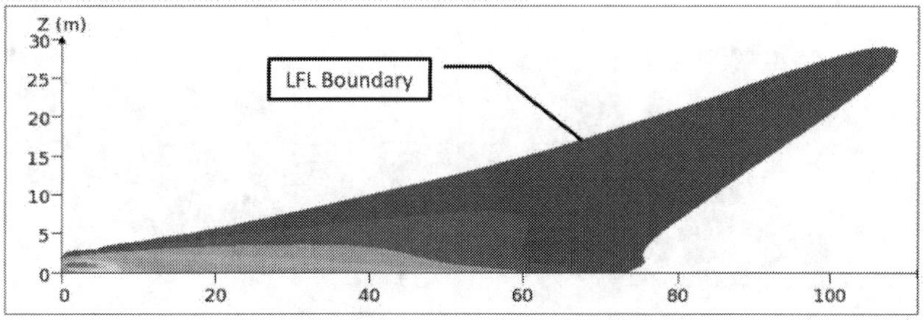

BAKERRISK

Dispersion of H_2 Releases (4 of 7)

- Congested process module
  - 37 m x 19 m x 12 m (8,120 m^3)

23

Dispersion of H_2 Releases (5 of 7)

- Flammable gas contours (8 kg/s)
  - Molar concentrations from LFL to 80% H_2

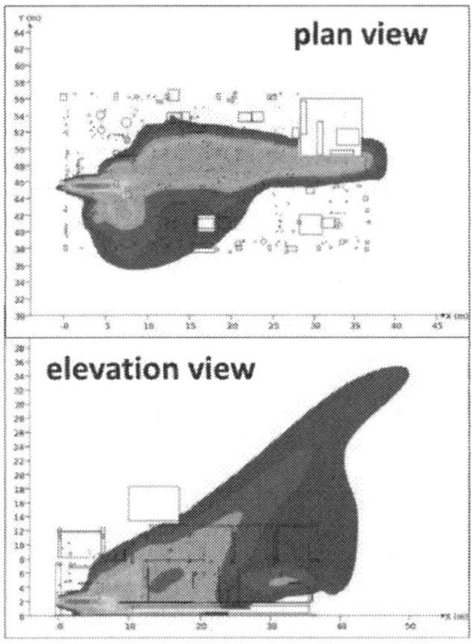

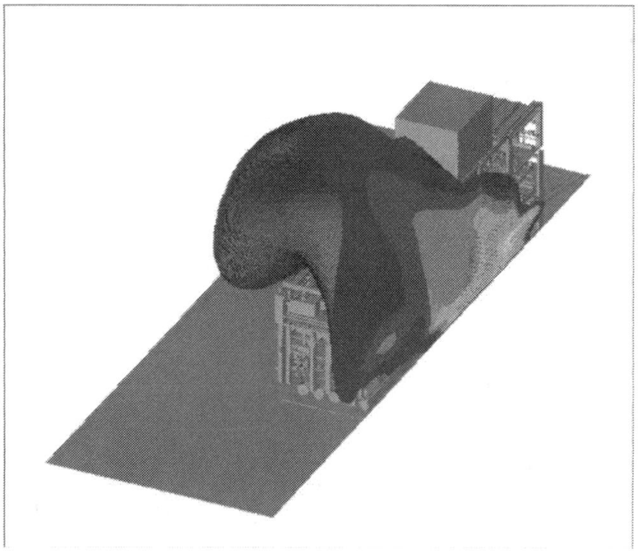

Dispersion of H_2 Releases (6 of 7)

- Results summary
  - Flammable cloud 15% to 46% of module volume at 8 kg/s (up to about 4,000 m^3)
    - Largest flammable cloud within module ≈ ½ flammable cloud without congestion
  - Flammable cloud 55% of module vol. at 16 kg/s
  - Total flammable cloud volume is roughly 3 to 7 times that within the module
    - Important for DDT, as detonation wave will propagate into flammable cloud outside module
    - Larger than cloud without congestion

BAKERRISK

25

Dispersion of H_2 Releases (7 of 7)

- Doesn't H_2 just float upwards, so that it can't form a significant flammable gas cloud within the process area (i.e., congested volume)?
- No, it doesn't.
- An H_2 release at significant rate will form large cloud with or without congestion
- H_2 will "just float upwards" only at small release rates, which are not of interest in typical hazard evaluations

BAKERRISK

26

Ignition of H_2 Releases

- Recent paper (2014 AIChE conf.) by Moosemiller and Galindo
  - Potential for delayed ignition
- Ordin (1974):
  - Hydrogen ignited in 62% of releases to environment (i.e., 38% did not ignite) & some ignitions were delayed (explosions)
- Potential for delayed ignition is real
- Consistent with occurrence of accidental unconfined H_2 VCEs

BakerRisk

27

H_2 VCE Blast Loads (1 of 3)

- So … how big would the consequence be?
- Blast load function of flame speed, energy & standoff
- Hydrogen flame speeds comparatively high
  - DDT expected for concentrations approaching stoichiometric in moderate levels of congestion (even with no confinement)
  - Flame speeds comparatively high even without a DDT

BakerRisk

28

H_2 VCE Blast Loads (2 of 3)

- Consider 4,000 m^3 H_2-air mixture
  - Based on congested module analysis
  - Not accounting for cloud outside module
  - Larger cloud for more severe release scenarios (> 2-inches or higher pressures) and/or larger congested volumes
- BST with M_f = Mach 0.5, 0.7 & 5.2
  - Expect DDT (M_f = 5.2) at moderate congestion

BAKERRISK

29

H_2 VCE Blast Loads (3 of 3)

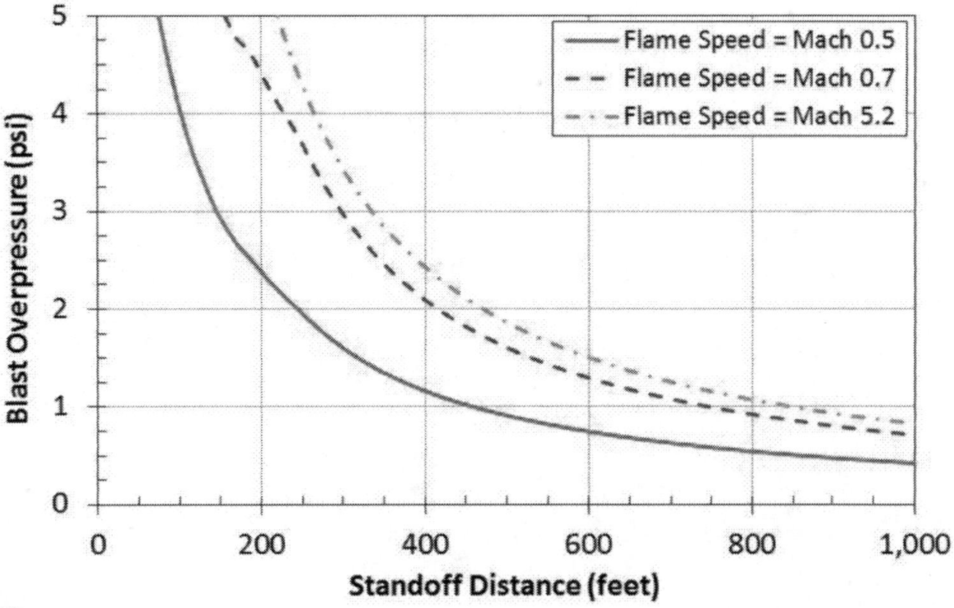

BAKERRISK

30

Conclusions (1 of 2)

- Doesn't H_2 just float upwards, so that it can't form a significant flammable gas cloud within the process area (i.e., congested volume)?
  - No, H_2 release can form a large flammable gas cloud near grade level and within a congested volume
- Doesn't H_2 always promptly ignite so that it just forms a jet fire?
  - No, delayed ignitions can and have occurred

BAKERRISK

31

Conclusions (2 of 2)

- Can unconfined H_2 VCEs actually occur?
  - Yes, and many such events have occurred
  - Should anticipate future events
- Blast loads from H_2 VCE can be large
- Recommend unconfined H_2 VCEs be treated as credible events

BAKERRISK

32

A Step-Wise Approach to Meeting the Growing Imbalance between Diesel/Gasoline Production

Mike Craig
Senior Hydroprocessing Technology Manager
Axens North America
Houston, Texas

Outline

- **Current Market Analysis**
- **Options Available**
- **Case Study: Step-Wise Approach**
- **Summary**

Crude and Natural Gas Prices

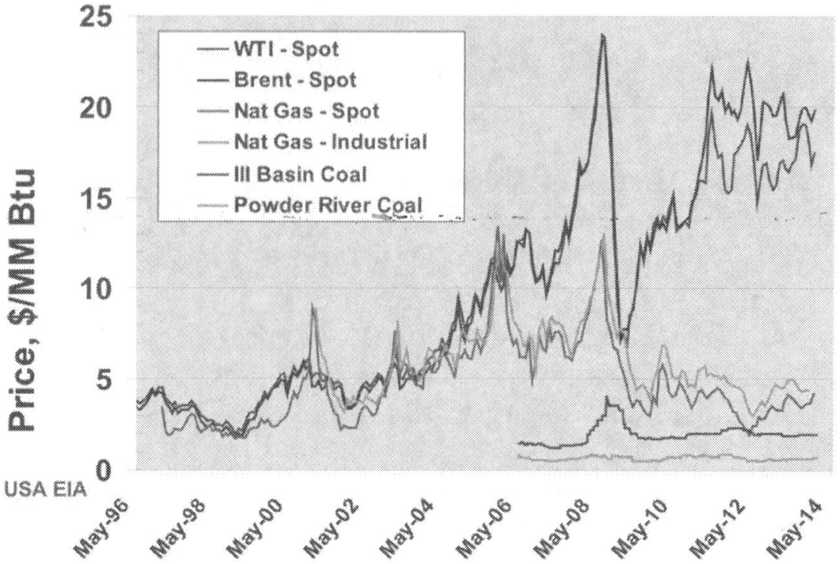

Source: USA EIA

Axens

3

Transportation Fuel Prices

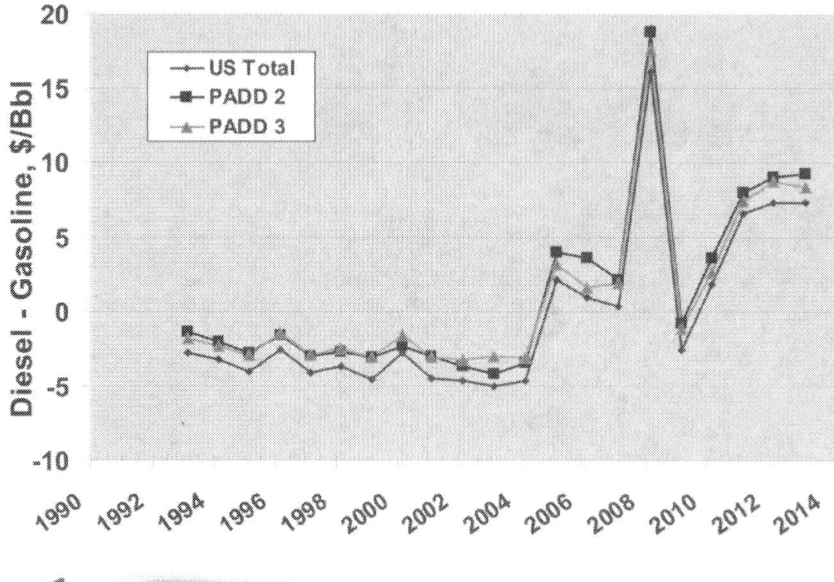

Axens

4

Market Observations

- Crude Oil Prices Staying High
- Low-Cost Hydrogen
- Diesel Production Favored over Gasoline
- 124 USA Refiners
  - Average Capacity is ~146,000 BPSD
  - 66 with Delayed Cokers
  - 88 with FCC Units
  - 47 with FCC Pre-treaters

Axens
5

Outline

- Current Market Analysis
- **Options Available**
- Case Study: Step-Wise Approach
- Summary

Axens
6

Options Available

- Changing Crude Feedstock
- Conversion of VGO
- Conversion of Vacuum Residue

Axens
IFP Group Technologies

7

Changing Crude Feedstock

Properties	WTI	Brent	Arab Light	Maya	Athabasca Bitumen
API	39	38.6	33.4	22	8.4
Sulfur, wt %	0.27	0.29	1.79	3.56	4.92
Distillation, vol %					
Naphtha	34	30	25	17	2
Kero & Diesel	29	29	28	22	14
VGO	21	29	30	27	34
Residue	14	9	15	33	50
D/G Ratio	0.85	0.94	1.12	1.28	7.0

Axens
IFP Group Technologies

8

Add VGO Hydrocracking Capacity

- Revamp FCC Pre-treater to MHC
  - Minimal Investment to Reaction Section
  - Up to 30 to 35% Conversion
  - Change out Catalyst
  - Requires new Fractionation Column
- Results
  - Increase in Volume Swell: 106.6 vol%
  - Improvement in Diesel Yield: D/G Ratio = 2.9

Axens
IFP Group Technologies

9

Single-Stage Hydrocracking

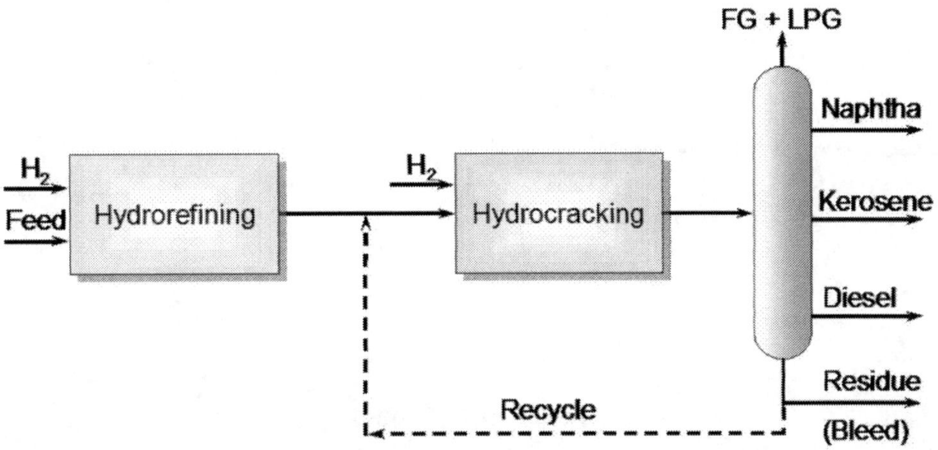

Axens
IFP Group Technologies

10

Selectivity vs. Conversion

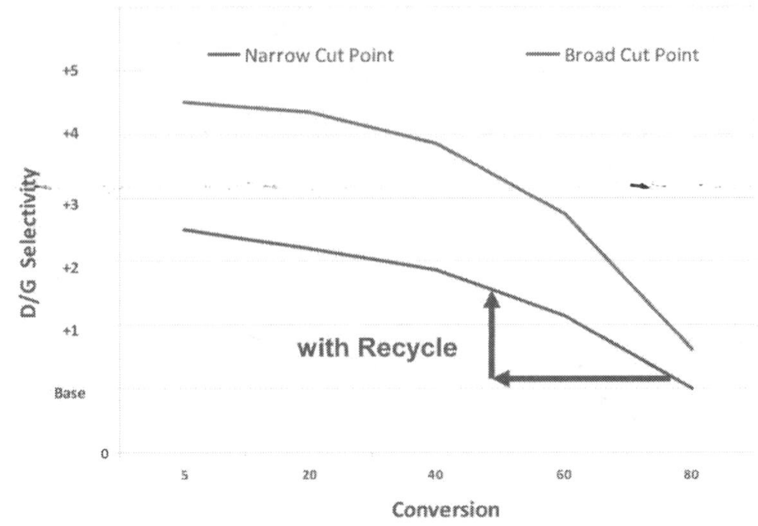

Axens
11

Two-Stage Hydrocracker with Recycle

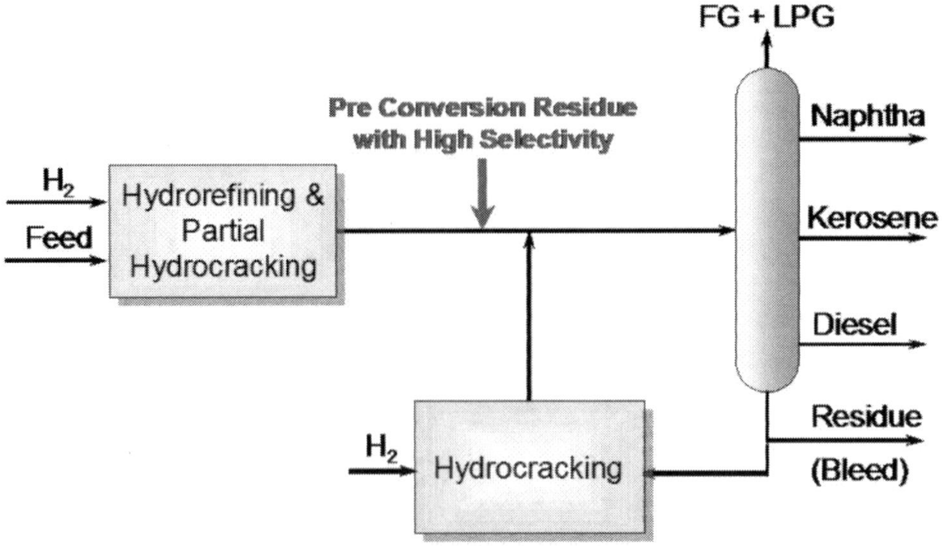

Axens
12

Axens Catalyst Portfolio

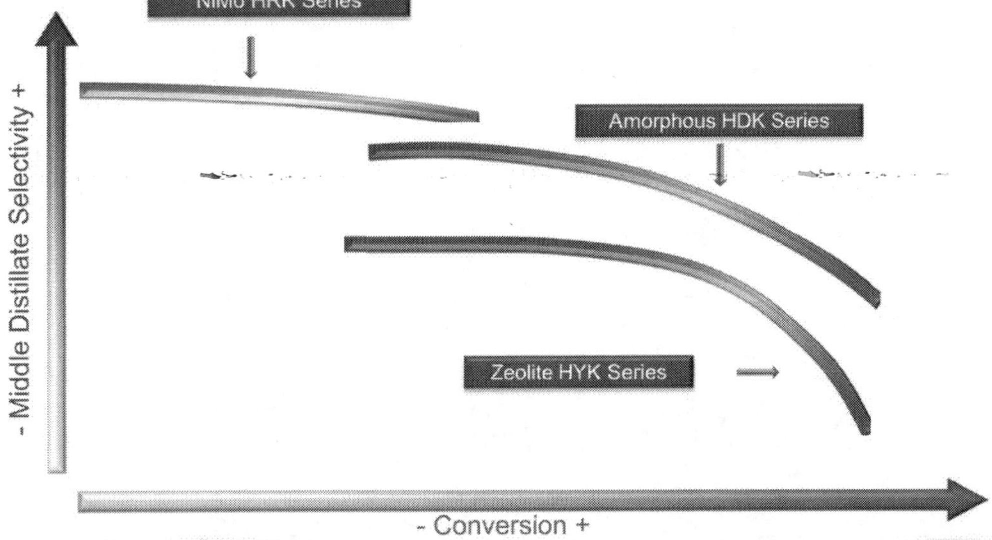

Axens

13

Conversion of Vacuum Residue

- Delayed Coking
- Residue Hydrocracking
- Residue Hydrocracking/Delayed Coking

Axens

14

Volume Yield vs. Conversion Unit Type

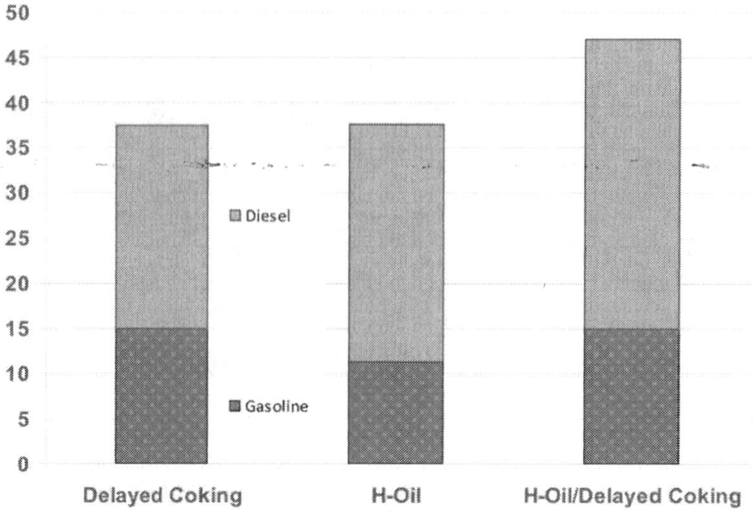

Axens

15

Outline

- Current Market Analysis
- Options Available
- **Case Study: Step-Wise Approach**
- Summary

Axens

16

Case Study

- Average Size Refinery: 150,000 BPSD
- Crude Blend WTI / WCS
- Conversion Units
  - FCC Unit
  - Delayed Coker

Axens

17

Base Case: Block Flow Diagram

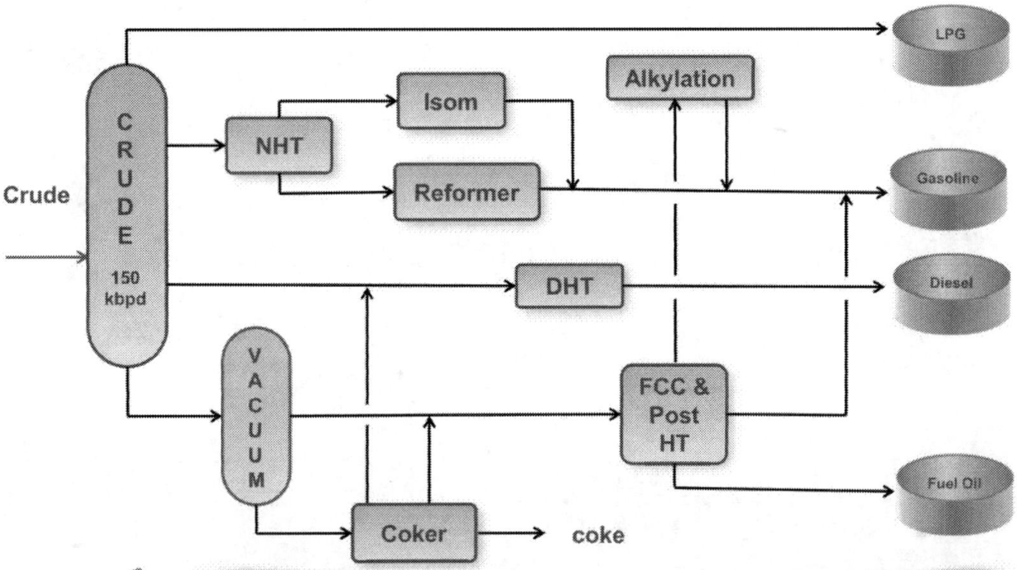

Axens

18

Crude Feedstock

	WTI / WCS	WTI / WCS / AB
Gravity, °API	34.6	20.2
Sulfur, wt%	1.2	3.1
Nitrogen, wppm	1,047	2,500
CCR, wt%	3.2	9.3
Metals, wppm	13	173
Distillation, vol %		
Naphtha	29.3	15.0
Diesel	30.8	22.4
VGO	24.5	29.2
Vacuum Residue	15.4	33.4
Crude D/G Ratio	1.1	1.5

Axens
IFP Group Technologies

19

Case 2: Block Flow Diagram

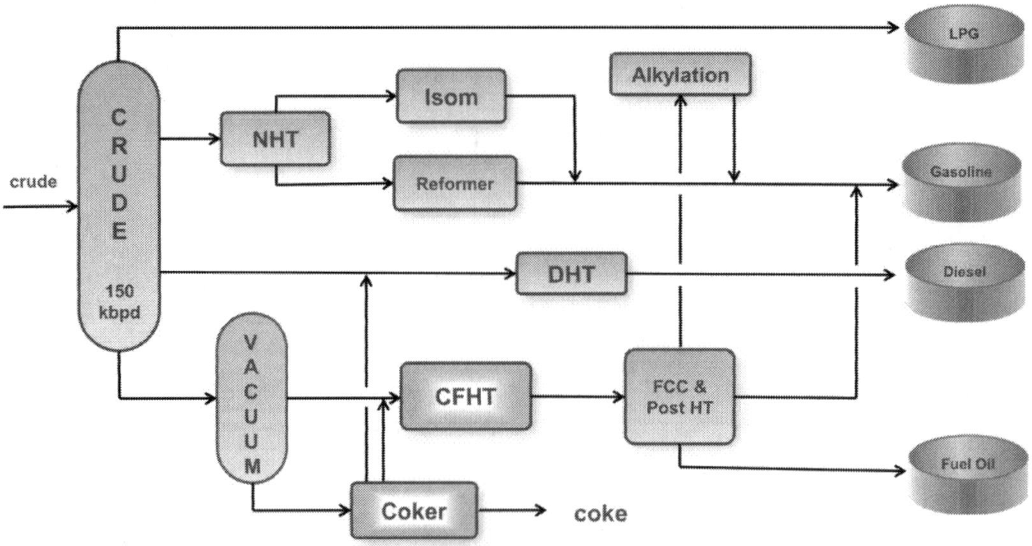

Axens
IFP Group Technologies

20

Cases Evaluated

- Case 2: add an Opportunity Crude
  - plus CFHT and Delayed Coker
- Case 2A: revamp CFHT to MHC
- Case 2B: add O-T Single-Stage Hydrocracking
- Case 2C: add Two-Stage w/Recycle Hydrocracking
- Case 3: add Residue Hydrocracking
  - use existing Delayed Coker from Base Case
  - add integrated VGO Hydrocracker

21

Relative Investment

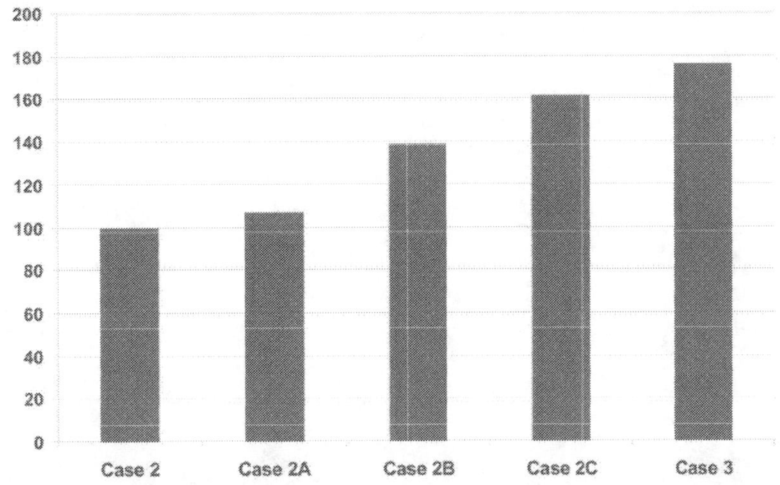

22

Yield of Transportation Fuels kBPSD

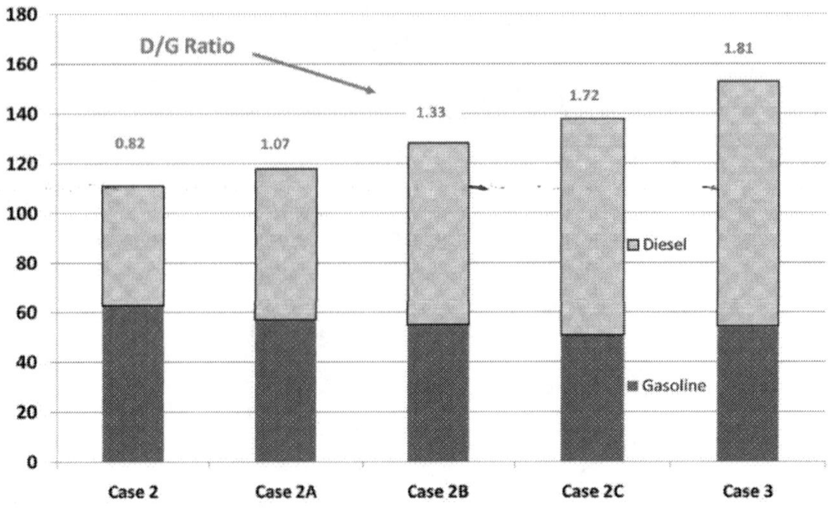

Axens
IFP Group Technologies

Outline

- Current Market Analysis
- Options Available
- Case Study: Step-Wise Approach
- **Summary**

Axens
IFP Group Technologies

Summary

- **Low Investment and Attractive Payout**
  - Revamping CFHT to Mild Hydrocracking
  - Revamping Distillation Column to broaden Diesel Cutpoint
- **Moderate Investment**
  - Adding Hydrocracking Capacity
    - Market Demand for Lube Oil
    - With or without FCC Unit
- **Highest Investment & Best Return**
  - Adding Residue Hydrocracking ahead of Delayed Coking
  - Integrate VGO Hydrocracking
  - Simple Payout < 3 years
  - Achieves highest D/G Selectivity

Axens IFP Group Technologies

25

Axens
IFP Group Technologies

A Step-Wise Approach to Meeting the Growing Imbalance between Diesel/Gasoline Production

Mike Craig
Senior Hydroprocessing Technology Manager
Axens North America
Houston, Texas

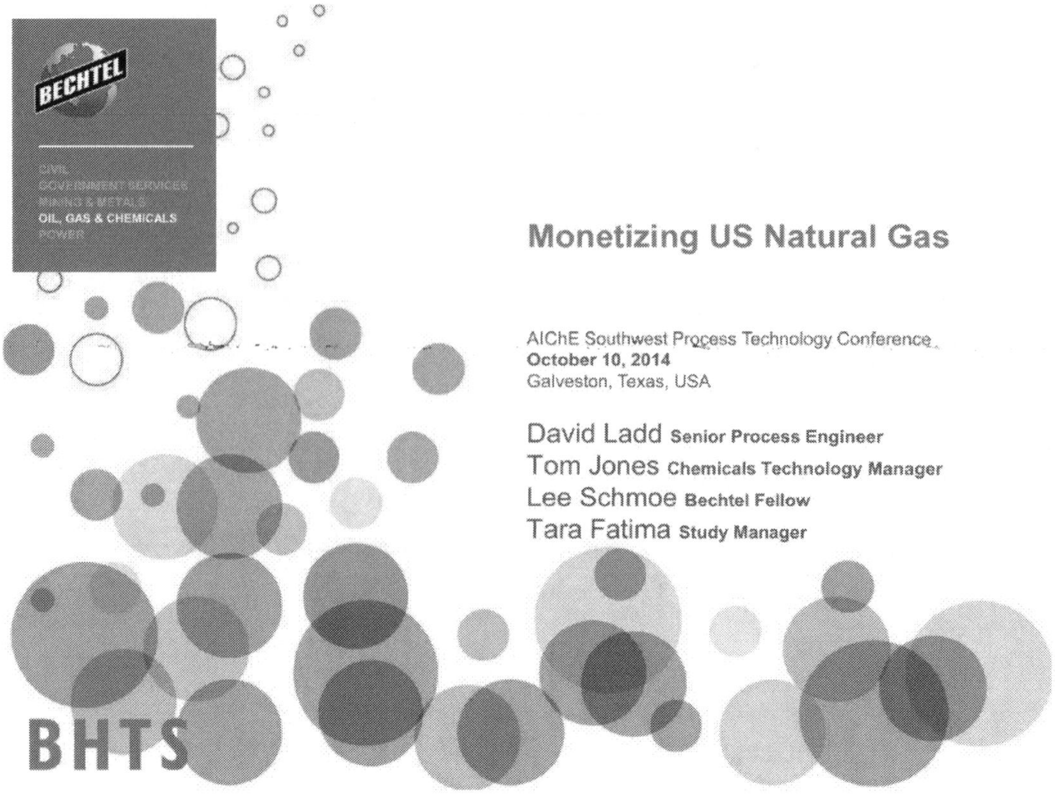
BECHTEL
CIVIL
GOVERNMENT SERVICES
MINING & METALS
OIL, GAS & CHEMICALS
POWER
Monetizing US Natural Gas
AIChE Southwest Process Technology Conference
October 10, 2014
Galveston, Texas, USA
David Ladd Senior Process Engineer
Tom Jones Chemicals Technology Manager
Lee Schmoe Bechtel Fellow
Tara Fatima Study Manager
BHTS

BECHTEL
Outline
Why economics matters for chemical engineers
Economics
Agriculture: Ammonia
Energy: LNG, GTL
Chemicals: Methanol, Ethylene
CHENIERE
Photo Cheniere Energy's Sabine Pass LNG terminal will be the first LNG Export terminal in the US since the last one was constructed at Kenai, AK in 1969.
© 2014 Bechtel Hydrocarbon Technology Solutions 2
MONETIZING US NATURAL GAS

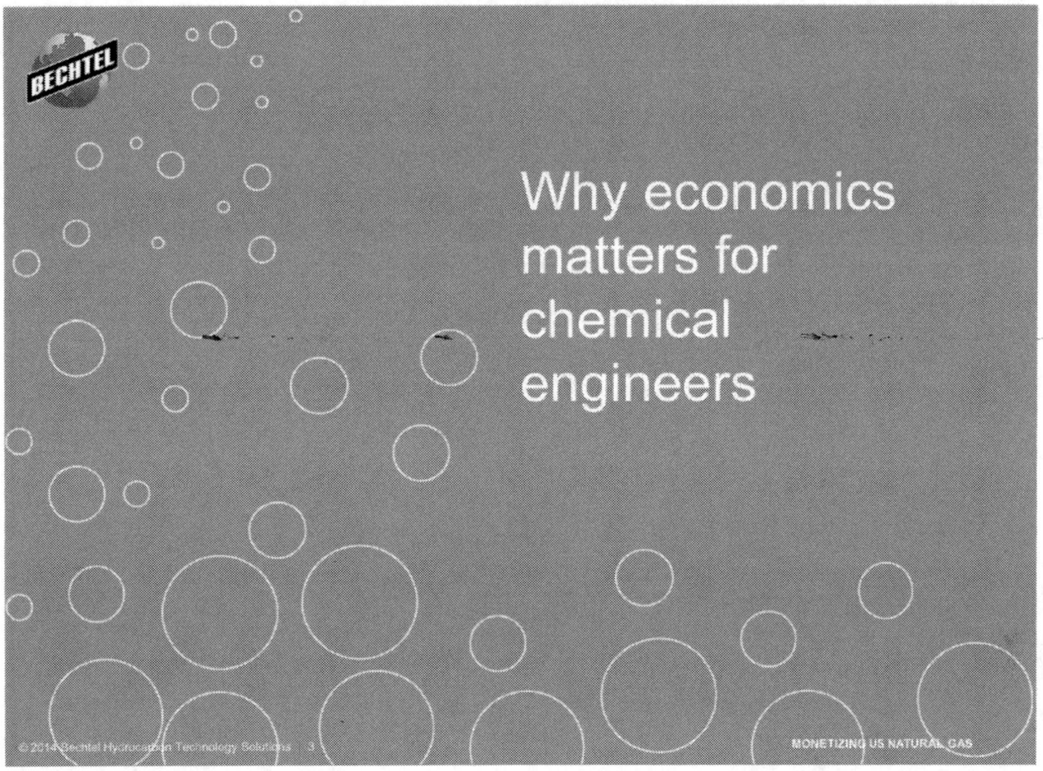
BECHTEL
Why economics matters for chemical engineers
© 2014 Bechtel Hydrocarbon Technology Solutions 3
MONETIZING US NATURAL GAS

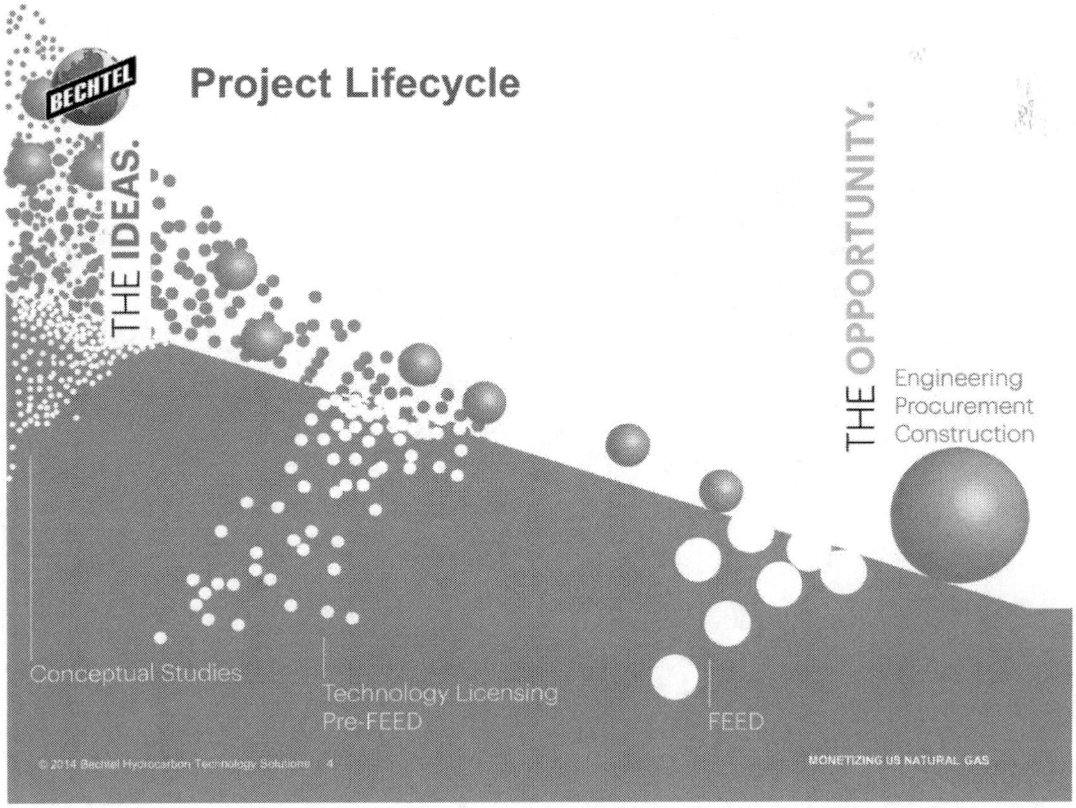
BECHTEL
Project Lifecycle
THE IDEAS.
THE OPPORTUNITY.
Engineering
Procurement
Construction
Conceptual Studies
Technology Licensing
Pre-FEED
FEED
© 2014 Bechtel Hydrocarbon Technology Solutions 4
MONETIZING US NATURAL GAS

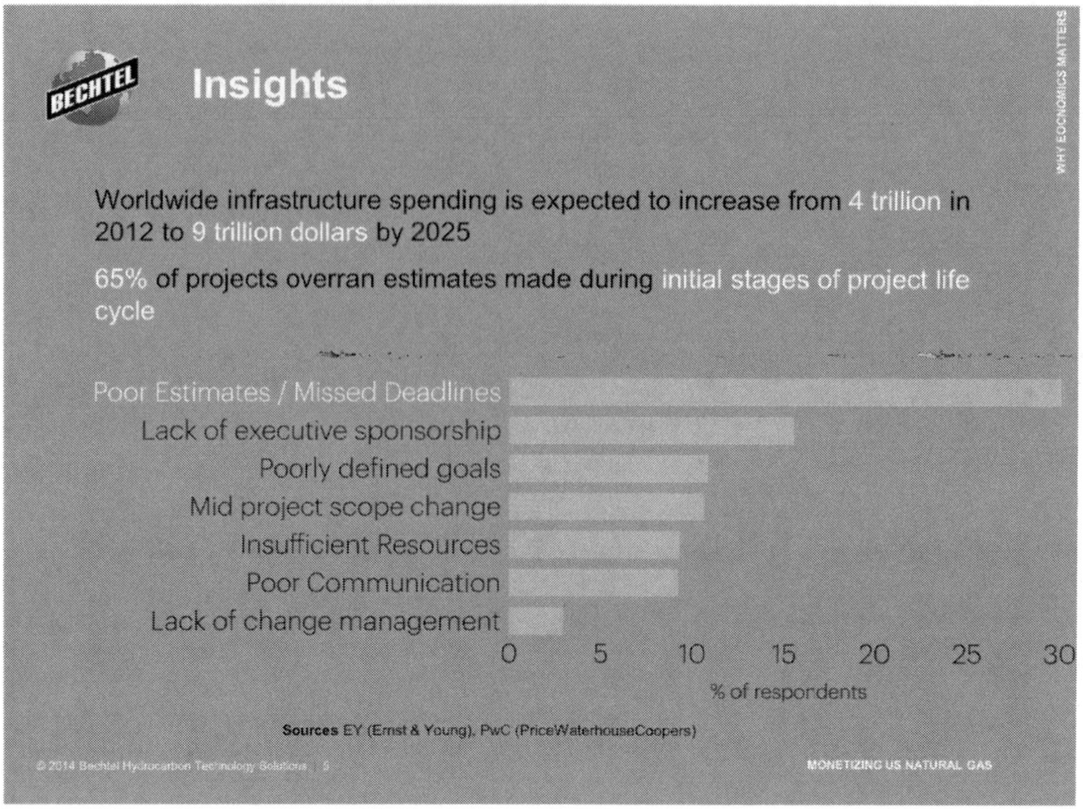

Insights
WHY EOCNOMICS MATTERS
Worldwide infrastructure spending is expected to increase from 4 trillion in 2012 to 9 trillion dollars by 2025
65% of projects overran estimates made during initial stages of project life cycle
Poor Estimates / Missed Deadlines
Lack of executive sponsorship
Poorly defined goals
Mid project scope change
Insufficient Resources
Poor Communication
Lack of change management
0
5
10
15
20
25
30
% of respondents
Sources EY (Ernst & Young), PwC (PriceWaterhouseCoopers)
© 2014 Bechtel Hydrocarbon Technology Solutions | 5
MONETIZING US NATURAL GAS

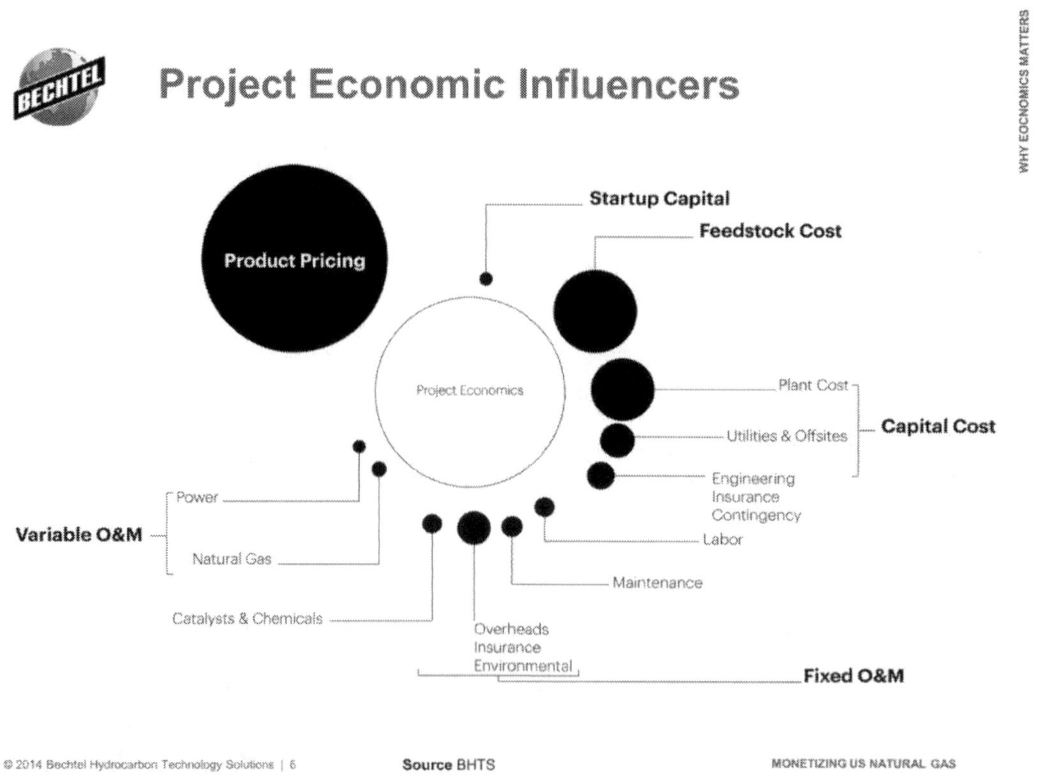

Project Economic Influencers
WHY EOCNOMICS MATTERS
Product Pricing
Startup Capital
Feedstock Cost
Project Economics
Plant Cost
Utilities & Offsites
Engineering
Insurance
Contingency
Capital Cost
Labor
Maintenance
Power
Natural Gas
Variable O&M
Catalysts & Chemicals
Overheads
Insurance
Environmental
Fixed O&M
© 2014 Bechtel Hydrocarbon Technology Solutions | 6
Source BHTS
MONETIZING US NATURAL GAS

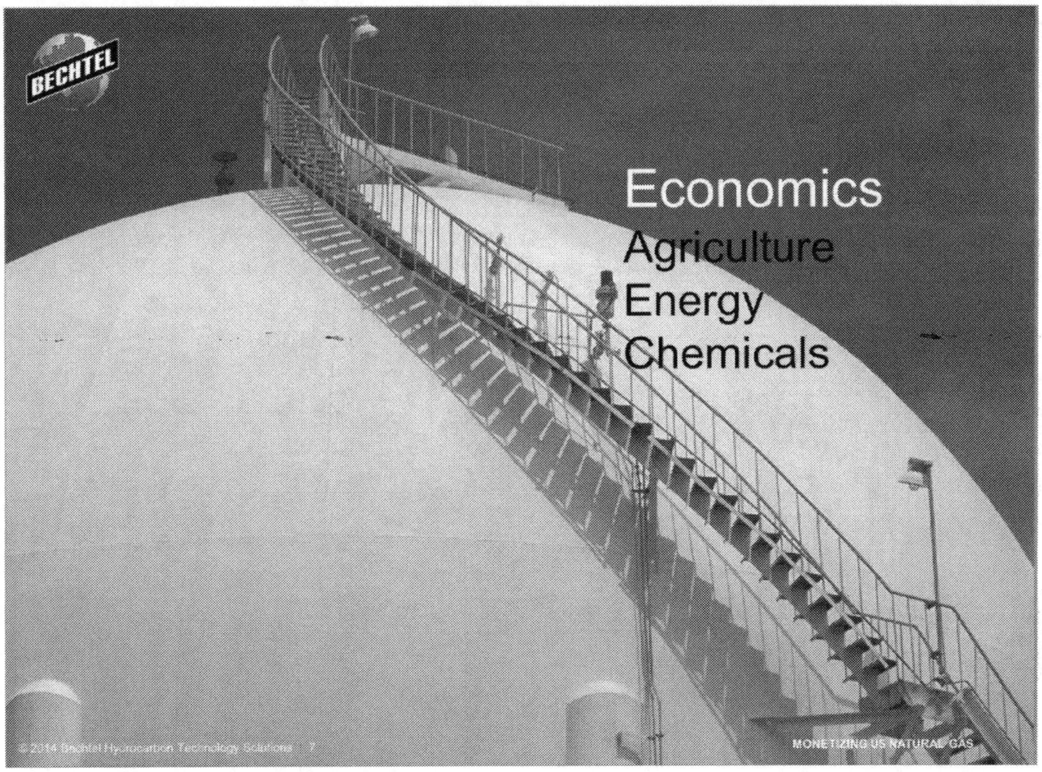
BECHTEL
Economics
Agriculture
Energy
Chemicals
© 2014 Bechtel Hydrocarbon Technology Solutions | 7
MONETIZING US NATURAL GAS

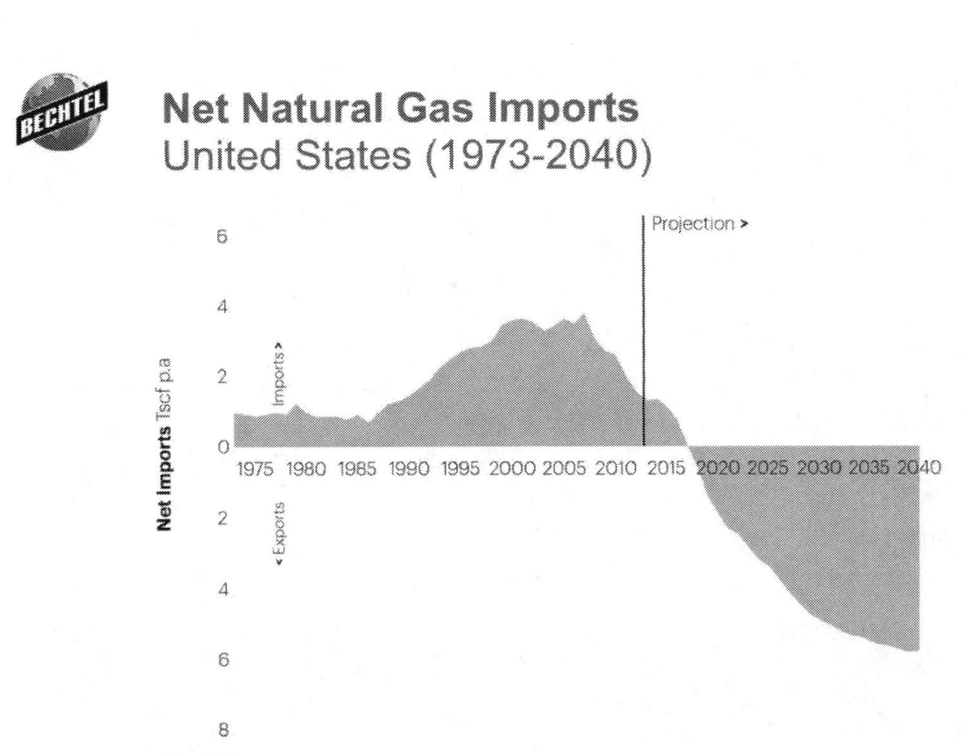
BECHTEL
Net Natural Gas Imports
United States (1973-2040)
Projection >
Net Imports Tscf p.a
Imports >
< Exports
6
4
2
0
2
4
6
8
1975 1980 1985 1990 1995 2000 2005 2010 2015 2020 2025 2030 2035 2040
Source EIA
© 2014 Bechtel Hydrocarbon Technology Solutions | 8
MONETIZING US NATURAL GAS

NATURAL GAS ECONOMICS

Natural Gas Price Forecasts
2014-2020

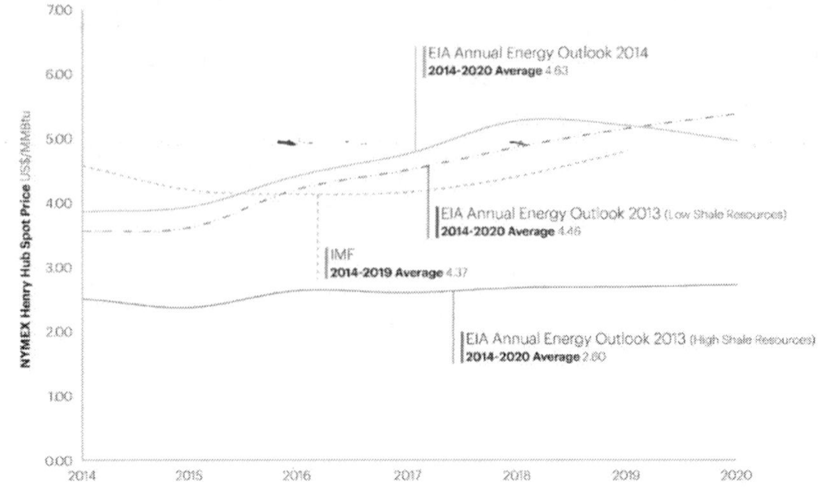

Sources EIA, IMF

MONETIZING US NATURAL GAS

Agriculture
Ammonia

MONETIZING US NATURAL GAS

Terminology

Feedstock Costs / Revenue

Feedstock Cost
Byproduct Revenue

Variable Operations & Maintenance

Utilities & Offsites — Power, Natural Gas

Fixed Operations & Maintenance

Catalysts
Chemicals
Labor
Maintenance
Overheads
Insurance
Environmental

+

Cash Cost

Capital Depreciation

+

Production Cost

Ammonia Economics
Becoming price competitive with Russia

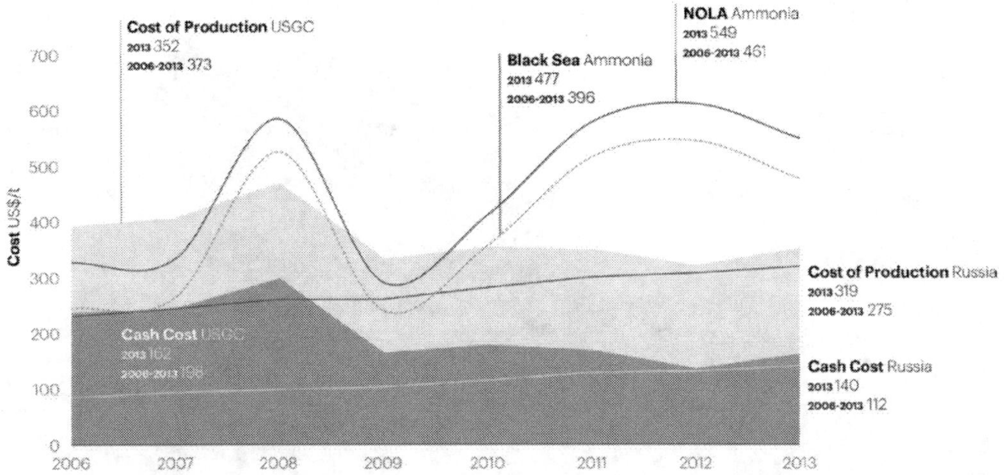

Pricing Sources Yara, ICIS
Economics Source BHTS

Energy

Liquefied Natural Gas (LNG)

Gas to Liquids (GTL)

World LNG Trade

Market dominated by Qatari export, Japanese imports

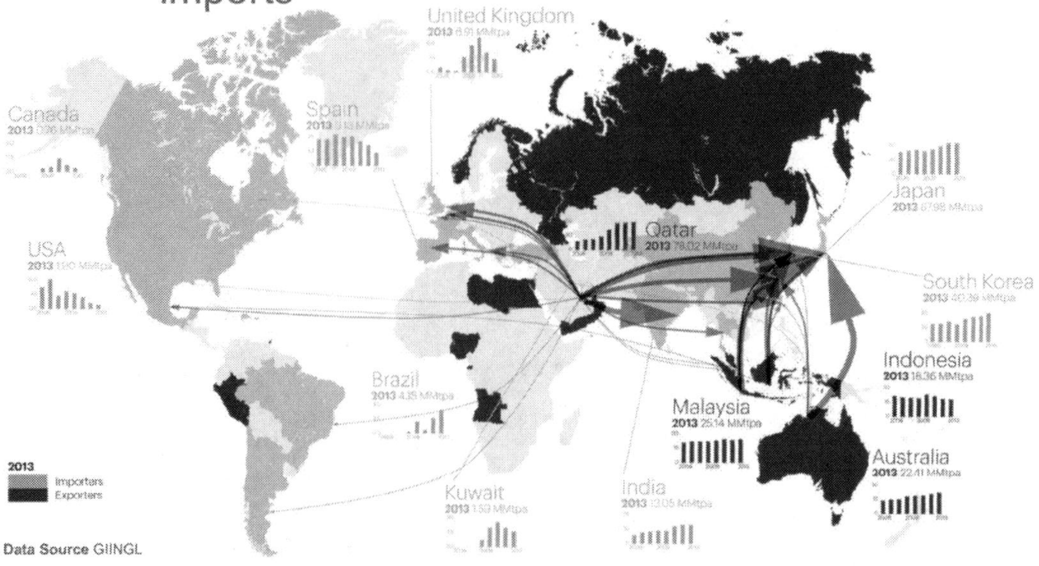

Data Source GIINGL

ENERGY LNG

Economics
North American LNG second lowest cash cost in the world

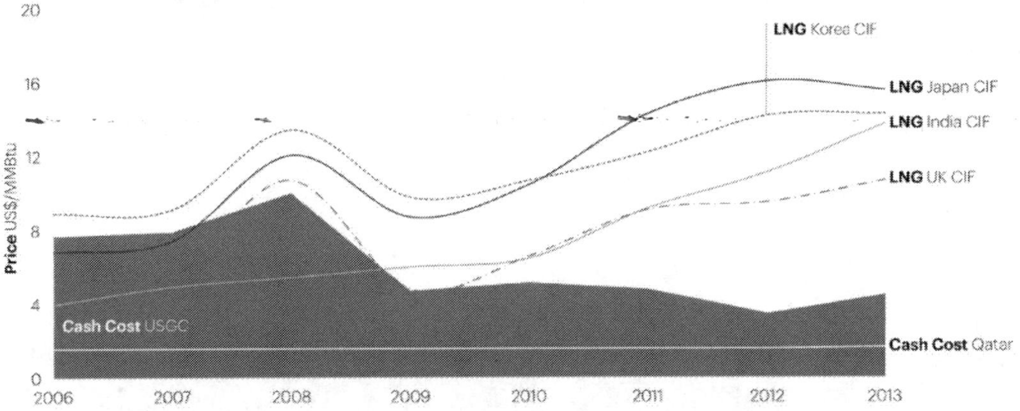

Pricing Source IHS, Bechtel Internal Project Intelligence
Economics Source BHTS

 MONETIZING US NATURAL GAS

ENERGY GTL

Capital Investment
Gas to Transportation Fuels (Diesel)

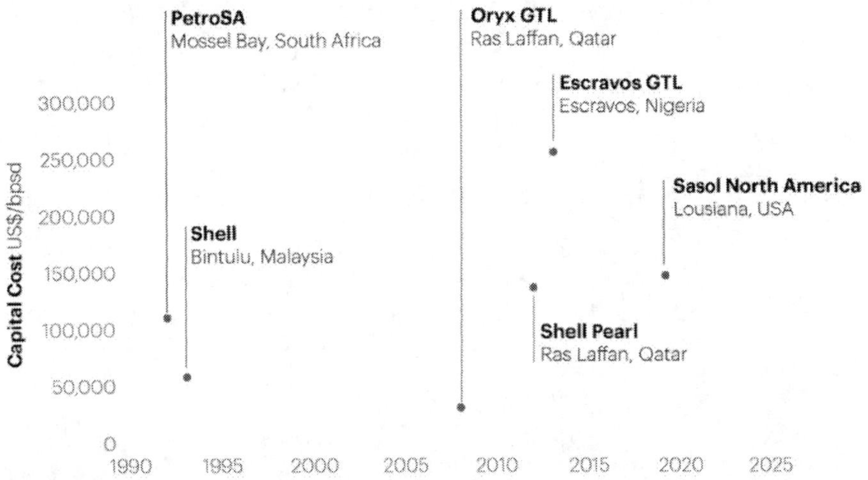

 Source BHTS Project Intelligence

MONETIZING US NATURAL GAS

ENERGY GTL

Gas to Fuels Economics

Spread between natural gas and crude creates an opportunity

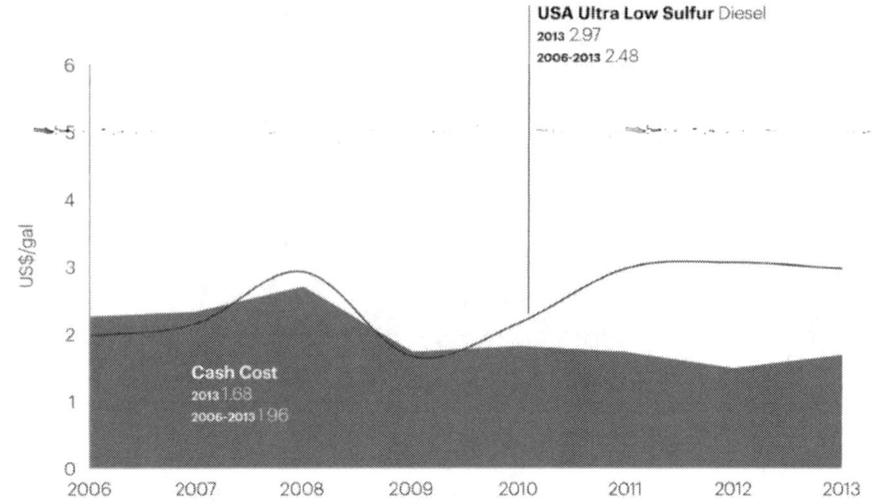

 Source BHTS MONETIZING US NATURAL GAS

Chemicals

Methanol

Ethylene

 MONETIZING US NATURAL GAS

Value Chain
Methanol

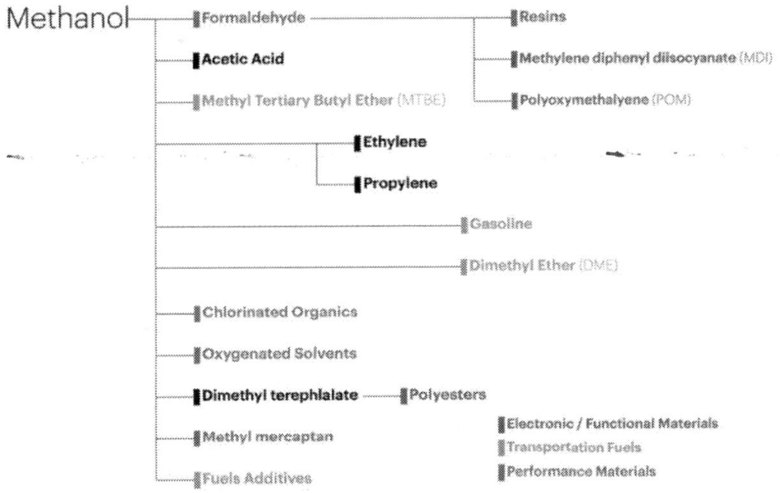

Source Various Sources

Economics
Methanol

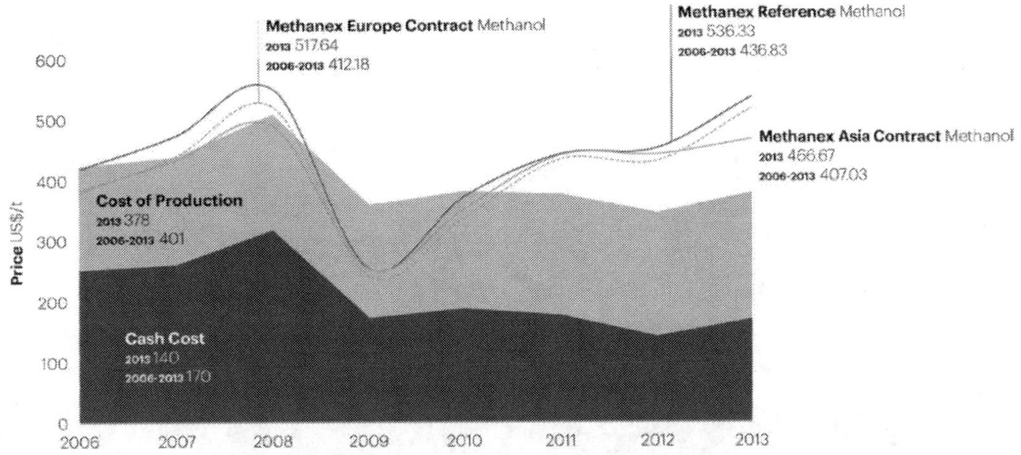

Pricing Source Methanex
Economics Source BHTS Internal Economic Model

Value Chain
Ethylene

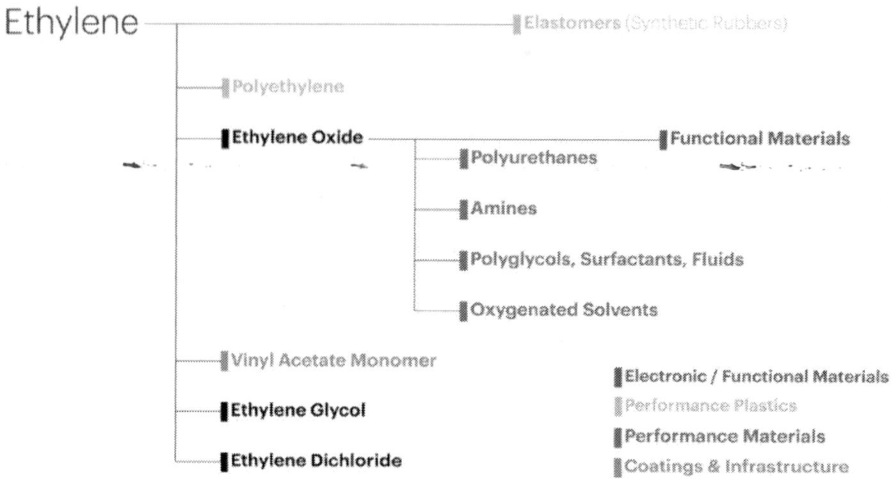

Source The Dow Chemical Company, BASF

Ethylene Production Economics
Two Strategic Shifts

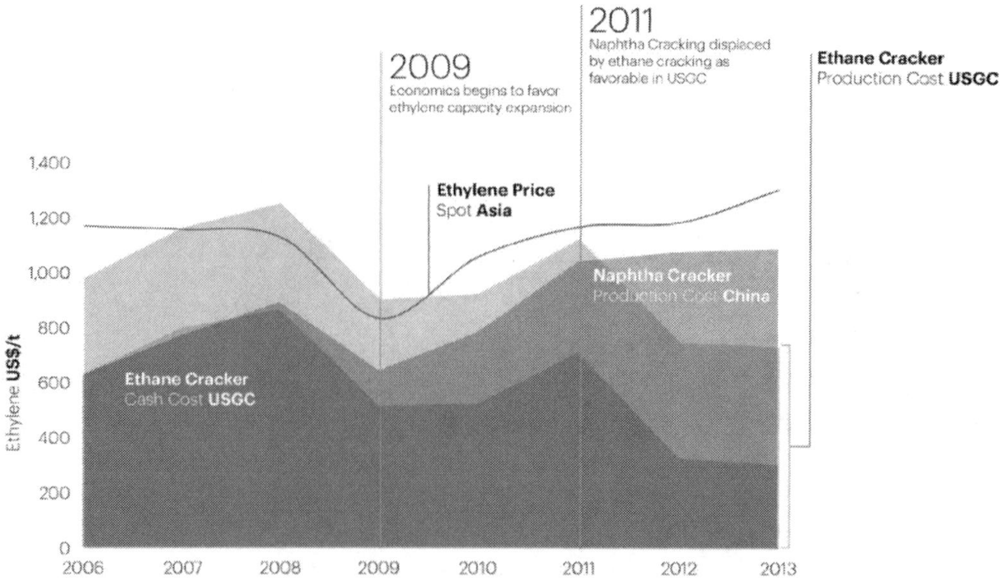

Summary

A structural shift in natural gas supply will lead to increased investment in gas monetization projects in a wave not seen in 30 years.

Conceptual engineering can significantly influence the outcome of major capital projects

Understanding what influences economics is key to maximizing project returns

Each technology is unique. Understanding chemical engineering + financial aspects ensures that the right variables are considered.

Important Notice

This presentation is the exclusive property of Bechtel Hydrocarbon Technology Solutions (BHTS). The conclusions and recommendations presented herein are preliminary in nature and shall not be used as a basis for final financial and management decisions or as a basis of any financial transaction. Any use of or reliance on this presentation or any information contained within is at the using party's sole risk and responsibility. BHTS, to the maximum extent permitted by law, disclaim all liability (including for losses or damages of whatsoever nature) to any third party, howsoever arising with respect to or in connection with the use of this presentation or any information contained within, including any liability caused by negligent acts or omissions.

Q&A

Amp Up Your Engineering Toolkit with Modeling Skills

DOW

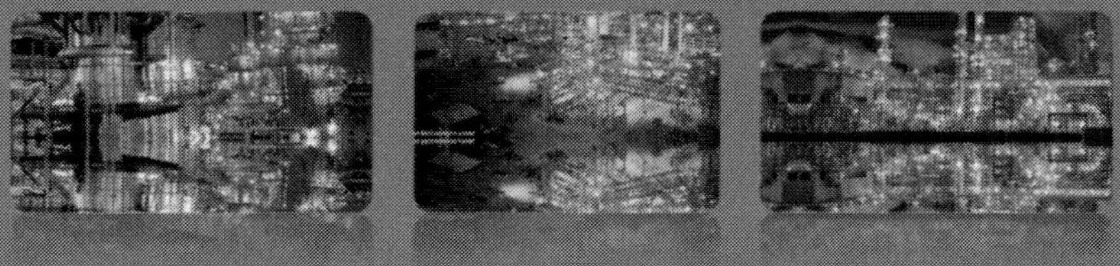

Intan M. Hamdan & Scott J. Bury

Friday, October 10, 2014

2014 AIChE Southwest Process Technology Conference

DOW

What Are Engineers?

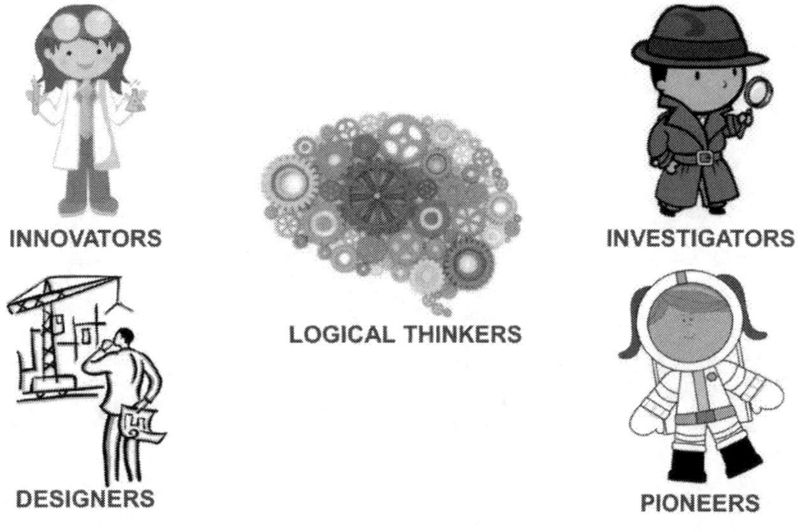

2

WHAT ARE ENGINEERS?

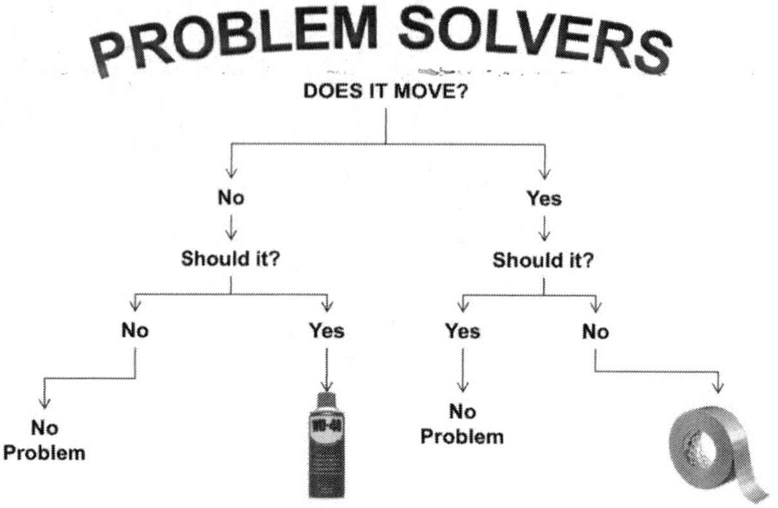

3

CHEMICAL ENGINEERING TRAINING CAMP

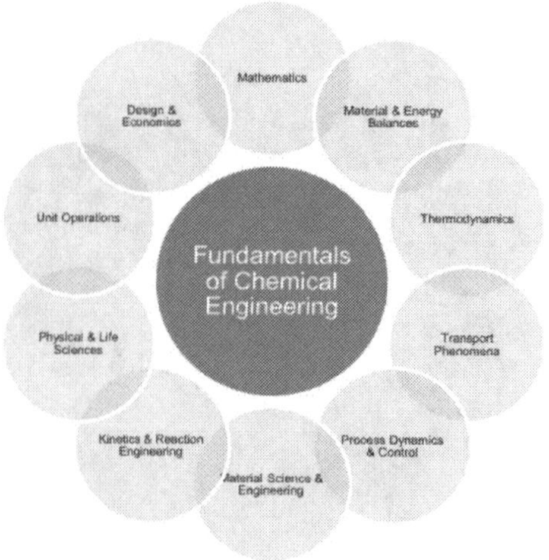

4

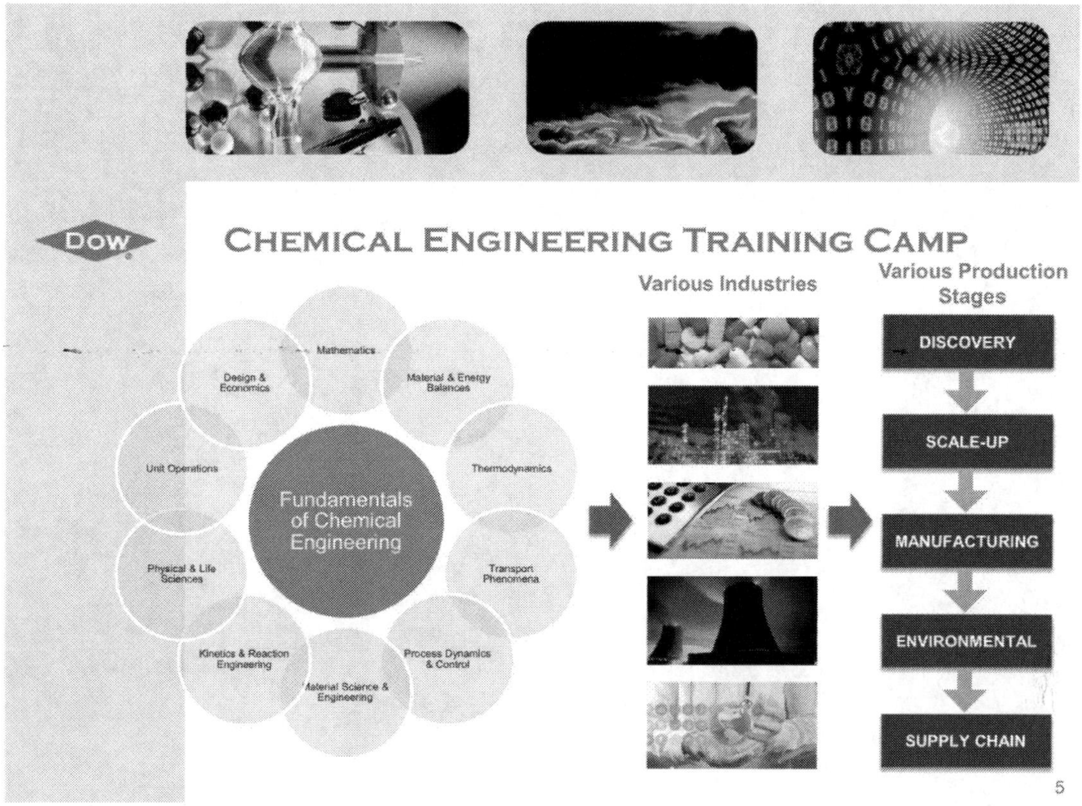
DOW
CHEMICAL ENGINEERING TRAINING CAMP
Various Industries
Various Production Stages
Mathematics
Material & Energy Balances
Thermodynamics
Transport Phenomena
Process Dynamics & Control
Material Science & Engineering
Kinetics & Reaction Engineering
Physical & Life Sciences
Unit Operations
Design & Economics
Fundamentals of Chemical Engineering
DISCOVERY
SCALE-UP
MANUFACTURING
ENVIRONMENTAL
SUPPLY CHAIN
5

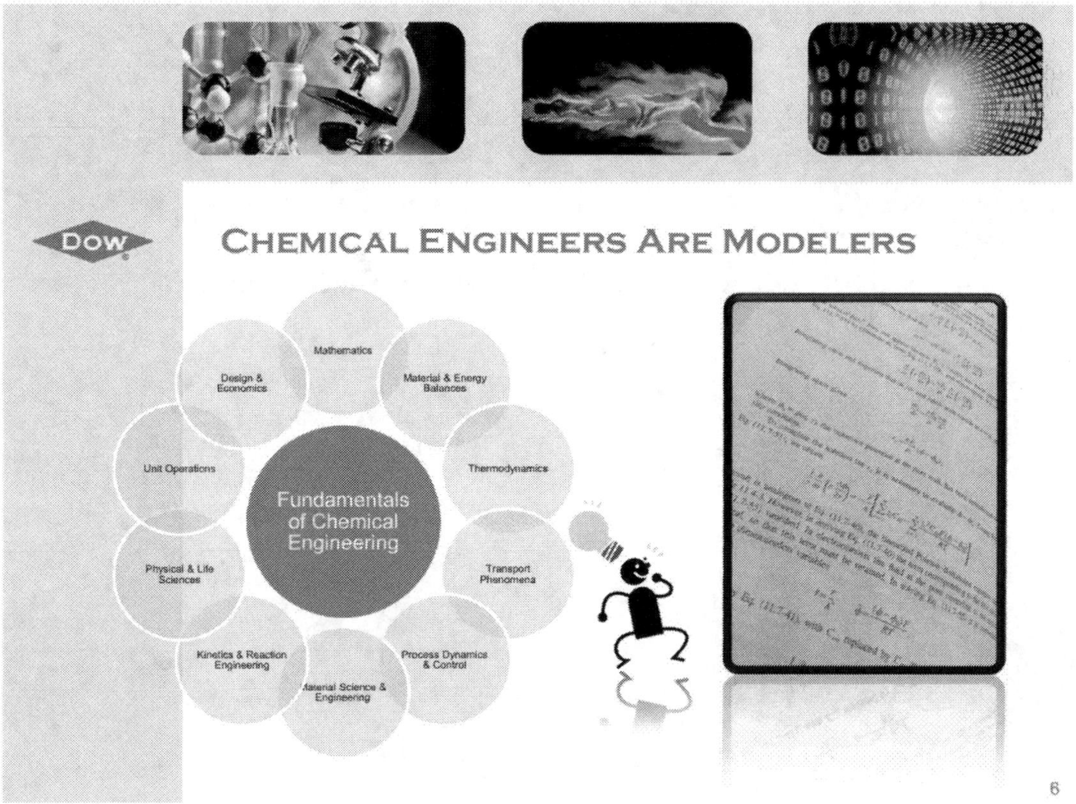
DOW
CHEMICAL ENGINEERS ARE MODELERS
Mathematics
Material & Energy Balances
Thermodynamics
Transport Phenomena
Process Dynamics & Control
Material Science & Engineering
Kinetics & Reaction Engineering
Physical & Life Sciences
Unit Operations
Design & Economics
Fundamentals of Chemical Engineering
6

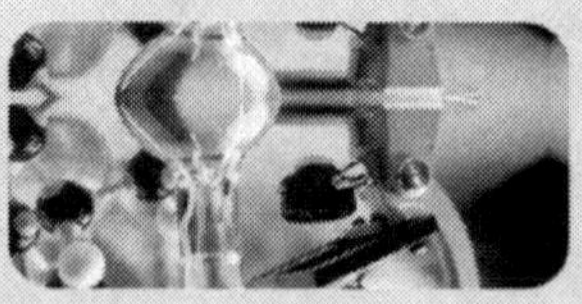
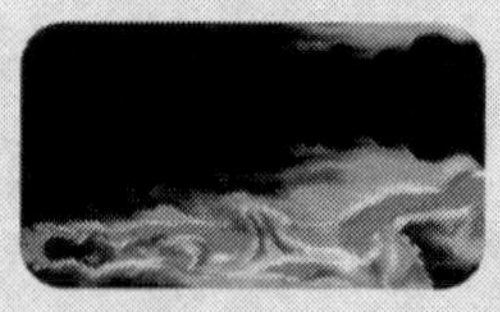

CHEMICAL ENGINEERS ARE MODELERS

Chemical engineers use models all the time:

- Calculating heat transfer coefficients
- Scaling-up a process from bench- to pilot-scale

OBJECTIVE: **Plant the seed for modeling and simulation.**

Increasing the sophistication of your modeling skills allows you to:

(1) Solve more engineering problems

(2) Increase your analytical skills when observing a system

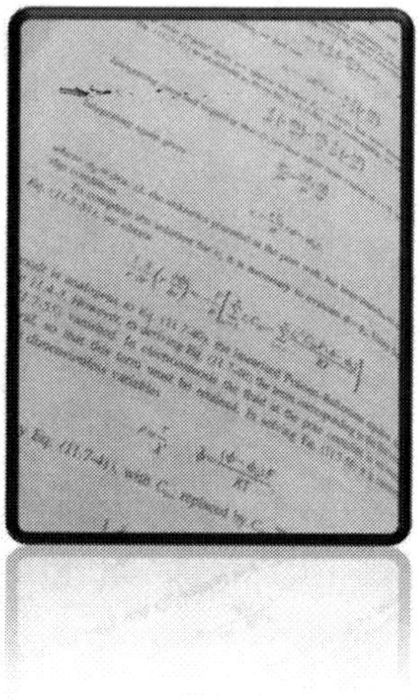

7

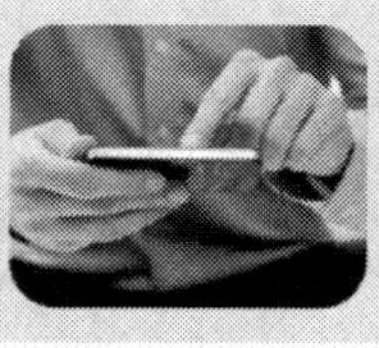

WHO WE ARE

Dow uses modeling and simulations in every step of the process

Combine the power of science and technology to passionately innovate what is essential to human progress

1897 Founded by Herbert H. Dow in Midland, Michigan

~160 Countries in which we serve customers

More than **5,000** products are manufactured at **188** sites in **36** countries

54,000 Approximate Employees worldwide

8

DOW'S DIVERSE & INTEGRATED PORTFOLIO

ELECTRONIC & FUNCTIONAL MATERIALS

- Dow Electronic Materials
  - Semiconductor Tech
  - Interconnect Tech
  - Display Technologies
  - Growth Technologies
- Functional Materials
  - Dow Wolff Cellulosics
  - Dow Consumer & Industrial Solutions
  - Dow Microbial Control
- JV: Dow Corning Corporation

COATINGS AND INFRASTRUCTURE SOLUTIONS

- Dow Building & Construction
- Dow Coating Materials
- Dow Water & Process Solutions
- Performance Monomers
- JV: Dow Corning Corporation

AGRICULTURAL SCIENCES

- Dow AgroSciences
  - Agricultural Chemicals
  - Seeds, Traits & Oils
  - Crop Protection

PERFORMANCE MATERIALS

- Amines
- Chlorinated Organics
- Dow Automotive Systems
- Dow Formulated Systems
- Dow Oil & Gas
- Dow Plastics Additives
- Epoxy
- Oxygenated Solvents
- Polyglycols, Surfactants & Fluids
- Polyurethanes
- Propylene Oxide/Propylene Glycol
- JV: Map Ta Phut
- JV: MEGLOBAL
- JV: Sadara Chemical Company

PERFORMANCE PLASTICS

- Dow Elastomers
- Dow Electrical & Telecommunications
- Dow Hygiene & Medical
- Dow Performance Packaging
- JV: EQUATE Petrochemical Company K.S.C
- JV: The Kuwait Olefins Company K.S.C.
- JV: The SCG-Dow Group
- JV: Sadara Chemical Company

FEEDSTOCK AND ENERGY

- Chlor-Alkali/Chlor-Vinyl
- Energy
- Ethylene Oxide/Ethylene Glycol
- Hydrocarbons
- JV: EQUATE Petrochemical Company K.S.C
- JV: The SCG-Dow Group
- JV: The Kuwait Olefins Company K.S.C

9

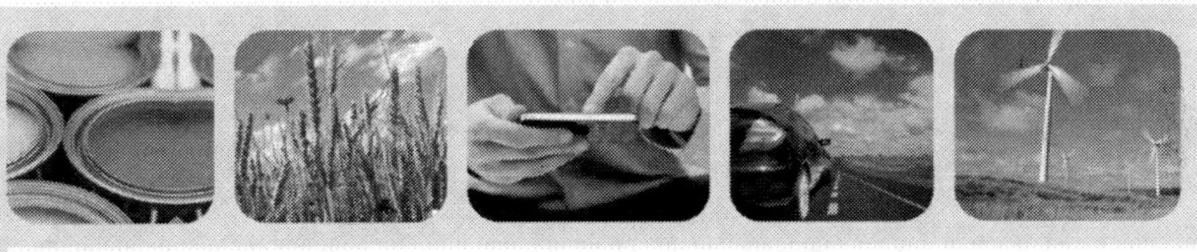

SIMULATION TOUCH POINTS IN CHEMICAL PRODUCTION

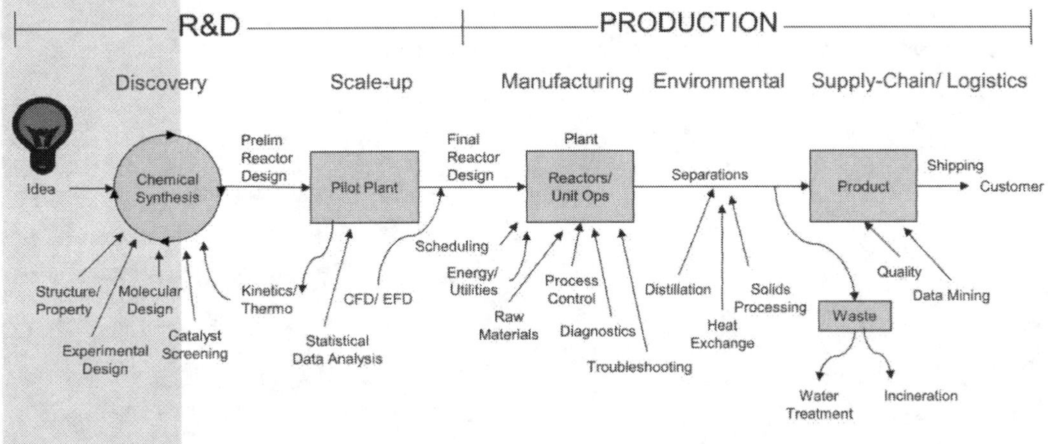

Adapted from *Modeling Challenges in Process Development (Chemical Industry Perspective)*, Alex Kalos, Presented at a two-day workshop organized by the Pharmaceutical Industry and the Council for Chemical Research, University of Maryland Biotechnology Institute (UMBI-Shady Grove, Rockville, MD), June 7-8 (2007). See CRI DOC 19723.

10

Simulation Touch Points in Chemical Production

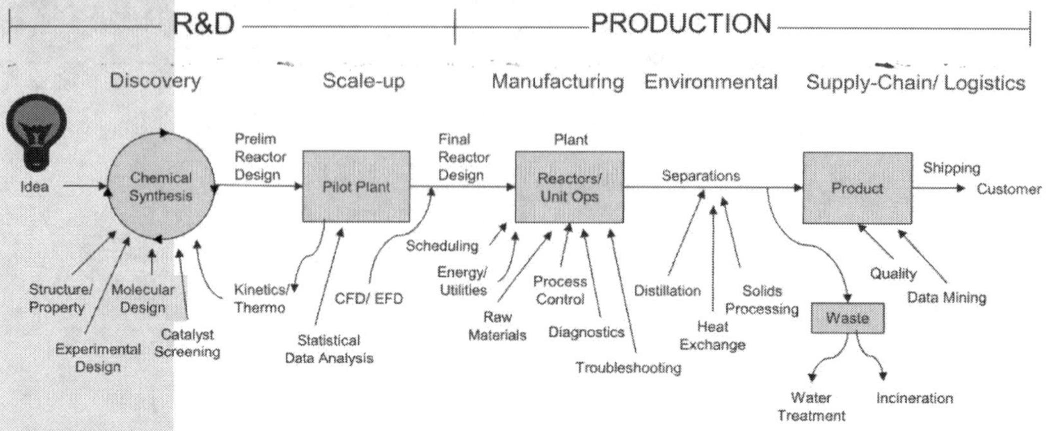

Adapted from *Modeling Challenges in Process Development (Chemical Industry Perspective)*, Alex Kalos. Presented at a two-day workshop organized by the Pharmaceutical Industry and the Council for Chemical Research, University of Maryland Biotechnology Institute (UMBI-Shady Grove, Rockville, MD), June 7-8 (2007). See CRI DOC 19723.

11

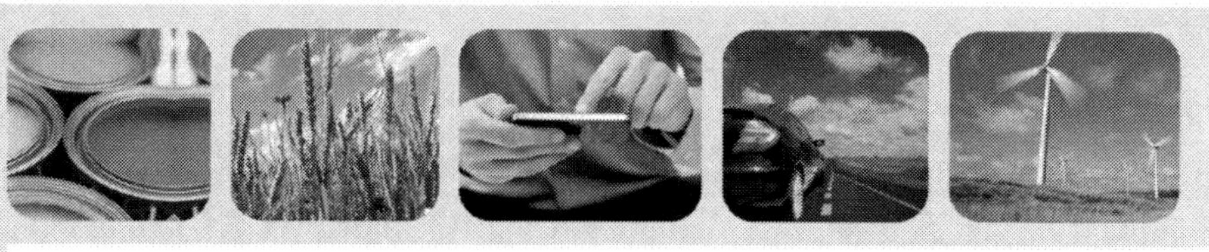

Computer-Aided Molecular Design

Objective: Develop, automate, and apply a systematic computer-aided molecular design framework that will identify molecules satisfying design specifications.

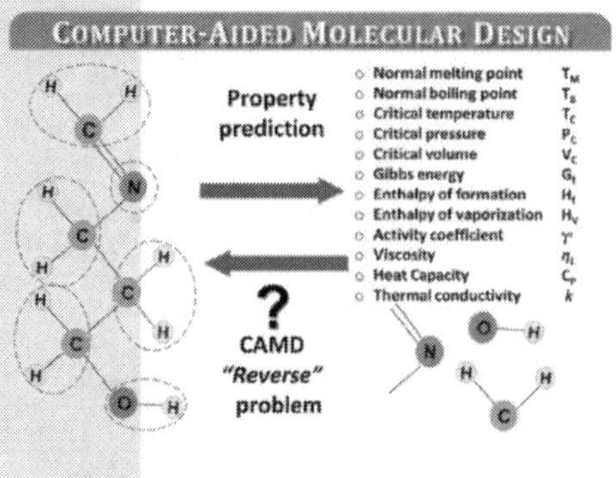

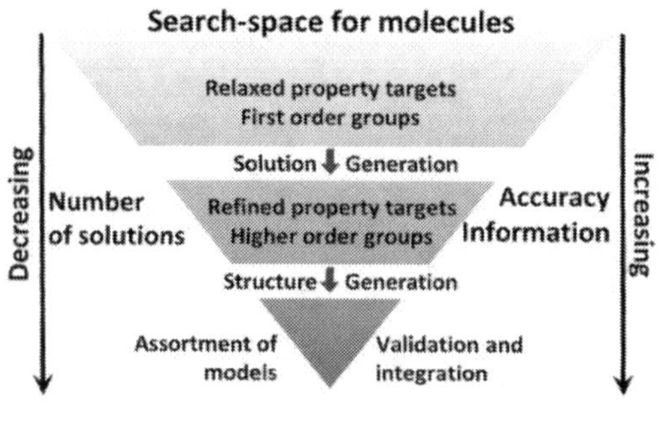

Austin, Samudra, Trahan, Sahinidis. *"A computational methodology for designing mixtures"*, AIChE Annual Meeting (2013)

12

SIMULATION TOUCH POINTS IN CHEMICAL PRODUCTION

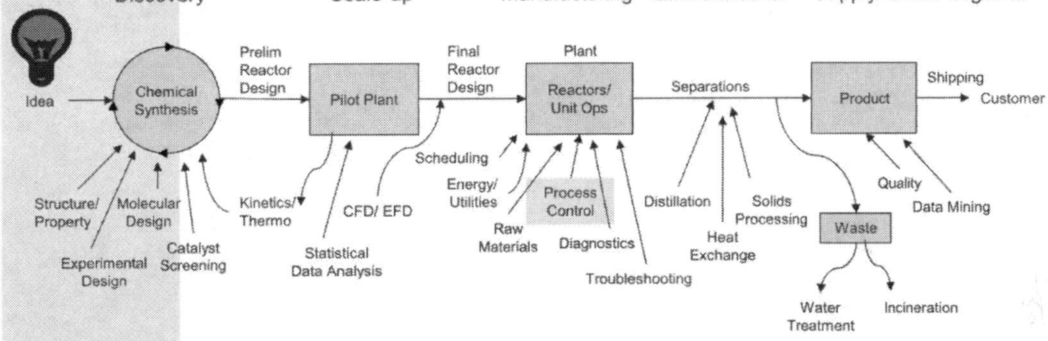

Adapted from *Modeling Challenges in Process Development (Chemical Industry Perspective)*, Alex Kalos, Presented at a two-day workshop organized by the Pharmaceutical Industry and the Council for Chemical Research, University of Maryland Biotechnology Institute (UMBI-Shady Grove, Rockville, MD), June 7-8 (2007). See CRI DOC 19723.

13

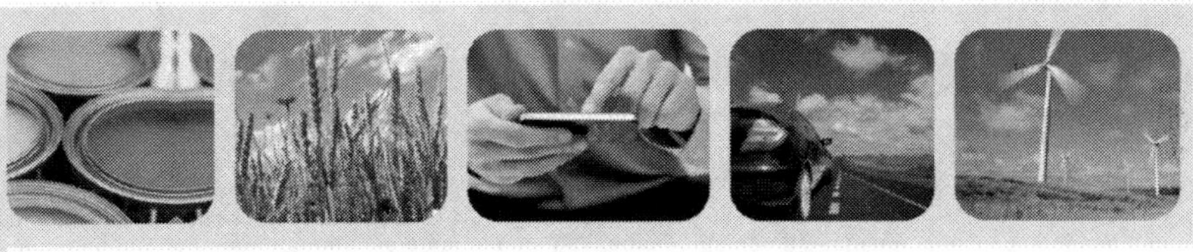

TRANSITION OPTIMIZATION OF SOLUTION POLYMERIZATION PROCESS

Objective: Develop an optimization framework that specifies optimal transition policy such that transition time and off-spec production are reduced.

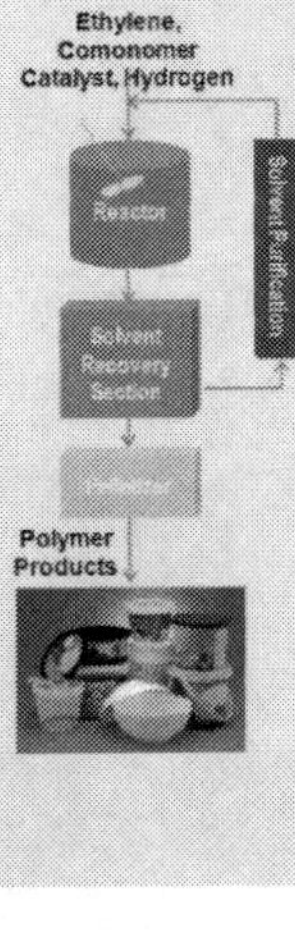

Min Off-spec + Trans. Time + Regularizing Term

Dynamic model

Grade A in-spec in Stage1

Grade B in-spec in Stage3

Continuity between stages

Initial and Final Conditions

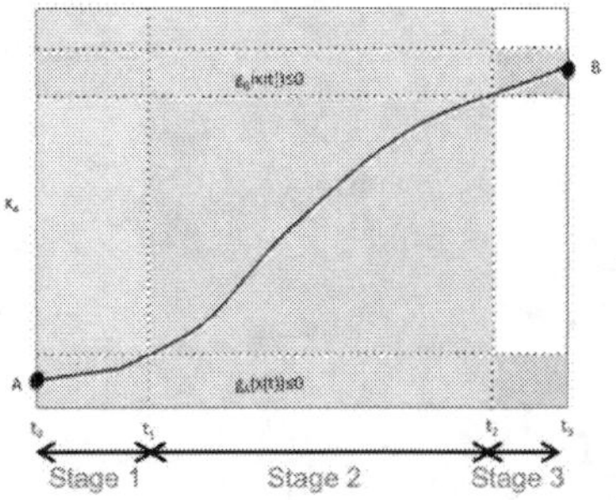

Shi, Hamdan, Munjal, Biegler. "*Model-based control for optimal transitions in polyethylene solution polymerization*", AIChE Annual Meeting (2013)

14

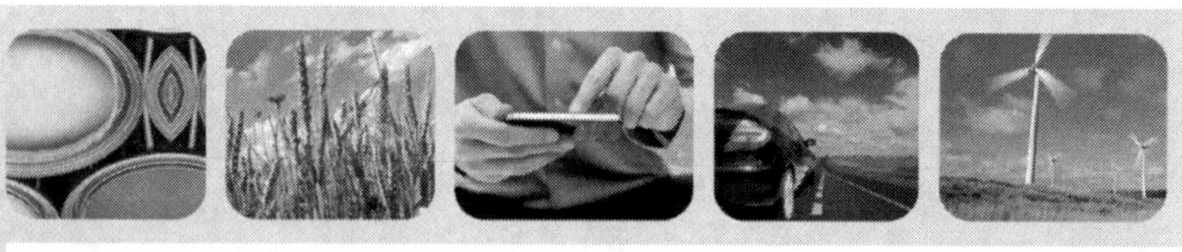

SIMULATION TOUCH POINTS IN CHEMICAL PRODUCTION

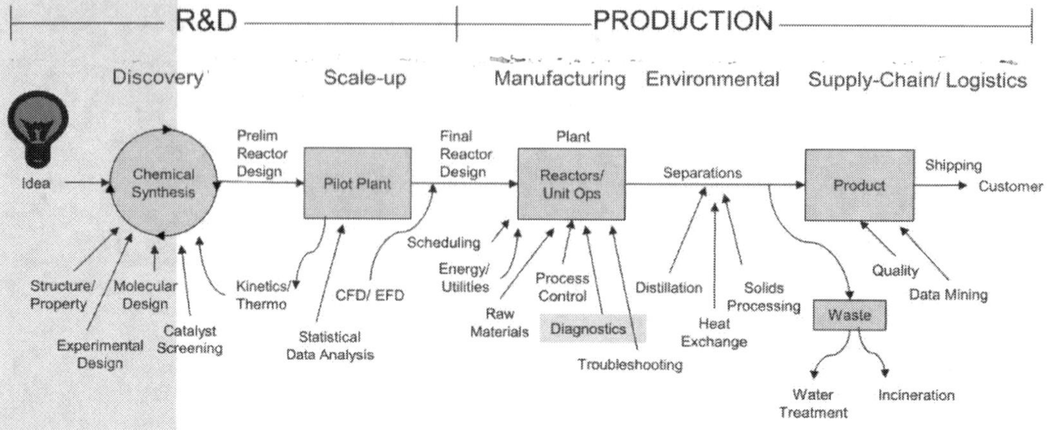

Adapted from *Modeling Challenges in Process Development (Chemical Industry Perspective)*, Alex Kalos. Presented at a two-day workshop organized by the Pharmaceutical Industry and the Council for Chemical Research, University of Maryland Biotechnology Institute (UMBI-Shady Grove, Rockville, MD), June 7-8 (2007). See CRI DOC 19723.

15

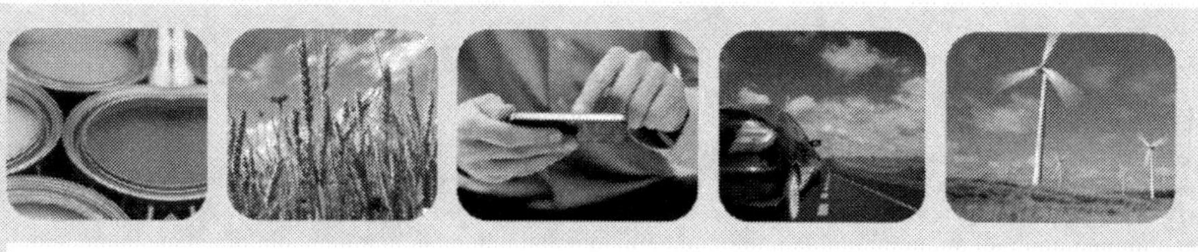

EXCEPTIONAL EVENTS MANAGEMENT

Objective: Develop EEM framework for real-time implementation on continuous tablet manufacturing process in order to mitigate incipient exceptional events and ensure normal operating conditions

Exceptional Events Management Framework Implemented with Moving Window

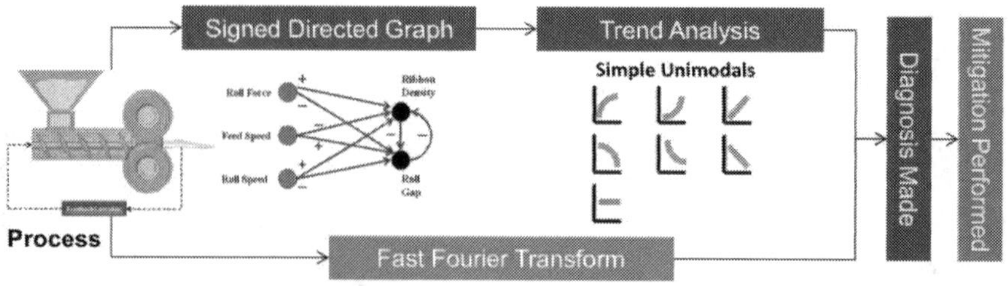

Hamdan, Reklaitis, Venkatasubramanian. "*Real-time exceptional events management for a partial continuous dry granulation line*", Journal of Pharmaceutical Innovation (2012)

16

Simulation Touch Points in Chemical Production

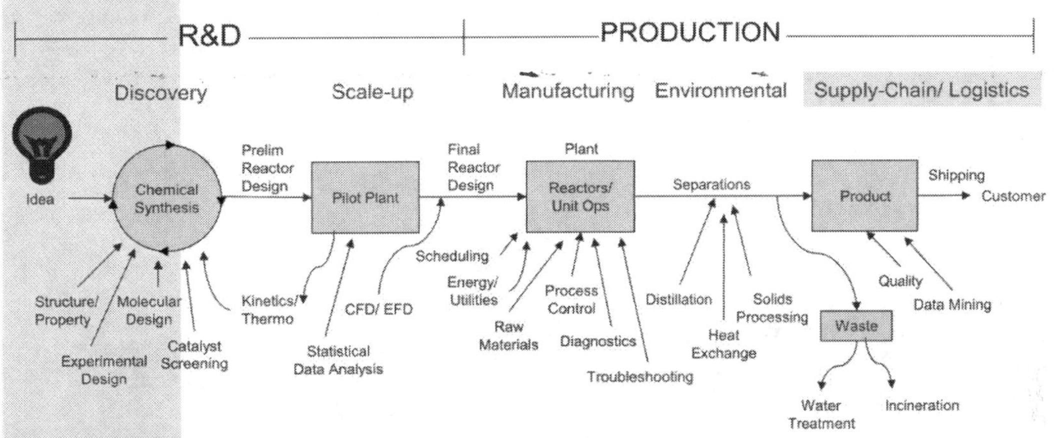

Adapted from *Modeling Challenges in Process Development (Chemical Industry Perspective)*, Alex Kalos, Presented at a two-day workshop organized by the Pharmaceutical Industry and the Council for Chemical Research, University of Maryland Biotechnology Institute (UMBI-Shady Grove, Rockville, MD), June 7-8 (2007). See CRI DOC 19723.

17

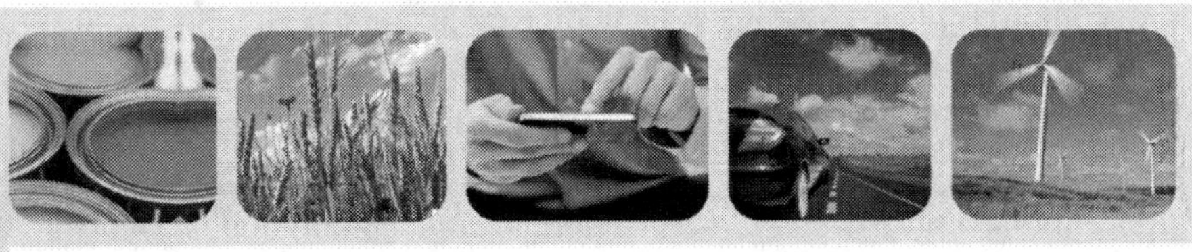

FMTO Model

Objective: Facilitate key capital decisions regarding marine terminal operations by evaluating the impact of various scenarios on terminal asset utilization.

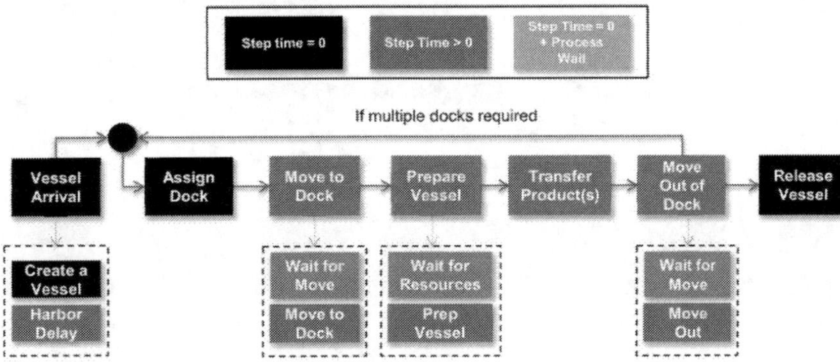

18

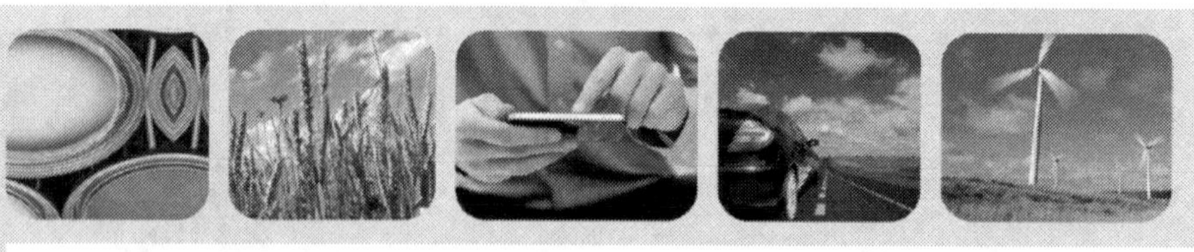

Modeling is Not Always the Right Tool

WHY YOU NO RUN?!!

99 little bugs in the code
99 little bugs in the code
Take one down, patch it around
117 little bugs in the code

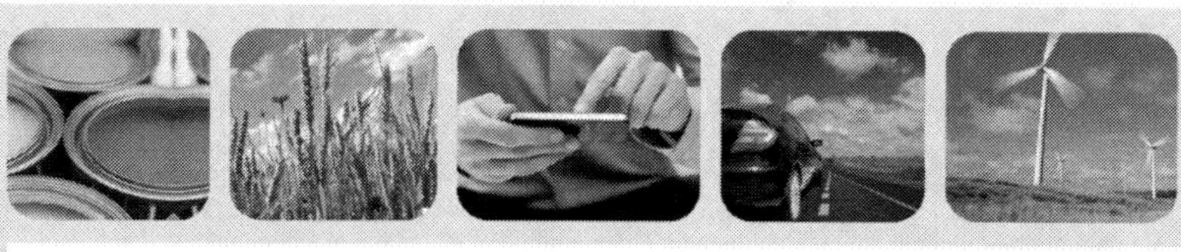

Flowchart to Help You Determine when Modeling and Simulation is Appropriate

"To a hammer, everything looks like a nail". **Don't be a hammer**

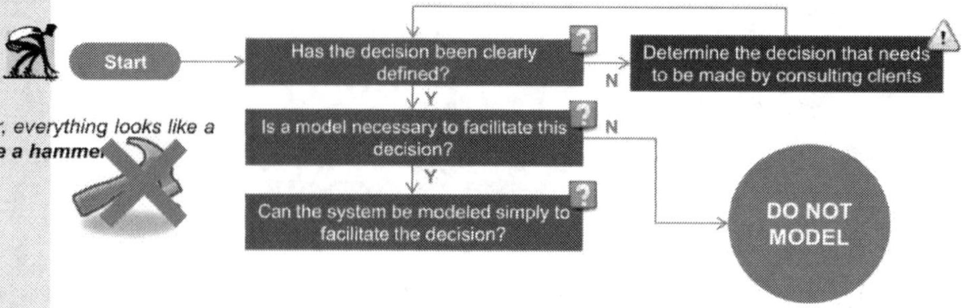

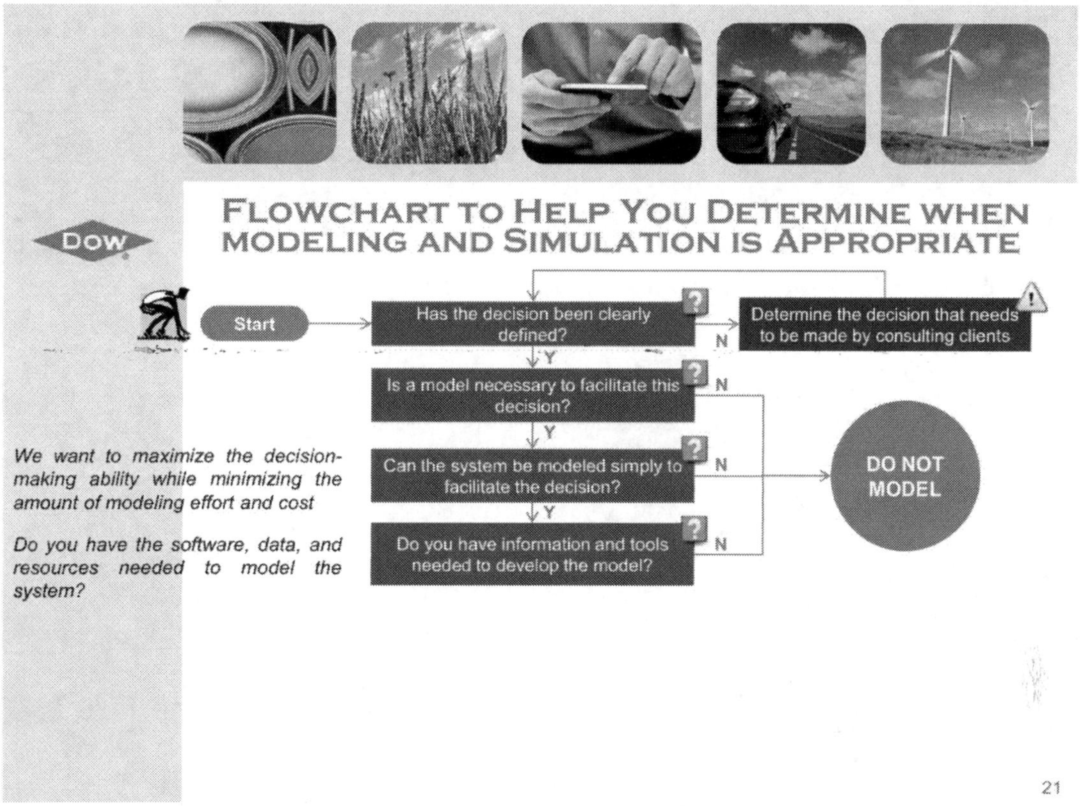
FLOWCHART TO HELP YOU DETERMINE WHEN MODELING AND SIMULATION IS APPROPRIATE
DOW
Start
Has the decision been clearly defined?
Determine the decision that needs to be made by consulting clients
N
Y
Is a model necessary to facilitate this decision?
N
Y
Can the system be modeled simply to facilitate the decision?
N
Y
Do you have information and tools needed to develop the model?
N
DO NOT MODEL
We want to maximize the decision-making ability while minimizing the amount of modeling effort and cost
Do you have the software, data, and resources needed to model the system?
21

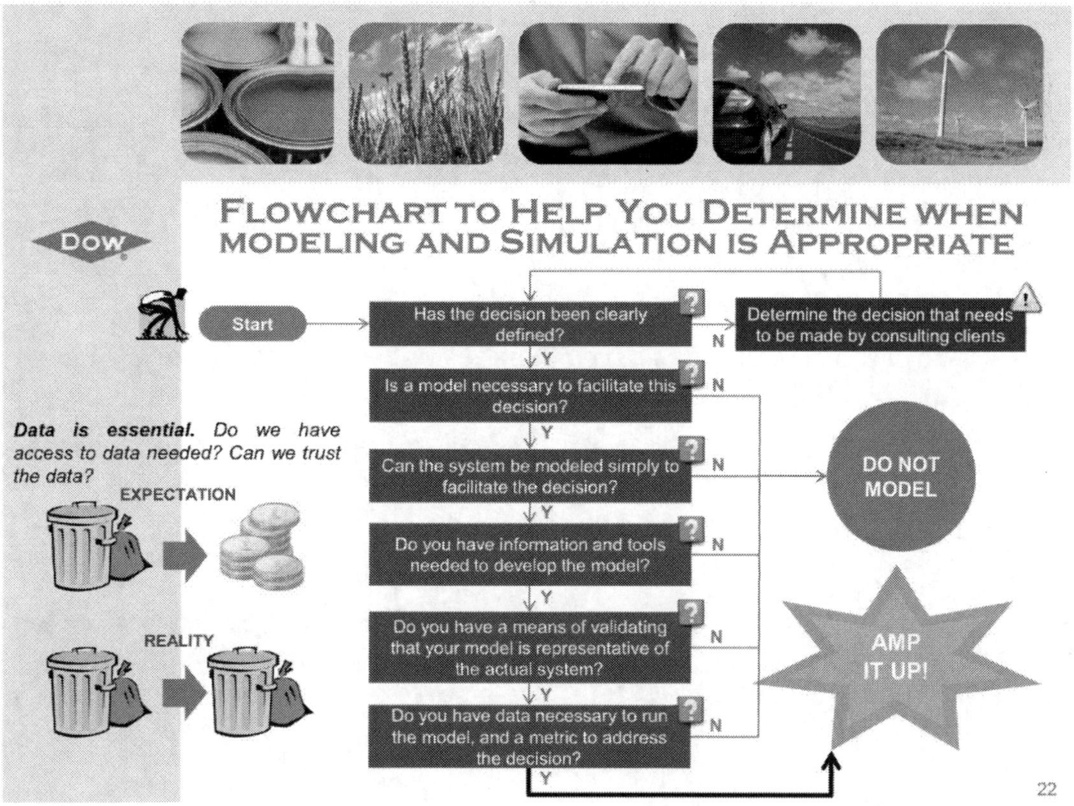
FLOWCHART TO HELP YOU DETERMINE WHEN MODELING AND SIMULATION IS APPROPRIATE
DOW
Start
Has the decision been clearly defined?
Determine the decision that needs to be made by consulting clients
N
Y
Is a model necessary to facilitate this decision?
N
Y
Can the system be modeled simply to facilitate the decision?
N
Y
Do you have information and tools needed to develop the model?
N
Y
Do you have a means of validating that your model is representative of the actual system?
N
Y
Do you have data necessary to run the model, and a metric to address the decision?
N
Y
DO NOT MODEL
AMP IT UP!
***Data is essential.** Do we have access to data needed? Can we trust the data?*
EXPECTATION
REALITY
22

THINGS TO KEEP IN MIND

1. The model objective must be clearly defined to enable correct scenario evaluations
2. Model must sufficiently capture the details that are to be evaluated… but not be too complex
3. Model validation is essential – "All models are bad, but some are useful"
4. Metrics chosen must be relevant to the decision being made or the question being answered
5. Meaningful interpretations of the results must be made, within the context of the decision

23

SUMMARY

- ❑ Engineers are modelers by training
- ❑ Models are developed across all industries for various stages of production
- ❑ Increasing the level of modeling sophistication enhances our engineering toolkit
- ❑ Modeling trains us to consider the details of the system that would otherwise be overlooked
- ❑ **CAUTION**: ***Modeling is not always the appropriate tool.*** Learning *when* to apply models is equally as important as knowing *how* to develop models.

24

Thanks To These Engineers with Amped Up Modeling Abilities...

whose models have been highlighted in this presentation

Apurva Samudra
Daniel Trahan
Intan Hamdan
Jun Shi
Larry Biegler
Naoko Akiya
Nick Austin
Nick Sahinidis
Scott Bury

25

Thank You!

For Your Attention

QUESTIONS

Process Safety in Engineering Design

Chaitanya Gottimukkala

RISKbytes Inc

Speaker Bio

- Chaitanya Gottimukkala is a Senior Risk Engineer at RISKbytes Inc.
- Over 6 years of progressive experience in technical safety, fire safety, and process engineering design activities for various Oil and Gas Clients involving Refining, Petrochemical and LNG facilities.
- M.E.S degree in Chemical Engineering from Lamar University, TX.
- An active STS-YP member for four years, and
- Vice-Chair of STS AIChE's Young Professionals Section for 2014.

RISKbytes

Presentation Overview

- What is Process Safety
- Process Safety – A Few Definitions
- Catastrophic Incident Process Safety Perspective
- Process Risk Management Design Categories
- Case Studies

RISKbytes

What is Process Safety

- A discipline that focuses on prevention of hazardous releases, fires and explosions throughout the lifecycle of a processing facility
- Applies formal assessments to critical areas of plant design, construction, installation, maintenance and operations

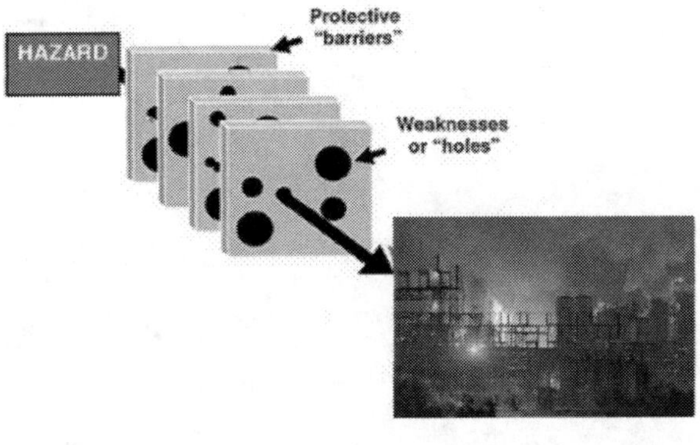

Process Safety – A Few Definitions

Term	Definition
Accident	A sequence of events occurring that produce unintended harm to people and property
Incident	The loss of containment of material or energy; not all incidents result in accidents
Hazard	An inherent chemical or physical property that has the potential to cause harm to people, property, or the environment
Risk	Measure of magnitude of loss times the likelihood of the incident to occur
Risk Analysis	Quantitative estimate of risk based utilizing engineering evaluation and mathematical techniques for calculating risk

Source: Chemical Process Safety Fundamentals with Applications, Daniel Crowl, Joseph Louvar, 3rd Edition

Process Safety Through the Facility Lifecycle

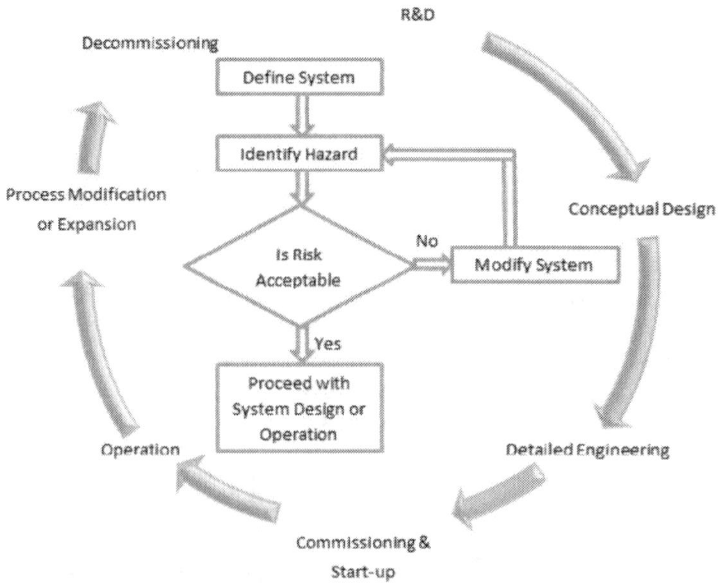

Process Safety throughout the Lifecycle of a Facility

Source: CCPS's Guidelines for Engineering Design for Process Safety, Second Edition

RISKbytes

Process Safety Through the Facility Lifecycle

Robust Management systems are critical in managing process safety through the facility lifecycle including, but not limited to:

- Conformance with Design Standards, Industry Codes and Standards
- Mechanical Integrity
- Understanding the Hazards of routine operations
  - Management of Change
  - Process Hazards Analysis
  - Incident and Near-Miss Investigation
- Safety Culture and Employee Involvement
- Management Review and Continuous Improvement

RISKbytes

Total Process Safety Incidents resulting in Injuries

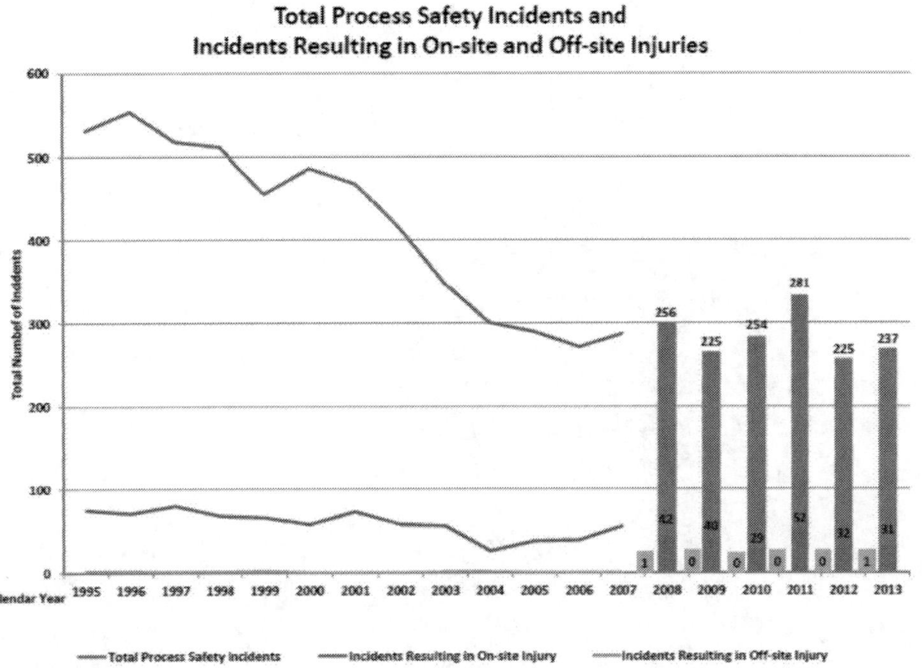

Source: American Chemistry Council Website

RISKbytes

Process Safety Incidents in the USA – Severity Index

Severity Level (Note 4)	Safety/Human Health (Note 5)	Fire or Explosion (including overpressure)	Potential chemical impact (Note 3)	Community/ environment impact (Note 5)
NA	Does not meet or exceed Level 4 threshold	Does not meet or exceed Level 4 threshold	Does not meet or exceed Level 4 threshold	Does not meet or exceed Level 4 threshold
4 (1 point used in severity rate calculations for each of the attributes that apply to the incident)	Injury requiring treatment beyond first aid to employee or contractors (or equivalent, Note 1) associated with a process safety incident (In U.S., incidents meeting the definitions of an OSHA recordable injury).	Resulting in $25,000 to $100,000 of direct cost	Chemical released within secondary containment or contained within the unit - see Note 2A	Short -term remediation to address acute environmental impact. No long term cost or company oversight. Examples would include spill cleanup, soil and vegetation removal.
3 (3 points used in severity rate calculations for each of the attributes that apply to the incident).	Lost time injury to employee or contractors associated with a process safety event	Resulting in $100,000 to $1 million of direct cost.	Chemical release outside of containment but retained on company property OR flammable release without potential for vapor cloud explosives - see Note 2B	Minor off-site impact with precautionary shelter-in-place OR Environmental remediation require with cost less than $1 million. No other regulatory oversight required. OR Local media coverage
2 (9 points used in severity rate calculations for each of the attributes that apply to the incident).	On-site fatality - employee or contractors associated with a process safety event; multiple lost time injuries or one or more serious offsite injuries associated with a process safety event.	Resulting in $1 million to $10 million of direct cost.	Chemical release with potential for injury off site or flammable release resulting in a vapor cloud entering a building or potential explosion site (congested/confined area) with potential for damage or casualties if ignited - see Note 2C.	Shelter-in-place or community evacuation. OR Environmental remediation required and cost in between $1 million - $ 2.5 million. State government investigation and oversight of process. OR Regional media coverage or brief national media coverage.
1 (27 points used in severity rate calculations for each of the attributes that apply to the incident).	Off-site fatality or multiple on-site fatalities associated with a process safety event.	Resulting in direct cost greater than $10 million.	Chemical release with potential for significant on-site or off-site injuries or fatalities - see Note 2D.	National media coverage over multiple days. OR Environmental remediation required and cost in excess of $2.5 million. Federal government investigation and oversight of process.

Source: American Chemistry Council Website

RISKbytes

Process Safety Incidents in the USA - 2013

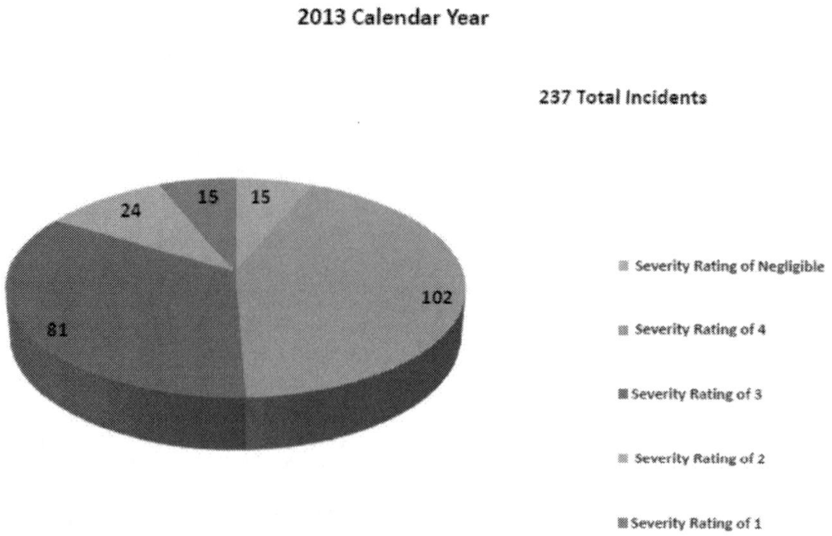

Source: American Chemistry Council Website

Catastrophic Accident: Process Safety Perspective - Bhopal

- December 3, 1984: Bhopal, Madhya Pradesh, India
- Poisonous cloud of 15 metric tons of methyl isocyanate (MIC), covering an area of more than 30 square miles leaked from Union Carbide India Limited
- More than 2000 people died instantly and over 300,000 injured

Catastrophic Accident: Process Safety Perspective - Bhopal

- Cause: Gas leaked as a result of an exothermic reaction of about 500 gallons (2,000 litres) of water entering MIC Tank 610
- 5 Process Safety Layers of Protection Failed:
  - refrigeration unit to minimize excess temperature,
  - high temperature alarm on Tank 610,
  - caustic soda scrubber for vent gases,
  - flare to combust MIC exiting the scrubber, and
  - water curtain to absorb any unburned MIC

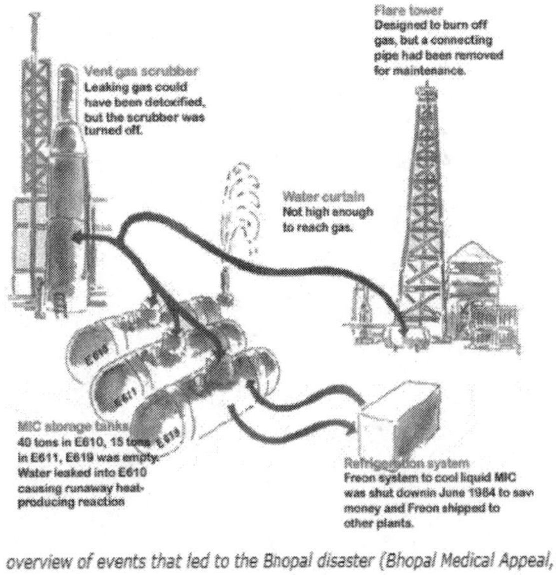

overview of events that led to the Bnopal disaster (Bhopal Medical Appeal, 2002)

Catastrophic Accident: Process Safety Perspective - Bhopal

- Key Findings from the Investigation
  - Inadequate Hazard Evaluation
  - Improper Planning in location of the facility storing large quantities of highly toxic chemical
  - Improper maintenance of Safety Critical Elements
  - No Management of Change
  - Lack of Technical Authorities
  - Personnel Competency
  - Inadequate Emergency Response

Process Risk Management Design Categories

Management of Process hazards and/or risks can be typically classified into the following categories:

- Inherently safe (Minimize, Substitute, Moderate, Simplify)
  - Substituting Hypochlorite for chlorine in water treatment
  - Minimizing hazardous chemical inventory
- Passive Mitigation
  - Increasing the design pressure of a vessel to operate below safe limits
  - Facility Layout and spacing
  - Dikes
- Active Mitigation
  - Pressure Relief Valve
  - Safety Instrumented Systems (SIS)
- Administrative Controls
  - Operating procedures
  - Emergency Response

Design Categories Cost Comparison

Comparison of Cost and Functional Attributes of Design Categories

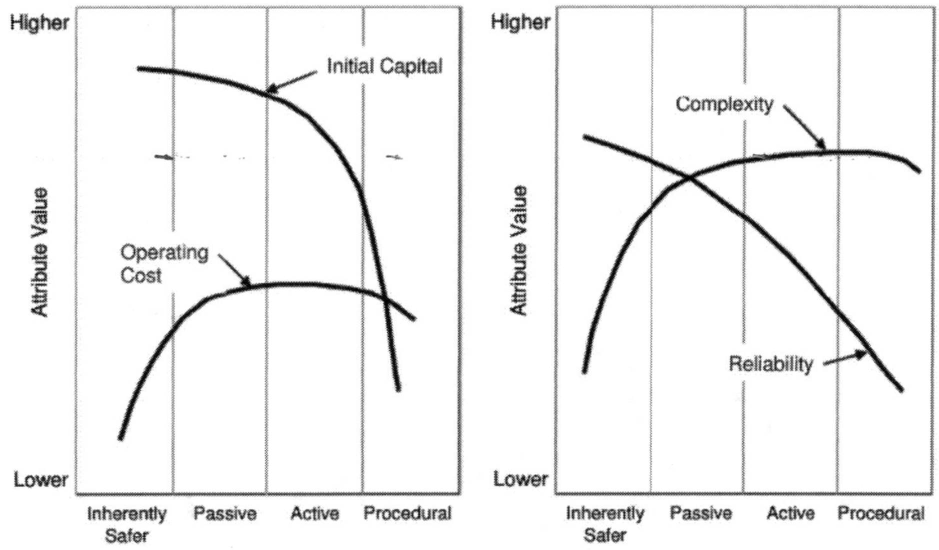

Source: CCPS Guidelines for Engineering Design for Process Safety, 2nd edition

Onion Model for Process Safeguards

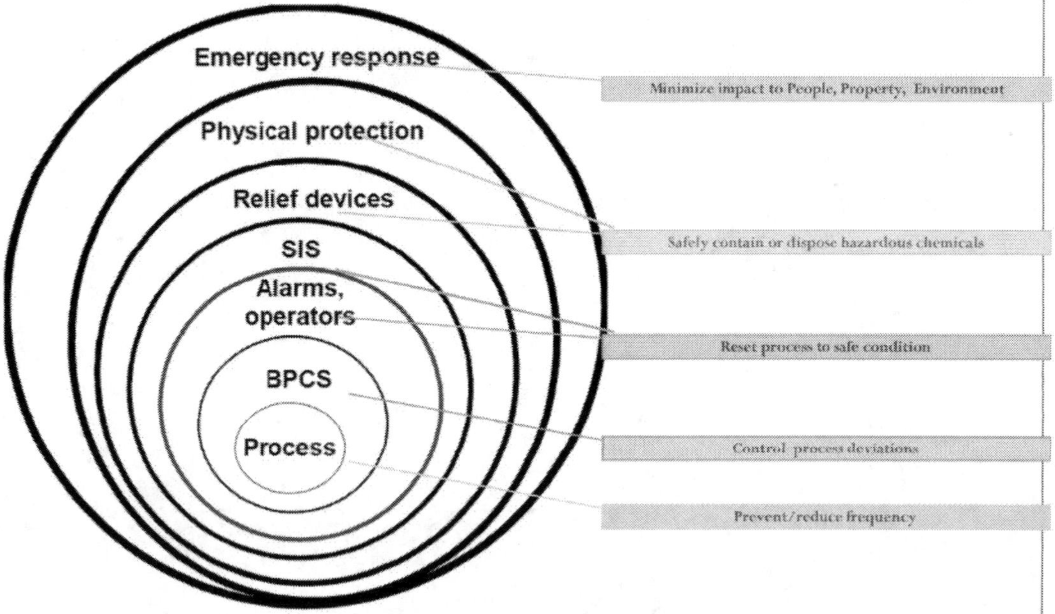

Source: ISA-84.00.01 Part 3

Case Study 1: Tank Overfill

Scenario: Overfill of Gasoline Tank (e.g. Buncefield Explosion, 2005)

Cause: Malfunction of Level Transmitter

Consequence: Potential loss of containment of flammable material leading to potential fire, explosion resulting in injury, property damage

Safeguards: Manual Block valve with operator action; SIS on high level

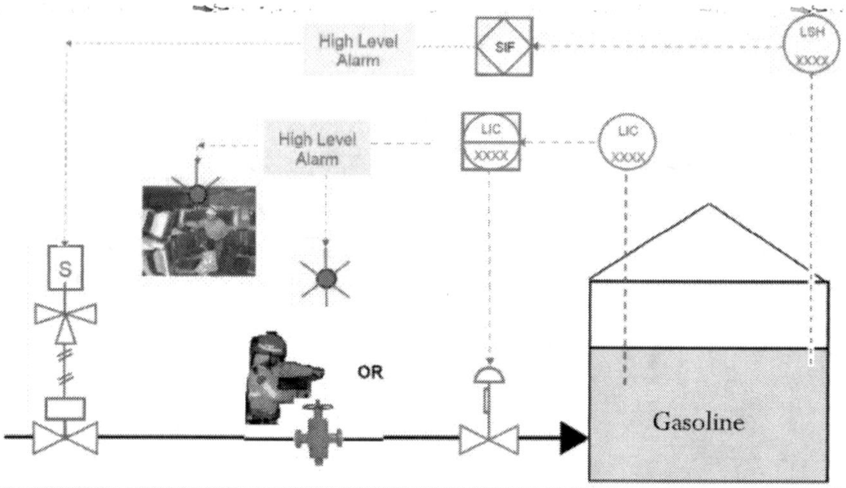

Case Study 2: Pump Flange Failure

Scenario: Pump Discharge Flange failure (e.g. Refinery Crude Unit)

Cause: Discharge Flange Failure

Consequence: Potential loss of containment of flammable material above AIT leading to potential fire, explosion resulting in injury, property damage

Safeguards: Manual Block valve with operator action

Recommended Safety Measures: Installing automatic suction isolation valve to shut off on flame detection; Install water spray system to cool adjacent equipment

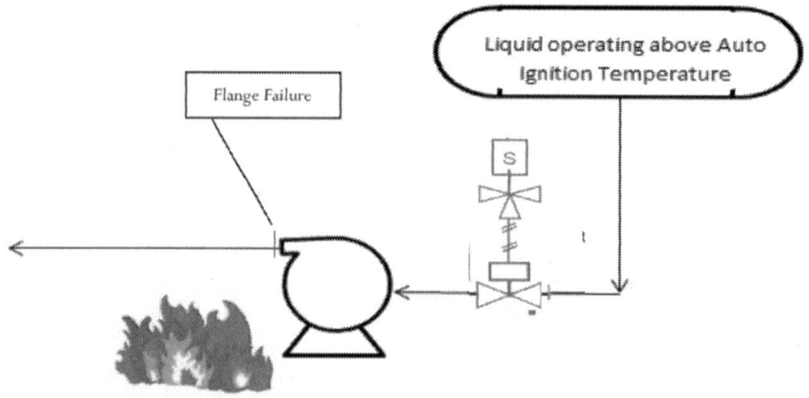

References

- CCPS Guidelines for Engineering Design for Process Safety, 2nd edition
- Chemical Process Safety Fundamentals with Applications, Daniel Crowl, Joseph Louvar, 3rd Edition
- American Chemistry Council, http://responsiblecare.americanchemistry.com
- http://www.lenntech.com/

Summary

- Learnt a little about Process Safety
- Process Safety – A Few Definitions
- Catastrophic Incident – Process Safety Perspective
- Process Risk Management Design Categories
- Case Studies

RISKbytes

Questions?

RISKbytes

Now,

Let us Go Design a Safer Industry !!!

Thank You!!

Should a chemical engineer consider B-school?

David Hill, Chemstations inc
Joseph Gomes, Chevron - Deepstar
10/10/14 Galveston, AIChE SPTC

Agenda

- Background on David and Joseph
- Career path in engineering
- How to get promoted
- Going beyond the concept of promotions
- What is B school? Is this for you?
- Case Study: Joseph the Engineer who went to B School

David

2

David Hill and Joseph Gomes – two different engineers

3

Where do you want to be?

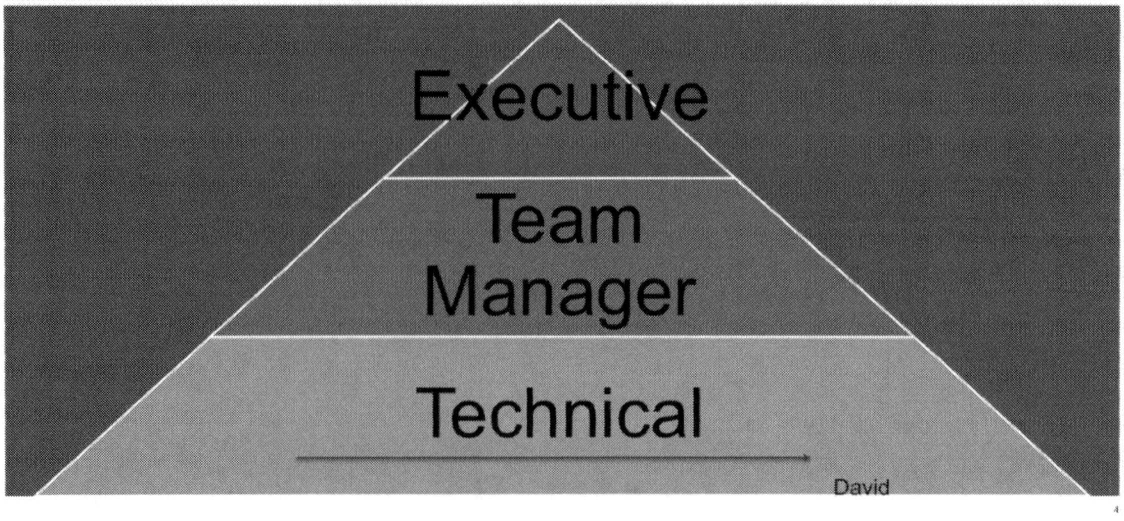

4

Determine Strengths and Weaknesses

David

5

Getting skills for the promotion

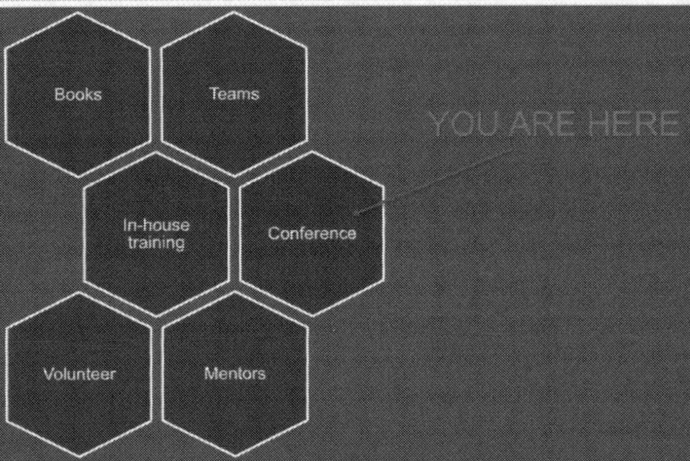

David

6

Are you looking for Next job, or looking to be more valuable asset?

- Do you want more than that?
- Promotion or long-term value?
- Job title, or contribution to the company?
- Goal to get a promotion or to be MORE – contribution
- "best at my job" or "best at contributing to my company?"

Joseph

Topics in Business school

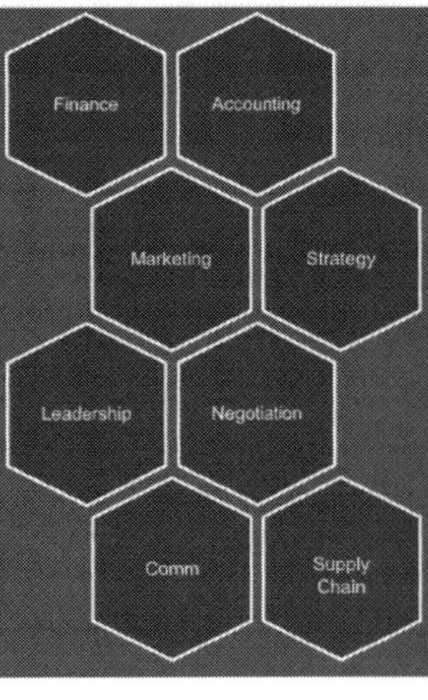

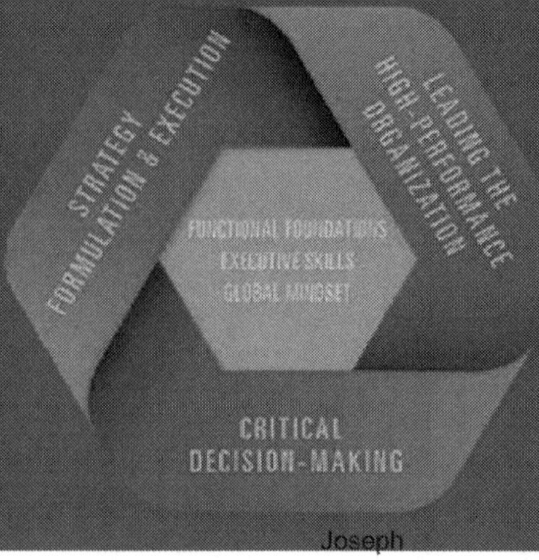

Joseph

Who goes to b school?

Joseph

Getting in

- Different programs
  - Full time: day classes
  - Professional: evening and/or weekend classes
  - Executive: weekend classes
  - Online…but…
- Application and cycle
  - Take the GMAT
  - Write some essays
  - *Admissions isn't just gpa*
- Go to the highest rank school you can get into

David

What to expect: Life in B-School

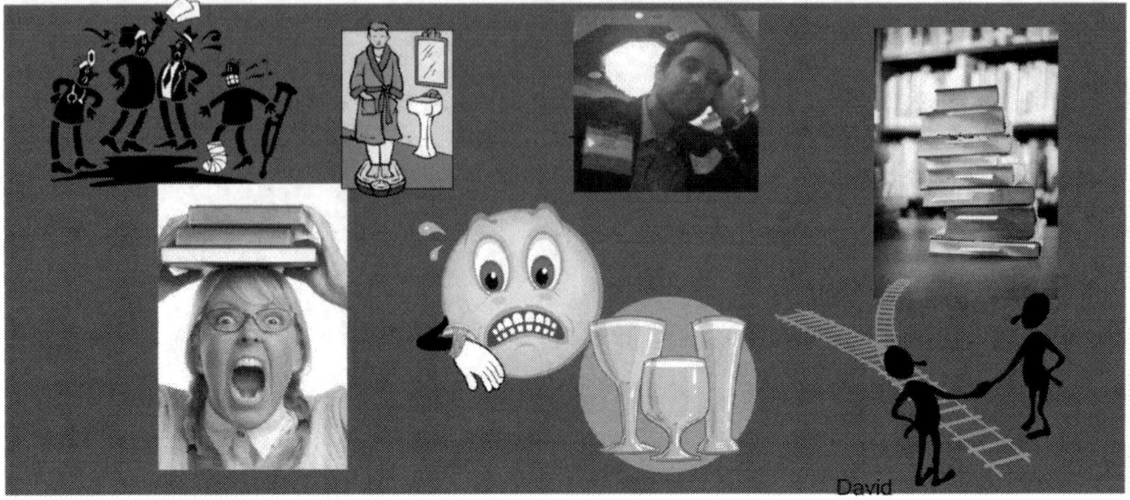

David

What you gain

BIG PICTURE

Compromise

Prioritize

Humility

Network

Service

Soft Skills

David

Is B School for you?

David and Joseph

Corporation Hierarchy

Executive
- Research Fellow

Management
- Technical Expert
- Team Lead Research Engineer
- Senior Research Engineer IV

Functional
- Research Engineer

Local
- Facilities Engineer

- CTO
- President / VP Technology Development
- Global BU Manager
- Functional Area Manager
- Regional Manager
- Technical Engineer
  - Business Law
  - Accounting
  - Finance

- COO
- President / VP Global Operations
- Regional Operations Manager
- BU Project Manager
- Project Manager

14

Process to Subsea

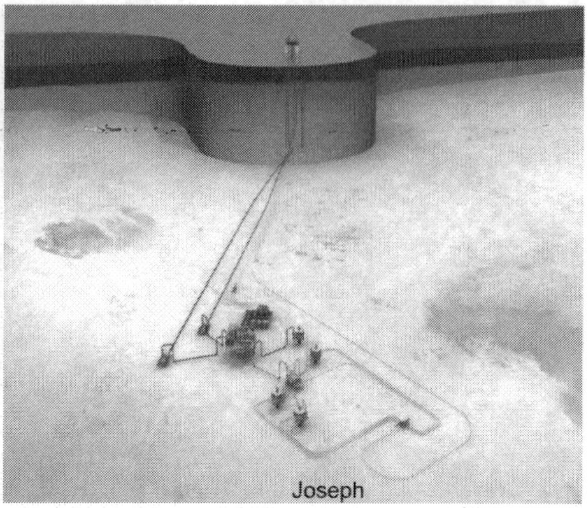

Chemical Engineering within Subsea Operations
Process to Subsea

Technology is a key enabler for success

Mission:

Develop and Deploy Flow Assurance Technology to Enable and Maximize Value of Deepwater Developments

Deepwater Production
Accurate Fluid Properties, Hydraulics, and Blockage Risk Modeling are important for system deliverability, availability, operability, and uptime

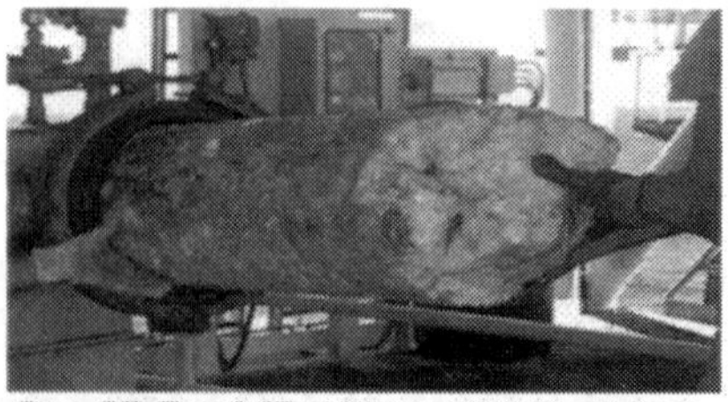

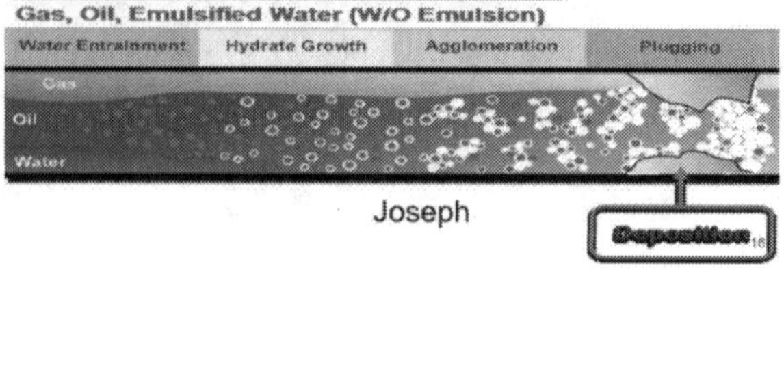

Joseph

DeepStar Members
Phase XII (January 2014 – December 2015)

Phase XII Participants

Chevron, TOTAL, Marathon Oil, bp, ConocoPhillips, MAERSK OIL, BG GROUP, woodside, PETROBRAS, Anadarko Petroleum Corporation, Statoil, nexen

Phase XII Contributors

2H Offshore Inc.	Doris Engineering	Oceaneering International
Advantek International	Exmar Offshore Company	Oil States Industries
Aker Solutions	Fluor Enterprises, Inc.	Saipem
American Bureau of Shipping	Fugro Chance, Inc.	SBM Offshore
Amog Consulting, Inc.	GE Oil & Gas (Vetco Gray)	Schlumberger
Asian Star Anchor Chain Co., LTD (ASAC)	General Marine Contractors, LLC (GMC)	Scoperta, Inc.
Assured Flow Solutions	Gibson Applied Technology & Engineering (GATE)	Sonomatic, Inc
ATI Metals (TDY Industries)	Granherne, Inc	Southwest Research Institute
Baker Peterolite Corporation	IntecSea, Inc.	Stress Engineering Services
Battelle Memorial Institute	Liquid Robotics Oil & Gas LLC	Trendsetter Engineering, Inc
Bechtel Oil, Gas and Chemicals, Inc.	Lockheed Martin Corp.	University of Houston
Bluewater Energy Services	Magma Global Limited	VWS Westgarth
Colorado School of Mines	Marintek USA, Inc	Weatherford
DNV-GL	Nalco Champion	Woodgroup Kenny

**Highlighted in yellow: New Contributor Members*

Joseph

17

DeepStar Management and DeepStar Committees

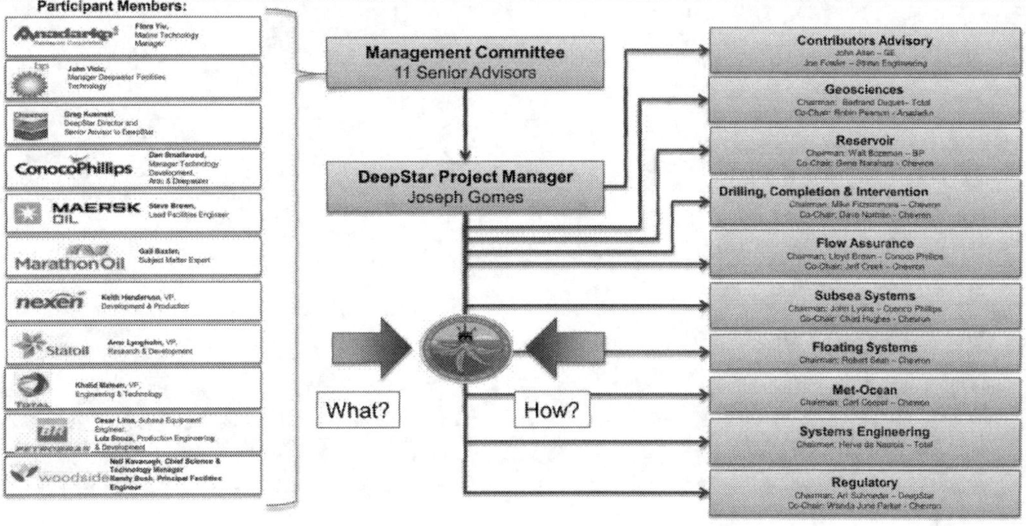

After B-school -- David

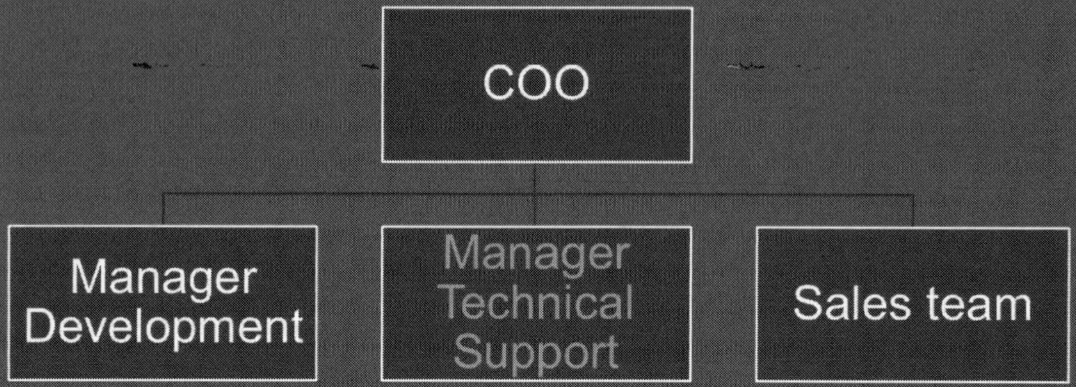

David

19

Closing thoughts

- Want a promotion? Take training
- B-school will change how you think, who you are
- This isn't a ticket to the C suite
- …it's not even an easy button
- Always looking for new opportunities
- Go Owls!

David

20

AUTHOR INDEX